EXS 62

Free Radicals and Aging

Edited by Ingrid Emerit
Britton Chance

Birkhäuser Verlag
Basel · Boston · Berlin

Editors' addresses:

Dr. Ingrid Emerit
Directeur de Recherche C.N.R.S.
Institut Biomédical des Cordeliers
Université Pierre et Marie Curie
15–21, Rue de l'Ecole de Médecine
F-75006 Paris
France

Prof. B. Chance
University of Pennsylvania
Department of Biochemistry
and Biophysics
D-501 Richards Building
Philadelphia, PA 19104-6089
U.S.A.

Library of Congress Cataloging-in-Publication Data

Free radicals and aging / edited by Ingrid Emerit, Britton Chance.
 p. cm.—(EXS; 62)
 Includes index.
 ISBN 978-3-0348-7462-5   ISBN 978-3-0348-7460-1 (eBook)
 DOI 10.1007/978-3-0348-7460-1
 1. Free radicals (Chemistry)—Physiological effect. 2. Aging—Molecular aspects.
 3. Free Radicals—metabolism—congresses.
 I. Emerit, Ingrid. II. Chance, Britton. III. Series.
 [DNLM: 1. Aging—physiology—congresses. W1 E65 v.62 / WT 104 F8529]
RB170.F75 1992
612.6′7—dc20

Deutsche Bibliothek Cataloging-in-Publication Data

Free radicals and aging / ed. by Ingrid Emerit; Britton
Chance.—Basel; Boston; Berlin: Birkhäuser, 1992
 (EXS; 62)
  ISBN 978-3-0348-7462-5

NE: Emerit, Ingrid [Hrsg.]; GT

©1992 Birkhäuser Verlag
Softcover reprint of the hardcover 1st edition 1992

  P.O. Box 133
  CH-4010 Basel
  Switzerland
  (FAX: + +41/61 271 76 66)

ISBN 978-3-0348-7462-5

Contents

Part II. Age-Related Diseases

Cardiovascular System

Brain

Preface

Many theories have been advanced to account for the process of aging. The argument that free radicals cause non-specific damage to macromolecules, such as DNA, lipids and proteins, has been called the free radical theory of aging. This theory proposed by D. Harman in 1956 has become quite persuasive in recent years due to developments in free radical biology. It is of practical interest, because it raises the possibility that the aging process could be influenced by interventions that modify free radical reactions. Since we are principally dealing with oxygen-derived free radicals, natural or synthetic antioxidants may have the power of retarding the aging process. The two major causes of death, atherosclerosis and cancer, are also related to increased oxygen radical production and can possibly be influenced by appropriate antioxidant treatment.

In our modern societies, average life expectancy now approaches a maximum of about 85 years. The risk of death drops after birth to a minimum figure around puberty and then rises almost exponentially at a standard rate so that few individuals reach 100 and none live beyond 115 years. Appropriate nutrition programs aimed at decreasing morbidity and mortality caused by degenerative diseases and non-specific age changes, and – if necessary – antioxidant supplements, should extend the span of healthy, productive life. It is indeed the primary objective of aging research to improve the quality of life and not simply extend it. This means that years can be added during which not only the individuals, but also society would benefit of the skills gained over a lifetime.

This volume assembles a selection of articles which have been critically refereed and whose authors were requested to write their respective contribution in the style of a small review. The carefully chosen group of authors who agreed to contribute to this volume emerged from the conference "Free Radicals and Aging" which took place in Paris at the end of September 1991 under the auspices of the Society for Free Radical Research, the University of Paris VI and the Centre National de la Recherche Scientifique.

The Editors

Free Radicals and Aging
ed. by I. Emerit & B. Chance
© 1992 Birkhäuser Verlag Basel/Switzerland

Free radical theory of aging: History

Denham Harman

University of Nebraska College of Medicine, Omaha, Nebraska 68198-4635, USA

Summary. Aging is the accumulation of changes responsible for the sequential alterations that accompany advancing age and the associated progressive increases in the chance of disease and death. These changes can be attributed to disease, environment, and the inborn aging process.

The aging process is now the major risk factor for disease and death after about age 28. The free radical theory of aging arose in 1954 from a consideration of aging phenomenon from the premise that a single common process, modifiable by genetic and environmental factors, was responsible for the aging and death of all living things. The theory postulates that aging is caused by free radical reactions, i.e., these reactions may be involved in production of the aging changes associated with the environment, disease and the intrinsic aging process. The origination of the theory and its application to the problem of increasing the functional life span are discussed.

Support for the free radical theory of aging has increased progressively and now includes: 1) studies on the origin of life and evolution, 2) studies on the effect of ionizing radiation on living things, 3) dietary manipulations of endogenous free radical reactions, 4) the plausible explanations it provides for aging phenomena, and 5) the growing numbers of studies that implicate free radical reactions in the pathogenesis of specific diseases.

The rapidly growing number of scientists involved in studies on the role of free radical reactions in biological systems should assure future significant increases in the healthy, useful, life span of man.

Introduction

Aging is the accumulation of changes responsible for the sequential alterations that accompany advancing age and the associated progressive increases in the chance of disease and death. These changes can be attributed to disease, environment, and to an inborn process, named the aging process. The latter produces aging changes at an apparently unalterable, exponentially increasing rate with advancing age. The contributions of the aging process to aging changes are small early in life but rapidly increase with age due to the exponential nature of the process (Harman, 1991).

As living conditions in a population approach optimum levels, the curve of the chance of death versus age shifts toward a limit (Harman, 1991). This is determined by the irreducible production in individuals of aging changes associated with environment and disease plus those produced by the aging process. Chances for death are now near limiting values in the developed countries. The aging process has now become the major risk factor for disease and death after around age 28 in the developed countries. The importance of the aging process to our health and well-being is obscured by the protean nature of its contributions, to nonspecific changes and to disease pathogenesis.

Many theories have been advanced to account for the aging process (Rockstein et al, 1974; Warner et al., 1987; Medvedev, 1990). No one theory is generally accepted. "This remarkable process remains a mystery" (Rothstein, 1986); it is doubtful that a single theory will explain all the mechanisms of aging (Schneider, 1987; Vijg, 1990). The importance attached to increasing the span of healthy productive life, i.e., the functional life span, dictates that aging hypotheses be explored for practical means of achieving this goal.

The free radical theory of aging shows promise of application today. This theory arose in 1954 from a consideration of aging phenomenon from the premise (Harman, 1981) that a single common process, modifiable by genetics and environmental factors, was responsible for the aging and death of all living things. The theory postulates that aging is caused by free radical reactions; i.e., these reactions may be involved in production of the aging changes associated with the environment, disease and the intrinsic aging process. The theory, applicable to all life, is based on the chemical nature of free radical reactions and their ubiquitous prominent presence in living things. The following is a brief account of the circumstances that led to this hypothesis and of my application of it.

Origin of the free radical theory of aging

On July 1, 1954 I joined the Donner Laboratory of Medical Physics on the Berkeley Campus of the University of California, as a research associate of Dr. John Lawrence. I spent four years at Donner: the first two years I was full time, part time during the second two years while I completed my residency requirements in internal medicine in San Francisco. My time was my own except for Wednesday mornings in the Hematology Clinic. I took advantage of this free time to pursue a long-time interest, the cause of aging. Because aging is universal I felt that aging might have a single basic cause.

My interest in aging had been sparked in December 1945 by an article (Laurence, 1945) in a popular magazine that my wife had called to my attention. The article, "Tomorrow you may be younger", was written by William L. Laurence, science editor of the New York Times. It was concerned with the work of Dr. Alexander A. Bogomolets, of the Institute of Gerontology in Russia, on an "anti-reticular-cytotoxic serum" (ACS). An account of these studies is contained in the book, Prolongation of Life, published in English in 1946 (Bogomolets, 1946).

In retrospect, I was probably uniquely qualified to approach the aging problem. I had the time, I had the interest, I had just completed a superb course in biology – medical school at Stanford and a rotating internship on the Standford Service at the San. Francisco City and County Hospital, and I had had over 15 years of almost continuous work in the chemistry laboratories of Shell Development Company, the chemical research division of Shell Oil Company. During my time at Shell I obtained my B.S. and Ph.D. degrees from the College of Chemistry at the University of California in Berkeley; my graduate work was concerned with mechanisms of organic reactions. During my last seven years at Shell I worked in the Reaction Kinetics Department, where I was involved almost entirely with free radical chemistry, largely of O_2 and compounds of phosphorus and sulfur.

The first four months at Donner was a period of progressively increasing frustration. Every possible lead I thought of led nowhere. The frustrations ended suddenly one morning during the first part of November, 1954. While reading in my office it suddenly occurred to me that free radical reactions, however initiated, could be responsible for the progressive deterioration of biological systems with time because of their inherent ability to produce random change (Scott, 1965; Pryor, 1966; Kochi, 1973; Nonhebel et al., 1974).

One day during the first part of December, 1954 I walked around the Berkeley campus and spoke with a number of people about the theory. Only two recognized that it might have merit: they were both chemists with biological interests – one in virology and the other in photosynthesis. The usual reaction was negative; it was too simple to explain such a complex process as aging. The theory was first published (Harman, 1955) on July 14, 1955 as a University of California Radiation Laboratory report, titled, "Aging: A theory based on free radical and radiation chemistry", and as a journal article (Harman, 1956a) in July 1956 in the Journal of Gerontology. The first talk, "Aging: The theory based on free radical and radiation chemistry with application to cancer and artherosclerosis", was presented on February 6, 1956 as a Donner Laboratory Seminar (Harman, 1956b).

Application of the theory

Shortly after the theory was first advanced it became obvious that it could not be directly proven or disproved and that it would be necessary to reduce it to practice. Almost simultaneously, studies were started in four areas: 1) action of catalase on H_2O_2, 2) cancer, 3) atherosclerosis, and 4) life span. The catalase study was initiated (Harman, 1956a) in an attempt to find evidence for the in vivo formation of the hydroxy radical $(HO\cdot)$ – by analogy with the Fenten reaction; the data were suggestive, but not conclusive. In 1955 an electron spin resonance spectrometer (ESR) became available in the Physics Department; this equipment failed to detect the presence of free radicals during the action of catalase on H_2O_2. Free radicals were first detected by ESR in a biological system in 1954 in yeast (Commenor et al., 1954). In 1964–

4

1965 while working for six months at Varian Associates in Palo Alto as an associate of Lawrence Piette, Ph.D., the presence of free radicals was detected in human serum (Harman et al, 1966).

The first paper based on the theory, "Reducing compounds as chemotherapeutic agent in cancer", was published as an abstract in 1956 in Clinical Research (Harman, 1956c). The first life span study (Harman, 1957a), with male AKR and female C3H mice, entitled "Prolongation of the Normal Life Span by Radiation Protection Chemicals", was published in 1957 as an abstract in Clinical Research, and subsequently as a paper in 1957 in the Journal of Gerontology (Harman, 1957b). A second life span paper "Prolongation of the normal life span and inhibition of spontaneous cancer by antioxidants" was published in 1961 in the Journal of Gerontology (Harman, 1961a). The term "free radical theory of aging" was first used in 1960 (Harman, 1960a) in the title of a paper on the decrease in serum mercaptan levels with age; these results were in accord with the fact that the chance of death rises exponentially with age.

During the early period of the free radical theory of aging another theoretical paper "Atherosclerosis: a hypothesis concerning initiating steps in pathogenesis" (Harman, 1957c), based on the free radical theory, was published in the Journal of Gerontology. This paper postulated that peroxidation of lipids in the serum and vessel wall was involved in atherogenesis. This prompted a letter to Lancet, published on November 30, 1957, entitled "Atherosclerosis: possible ill-effects of the use of highly unsaturated fats to lower serum cholesterol levels (Harman, 1957d). The first experimental paper (Harman, 1958) based on the atherosclerosis hypothesis was an abstract, "Atherosclerosis: effect of an antioxidant-cysteine" published in 1958 in Clinical Research.

In July, 1958, I moved to the University of Nebraska College of Medicine in Omaha, Nebraska to occupy the chair of Cardiovascular Research. For a number of years thereafter I was mainly involved with atherosclerosis research. Experiments were largely concerned with the possibility that the initiating steps in atherogenesis were: 1) oxidative-polymerization of constituents of serum lipoproteins; 2) anchoring of the oxidized materials in the arterial wall; and 3) an inflammatory reaction induced in the arterial wall by these condensed products (Harman, 1957c). These studies included: 1) Atherosclerosis: Oxidation of serum lipoprotein and its relationship to pathogenesis (Harman, 1960b); 2) Atherosclerosis: Possible role of phagocytosis (Harman, 1960c); 3) Atherosclerosis; Effect of the rate of growth (Harman, 1962a); 4) Atherosclerosis: Inhibiting effect of an antihistaminic drug, chlorpheniramine (Harman, 1962b); 5) Role of serum copper in coronary atherosclerosis (Harman, 1963); 6) Atherosclerosis: Possible role of drinking water copper (Harman, 1965a); 7) The free radical theory of aging: Effect of age on serum copper levels (Harman, 1965b); 8) Pigatherosclerosis: Effect of the antihistamine, chlorpheniramine, on atherogenesis and serum lipids (Harman, 1969), and 9) Atherogenesis in minipigs: Effect of dietary unsaturation and of copper (Harman, 1970).

Research emphasis gradually shifted so that by the late 1960's it was mainly concerned with aging and disease processes. By the mid 1960's it had become apparent that decreasing endogenous free radical reactions by antioxidants or altering the composition of the diet, increased average life expectancy, but had little, if any, effect on the maximum life span. Consideration of this fact led to the publication in 1972 of a short paper in the Journal of the American Geriatrics Society entitled, "The Biologic Clock: The Mitochondria?" (Harman, 1972). This paper was expanded in 1983; "Free radical theory of aging: consequences of mitochondrial aging" (Harman, 1983). Around this time Dr. Jaime Miquel (Fleming et al., 1982; Miquel et al., 1980, 1982) and his associates published on the role of mitochondria in aging. Recently others have published mitochondrial theories of aging based on the high mutation rate of mitochondrial DNA (Linnane et al., 1989; Bandy et al., 1990).

A study of the effect of antioxidants on the life span of male LAF_1 mice in 1968 (Harman, 1968) indicated that the observed increases in life span were probably largely due to inhibition of amyloid in this susceptible stain by the antioxidants. A subsequent study (Harman et al., 1976a) of casein-induced amyloidosis showed that the antioxidants increased the serum concentration of a component in the α-glycoprotein region. It is possible that the antioxidants inhibited the rate of oxidative formation of amyloid fibrils to that there was time for them to be further degraded before they could aggregate and be stabilized by substance P.

The effect of the amount and degree of unsaturation of dietary fat was evaluated in 1971 (Harman, 1971). The data were compatible with the possibility that lipid peroxidation contributed to biological degradation but was not a major factor determining the mortality rate when the diet contains amounts of antioxidants in the range ordinarily employed to prevent signs of vitamin E deficiency.

Chronic organic brain syndrome (COBS) is a major health problem. The majority of cases of COBS are due to senile dementia of the Alzheimer's type (SDAT) (Kay, 1972). Because lipid peroxidation might play a role in these disorders a number of studies were directed to this area. They include: 1) Free radical theory of aging: effect of dietary fat on central nervous system function (Harman et al., 1976b) – the number of errors made by rats in a Hebb-Williams maze was found to be higher as the amount or degree of unsaturation of the dietrary fat was increased; and 2) Free radical theory of aging: effect of age, sex and dietary precursors on rat-brain docosahexanoic acid (Eddy et al., 1977) – female rats were found to have a greater capacity than males to convert $\omega3$ fatty acids to brain docosahexanoic acid.

Antioxidants were shown in 1977 (Harman et al., 1977) to enhance cellular and humoral immune responses in mice. Thus, the immune deficiency of age (Kay et al., 1976) can be attributed, at least in part, to the increase in the free radical reaction level with age (Harman, 1986; Noy et al., 1985).

The possibility was evaluated in 1979 (Harman, 1979) that the expected free radical reaction damage associated with the high metabolic activity in the first stages of life might be reflected in a shortened life span. This proved to be the case. Mice whose mothers had received one of several antioxidants during pregnancy and until their offspring were weaned – their offspring were then all placed on the same normal mouse chow – lived longer than mice born of a mother on the control diet. Unexpectantly, the increase was greater for male offspring (14.7%) than for females (8.6%). Consideration of the data of this study in the light of early events in embryogenesis, led to the suggestion that the usual greater longevity of females is due, at least in part, to greater protection of female embryos from free radical damage during the period before random inactivation of one of their two X-chromosomes.

The fact that the suppressor cells are more radiosensitive than the other cells involved in the immune respnse prompted a study, published in 1980 (Harman, 1980a), of the effect of antioxidants on the autoimmunity of male NZB mice. Ethoxyquin almost completely suppressed the autoimmunity of these mice; this study shed some light on the etiology of systemic lupus erythematosus.

A paper in 1980 addressed the question of why free radical reactions are ubiquitous in living things (Harman, 1980b). It would appear that life originated as a result of free radical reactions, selected free radical reactions to play major metabolic roles, and utilized them to provide for mutation and death.

There has long been interest in the association between aging and disease. This was discussed in 1984 in the paper, "Free radical theory of aging: the "free radical" diseases" (Harman, 1984). The free radical theory of aging provides a plausible explanation for the relation between disease and aging. A disease is a combination of changes, usually forming a readily recognized pattern, that have detrimental effects on function and that in some cases may lead to death. The ubiquitous free radical reactions would be expected to produce progressive adverse changes that accumulate with age throughout the body. The "normal" sequential alterations with age can be attributed to those changes more-or-less common to all persons. Superimposed on this common pattern of change are patterns that should differ from individual to individual owing to genetic and environmental differences that modulate free radical reaction damage. The superimposed patterns of change may become progressively more discernable with time and some may eventually be recognized as diseases at ages influenced by genetic and environmental risk factors. Aging may also be viewed as a disease, differing from others in that the aging pattern is universal. The probability of developing any one of the "free radical" diseases should be decreased by lowering the free radical reaction level by any means (e.g., food restriction, antioxidants) and in the case of a specific disease, lowered further by decreases in contributing environmental factors (e.g., cholesterol in atherosclerosis). The growing number (Harman, 1984, 1986; Halliwell et al., 1989) of free radical diseases includes the two major causes of death, cancer and atherosclerosis, as well as other common degenerative disorders. Data on the beneficial effects of antioxidants on the free radical diseases are accumulating rapidly (Slater et al., 1991).

The free radical theory of aging has been summarized a number of times, first in 1961 in the Lancet (Harman, 1961b) and secondly in 1962 in Radiation Research (Harman, 1962c). In the latter paper the data at the time supporting the probable presence of free radicals in living things was presented – this took less than a page. Although studies by others on the role of free radicals in biological systems began to increase in the late 1950's, the increase was slow until the discovery by McCord and Fridovich of superoxide dismutase (SOD) in 1969 (McCord et al., 1969). There was then an explosion of studies so that today around 200 articles are abstracted each month (CA Selects, 1991). The theory was summarized again in 1981 (Harman, 1981) in "The Aging Process", published in the Proceedings of the National Academy of Sciences (PNAS) and in 1986 (Harman, 1986) in a book chapter entitled "Free radical theory of aging: role of free radicals in the origination and evolution of life, aging, and disease processes"; a significant portion of this paper was devoted to atherogenesis, the major cause of death.

Interest in formation of lipofuscin (age pigment) resulted in a paper in 1990 (Harman, 1990) entitled "Lipofuscin and ceroid formation: the cellular recycling system". This system probably arose early in evolution to reuse cellular components damaged early on primarily by ionizing radiation and later by free radical reactions involving O_2.

Continuing concern for the relatively low level of funding for basic biomedical aging research prompted publication in the June 15, 1991 issue of the PNAS of a paper entitled, The Aging Process: major risk factor for disease and death (Harman, 1991). Average life expectancies at birth in the developed countries are close to plateau values. Efforts to further increase the functional life span by conventional biomedical research are now almost futile. This paper contains a brief discussion of the free radical theory of aging to bolster the assertion that prospects for significantly increasing the functional life span through biomedical aging research are bright.

Comment

In this brief history of the free radical theory of aging the focus has been primarily on my own work. These studies, plus those of many others

8

(Halliwell et al., 1989; Freeman et al., 1982; Pryor, 1976–1984), provides support for the free radical theory of aging. This now includes: 1) studies on the origin of life and evolution, 2) studies of the effect of ionizing radiation on living things, 3) life span experiments in which adverse endogenous free radical reactions were expected to be lowered by dietary manipulation, 4) the plausible explanation it provides for aging phenomena, and 5) the growing number of studies that implicate free radical reactions in the pathogenesis of specific diseases.

It is apparent that the free radical theory of aging can serve as a useful guide for efforts to increase the healthy life span. This is not to say that other approaches to doing so may not be found in the future, but only that it now seems feasible to augment it to some degree by interfering with what appears to be the natural process which leads to death. A rapidly growing number of scientists are now involved in aging and in the role of free radical reactions in biological systems. Their efforts should assure future significant increases in the healthy, useful, life span of man.

Bandy, B., and Davison, A. J. (1990) Mitochondrial mutations may increase oxidative stress: Implications for carcinogenesis and aging? Free Rad. Bio. Med. 8: 523–539.

Bogomolets, A. A. (1946) The Prolongation of Life. Duell, Sloan, and Pearse, Inc., New York.

CA Selects (1991) Free radicals (biochemical aspects). Amer. Chem. Soc., Chem. Abstr. Ser., Columbus, Ohio.

Commoner, B., Townsend, J., and Pake, G. E. (1954) Free radicals in biological materials. Nature 174: 689–691.

Eddy, D. E., and Harman, D. (1977) Free radical theory of aging: Effect of age, sex and dietary precursors on rat-brain docosahexanoic acid. J. Amer. Geriatrics Soc. 25: 220–229.

Fleming, J. E., Miquel, J., Cottrell, S. F., Yengoyan, L. S., and Economos, A. C. (1982) Is cell aging caused by respiratory-dependent injury to the mitochondrial genome? Gerontology 28: 44–53.

Freeman, B. A., and Crapo, J. D. (1982) Biology of Disease: Free radicals and tissue injury. Lab. Invest. 471: 412–426.

Halliwell, B., and Gutteridge, J. M. C. (1989) Free Radicals in Biology and Medicine, 2nd edition, Clarendon, Oxford, U.K., pp. 418–419.

Harman, D. (1955). Aging: A theory based on free radical and radiation chemistry. Univ. Calif. Rad. Lab. Report No. 3078, July 14.

Harman, D. (1956a) Aging: A theory based on free radical and radiation chemistry. J. Gerontol. 11: 298–300.

Harman, D. (1956b) Aging: The theory based on free radical and radiation chemistry with application to cancer and atherosclerosis. Rad. Lab. Calender: Univ. Calif.-Berkeley, 3: No. 3, February 2.

Harman, D. (1956c) Reducing compounds as chemotherapeutic agents in cancer. Clin. Res. 4: 54–55.

Harman, D. (1957a) Prolongation of the normal life span by radiation protection chemicals. Clin. Res. 5: 46.

Harman, D. (1957b) Prolongation of the normal life span by radiation protection chemicals. J. Gerontol. 12: 257–263.

Harman, D. (1957c) Atherosclerosis: A hypothesis concerning the initiating steps on pathogenesis. J. Gerontol. 12: 199–202.

Harman, D. (1957d) Atherosclerosis: Possible ill-effects of the use of highly unsaturated fats to lower serum cholesterol levels. Lancet II: 116–117.

Harman, D. (1958) Atherosclerosis: effect of an antioxidant-cysteine. Clin. Res. 6: 50–51.

Harman, D. (1960a) The free radical theory of aging: The effect of age on serum mercaptan levels. J. Gerontol. 15: 38–40.

Harman, D. (1960b) Atherosclerosis: oxidation of serum lipoproteins and its relationship to pathogenesis. Clin. Res. 8: 108.

Harman, D. (1960c) Atherosclerosis: Possible role of phagocytosis. Circulation 22: 681.

Harman, D. (1961a) Prolongation of the normal life span and inhibition of spontaneous cancer by antioxidants. J. Gerontol. 16: 245–254.

Harman, D. (1961b) Mutation, cancer, and aging. Lancet I: 200–201.

Harman, D. (1962a) Atherosclerosis: Effect of rate of growth. Circulation Res. 10: 851–852.

Harman, D. (1962b) Atherosclerosis: Inhibiting effect of an antihistaminic drug, chlorpheniramine. Circulation Res. 11: 227–282.

Harman, D. (1962c) Role of free radicals in mutation, cancer, aging, and the maintenance of life. Rad. Res. 16: 753–763.

Harman, D. (1963) Role of serum copper in coronary atherosclerosis. Circulation 28: 658.

Harman, D. (1965a) Atherosclerosis: possible role of drinking water copper. Clin. Res. 13: 91.

Harman, D. (1965b) The free radical theory of aging: Effect of age on serum copper levels. J. Gerontol. 20: 151–153.

Harman, D., and Piette, L. H. (1966) Free radical theory of aging: Free radical reactions in serum. J. Gerontol. 21: 560–565.

Harman, D. (1968) Free radical theory of aging: Effect of free radical reaction inhibitors on the mortality rate of male LAF_1 mice. J. Gerontol. 23: 476–482.

Harman, D. (1969) Pig atherosclerosis: Effect of the antihistamine, chlorpheniramine, on atherogenesis and serum lipids. J. Atheroscler. Res. 10: 77–84.

Harman, D. (1970) Atherogenesis in minipigs: Effect of dietary fat unsaturation and of copper, in: Jones, R. J. ed. Atherosclerosis: Proc. of the Second Intrnl. Symposium, Springer-Verlag, New York, pp. 471–475.

Harman, D. (1971) Free radical theory of aging: Effect of the amount and degree of unsaturation of dietary fat on mortality rate. J. Gerontol. 26: 451–457.

Harman, D. (1972) The biological clock: The mitochondria? J. Amer. Geriatrics Soc. 20: 145–147.

Harman, D., Eddy, D. E., and Noffsinger, J. (1976a) Free radical theory of aging: Inhibition of amyloidosis in mice by antioxidants; possible mechanisms. J. Amer. Geriatrics Soc. 24: 203–210.

Harman, D., Hendricks, S., Eddy, D. E., and Seibold, J. (1976b) Free radical theory of aging: Effect of dietary fat on central nervous system function. J. Amer. Geriatrics Soc. 24: 301–307.

Harman, D., Heidrick, M. L., and Eddy, D. E. (1977) Free radical theory of aging: Effect of free-radical-reaction inhibitors on the immune response. J. Amer. Geriatrics Soc. 25: 400–407.

Harman, D. (1979) Free radical theory of aging: Beneficial effect of adding antioxidants to the maternal mouse diet on the life span of offspring; possible explanation of the sex difference in longevity. Age 2: 109–122.

Harman, D. (1980a) Free radical theory of aging: Beneficial effect of antioxidants on the life span of male NZB mice; role of free radical reactions in the deterioration of the immune system with age and in the pathogenesis of systemic lupus erythematosus. Age 3: 64–73.

Harman, D. (1980b) Free radical theory of aging: Origin of life, evolution, and aging. Age 3: 100–102.

Harman, D. (1981) The aging process. Proc. Natl. Acad. Sci. USA 78: 7124–7128.

Harman, D. (1983) Free radical theory of aging: Consequences of mitochondrial aging. Age 6: 86–94.

Harman, D. (1984) Free radical theory of aging: The "free radical" diseases. Age 7: 111–131.

Harman, D. (1986) Free radical theory of aging: Role of free radicals in the origination and evolution of life, aging, and disease processes, in: Free Radicals, Aging, and Degenerative Diseases. Johnson, Jr., J. E., Walford, R., Harman, D., and Miquel, J. eds. Liss, New York, pp. 3–49.

Harman, D. (1990) Lipofuscin and ceroid formation: The cellular recycling system, in: Lipofuscin and Ceroid Pigments. Porta, E. A. ed. Plenum Press, New York.

Harman, D. (1991) The aging process: Major risk factor for disease and death. Proc. Natl. Acad. Sci. USA 88: 5360–5363.

Kay, D. W. K. (1972) Epidemiological aspects of organic brain disease in the aged, in: Aging and the Brain. Gaitz, G. M., ed. Plenum Press, New York, pp. 15–27.

10

Kay, M. M. B., and Makinodan, T. (1976) Immunobiology of aging: Evaluation of current status. Clin. Immunol. Immunopathol. 6: 394–413.

Kochi, J. K., ed. (1973) Free Radicals, Vol. 2, Wiley, New York.

Laurence, W. L. (1945) Tomorrow you may be younger. Ladies Home Journal 62: No. 12, 22–23, 135–137.

Linnane, A. W., Marzuki, S., Ozawa, T., and Tanaka, M. D. (1989) Mitochondrial DNA mutations as an important contributor to ageing and degenerative diseases. Lancet I: 642–645.

McCord, J. M., and Fridovich, I. (1969) Superoxide dismutase: An enzymatic function for erythrocuprein (Hemocuprein). J. Biol. Chem. 244: 6049–6055.

Medvedev. Z. A. (1990) An attempt at a rational classification of theories of aging. Biol. Rev. 65: 375–398.

Miquel, J., Economos, A. C., Fleming, J., and Johnson, Jr., J. E. (1980) Mitochondrial role in cell aging. Exper. Gerontol. 15: 575–591.

Miquel, J., and Fleming, J. E. (1982) Experimental support for the mitochondrial mutation theory of aging. Age 5: 142–143.

Nonhebel, D. C., and Walton, J. C. (1974) Free Radical Chemistry. University Press, Cambridge, U.K.

Noy, N., Schwartz, H., and Gafni, A. (1985) Age-related changes in the redox status of rat muscle cells and their role in enzyme-aging. Mech. Ageing Devel. 29: 63–69.

Pryor, W. A. (1966) Free Radicals. Raven Press, New York.

Rockstein, M., Sussman, M. L., and Chesky, J., eds. (1974) Theoretical Aspects of Aging. Acad. Press, New York.

Rothstein, M. (1986) Biochemical studies of aging. Chem. Eng. News 64: (32), 26.

Schneider, E. L. (1987) Theories of aging: a perspective, in: Modern Biological Theories of Aging. Warner, M. R., Butler, R. N., Sprott, R. L. and Schneider, E., eds. Raven Press, pp. 1–4.

Scott, G. (1965) Atmospheric Oxidation and Antioxidants. Elsevier Publishing Co., New York.

Slater, T. F., and Block, G., eds. (1991) Antioxidants vitamins and β-carotene in disease prevention. Amer. J. Clin. Nutr. 53: Suppl. 1, 189S–396S.

Vijg, J. (1990) Searcing for the molecular basis of aging: the need for life extension models. Aging 2: 227–229.

Warner, H. R., Butler, R. N., Sprott, R. L., and Schneider, E. L., eds. (1987) Modern Biological Theories of Aging. Raven Press, New York.

Free Radicals and Aging
ed. by I. Emerit & B. Chance

Free radical theory of aging: View against the reliability theory

Vitali K. Koltover

Institute of Pharmacology and Toxicology, Veterinary University of Vienna, Vienna, Austria

Summary. The reliability theory approach to the free radical theory of aging is presented. First, the structural and functional inhomogeneity – hierarchy – of living systems has been taken into account. Hence, the existence of the finite number of critical structures – supervisors – in the system has been postulated. Second, I have postulated the stochastic nature of age-changes in the supervisors, paying particular attention to oxygen free radicals as the by-products of normal oxidative metabolism. Third, according to the cooperativity of main processes in biological systems, the existence of the threshold values of the reliability parameters of the supervisors has been postulated. In the terms of this reliability free radical theory of aging it becomes possible to explain the character of age-changes in mortality data for humans and animals; to explain how the linear kinetics of accumulation of free radical damages in the cellular targets (supervisors) can lead to the exponential kinetics of the mortality rate growth with age, i.e., to deduce the so-called Gompertz law of mortality; to explain the nature of the well-known empiric interspecies correlations between the maximum life span values and metabolic factors; to estimate species-specific life span potentials of humans and animals. The chromosomes of neural postmytotic cells have been suggested to function as the supervisors. In the terms of the theory it has been estimated that the life span of primates could run up 250 years but for the damages due to the oxygen free radicals.

Introduction

The free radical hypothesis of aging was suggested more than 35 years ago (Harman, 1981). Up to date this hypothesis has been supported by a great deal of indirect evidence (Nohl, Hegner, 1978; Chance et al., 1979; Fridovich, 1986; Fraga et al., 1990; Khan et al., 1990). However, direct measurements of concentrations of instable free radicals in situ are still impossible. Apart from that, there is an important kinetic problem. If oxygen free radicals do play a role in aging, the accumulation of oxidative damage in cellular targets with time does not seem to proceed at a quicker pace than a linear kinetic process (Koltover, 1983). Radical chain reactions as combustions or outbreaks seem unthinkable in a living system. Meanwhile, it is well-known that the kinetics of mortality growth for animals and humans (for ages from approximately 35 to 85 years) obeys the exponential law (the so-called Gompertz's law of mortality (Strehler, 1977)). The free radical theory must explain this exponential growth of the mortality rate.

12

In order to find answers to this and some other questions concerning the mechanisms of aging, the reliability theory approach to the problem of molecular mechanisms of aging has been developed. (Koltover, 1981; 1982; 1983; 1985; 1986; Grodzinskij et al., 1987; Koltover et al., 1991).

The principles of the approach

First of all, account should be taken of the structural and functional inhomogeneity and resulting hierarchy of living systems. Hence, the first postulate of the theory states that there is a finite number (Q) of critical supervisory structures and numerous subordinate structures in such a system. Each supervisor performs a unique, vitally important, function. Hence, a failure of any of them results in the organism's death.

With time stochastic changes occur to the supervisors resulting in the irreversible growth in nonoptimality of their functional characteristics. That is the second postulate of the theory. Let the random variable m_j be a quantitative characteristic (a parameter) of an initial nonoptimality (disfunction) of the j-th supervisor ($j = 1, 2, \ldots Q$) at $t = 0$. Then the set of values m_j represents a random sample which obeys a particular distribution function $F(m) = P(m_j < m)$. The type of the function should be founded on experiments but we can try to "guess" it.

With this aim in view, account shall be taken of another fundamental property of biological systems. I mean the cooperativity of biological processes, i.e., the existence of threshold values of the most important functional parameters. Hence the third postulate of the theory: There is a threshold value of the supervisor's nonoptimality at which the supervisor fails. Thus, the random values m_j should be limited: $0 \leqslant m_j \leqslant m_c$, where m_c is the limit value. For the sake of simplicity this value is taken to be the same for every supervisor. Hence, the distribution function $F(m)$ should be truncated. The simplest asymmetrical distribution known from the mathematical reliability theory is the exponential function (Lloyd and Lipov, 1964). Therefore, the m_j can be assumed to represent a random sample of the exponential, though truncated, distribution:

$$F(m) = P(m_j < m) = \frac{1 - \exp(-am)}{1 - \exp(-am_c)} \tag{1}$$

where a is the parameter of the distribution (Koltover, 1981).

Let $\tau_j = b(m_c - m_j)$ be the time-to-failure value for the j-th supervisor – a random value with the distribution function $G(t)$. We have defined death to occur as soon as anyone of the supervisors develops the threshold nonoptimality, i.e. the expected life span $\tau = \min \tau_j$ ($j = 1, 2, \ldots Q$). Then the survival function – probability of being alive at the moment of time t – is given by the smallest sample value of the random sample of size Q:

$$R(t) = P(\tau \geqslant t) = [1 - G(t)]^Q = \left[1 - \frac{\exp(\gamma t) - 1}{\exp(\gamma T) - 1}\right]^Q \qquad (2)$$

where $\gamma = a/b$; $T = b \cdot m_c$ (Koltover, 1981; 1982).

Results and discussion

At not too high values of time it is easy to derive the following approximate expression from eq. (2):

$$R(t) \approx \exp[-QG(t)] = \exp[-\beta(\exp(\gamma t) - 1)] \qquad (3)$$

where $\beta = Q/(\exp(\gamma T) - 1)$.

From eq. (3) it is easy to obtain the mortality rate function – the logarithmic derivative of $R(t)$;

$$h(t) = -\mathrm{d}R/\mathrm{d}t/R = h_0 \exp(\gamma t) \qquad (4)$$

where γ and $h_0 = \beta \cdot \gamma$ are the constants. According to eq. (4), the mortality rate should increase exponentially with time. Thus, the theory explains the basic quantitative law of gerontology – the Gompertz law of mortality.

In order to test the theory, we used life tables for men and women in Sweden (Statistisks Arsbok för Sverige, 1957; 1975; 1977). These life tables are usually used for such simulations because they represent survivorship data with much higher accuracy (1-year age intervals) than those of other countries. From these tables we calculated the values of the functions $R(t) = n(t)/n(0)$ and $h(t) = -\Delta n(t)/n(t)/\Delta t$, where $n(t)$ is the number of alive individuals of the age t, Δn is the number of those who died during the time interval Δt taken to be 1 year. As real populations are inhomogeneous we averaged the survival functions over the ensemble assuming the normal distribution of the values of T and γ around their means T_0 and γ_0 with the respective standard deviations σ_T and σ_γ (Koltover et al., 1991).

Figure 1 and Table 1 show the results of the simulations. One can see that the theory fits the observable data (including the Gompertz interval of ages as well as geriatric ages) with the agreement function values not worse than 13 percent.

The equations (2)–(4) contain the limit life span T. It has appeared as the direct result of the third postulate of the theory – of the limit values of the supervisors' nonoptimality: $T = b \cdot m_c$, where b means the reciprocal rate of age-changes in the nonoptimality of the supervisors. Since this parameter has been suggested to be independent of time, the age-changes obey linear kinetics. Nevertheless, these linear changes turn into the exponential increase of the mortality – eq. (4). From the eq.

14

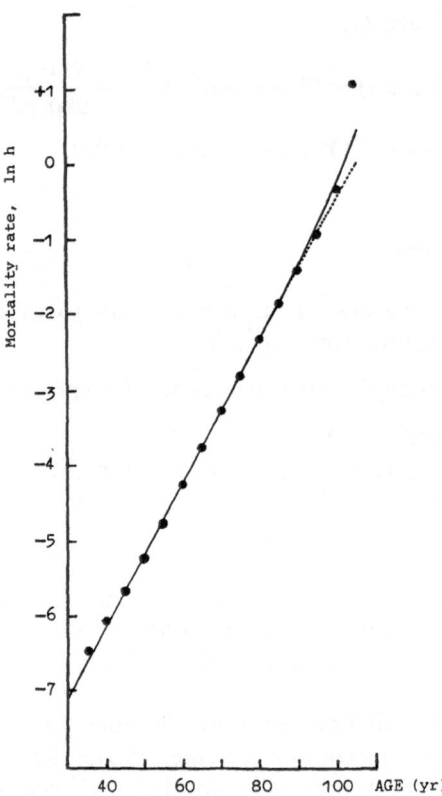

Figure 1. Dynamics of age-mortality rate for the male population of Sweden (Gompertz period and geriatric period) for the 1971/1975 calendar years. The solid curve was generated from the mathematical model with the values $Q = 46$, $T_0 = 120.0\ y$, $\sigma_T = 0.34\ y$, $\gamma_0 = 0.09505\ y^{-1}$, $\sigma_\gamma = 8.7 \times 10^{-4}\ y^{-1}$. The dashed line approximates the Gompertz linear area. The circles represent the experimental data obtained from the life tables (Koltover et al., 1991).

(2)–(3) it can be seen that if there were no threshold values (if the m_c value was infinitely large), the probability of being alive would be equal to 1 at all moments of time. Thus, due to the existence of such threshold effects, linear accumulation of damages in cellular structures results in the exponential mortality increase with time.

Now, let us consider the case of free radical damages initiated by superoxide radicals. In the first approximation, the rate of the age-changes in the supervisors should be proportional both to probability w of a damage and probability u of the damage contribution in the supervisor nonoptimality: $1/b \sim w \cdot u$. As superoxide radicals are "toxic", their appearance in a cell can be regarded – from the point of view of the reliability theory – as results of malfunctions (recurrant failures) of cellular redox-enzymes. The frequency of such malfunctions should be proportional to the intensity of oxidative metabolism

Table 1. The sets of the parameters which give the best agreement between the mathematical model and the experimental data (the life-tables of Sweden). The function $F = 0$ in the case of "ideal fitness" (Koltover et al., 1991)

Population (calendar years)	Q	$T_0 \pm \sigma_T$ (yr)	$\gamma_0 \pm \sigma_y$ (yr^{-1})	F (%)
men (1951–1955)	46	119.8 ± 0.3	0.0930 ± 0.0001	2.05
men (1969–1973)	46	120.0 ± 0.3	0.0951 ± 0.0010	12.36
men (1971–1975)	46	120.0 ± 0.3	0.0950 ± 0.0009	13.32
women (1951–1955)	46	118.7 ± 0.3	0.1009 ± 0.0001	2.16
women (1969–1973)	46	119.4 ± 0.3	0.1073 ± 0.0001	13.27
women (1971–1975)	46	119.8 ± 0.3	0.1060 ± 0.0001	13.55

(Koltover, 1981; 1982; 1983). Hence, the rate of superoxide production $I = qV$, where V is the oxidative metabolism rate and q is the probability of the malfunctions of the redox-enzymes resulting in the superoxide emergence.

Thus, in the case of damages owing to superoxide radicals we can rewrite the expression for the limit life span potential as follows:

$$T = b \cdot m_c \sim m_c / w \cdot u \sim m_c / u(qV/E) + D \qquad (5)$$

where E is the activity of the protective enzyme (superoxide dismutase) and D is the index to incorporate other injuries which are not associated with the superoxides (Koltover, 1982).

Thus, the theory predicts the linear equation

$$1/T = A(V/E) + B \qquad (6)$$

between the reciprocal maximum life span value and the ratio of V/E for animals. The activity of superoxide dismutase in cytoplasmic fractions of brain, liver and heart of 13 species of animals and of humans were measured at R. Cutler's laboratory (Tolmasoff et al., 1980). According to eq. (6), I have plotted the values $1/T$ as a function of the ratio of specific metabolic rate to the superoxide dismutase activity data for these species. As a result the straight lines with the correlation coefficients of 0.997 (for brain and liver tissues) and 0.981 (for heart tissues) were obtained (Fig. 2). It was calculated, using the free coefficients of these linear equations for heart, liver and brain tissues, that the life span of primates could reach 100, 200 and 250 years, respectively were it not for the damaging effects of the superoxide radicals (Koltover, 1982; 1983). The theory is also in good agreement with other interspecies correlations of this kind (Grodzinskij et al., 1987).

16

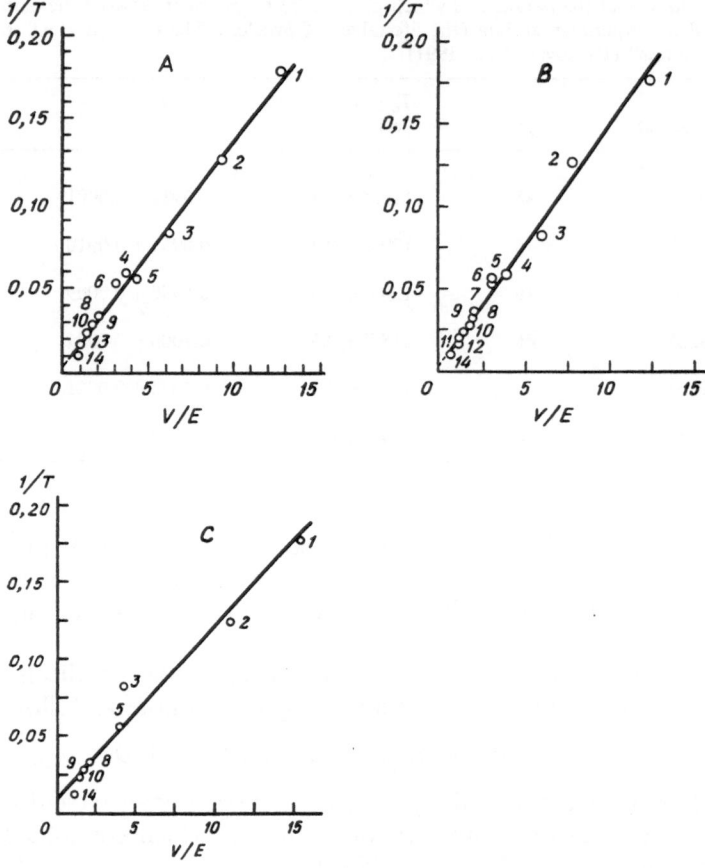

Figure 2. Correlation graphs between the reciprocal maximum life span potential ($1/T$) of different mammalian species and the ratio of specific metabolic rate to superoxide dismutase activity (V/E) for brain (A), liver (B) and heart (C), plotted according to our mathematical model. Experimental data were obtained from a paper by R. Cutler's group (Tolmasoff et al., 1980). Numbers correspond to the species: *Mus musculus* (1), *Peromyscus maniculatus* (2), *Tupaii glis* (3), *Saimarii scuireus* (4), *Galago crassicaudatus* (5), *Saguinus mystak* (6), *Lemur macaca fulvus* (7), *Cercopithecus aethiops* (8), *Macaca mulatta* (9), *Papio anubis* (10), *Gorilla gorilla* (11), *Pan troglodytes* (12), *Pongo pygmaeus* (13), *Homo sapiens* (14) (Grodzinskii et al., 1987).

The parameters Q, m_c, a, b are predetermined at the initial moment of time ($t = 0$). This means that – in the terms of the theory – aging is actually a programmable process. The longevity parameters have been incorporated into the original design of the system. However such intracellular processes as superoxide production, DNA damages, etc., have a stochastic nature. Hence, realization of the aging program proceeds stochastically.

From the postulated threshold value m_c for the supervisors nonoptimality, the limit life span value appeared in the formulas of the model $(T = b \cdot m_c)$. Indeed, it is well-known that there are no mice exceeding 3 or 4 years of age and a human life span does not exceed 120 years (provided we take reliable data into account, not sensational press reports or legends) (Olshansky et al., 1990). The limited life span of normal human fibroblasts is also a well-known phenomenon (Phillips and Cristofalo, 1990). Such a limit is not rare in nature (the limit speed of light in vacuum might be exemplified). In any case, due to the existence of the threshold of the supervisors' nonoptimality it is possible to answer the question as to how the linear accumulation of free radical damages in cells turns up in the exponential increase in mortality rate.

In conclusion, it is necessary to make some remarks on the nature of the supervisory structures. It is clear that such a structure must satisfy the following condition: its longevity should be comparable to the organism's life span. Perhaps, only genetical structures of postmytotic cells can meet this condition. Having analyzed the age-dynamics of human mortality we obtained the mean value of the limit life span potential $T = 120$ years at the value of the number of the supervisors $Q = 46$ (Table 1). It is hardly possible to consider these estimations as random figures. Indeed, the life span of humans does not not exceed 120 years. As to the estimated value of Q, it coincides with the number of chromosomes in human diploid cells. Considering the values of the Gompertz mortality parameters (h_0 and γ, available from the literature) and taking into account the chromosome numbers we estimated the life span limits for mouse, rat, domestic sheep and African buffalo. The estimations are in good agreement with the values known from the gerontological literature (Table 2).

Oxygen free radicals seem to be capable of triggering (directly or indirectly) functional changes in chromosomes. For example, we have shown that in a mouse's liver in vivo the appearance of nitric oxide

Table 2. The numbers of the supervisory structures (Q) and the life span potentials (T) for some mammalian species estimated according to the reliability theory model of aging (Koltover et al., 1991)

Species	Number of chromosomes in a diploid cell	Q	T (yr)
human	46	46	120.0
mouse (NZB males)	40	40	3.0
rat (Wistar males)	42	42	3.8
domestic sheep	54	54	19.5
African buffalo	60	60	29.8

18

radicals (NO) is followed by intensification of transcription (Grigoriev and Koltover, 1987). The life-prolongation antioxidant butylated hydroxytoluene operates on adenohypophysis (Frolkis et al., 1990). The neuro-endocrine system plays the crucial role in aging (Medvedev, 1990). Therefore, DNA of some neuronal cells might pretend to serve as the supervisors. Now it is difficult to bind the parameters m to a particular type of the genome nonoptimality. In recent years molecular biologists have been paying special attention to the so-called mobile genetic elements (transposons) and their role in ontogenesis (Ayala, Kiger, 1984). The threshold values could characterize a critical arrangement of the transposon topology at which the organism's vitality reaches zero. The suggestions concerning the nature of "supervisors" are, of course, speculative. But still they can be checked by doing experiments.

Acknowledgment. This contribution was supported by the Institute of Pharmacology and Toxicology of the Veterinary University of Vienna. I am greatly indebted to Prof. H. Nohl for discussions and comments on the contribution.

Permanent address: Institute of Chemical Physics, Acad. of Sci., Chernogolovka, Moscow District, Russia.

Ayala, F. J., and Kiger, J. A. (1984) Modern genetics (2nd edn), Benjamin Cummings Publ. Co., Menlo Park.

Chance, B., Sies, H., and Boveris, A. (1979) Hydroperoxide metabolism in mammalian organs. Physiol. Rev. 59: 527–605.

Fraga, C. G., Shigenaga, M. K., Park, J. W., Degan, P., and Ames, B. N. (1990) Oxidatible damage to DNA during aging: 8-hydroxy-2′-deoxyguanosine in rat organ and urine. Proc. Natl. Acad. Sci. USA 87: 4533–4537.

Fridovich, I. (1986) Biological effects of superoxide radical. Archs. Biochem. Biophys. 247: 1–11.

Frolkis, V. A., Gorban, E. N., and Koltover, V. K. (1990) Effects of antioxidant butylated hydroxytoluene (BHT) on hormonal regulation and ESR signals in adult and old rats. Age 13: 5–8.

Grigoriev, M. Y., and Koltover, V. K. (1987) Stimulation of transcription in mouse liver cells by nitric oxide free radicals. Biochemistry (Moscow) 52: 758–760.

Grodzinskij, D. M., Vojtenko, V. P., Kutlakhmedov, Y. A., and Koltover, V. K. (1987) Reliability and aging of biological systems. Naukova dumka, Kiev (in Russian).

Harman, D. (1981) The aging process. Proc. Natl. Acad. Sci. USA 78: 7124–7128.

Khan, Sh. H., Emerit, I., and Feingold, J. (1990) Superoxide and hydrogen peroxide production by macrophages of New Zealand black mice. Free Radical Biol. Med. 8: 339–345.

Koltover, V. K. (1981) Reliability of enzymatic protection of cells from superoxide radicals and aging. Doklady Biophysics (A translation of the Biophysics section of Doklady Akademii Nauk SSSR) 256: 3–5.

Koltover, V. K. (1982) Reliability of enzyme systems and molecular mechanisms of ageing. Biophysics (A translation of Biofizika) 27: 635–639.

Koltover, V. K. (1983) The reliability theory, superoxide radicals and aging. Inst. Chem. Physics Acad. Sci. Chernogolovka, Russia.

Koltover, V. K. (1985) Deterministic reliability of enzymes and stochastic nature of aging. Reliability of Biol. Systems, Naukova dumka, Kiev, pp. 148–161 (in Russian).

Koltover, V. K. (1986) Genetic determinism of cellular molecular constructions and stochastic nature of processing of the aging programme. Reliability and elementary events in processes of aging of biological systems. Naukova dumka, Kiev, pp. 38–52 (in Russian).

Koltover, V. K., Andrianova, Z. S., and Ivanova, A. N. (1991) Simulation of survival and mortality curves of human populations by the reliability theory method. Izvestiya Akademii Nauk SSSR, Biol. Ser., No. 4 (in press).

Lloyd, D. K., and Lipov, M. (1964) Reliability: management, methods and mathematics. Prentice Hall, New York.

Medvedev, Z. A. (1990) An attempt at a rational classification of theories of ageing. Biol. Rev. 65: 375–398.

Nohl, H., and Hegner, D. (1978) Do mitochondria produce oxygen radicals in vivo? Eur. J. Biochem. 82: 563–567.

Olshansky, S. J., Carnes, B. A., and Cassel, C. (1990) In search of Methuselach – estimating the upper limits to human longevity. Science 250: 634–640.

Phillips, P. D., and Cristofalo, V. J. (1990) Recent advances in cellular aging research: understanding the limited lifespan of normal human fibroblasts. Rev. Biol. Res. Aging 4: 265–279.

Statistisks Arsbok för Sverige. 1957. Statistiska Centralbyran, Stockholm, 1957.

Statistisks Arsbok för Sverige. 1975. Statistiska Centralbyran, Stockholm, 1975.

Statistisks Arsbok för Sverige. 1977. Statistiska Centralbyran, Stockholm, 1977.

Strehler, B. L. (1977) Time, cells and aging (2nd edn), Acad. Press, New York.

Tolmasoff, J., Ono, T., and Cutler, R. (1980) Superoxide dismutase: correlation with life span and specific metabolic rate in primate species. Proc. Natl. Acad. Sci. USA 77: 2777–2781.

Free Radicals and Aging
ed. by I. Emerit & B. Chance

The rate of DNA damage and aging

Michael G. Simic

Department of Pharmacology and Toxicology, School of Pharmacy, University of Maryland, Baltimore, MD 21201, USA

Summary. A new theory of aging based on the rate of DNA damage is presented, and the relationship between the rate of oxidative DNA damage and maximum life span (MLS) of mammalian species is explored. In humans the level of oxidative DNA damage, as measured by urinary biomarkers, can be modulated by caloric restriction and dietary composition. Consequently, longevity may depend not only on the basal metabolic rate but also on dietary caloric intake and the type of diet. The theory may provide the basis for a practical approach for reduction of degenerative diseases in general, extension of life expectancy, and optimization of individual lifestyles.

Introduction

Evolutionary forces have favored extension of the maximum life span (MLS), allowing longer gestation and a single offspring as well as development of larger brains and greater intelligence. The extension of MLS in mammalian species was achieved by substantial reduction of specific metabolic rate (SMR).

The inverse correlation of MLS and SMR was first defined by Rubner (1908), who proposed that mammals have a limited capacity for metabolic processing of energy. This capacity, expressed as lifetime energy intake (LEI), was found to be about 200 kcal/g of soft tissue. This concept of shorter life span for species with higher metabolic rates was further extended and refined by the works of Pearl (1928), Sacher (1977), Cutler (1985), and Ames (1989).

The inverse relationship between MLS and SMR holds only as a general rule with notable exceptions. Values for a few species are given in Table 1. It appears from Table 1 that there are factors other than SMR or LEI that determine longevity of a species. Because the mechanisms of aging are not firmly established, this issue is not easily resolved.

An attractive explanation of the causes of aging is damage to biosystems by free radical processes (Harman, 1991). If free radical reactions were the major cause of aging, a reduction in their levels by antioxidants or antioxienzymes (e.g., catalase, superoxide dismutase, and glutathione peroxidase) should, in principle, retard aging, with a concomitant extension of MLS. Levels of antioxidants and antioxienzymes in different

Table 1. Specific metabolic rate (SMR), maximum life span (MLS), and life energy intake (LEI) for selected mammalian species

Species	SMR cal/g/day	MLS years	LEI(LEP)[a] kcal/g
Mouse	180	3.5	232
Cow	15	30	253
Horse	14	49	152
Chimpanzee	27	50	450
Elephant	13	70	220
Human	25	100	815

[a]Life energy potential (LEP) as defined by Cutler (1985).

mammalian species, however, appear to be constant. Another attempt to explain this relationship considered concentrations of antioxidants and antioxienzymes relative to SMR rather than absolute concentrations (Cutler, 1985). Although applicable to mouse, chimpanzee, and human, this proposition does not hold for horse, cow, and elephant.

More recently, with the development of sensitive analytical instrumentation, it became possible to measure less than nmol levels of products generated by oxidative damage to DNA in living organisms (Dizdaroglu and Bergtold, 1986; Dizdaroglu, 1991). These measurements were made either directly on DNA or on excised damage products found in the urine (Cathcart et al., 1984; Bergtold et al., 1988; Ames, 1989; Shigenaga et al., 1991). The yield of urinary biomarkers of oxidative DNA base damage is related to SMR (Ames, 1989) or, more justifiably, to MLS (Bergtold et al., 1988; Simic and Bergtold, 1991a; Simic, 1991; Bergtold et al., 1992).

In addition to the general correlations of MLS with certain physiological factors for various species, an important relationship between MLS and dietary caloric intake (DCI) within a species was discovered by McCay and co-workers (1935; 1943) for mice and rats. More recently other investigators (for review, see Weindruch and Walford, 1988; Fishbein, 1991) have confirmed the pioneering observations of McCay and concluded that the MLS of animals fed calorically restricted diets may be about one third longer than that of animals fed ad libitum. No experiments in humans have been conducted for obvious reasons.

Comprehensive mechanisms of caloric restriction (CR) are still missing, although diverse biological effects have been observed (Masoro et al., 1982; Weindruch and Walford, 1988). A new hypothesis for the molecular mechanisms of caloric restriction effects based on the rate of DNA damage was proposed recently (Simic and Bergtold, 1991a; 1991b) and was extended further to the mechanisms of aging (Simic, 1991). Some of these concepts are described in this review and their relevance to the theory of aging assessed.

Generation of free radicals in vivo

There are numerous mechanisms for generation of free radicals in vivo (see, e.g., Simic et al., 1988; Emerit et al., 1990), and mechanisms pertinent to aging have been reviewed recently (Simic, 1991; Karam et al., 1991). Two major mechanisms of free radical generation in vivo are described below.

The mitochondrial electron-transport chain may leak an electron to oxygen via the semiquinone radical of ubiquinone (Chance et al., 1979; Nohl et al., 1986),

$$O_2 + \cdot UQ^- \rightarrow \cdot O_2^- + UQ \tag{1}$$

The resulting superoxide radical may give hydrogen peroxide via the following SOD-catalyzed reaction (McCord and Fridovich, 1969)

$$\cdot O_2^- + \cdot O_2^- + 2H^+ \rightarrow H_2O_2 + O_2 \tag{2}$$

Hydrogen peroxide generated in reaction (2) as well as by any other metabolic process may undergo the well-known Haber-Weiss reaction in which ferrous iron and some of its complexes (e.g., with DNA) act as electron donors,

$$H_2O_2 + Fe(II) \rightarrow \cdot OH + OH^- + Fe(III) \tag{3}$$

In a biosystem where reducing conditions prevail, Fe(II) can be regenerated easily by any electron donor,

$$Fe(III) + e^- \rightarrow Fe(II) \tag{4}$$

Hence, an iron ion attached to DNA may induce multiple damages (Chevion, 1988; Imlay et al., 1988).

A high metabolic rate and high dietary caloric intake should generate higher yields of $\cdot O_2^-$, H_2O_2, and $\cdot OH$. Although $\cdot O_2^-$ and $\cdot OH$ are extremely difficult, if not impossible, to detect and monitor in vivo, H_2O_2 has been observed and quantitated in body fluids (e.g., blood plasma) and cells (Szatrowski and Nathan, 1991).

Biomarkers of oxidative DNA damage

The $\cdot OH$ radicals generated by reaction (3) on DNA react rapidly with deoxyribose and DNA bases. As a consequence numerous products are generated (von Sonntag, 1987). From mechanistic considerations the same types and fractions of products are expected in most living organisms. Hence, in principle, any of these products can be used as a biomarker of oxidative DNA damage provided no other processes contribute to its genesis.

Two of the products of ·OH radical reaction with DNA bases were shown to be the most suitable sources of DNA biomarkers – thymine glycol (Tg) and 8-hydroxyguanine (8-G-OH). These two compounds are the major products of ·OH radical reaction with thymine (T) and guanine (G), respectively. Neither of these products was shown to result from enzymatic hydroxylation processes.

Being foreign to DNA, they are eliminated by repair enzymes (excision enzymes and glycosylases) as they are generated. An equilibrium level of these products on DNA, however, is always present (Bergtold et al., 1988; Richter et al., 1988; Fraga et al., 1990). Enzymatic elimination would generate both free base products (Tg and 8-G-OH) and their nucleoside moieties (dRTg and 8-dRG-OH). For example,

$$\text{DNA-Tg} \xrightarrow{\text{Ez}} \text{DNA (or DNA')} + \text{Tg} + \text{dRTg} + \text{N} + \text{P} \qquad (5)$$

where DNA' is mutated DNA, N represents unaltered nuclear material (free bases, nucleosides, nucleotides, and their polymers), and P stands for a variety of other products. Once detached from DNA some of the products of this reaction will be catabolized. The catabolized and unchanged products are taken up into the blood plasma and excreted in the urine. Tg and some other products of thymine were first observed in urine by Ames and co-workers (Cathcart et al., 1984). Since then there have been numerous other reports of urinary DNA base products (Bergtold et al., 1988; Ames, 1989; Shigenaga et al., 1991; Bergtold and Simic, 1991). Formation, measurement, and application of these products to aging and carcinogenesis have been reviewed comprehensively (Simic, 1991, 1992).

Despite their specificity, the free base products of ·OH radical reactions (Tg and 8-G-OH) do not qualify as biomarkers of oxidative DNA damage, because they may be absorbed from dietary sources through the digestive tract. Because their nucleosides are not absorbed through the gut (Cathcart et al., 1984), they qualify as true biomarkers. The structures of thymidine glycol (dRTg) and 8-hydroxydeoxyguanosine (8-dRG-OH) are shown below

thymidine glycol   8-hydroxydeoxyguanosine

(dRTg)   (8-dRG-OH)

Rate of oxidative DNA damage

Although Tg and 8-G-OH are found in DNA, and dRTg and 8-dRG-OH in urine, they cannot be used to assess the rate of oxidative DNA damage before certain facts and mechanisms are clearly established. The level of Tg and 8-G-OH in DNA depends on the rate of formation and rate of elimination of the compounds by repair enzymes. Consequently, the equilibrium levels on DNA cannot be compared quantitatively in different organisms until these two rates are determined for each organism.

Daily urinary output of dRTg and 8-dRG-OH could, in principle, represent the rate of oxidative DNA damage, provided no loss occurs from the moment of excision to the time of urine collection and quantitation by GC/MS methodologies (Bergtold et al., 1988; Bergtold and Simic, 1991). This equivalency was clearly demonstrated by irradiation experiments in mice and humans (Bergtold and Simic, 1991; Simic and Bergtold, 1991a; Simic, 1991) using ionizing radiations (e.g., γ-rays and high-energy electrons). Ionizing radiations were used because they generate a known quantity of ·OH radicals by splitting water,

$$H_2O \xrightarrow{\nu} H_2O^{·+} + e^- \tag{6}$$

$$H_2O^{·+} + H_2O \rightarrow ·OH + H_3O^+ \tag{7}$$

The quantity of ·OH radicals generated is proportional to the water content in the irradiated tissue and the energy input (E),

$$E(J) = Dose(Gy) \times mass(kg) \tag{8}$$

The results are shown in Table 2.

Increments of daily output of urinary biomarkers (difference in daily output before irradiation and one day after irradiation) were proportional to the energy input, i.e., to the number of ·OH radicals for both the mouse and human. It was concluded therefore, that these products are not being catabolized and *represent the rate of oxidative DNA damage*.

Table 2. Daily increment of urinary biomarkers of oxidative DNA damage induced by ionizing radiation (From Bergtold and Simic, 1991)

Species	Δ dRTg (nmol/d/J)	Δ 8-dRG-OH (nmol/d/J)
Human	3.1 ± 0.8	6.7 ± 1.5
Mouse	3.0 ± 0.6	6.9 ± 1.3

Factors of oxidative DNA damage

Having established the proportionality between the number of ·OH radicals and the daily output of dRTg and 8-dRG-OH in the urine, these two biomarkers can be used to measure the rate of oxidative DNA damage. There are a few parameters that alter the rate of DNA damage. In this paper only the relationship with metabolic rate and dietary caloric intake are presented because they provide sufficient information for development of the rate of DNA damage theory of aging.

Metabolic rate

Species with a high metabolic rate also have a high rate of oxidative DNA damage (Ames, 1989; Bergtold et al., 1988). Humans have the lowest rate of DNA damage among the 20 mammalian species investigated (Bergtold et al., 1992). A comparison of the rates of mouse and human oxidative DNA damage is shown in Table 3.

The yields of biomarkers are not proportional to SMR, indicating participation of other parameters, e.g., efficacy of the antioxidant defense system. In this particular case the rate of DNA repair is not a parameter.

Dietary caloric intake

Dietary caloric intake (DCI) was found to play a major role in the rate of oxidative DNA damage (Simic and Bergtold, 1991a; Simic and Bergtold, 1991b; Simic, 1991). Human data are shown in Table 4.

The effect of DCI on the rate of damage in one species (Table 4) is not as large as the effect of SMR when values for different species are compared (Table 3). Hence, the effect of DCI on MLS is not as pronounced as the effects of SMR. For mice and rats a 40% reduction of

Table 3. Yields of urinary biomarkers of oxidative DNA damage[a] for organisms with low and high metabolic rate (From Bergtold et al., 1988)

Species	Specific metabolic rate (cal/g/d)	Thymidine glycol dRTg (nmol/kg/d)	8-Hydroxy-2'-deoxy-guanosine (8-dRG-OH) (nmol/kg/d)
Human	25	0.3 ± 0.1	0.3 ± 0.1
Mouse	180	7.3 ± 1	11 ± 2

[a]These are the mean values for 16 mice and 23 humans; errors given are standard deviations, reflecting variations between individuals.

Table 4. Yields of urinary thymidine glycol in human as a function of dietary caloric intake of same type of diet (From Simic and Bergtold, 1991b)

Daily caloric intake (kcal/d)	dRTg[a] (nmol/kg/d)
2100	0.27
1200	0.11

[a]Average of 4 CR cycles.

DCI extends MLS by one third (Weindruch and Walford, 1988). DCI may also affect the fidelity of DNA repair (Turturro et al., 1991) and general free radical damage observed in organisms (Yu, 1991). Hence, a reduced rate of oxidative damage to DNA, increased fidelity of DNA repair, and decreased overall free radical damage constitute a molecular basis for extension of MLS in organisms subjected to optimal CR.

Rate-of-damage theory

A proven relationship between cumulative effects, such as onset of cancer and death (MLS), and rate (intensity) of damage would greatly simplify and shorten mechanistic studies of aging and toxicological testing. An attempt to relate such factors is presented in Table 5.

As shown in Table 5, there is an apparent relationship between the endogenous rate of oxidative DNA damage (daily output of dRTg in urine per unit mass), MLS, and the probability of cancer onset for the mouse and human. In general, an inverse relationship exists between SMR and MLS. As shown in Table 1, there are numerous exceptions to that rule. Because the cow, horse, and elephant have much lower SMR than the human, they would be expected to live longer. In contrast an inverse relationship between SMR and the rate of oxidative DNA damage (Simic, 1991) better explains these exceptions (Bergtold et al., 1992).

Table 5. Specific metabolic rate (SMR), maximum life span (MLS), cancer onset, the yield of thymidine glycol (dRTg) in urine under normal conditions, and the increase ΔdRTg induced by ionizing radiations (x- and γ-rays) (From Simic and Bergtold, 1991a, 1991b)

Species ($n = 5$)	SMR (cal/g/d)	MLS years	Cancer onset (2% probability) years	dRTg (nmol/kg/d)	Δ dRTg (nmol/d/J)
Mouse	180	3.5	~1	7.2 ± 1	3.0 ± 0.6
Human	25	100	~50	0.29 ± 0.1	3.1 ± 0.8

For a constant rate of DNA damage ($\Delta D/\Delta t$), the total life span damage (TLSD) is a function of time (t),

$$\text{TLSD} = (\Delta D/\Delta t) \times t \qquad (9)$$

For a changing rate of damage, a more likely occurrence due to dietary (Table 4) and other parameters,

$$\text{TLSD} = \int_0^{\text{LS}} \frac{dD(t)}{dt}\, dt \qquad (10)$$

Although TLSD and LEI (or LEP) are related, they may differ substantially because of variability of the efficacy of bioenergetics, inhibition of damaging species, and chemical repair processes (see Table 1). Hence, the yield of urinary biomarkers of DNA damage represents net $\Delta D/\Delta t$, which cannot be quantitatively correlated to energy input. On the other hand, although biological consequences are a function of $\Delta D/\Delta t$, they depend on the efficacy and fidelity of enzymatic repair (Friedberg, 1985), polymerase efficacy (Shibutani et al., 1991), individual (or species) physiology (Cutler, 1985), and the genetic constitution of an organism. Therefore, rate-of-damage correlations can be observed only by strictly controlling or taking into account relevant parameters.

Cancer and aging, the two biological consequences with the most complex etiology, also may be a function of other types of endogenous DNA damage not yet understood. Additional sources of DNA damage, such as alkylation (Lindahl et al., 1988) or biochemical control processes, however, would not alter the proposed concepts, but would add some additional parameters, provided these parameters made a significant contribution.

Despite enzymatic repair and other defenses, continuous DNA damage may progressively alter the genomic sequence (Basu et al., 1989; Shibutani et al., 1991; Moraes et al., 1990), as indicated by aging (Vijg, 1990) to the extent that it contributes to the development of cancer, e.g., by activation of oncogenes or inactivation of suppressor genes. Consequently, the probability of mutation and concomitant development of tumors and reduction of MLS increases with the rate of oxidative DNA damage.

The rate of oxidative DNA damage (ΔdRTg) induced by an exogenous, free radical-generating agent (radiation) under the same conditions (Table 5) is the same in both organisms (Bergtold and Simic, 1991). Consequently, there is no proportionality between dosage of an exogenous agent and the total (the sum of endogenous and exogenous) rate of damage. These observations are crucial to interspecies extrapolation for exogenous toxicants.

28

Conclusions

The lack of understanding of aging mechanisms makes the problem of life extension speculative and challenging. If aging represents accumulation of damage, then deleterious or beneficial effects of an agent or process on aging may be assessed through biomarkers that reflect either cumulative effects or the rate of damage. The identification of a biomarker requires a mechanistic understanding of its genesis both in model systems and in vivo.

Free radicals or their precursors are generated in vivo by endogenous processes and by numerous exogenous agents. These free radical processes may lead to permanent damage, defined as DNA sequence changes, i.e., mutations. The accumulation of damage contributes to development of degenerative diseases (cardiovascular, cancer) and aging.

Studies of factors affecting damage indicate that the rate of bio-damage may be modulated, which may be critical to cancer prevention and retardation of aging.

The apparent relationship between the rate of DNA damage and MLS suggests two major conclusions: (1) agents or processes that increases the rate of damage may accelerate the rate of aging and decrease life span, and (2) the rate of damage apparently may be decreases by lowering specific metabolic rate, decreasing dietary caloric intake, inhibiting damaging species, or increasing fidelity of DNA repair.

The measurement of rate of accumulation of damage (intensity factors) that contributes to the appearance of endpoints through direct or indirect monitoring becomes highly attractive because it substantially reduces the required experimental time. Rate studies therefore become an economical approach to the investigation of long-term cumulative processes. The rate-of-damage theory, which emphasizes DNA as the relevant and critical biological target of damage, properly validated through classical testing and implemented, could save society enormous amounts of time and money on research and tests. In addition, rapid preliminary assessment of rate parameters would contribute critically to better experimental design and would strengthen the final conclusions based on animal and human studies.

Acknowledgments. Partial support of this work was provided by DNA and the ILSI Risk Science Institute. Special thanks are due to Gloria Wiersma for technical assistance. Comments and suggestions by Lorraine E. Twerdok, Miro Radman, Richard Cutler, and Maurice Fox are gratefully acknowledged.

Ames, B. N. (1989) Mutagenesis and carcinogenesis: Endogenous and exogenous factors. Environ. Mol. Mutagen. 14: 16, 66–77.

Basu, A. K., Loechler, E. L., Leadon, S. A., and Essigman, J. M. (1989) Genetic effects of thymine glycol: Site-specific mutagenesis and molecular modeling studies. Proc. Natl. Acad. Sci. USA 86: 7677–7681.

Bergtold, D. S., Simic, M. G., Alessio, H., and Cutler, R. G. (1988) Urine biomarkers for oxidative DNA damage, in: Oxygen Radicals in Biology and Medicine, pp. 483–490. Eds M. G. Simic, K. A. Taylor, J. F. Ward and C. von Sonntag. Plenum Press, New York.

Bergtold, D. S., and Simic, M. G. (1991) Hydroxy radical in radiation dosimetry and metabolism: Dietary caloric effect, in: Trends in Biological Dosimetry, pp. 21–32. Eds B. Gledhill and F. Mauro. J. Wiley & Sons, New York.

Bergtold, D. S., Cutler, R. G., and Simic, M. G. (1992) to be published.

Cathcart, R., Schwiers, E., Saul, R. L., and Ames, B. N. (1984) Thymine glycol and thymidine glycol in human and rat urine: A possible assay for oxidative DNA damage. Proc. Natl. Acad. Sci. (USA) 81: 5633–5637.

Chance, B., Sies, H., and Boveris, A. (1979) Hydroperoxide metabolism in mammalian organs. Physiol. Rev. 59: 527–605.

Chevion, M. (1988) A site-specific mechanism for free radical induced biological damage: The essential role of redox-active transition metals. Free Radical Biol. Med. 5: 27–37.

Cutler, R. G. (1985) Antioxidants and longevity of mammalian species, in: Molecular Biology of Aging, pp. 15–73. Eds A. D. Woodhead, A. D. Blackett and A. Hollaender. Plenum Press, New York.

Dizdaroglu, M., and Bergtold, D. S. (1986) Characterization of free radical-induced base damage in DNA at biologically relevant levels. Analyt. Biochem. 156: 182–188.

Dizdaroglu, M. (1991) Chemical determination of free radical-induced damage to DNA. Free Radical Biol. Med. 10: 225–242.

Emerit, I., Packer, L., and Auclair, C. (1990) Antioxidants in therapy and Preventive Medicine. Plenum Press, New York.

Fishbein, L. (Ed.) (1991) Biological Effects of Dietary Restrictions. Springer-Verlag, New York.

Fraga, C. G., Shigenaga, M. K., Park, J. W., Degan, P., and Ames, B. N. (1990) Oxidative damage to DNA during aging. Proc. Natl. Acad. Sci. 87: 4533–4537.

Friedberg, E. C. (1985) DNA Repair. W. H. Freeman & Co., New York.

Harman, D. (1981) The aging process. Proc. Natl. Acad. Sci. USA 78: 7124–7128.

Imlay, J. A., Chin, S. M., and Lynn, S. (1988) Toxic DNA damage by hydrogen peroxide through the Fenton reaction in vivo and in vitro. Science 240: 640–642.

Karam, L. R., Bergtold, D. S., and Simic, M. G. (1991) Biomarkers of OH radical damage in vivo. Free Rad. Res. Comms. 12–13: 11–16.

Lindahl, T., Sedgwick, B., Sekiguchi, M., and Nakabeppu, Y. (1988) Regulation and expression of the adaptive response to alkylating agents. Ann. Rev. Biochem. 57: 133–157.

Masoro, E. J., Yu, B. P., and Bertrand, H. A. (1982) Action of food restriction in delaying the aging process. Proc. Natl. Acad. Sci. (USA) 79: 4239–4241.

McCay, C. M., Crowell, M. F., and Maynard, L. A. (1935) The effect of retarded growth upon the length of the life span and upon the ultimate body size. J. Nutr. 10: 3.

McCay, C. M., Sperling, G., and barnes, L. L. (1943) Growth, aging, chronic diseases and life span in rats. Arch. Biochem. Biophys. 2: 469.

McCord, J. M., and Fridovich, I. (1969) Superoxide dismutase. J. Biol. Chem. 244: 6049–6055.

Moraes, E., Keyse, S., and Tyrell, R. (1990) Mutagenesis by hydrogen peroxide treatment of mammalian cells: A molecular analysis. Carcinogenesis 11: 283–293.

Nohl, H., and Jordan, W. (1986) The mitochondrial site of superoxide formation. Biochem. Biophys. Res. Comm. 138: 533–539.

Pearl, R. (1928) The Rate of Living. Alfred Knopf, New York.

Richter, C., Park, J. W., and Ames, B. N. (1988) Normal oxidative damage to mitochondrial and nuclear DNA is extensive. Proc. Natl. Acad. Sci. USA 85: 6465–6467.

Rubner, M. (1908) Das Problem der Lebensdauer und Seine Beziehungen Zun Wachstum und Ernahrung, Oldenbourg, Munich.

Sacher, G. A. (1977) Life table modification and life prolongation, in: Handbook of the Biology of Aging, pp. 582–638. Eds C. E. Finch and L. Hayflick. Van Nostrand Reinhold, New York.

Shibutani, S., Takeshita, M., and Grollman, A. P. (1991) Insertion of specific bases during DNA synthesis past the oxidation-damaged base 8-oxodG. Nature 349: 431–434.

Shigenaga, M. K., and Ames, B. N. (1991) Assays for 8-hydroxy-2'-deoxyguanosine: A biomarker of in vivo oxidative DNA damage. Free Radical Biol. Med. 10: 211–216.

30

Simic, M. G., Taylor, K. A., Ward, J. F., and von Sonntag, C. (Eds.) (1988) Oxygen Radicals in Biology and Medicine. Plenum Press, New York.

Simic, M. G., and Bergtold, D. S. (1991a) Dietary modulation of DNA damage in humans. Mutat. Res. 250: 17–24.

Simic, M. G., and Bergtold, D. S. (1991b) Urinary biomarkers of oxidative DNA base damage and human caloric intake, in: Biological Effects of Dietary Restriction, pp. 217–225. Ed. L. Fishbein. Springer-Verlag, New York.

Simic, M. G. (1992) Urinary biomarkers and the rate of DNA damage in carcinogenesis and anticarcinogenesis. Mutat. Res. (in press).

Szatrowski, T. P., and Nathan, C. F. (1991) Production of large amounts of hydrogen peroxide by human tumor cells. Cancer Res. 51: 794–798.

Turturro, A., and Hart, R. W. (1991) Caloric restriction and its effects on molecular parameters especially DNA repair, in: Biological Effects of Dietary Restriction, pp. 185–190. Ed. L. Fishbein. Springer-Verlag, New York.

Vijg, J. (1990) DNA sequence changes in aging: How frequent, how important? Aging 2: 105–123.

von Sonntag, C. (1987) The Chemical Basis of Radiation Biology. Taylor and Francis, New York.

Weindruch, R., and Walford, R. L. (1988) The retardation of aging and disease by dietary restriction. Charles C. Thomas Publishers, Springfield, Illinois.

Yu, B. P. (1991) Free radicals and modulation by dietary restriction. Age & Nutrition 2: 84–89.

Free Radicals and Aging
ed. by I. Emerit & B. Chance
© 1992 Birkhäuser Verlag Basel/Switzerland

Genetic stability and oxidative stress: Common mechanisms in aging and cancer

Richard G. Cutler

Gerontology Research Center, National Institute on Aging, 4940 Eastern Avenue, Baltimore, MD 21224, USA

Summary. Much evidence indicates that age-related diseases as well as the aging process itself may be a result of genetic instability, resulting in cells changing their proper state of differentiation. Such genetic changes could be of an epigenetic or genetic nature (or both) and suggest that mechanisms acting to stabilize the differentiated state of cells could be major determinants of animal species' general health and longevity. There is also much data indicating that a causative factor in cancer is genetic instability, suggesting the importance of genetic stabilized factors governing the frequency of this disease. Possibly related to these two observations is the good positive correlation between rate of cancer and the rate of aging of different mammalian species. In addition, cells from longer-lived species appear to have a higher intrinsic ability to maintain their proper differentiated state and animals whose life span has been extended by food restriction also have a postponed onset of cancer. These data suggest the possibility that genetic instability may represent a common causative mechanism for both aging and cancer. This suggestion is supported by evidence indicating that oxidative stress can increase cancer frequency and that longer-lived species appear to have a lower oxidative stress state. However, there appears to be a serious lack of evidence indicating that physical and chemical mutagenic agents accelerate aging as they do cancer. For these and other considerations, it is proposed that aging is not a result of genetic instability as are many different age-related diseases. Thus, physiological aging and some age-related diseases could have separate and distinct mechanisms of causation and control.

One of the key issues in biogerontology today concerns the complexity of the aging process and whether it might ever be possible to decrease the rate of human aging (Weiss, 1966). Health care systems appear to be based on the concept that either human aging does not play a critical role in the health status of individuals and/or little can be done about the aging process. Consequently, it may be concluded that it is best to focus on specific diseases of aging rather than the aging process per se.

There is, however, some evidence based on comparative and evolutionary data indicating that, in spite of the vast complexity of the ageing processes, relatively few mechanisms might exist that govern aging rate (Cutler, 1976a,b; 1982; 1984a). Such a possibility should not be left unexplored because, if true, it may prove possible to significantly increase the healthy years of human life span through relatively inexpensive means.

This concept forms the basis of the longevity determinant hypothesis of aging and longevity and will be reviewed now along with the presentation of recent experimental data testing its validity. Here, I shall place emphasis on experiments from our laboratory examining if (1) cell dysdifferentiation represents a primary aging process, (2) if reactive oxygen species cause dysdifferentiation and (3) if mechanisms acting to stabilize the proper state of differentiation, such as antioxidants, represent a class of longevity determinants.

The goal of reducing the overall aging rate in humans runs immediately into the problem of the vast complexity of normal human biology and what little we presently know in this area of science. Indeed, it is usually argued that the causes of aging are likely to be so complex that we can only begin to understand aging after much more is learned of normal biology. There is also the argument that aging is likely to have multiple causes and it is unwise to speculate that relatively few genetic or biochemical means of intervention are possible or could have any significant impact on decreasing human aging rate.

This all may be true, but these pessimistic arguments are not too convincing to me when critically evaluated. The fact is that the problem of the complexity of the aging process and of the possibility of developing intervention strategies to increase the healthy years of life span and productivity by reducing human aging rate has simply not received serious scientific attention or evaluation. I have presented an alternative argument suggesting that, in spite of the vast complexity of aging, the processes governing aging rate or life span may be much less complex and therefore subject to intervention in the near future (Cutler, 1982; 1984a,b).

The basis of this prediction comes from evolutionary and comparative studies of mammalian species closely related to one another biologically and evolutionarily but having substantial differences in their life span and aging rate. These studies led to the formulation of the 'longevity determinant gene hypothesis' (Cutler, 1975; 1976a,b; 1978). This hypothesis predicts that aging is a result of normal biological processes necessary for life but which have long term negative or aging effects on the organism. These normal biological processes have been divided into two major categories: (1) developmentally-linked biosenescent processes and (2) the continually-acting biosenescent processes. According to this hypothesis, longevity of a species is related to how efficiently the aging effect of these common problems are reduced.

I will not deal with the first category today, but will note that they include many of the hormones associated with development. For example, most pituitary, thyroid, adrenocorticoid, ovarian and testicular hormones have been found to have long term aging effects. An important exception may be growth hormone, which does appear to have many rejuvenative effects when administered to old animals (Everitt and

Meites, 1989). However, long-term studies of growth hormone administration have not yet been conducted to determine possible side effects or if healthy years of life span may be increased.

Our research has focused on the second category, the continuously-acting biosenescent processes, in testing the longevity determinant gene hypothesis. Here, we have examined the possible aging effects of oxygen metabolism, which is known to produce what are called 'reactive oxygen species' during normal energy production. Reactive oxygen species can interact with the cell's genetic apparatus and alter its proper state of differentiation. These changes in differentiation, as they might occur during normal aging, have been called dysdifferentiation and could explain many aspects of the normal aging process.

The unique feature of the longevity determinant gene hypothesis is that it suggests that all animal species share common aging causes and common mechanisms of regulating aging rate. Although the processes causing aging may be very complex, it is predicted that less complex key processes exist that regulate aging rate or life span. For example, human and chimpanzee are closely related to one another, both evolutionarily (about 3–5 million years from a common ancestor) and biologically (DNA sequences are about 98% common). Because of the remarkable biological similarities between human and chimpanzee, we would expect the aging processes to be similar qualitatively and, as far as we know, they are. Yet, in spite of these remarkable biological similarities, the chimpanzee has a life span about one half that of human (50 y vs 100 y) and accordingly appears to age about twice as fast in all biological aspects and exhibits the same diseases of aging but in half the time.

Unfortunately, there has been some confusion with the term "Longevity Determinant Genes" or "LDGs". Clearly, most genes of an animal's genome are likely to be involved directly or indirectly in determining duration of health and longevity. In fact, mutations of most genes are likely to be deleterious and shorten life span. Longevity determinant genes (LDGs) however represent a new and different concept, a specific class of genes which are proposed to have evolved and are specialized in determining longevity or health duration of an animal. In addition, as previously mentioned, such a class of LDGs may be relatively small in number (or information content) with respect to the size of the genome itself. Indeed, only a few key regulatory genes governing expression of LDGs may exist.

These LDGs were first called "Life Maintenance Genes" governing Life Maintenance Processes (Cutler, 1972; 1976a). They were later renamed Longevity Determinant Genes (Cutler, 1982) for clarity of terminology. The term Longevity Assurance Genes has also been suggested (Sacher, 1978).

So the central question we ask is what is so unique about human biology as compared to the chimpanzee or other shorter-lived animals that allows human to be so long-lived? Because of the strikingly similar biology between human and chimpanzee as well as their close evolutionary relatedness, it is argued that the genetic and biochemical differences governing their life span differences are also not likely to be very great. Thus, we ask, how complex biologically speaking are the processes governing human aging rate?

Biological nature of human aging and longevity

Human life expectancy (average survival time) for most of the time *Homo sapiens* has been on this earth ($\approx$ 100,000 y) is generally believed to be about 20 to 30 years (Acsádi and Nemeskéri, 1970). Only recently over the last 400 years or so in developed nations has life span expectancy increased, and this increase has been rather dramatic (about 40 to 50 years). The reason behind this increase is not altogether clear, but it certainly includes reducing environmental hazards and better nutrition and health care. The important point to be made here, however, is that in primitive cultures where life expectancy was about 30 years, few people lived long enough to suffer appreciably from the processes of senescence or aging. Death was almost entirely due to environmental hazards and infectious diseases. Today in developed countries, although people have life expectancies in the range of 70 to 80 years, they are not living younger longer but are instead actually living older longer. This is because the increased life expectancy occurred with life span remaining essentially constant. That is, the increase of life expectancy did not occur as a result of slowing down normal aging processes but instead by reducing the major external environmental hazards to life. Thus, the present day problem of an older population in developed nations is actually an artifact of our civilization.

Normal human aging results in a steady decline in functional capacity of essentially every physiological and mental processes, beginning shortly after the age of sexual maturation (Bafitas and Sargent, 1977). The slope of the decrease in function with age could be taken to generally reflect an average of overall aging rate of the body. A fundamental question of biological research in gerontology today is to understand the mechanisms leading to this age-dependent decrease in function. With increasing age there also occurs an increased incidence of disease, which finally results in the death of the individual. An example of such a disease is cancer (Kohn, 1978; 1982).

It is important to emphasize here the apparently great uniformity of aging between different individuals and in the biological functions that are affected. With increasing age, it is expected theoretically that essentially all body functions would decline with increasing chronological age synchronously (Maynard Smith, 1966). At least nothing biologically is expected to improve with age. Persons that are able to live past 100 years do so not because they age more slowly but instead because they age more uniformly. That is, they appear to not suffer from any particular weak link in their body such as from heart disease or diabetes. This information may be extremely important in the development of means to increase the healthy years of life span. The theoretical point can be made that, even if the major diseases causing death today could somehow be completely eliminated, the impact on increased life expectancy over the entire country is not nearly as great as one might believe. For example, elimination of all cancer results in an increase of only about 2 years of life expectancy. The reason for this result is that most people suffer from the major killer diseases today are 65 years of age or older. Thus, the removal of a major disease causing death of the elderly would simply result in uncovering a new disease or problem since essentially every aspect of body function is being reduced through the aging process.

Such studies indicate that significant increase in the healthy years of life span of the general population of a nation is only likely to be achieved by actually reducing uniformly the rate of aging of the entire body, not by a piecemeal approach of reducing or eliminating specific disease processes. Thus, efforts being made in the field of cancer or even Alzheimer's disease may not achieve dramatic reduction in medical costs if effective therapy uncovers new diseases characteristic of many patients as they grow older.

The longevity determinant gene hypothesis

If significant gain to reduce the high costs of human aging to society does indeed require a uniform reduction in human aging rate, then how might this objective be achieved? The answer to this question of course depends on how complex the problem is. There is unfortunately very little hard data in this area but a few studies based on an evolutionary and comparative biology of species closely related to one another but having differences in aging rate have suggested that key longevity determinant processes may indeed exist.

The basis of such studies is the fact that different mammalian species do indeed have different aging rates and that they age qualitatively in a similar way. In addition, estimates have been made as to how fast life span evolved along the hominid ancestral-descendant sequence leading to the *Homo sapiens* species. It was found to increase at about 14 years per 100,000 years about 100,000 years ago (Cutler, 1975; 1976b). Such a high rate suggests that alterations in gene sequence are not likely to be involved in the rapid evolution of life span. Instead, these data support the concept that few changes in gene regulatory processes occurred during hominid evolution, resulting in a uniform decrease in aging rate.

These evolutionary comparative studies led to the proposal that key longevity determinant genes of a regulatory nature might exist that are capable of governing the aging rate of the entire body. This concept is in contrast to what has been generally believed to be the case. This is where aging is a result of biological functions equally as complex as the organism itself and that life span or aging rate is determined by thousands of genes operating in highly complex mechanisms unique for each cell or tissue in the organism.

Another prediction of the longevity determinant gene hypothesis is that aging is not genetically programmed but instead is the result of normal biological processes. All mammalian species have remarkably similar biological processes (particularly chimpanzee and human). The question then arises of what normal biological processes are most responsible for causing aging and what key mechanisms evolved to control their rate of their expression?

To answer this question, we began a comparison of human biology with that of other closely related species as a function of their life span with the hope of discovering small differences that could help explain why humans are so long-lived. One such comparative study is a plot of life span vs specific metabolic rate (Cutler, 1983). These data indicate that aging rate of different species is related to their metabolic rate. Thus, it appears that by-products of oxygen metabolism may have a role in causing aging. These data also indicate that most animals consume a constant amount of energy over their life span but that humans are exceptional in consuming about four times the energy over their life span.

Oxidative stress as a primary aging process

It is well known now that utilization of oxygen represents an efficient mechanism for aerobic organisms to generate energy, but during this process, by-products called reactive oxygen species are also created that could damage a cell (Halliwell and Gutteridge, 1989). Indeed, all aerobic organisms require a vast complex network of defense mechanisms to reduce the toxic effects of the by-products of oxidative energy metabolism. There are a number different sources of reactive oxygen species and a number of defense mechanisms have already been identified.

Many different diseases, some related to aging, appear to be reduced by dietary antioxidant levels, indicating the biological significance of antioxidants to protect against age-related diseases (Am. J. clin. Nutr. 1991). There is also interesting data indicating that the onset frequency of cancer incidence leading to death is related to the aging rate of a

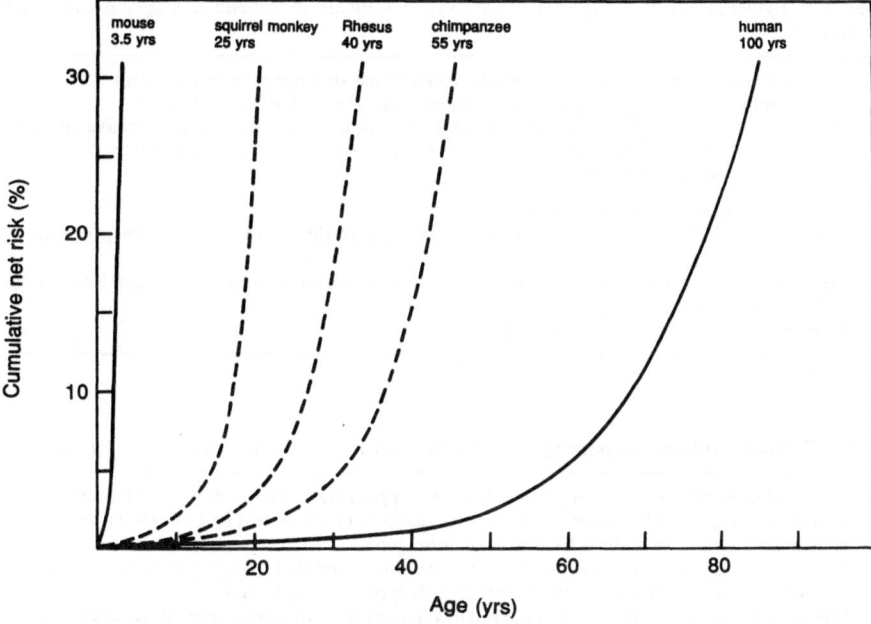

Figure 1. Cumulative net risk of death from cancer as a function of age and maximum life span potential. This family of curves represents typical data describing the onset frequency of all types of cancer with age in different mammalian species. The data are plotted as Probability (P) being proportional to (age)[4]. Curves for mouse and human are most reliable (solid curves based on the species' life span and data indicating relative rank order of longer-lived mammals having less cancer incidence with age. Figure from Cutler and Semsei, 1989.

species. These data imply that not only does cancer incidence increase exponentially with chronological age but also that the rate of this increase is related to the aging rate of the species (Fig. 1). This important results suggests that whatever mechanism protects human from cancer so efficiently as compared to other species may also be responsible for the slow aging rate humans possess.

Much evidence indicates that cancer is caused by genetic alterations, resulting in an improperly differentiated cell (Cutler, 1991). Such genetic alterations could be caused by reactive oxygen species (Table 1). Thus, a common mechanism which could cause both cancer and aging is reactive oxygen species causing a genetic instability. Such a idea is consistent with the dysdifferentiated hypothesis of aging, where aging is predicted to be a result of cells drifting slowly away from their proper state of differentiation (Ono and Cutler, 1978; Cutler, 1985a). There is considerable evidence supporting dysdifferentiation as a primary cause of aging (Table 2) and reactive oxygen species to be related to causes of

Table 1. Evidence for the importance of oxygen radicals in genetic damage, disease and cancer

---

1. *Dietary carcinogens and anti-carcinogens, oxyradicals and degenerative diseases*
   Many mutagens and carcinogens act through generation of oxygen radicals.
   Oxygen radicals also play a major role as endogenous initiators of degenerative diseases (cancer, heart disease) through DNA damage, mutation and promotion effects.
   Bruce Ames, Science 221: 1256, 1983.

2. *Prooxidant state and tumor promotion*
   There is convincing evidence that active oxygen, peroxides and radicals can promote or initiate cells to neoplastic growth.
   Prooxidant state can be caused by xenobiotics, metabolites, inhibitors of antioxidant state, and membrane-active agents.
   Peter A. Cerutti, Science 227: 375, 1985.

Table 2. Some evidence supporting dysdifferentiation as a possible causative factor in aging

---

1. Cells in tissue culture from longer-lived species appear more stable in terms of spontaneous mutation, greater cell-doubling number to reach a crisis state (Röhm, 1981) and lower sensitivity to transformation by chemical mutagens.
2. Rate of accumulation of chromosomal aberrations is related to aging rate of the species.
3. Age-dependent accumulation of abnormal cell types (metaplasias).
4. Qualitative and quantitative age-dependent changes in gene expression, structural proteins and enzyme levels have been found.
5. Basic housekeeping functions of a cell appear to be adequate in older organisms.

Table 3. Evidence supporting active oxygen species as a possible causative factor in dysdifferentiation

---

1. Low concentration of mutagenic/carcinogenic agents induce improper gene regulation, some by induction of transposable elements involved in control of gene expression.
2. Active oxygen species react with chromatin, oxidize specific DNA bases, and cause single-strand DNA breaks.
3. Active oxygen species induce chromosomal aberrations, and longer-lived species have a slower rate of accumulation of chromosomal aberrations.
4. Longer-lived species show a slower rate in accumulation of lipofuscin age pigments.
5. Aging rate is proportional to metabolic rate (cal/g/day) for many different mammalian species.

---

aging (Table 3). Taking these data together provides more support that aging may be a result of genetic instability leading to dysdifferentiated cells, which finally causes decline in physiological function.

An important aspect of this hypothesis is that it predicts that life span is a result of factors acting to stabilize proper differentiated state of cells. That is, with human being the longest-lived mammalian species, this hypothesis would predict that human cells should be the most stable in maintaining their proper state of differentiation. If this is true, then it would be important to know what mechanism acts to stabilize cellular

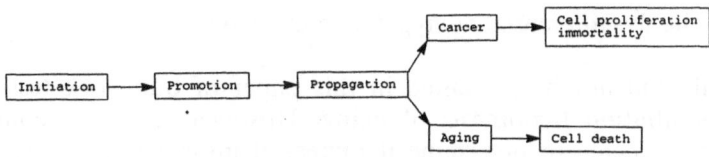

Figure 2. Multistep model of aging and cancer. This model represents the proposed multistep phenomena of both cancer and aging in their progress to increasing stages of dysdifferentiation. Ultimate end result of transformed cancer cell is cell proliferation immortality and of the aging process (other than cancer) is cell death. Figure from Cutler and Semsei, 1989.

differentiation. Essentially nothing is known of mechanisms acting to stabilize the proper differentiated state of cells once the genetic program of development has been completed. Thus, in light of the longevity determinant hypothesis, this area of research should deserve increased attention. Also, if reactive oxygen species do play a role in destabilizing the genetic apparatus of cells, the antioxidants and other defense mechanisms acting to lower cellular oxidative stress may be an important component determining genetic stability.

These and other data suggest the hypothesis that oxygen radicals may cause aging. This model can, however, be expanded into what we have called the multistep model of cancer and aging, as shown in Figure 2. The important feature of this model is the possiblity of taking advantage of much of what has already been learned in cancer research (such as to the multistep model of carcinogenesis) and applying it towards an advancement of an integrated model for research in aging and cancer mechanisms. For example, factors governing aging rate could act at the initiation stage (mutagens) or the progressive stage (epigenetic). Thus, dietary factors known to be possible causes of cancer (initiators or promotors) may also be acting to accelerate aging rate, and in turn those agents found to protect against cancer (dietary antioxidant) may also be of some benefit in protecting against abnormally high aging rate.

These data briefly reviewed here have suggested the working hypothesis shown in Table 4 that has guided much of our recent experimental work. Most recently, an extension of the dysdifferentiation model has been proposed. It is now recognized that small changes in the proper differential state of cells may have their greatest impact in certain areas of the brain involved in regulating homeostatic functions. Such tissues are represented by the various areas of the pituitary gland and the hypothalamus. Such localized brain tissues could be under high oxidative stress and often consist of a relatively few number of cells. Thus, alternations of regulatory tissues are predicted to be the most important targets in the dysdifferentiation model of aging.

Testing the dysdifferentiative hypothesis of aging

Details will not be presented in this paper of our work testing the dysdifferentiation hypothesis of aging. However, we have conducted some experiments to determine if increased improper gene expression does occur with age (Ono and Cutler, 1978; Cutler, 1985a). These results are briefly summarized in Table 5. Because of evidence that an increase of c-*myc* expression may be caused by genetic alterations, it appears possible that such genetic alterations may also be accumulating with age in normal-appearing tissue, increasing the probability steadily with age of the transformation of every cell in the tissue (Semsei et al., 1989; Cutler and Semsei, 1989).

Table 4. The working hypothesis being tested in recent experimental work

1. Aging is in part a result of changes in proper gene expression and regulation. This process has been called dysdifferentiation.
2. Active oxygen species contribute to the dysdifferentiation process.
3. Mechanisms that act to reduce the dysdifferentiation effects of active oxygen species prolong the proper state of differentiation and thus the longevity of the animal.
4. Human longevity is a result of unusually high efficiency of these stabilizing mechanisms.
5. Aging and cancer are a result of the same dysdifferentiation process.

Table 5. Summary of investigation of possible age-dependent changes of specific genes

No change with age: (brain, liver, kidney)
1. alpha-fetoprotein (mouse)
2. casein (mouse)
3. alpha and beta hemoglobin (human WI-38 cells)

Increase with age: (brain, liver, kidney)
1. alpha and beta hemoglobin, quantitative (mouse and human)
2. mouse leukemia virus (MuLV), qualitative (mouse)
3. mouse mammary tumor virus (MMTV), qualitative (mouse)
4. c-*myc* oncogene, quantitative (3rd Extron, mouse)

Table 6. Summary of antioxidant comparison results

Positive correlation	No correlation	Negative correlation
1. Cu/Zn SOD	1. Ascorbate	1. Catalase
2. Mn SOD	2. Retinol	2. Glutathione
3. Carotenoids		3. Glutathione peroxidase
4. Alpha tocopherol		
5. Urate		
6. Ceruloplasmin		

Testing the role reactive oxygen species may have in aging

Our first approach in testing the concept that aging may in part be caused by reactive oxygen species was carried out by determining if longer-lived species had superior defense mechanisms (antioxidants). According to our hypothesis, longer-lived species would not be predicted to have superior types of protective mechanisms (a qualitative difference) but rather more of the same type of protective mechanisms (a quantitative difference) resulting from genetic changes occurring in regulatory genes.

Results of some of these experiments are summarized in Table 6 and shown specifically for superoxide dismutase in Figure 3. These data generally indicate that longer-lived species, particularly human, do indeed have higher levels of antioxidants per amount of oxygen utilized. A simple but convincing experiment clearly demonstrating this conclusion was carried out by comparing the spontaneous autoxidation rate in air of whole tissue homogenates taken from animals having different life spans. Rate of autoxidation was measured by the amount of peroxides being produced (TBA assay). A typical result is shown in Figure 4 for brain tissue, indicating an increased resistance of autoxidation with

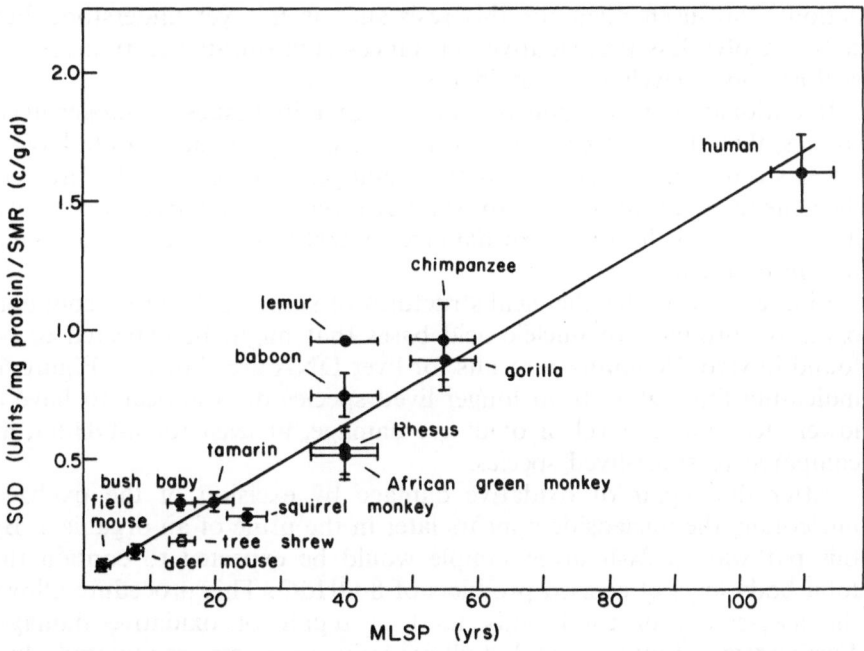

Figure 3. Superoxide dismutase (SOD) concentration per specific metabolic rate (SMR) in liver of mammals as a function of maximum lifespan potential (MLSP). Figure from Cutler, 1985b.

42

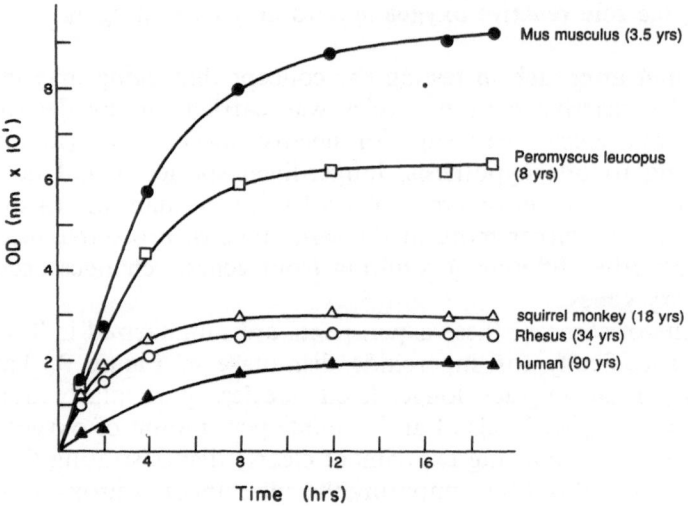

Figure 4. Kinetics of autoxidation of brain homogenate. Figure from Cutler, 1985c.

increasing life span of the species. Thus, in this experiment, human brain tissue was found to be the most resistant to spontaneous autoxidation. The mechanism for this resistance is not yet understood but could involve less peroxidative substances (unsaturated fatty acids) as well as higher levels of antioxidants.

If antioxidant protection is indeed higher in tissues of longer-lived species, then it would be expected that these same tissues should have a lower steady state level of oxidative damage. Since our model predicts the genetic apparatus of cells to be a key target to oxidative damage, we have compared the oxidative damage in DNA as a function of aging rate in different species.

Figure 5 shows the chemical structures of some of the more common oxidative products of nucleic acid bases that might be expected to be found in vivo. Preliminary results for liver DNA are shown in Figure 6, indicating that DNA from longer-lived species does appear to have a lower steady state level of oxidative damage, at least for 8-OHdG, as compared to short-lived species.

After the repair of oxidative damage by excision of the oxidized nucleotide, the nucleoside appears later in the urine of an organism. By this pathway, a 24-h urine sample would be expected to contain the total body level of excised products of 8-OHdG. This procedure allows the assessment of total body level of repair of oxidative damage. Furthermore, if we assume that all oxidative products are repaired, then the amount of 8-OHdG detected could represent rate of DNA damage as well as rate of repair.

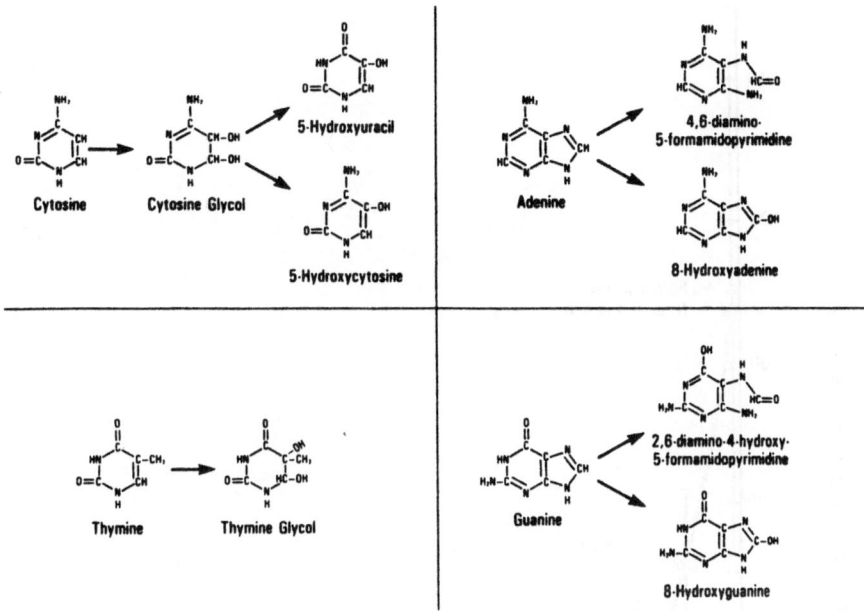

Figure 5. Major modifications of nucleic acid bases formed by the superoxide radical. Figure from Jackson et al., 1989.

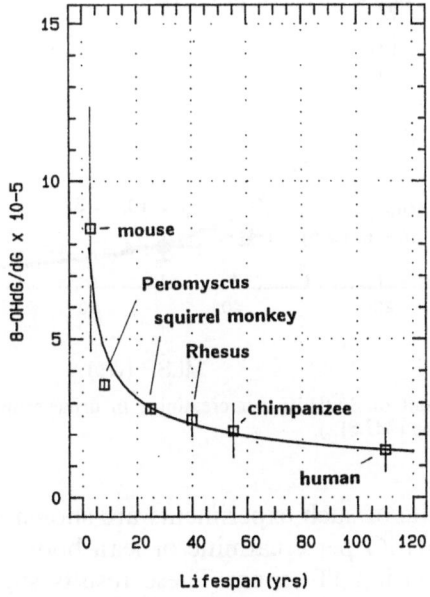

Figure 6. Steady state level of 8-OHdG in liver DNA in species having different life spans (MLSPs).

44

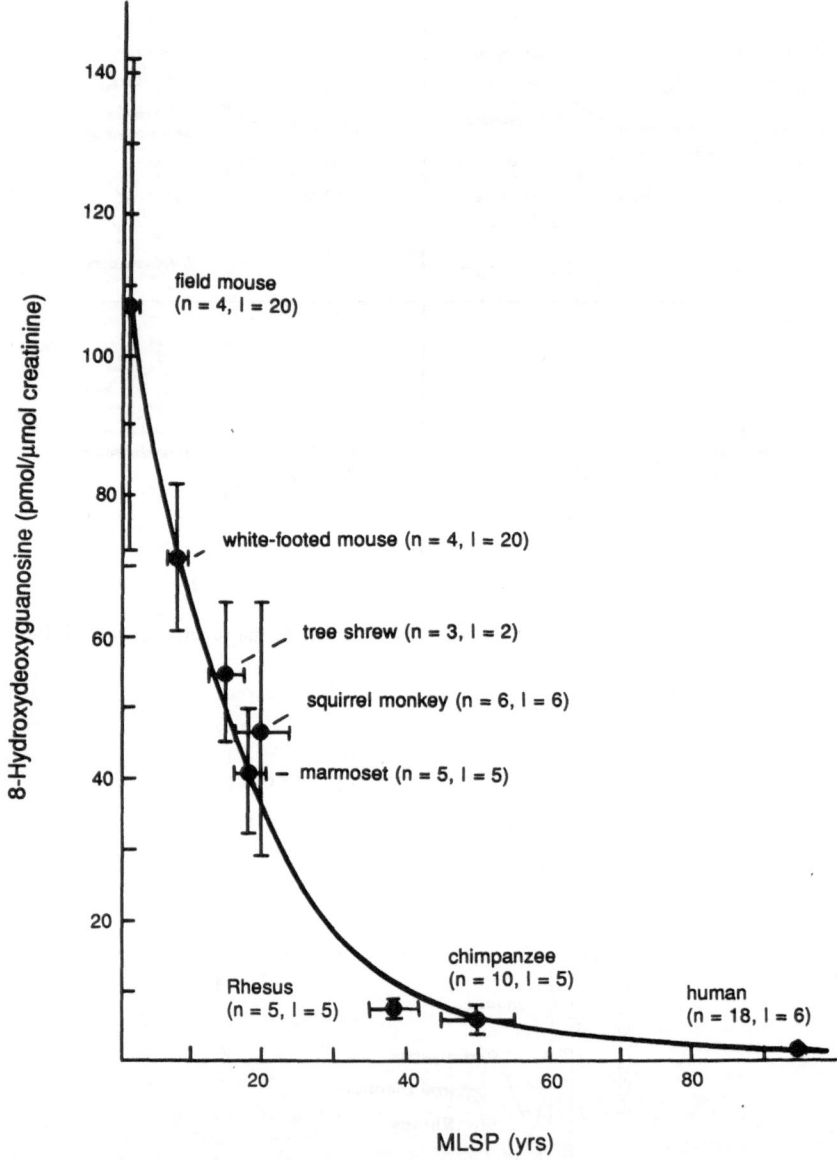

Figure 7. Relative amount of 8-OHdG per creatinine in urine samples taken from species having different life spans (MLSPs).

Preliminary results of such experiments are shown in Figure 7, where the amount of 8-OHdG per creatinine or lean body mass was found to decrease with increasing life span. These results suggest that the low levels of 8-OHdG found in human urine are likely to be the result of less 8-OHdG in the DNA available to be repaired. The reason for the low level of oxidative damage in human DNA may in turn be the result of the unusually high levels of antioxidants that have been found in human tissue. Thus, the comparative results of antioxidant levels in tissues, DNA damage in tissues and urine levels of oxidized nucleosides are mutually complementary and together support the importance of oxidative damage as a cause of aging and of anti-oxidants in governing aging rate.

Summary and conclusion

A brief overview has been given of the biological nature of human aging processes, where it has been emphasized that, in addition to the diseases of aging, there is also great economic loss as a result of human aging processes that began many years before medical costs related to aging escalated. Because of the ubiquitous nature of aging, reducing the function of essentially all physiological processes, it appears that the only long-term solution to human aging problems is to decrease uniformly the aging rate of the entire body.

Although the means to decrease human aging processes has usually been considered impossible, there are theroetical arguments and experimental data suggesting that progress could be made in achieving this goal. This idea is based on the longevity determinant gene hypothesis predicting the existence of a relatively few key regulatory genes governing the aging rate of the entire organism. If this hypothesis is not true, then indeed the prospect for significant intervention into human aging would appear impossible to achieve in the near future.

Experiments have been reviewed testing the longevity determinant gene hypothesis, the possibility that aging may be a result of dysdifferentiation and if aging rate is determined by mechanisms acting to stabilize the differentiated state of cells.

In testing the dysdifferentiation hypothesis of aging, there is not yet sufficient data available to make an assessment. There is however good evidence now that changes in gene expression with age do occur, sometimes involving endogenous retroviruses or oncogenes and an increase with age of unusual cell types such as metaplasia cells.

There is however considerably more evidence suggesting that aging involves genetic instability and that longer-lived species have a more stable genetic apparatus and superior protection against reactive oxygen species. There is a striking similarity in this model and models of cancer, and much might be gained in bringing together these two fields of research.

Acknowledgments. I thank Edith Cutler for technical and clerical support. Her help was made possible by a grant from the Paul Glenn Foundation for Medical Research.

46

Ascádi, G., and Nemeskéri, J. (1970) History of Human Lifespan and Mortality. Akadémiai Kiadó, Budapest.

Ames, B. N. (1983) Dietary carcinogens and anticarcinogens. Science 221: 1256–1264.

Bafitis, H., and Sargent, F. (1977) Human physiological adaptability through the life sequence. J. Gerontology 32: 402–410.

Cerutti, P. A. (1988) Prooxidant states and tumor promotion. Science 227: 375–381.

Comfort, A. (1978) The Biology of Senescence. Elsevier, New York.

Cutler, R. G. (1972) Transcription of reiterated DNA sequence classes throughout the life-span of the mouse in: Advances in Gerontological Research, vol. 4, pp. 219–321. Ed. B. L. Strehler.

Cutler, R. G. (1975) Evolution of human longevity and the genetic complexity governing aging rate. Proc. natl Acad. Sci. USA 72: 4664–4668.

Cutler, R. G. (1976a) Nature of aging and life maintenance processes, in: Interdisciplinary Topics in Gerontology, vol. 9, pp. 83–133. Ed. R. G. Cutler. Karger, Basel.

Cutler, R. G. (1976b) Evolution of longevity in primates. J. hum. Evol. 5: 169–202.

Cutler, R. G. (1978) Evolutionary biology of senescence, in: The Biology of Aging, pp. 311–360. Eds J. A. Behnke, C. E. Finch and G. B. Moment. Plenum Press, NY.

Cutler, R. G. (1982) Longevity is determined by specific genes: Testing and hypothesis, in: Testing the Theories of Aging, pp. 25–114. Eds R. Adelman and G. Roth. CRC Press, Boca Raton, FL.

Cutler, R. G. (1983) Superoxide dismutase, longevity and specific metabolic rate. Gerontology 29: 113–120.

Cutler, R. G. (1984a) Evolutionary biology of aging and longevity, in: Aging and Cell Structure, vol. 2, pp. 371–428. Ed. J. E. Johnson. Plenum Press, New York.

Cutler, R. G. (1984b) Antioxidants, aging, and longevity, in Free Radicals in Biology, vol. VI, pp. 371–428. Ed. W. A. Pryor. Academic Press, New York.

Cutler, R. G. (1985a) Dysdifferentiation and aging, in: Molecular Biology of Aging: Gene Stability and Gene Expression, pp. 307–340. Eds R. S. Sohal, L. Birnbaum and R. G. Cutler. Ravan Press, New York.

Cutler, R. G. (1985b) Antioxidants and longevity of mammalian species, in: Molecular Biology of Aging, pp. 15⊢73. Eds A. D. Woodhead, A. D. Blackett, A. Hollaender. Plenum Press, New York.

Cutler, R. G. (1985c) Peroxide-producing potential of tissues: Correlation with the longevity of mammalian species. Proc. natl. Acad. Sci. USA 82: 4798–4802.

Cutler, R. G., and Semsei, I. (1989) Development, cancer and aging: Possible common mechanisms of action and regulation. J. Gerontol. 44: 25–34.

Cutler, R. G. (1991) Human longevity and aging: Possible role of reactive oxygen species. Ed. W. Pierpaoli. N.Y. Acad. Sci. 621: 1–28.

Everitt, A., and Meites, J. (1989) Aging and anti-aging effects of hormones. J. Gerontol. 44: B139–B141.

Halliwell, B., and Gutteridge, J. M. C. (1989) Free Radicals in Biology and Medicine. Clarendon Press, Oxford.

Jackson, J. H., Gajewski, E., Schraufstatter, I. U., Hyslop, P. A., Fuciarelli, A. F., Cochrane, C. G., and Dizdaroglu, M. (1989) Damage to the bases in DNA induced by stimulated human neutrophils. J. clin. Invest. 84: 1644–1649.

Kohn, R. R. (1978) Principles of Mammalian Aging. Prentice-Hall, Inc., Englewood Cliff, New Jersey.

Kohn, R. R. (1982) Cause of death in very old people. J. Am. med. Assoc. 247: 2793–2797.

Maynard Smith, J. (1966) Theories of aging, in: Topics in the Biology of Aging, pp. 1–35. Ed. P. L. Krohn. Interscience, New York.

Ono, T., and Cutler, R. G. (1978) Age-dependent relaxation of gene repression; Increase of globin and endogenous murine leukemia virus related RNA in brain and liver of mouse. Proc. natl Acad. Sci. USA 75: 4431–4435.

Röhm, D. (1981) Evidence for a relationship between longevity of mammalian species and life spans of normal fibroblasts in vitro and erythrocytes in vivo. Proc. natl. Acad. Sci. USA 78: 5009–5013.

Sacher, G. A. (1978) Longevity, aging and death: An evolutionary perspective. The Gerontologist 18: 113–119.

Semsei, I., Ma, S., and Cutler, R. G. (1989) Tissue and age specific expression of the *myc* protooncogene family throughout the life span of the C57BL/6J mouse strain. Oncogene 4: 465–470.

Weiss, P. (1966) Aging: A corollary of development in: Perspectives in Experimental Gerontology, pp. 314–322. Ed. C. C. Thomas, Springfield, III.

Free Radicals and Aging
ed. by I. Emerit & B. Chance
© 1992 Birkhäuser Verlag Basel/Switzerland

Oxygen-induced mitochondrial damage and aging

Jaime Miquel[a,b], Emilio de Juan[a] and Inmaculada Sevila[a]

[a]*Laboratorio de Neurogerontología, Facultad de Medicina, Aptdo. Correos 374, San Juan, Alicante, Spain, and* [b]*Linus Pauling Institute of Science and Medicine, Palo Alto, California 94306, USA*

Summary. Although the views of Harman and Gerschman provide a reasonable explanation for many of the effects of aging, they fail to explain why many cell types, from amoebae to mammalian spermatogonia, do not show a time-related involution, while other cells (especially the neurons) change with age. We feel that a better understanding of senescence (from the molecular to the organ and organismic levels) can be gained by integrating the free radical theory of aging with the classic concepts of Minot and Pearl on the role of cell differentiation and metabolic rate in, respectively, triggering and pacing senescence.

In agreement with the above, we maintain that aging is the non-programmed but unavoidable "side effect" of oxy-radical damage to the membrane and genome of the mitochondria of irreversibly differentiated cells. If oxy-radical damage to mtDNA occurs, it will block the rejuvenation of the mitochondrial population by the process of organelle division, thus leading to bioenergetic decline and cellular death.

Introduction

The concept that mitochondria are involved in aging derives from the views of Harman (1956) and Gerschman (1962) linking senescence to the injurious effects of free radicals, especially those arising from the univalent reduction of oxygen in cellular metabolism. However, when first proposed by Harman (1972), the concept that mitochondria play a role in cell senescence did not gain much acceptance. The most popular opinion was that, since mitochondria are self-replicating organelles, they should be able to counteract any age-related loss. More recent research has shown that aging indeed results in some decrease in the number of mitochondria and that the organelles found in old cells may undergo some degree of biochemical alteration. Further, it has been experimentally shown that the reduction of oxygen in the respiratory chain of the inner mitochondrial membrane is accompanied by the formation of superoxide radical at concentrations that depend on the metabolic state of the organelles (Chance et al., 1979; Flohe, 1982).

Presently, there is a surge of interest in mitochondrial aging, as shown by a monographic symposium on this subject at the 1990 meeting of the "American Aging Association" and by the publication of a series of hypotheses on the role of these organelles in cell aging, which focus on senescent alteration of the mitochondrial genome (Table 1). Probably,

48

Table 1. Gerontological hypotheses related to mutations of the mitochondrial genome*

Proposed cause of aging	References
Free radical injury to mitochondria, including their DNA	Harman, 1972
Free radical or lipid peroxide induced inactivation of the *mitochondrial DNA of fixed postmitotic cells*	Miquel et al., 1980
Irreversible injury to mtDNA	Fleming et al., 1982
Changes in the inner membrane because of free radical effects on both mitochondrial and nuclear DNA	Harman, 1983
Intrinsic mitochondrial mutagenesis in terminally differentiated cells	Miquel and Fleming, 1984
Nuclear accumulation of mitochondrial DNA fragments	Richter, 1988
Accumulation of mitochondrial mutations and subsequent cytoplasmic segregation of these mutations	Trounce et al., 1989
Injury to mitochondria by free radicals	Shapira et al., 1989

*Modified from J. Miquel, 1991.

the finding by Shapiro et al. (1989) of a deficiency of respiratory complex I in the mitochondria of the substantia nigra in Parkinson disease will further spur the interest in the theoretical and biomedical implications of age-related changes in these subcellular organelles.

Age pigment accumulation and the mitochondrial damage hypothesis of cell aging

In our research on *Drosophila* aging, we were the first to show that the aging process of differentiated cells of invertebrates is linked to an accumulation of age pigment (Fig. 1. Takahashi et al., 1970; Miquel, 1971; Miquel et al., 1974), similar to the lipofuscin increase that is one of the most consistent effects of mammalian nerve cell aging (see reviews by Miquel et al., 1977, 1983). Our laboratory also showed evidence of age-related accumulation of age pigment in two other types of fixed postmitotic cells, namely the Leydig and Sertoli cells of the mammalian testis (Miquel et al., 1978, Fig. 1). These data and our findings that in both insects and mammals, a great amount of age pigment originates from degenerating mitochondria (Miquel et al., 1980) led us to propose that cell aging is linked to a process of mitochondrial destruction that only occurs in irreversibly differentiated cells.

In our first presentation of the concept (Miquel et al., 1980), we maintained that cell aging may derive from injury caused to the mito-chondrial genome (mtDNA) by free radicals from the inner mitochon-

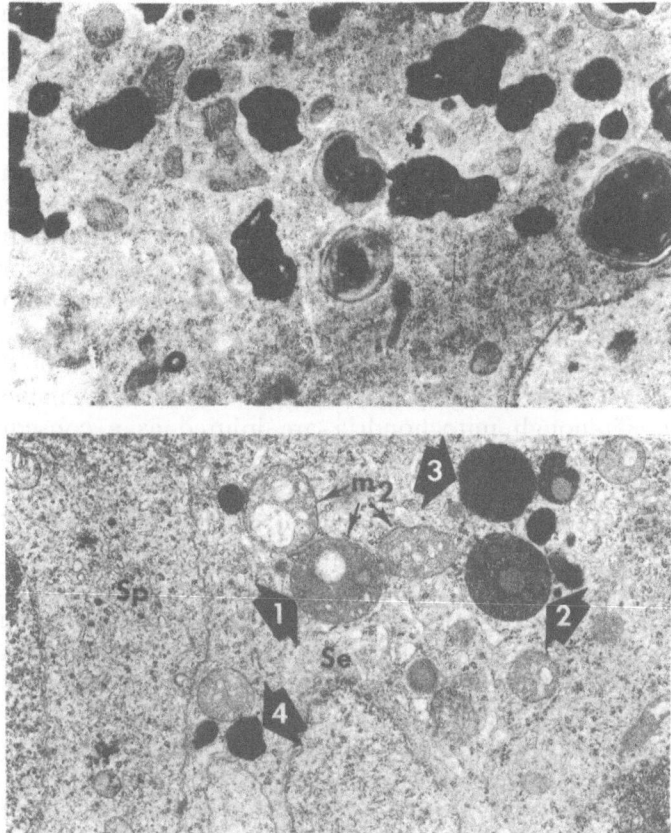

Figure 1. *Upper electron micrograph*: abundant granules of age pigment, many of which derive from altered mitochondria, in an irreversibly differentiated cell from an old insect (*Drosophila melanogaster*). ×15,000 (From Takahashi et al., 1979). *Lower electron micrograph*: testis cells from a 4-month-old mouse. Even at this young age, some irreversibly differentiated Sertoli cells (Se) contain abundant pigment granules (arrows 3 and 4) in the vicinity of normal mitochondria (arrow 1). The fast-replicating spermatocytes (Sp), on the other hand, do not show any pigment. It seems that Sertoli cells and other types of differentiated cells can suffer a process of mitochondrial degeneration starting with increased lipid storage (shown by a high degree of osmiophilia: arrow 2) and leading to mature age pigment (arrows 3, 4). Probably, because of mtDNA mutation, the mitochondrial damage is more striking in old animals. ×18,200 (From Miquel et al., 1978).

drial membrane. In our concept, fast-replicating cells do not suffer free radical attack because of their lower levels of oxygen utilization. Conversely, irreversibly differentiated cells, such as neurons, suffer mitochondrial aging because of the high levels of oxygen utilization required by their specialized work. Since the mitochondrial genome is needed for the division of these organelles, damage to mtDNA would block mitochondrial turnover and replication, with concomitant peroxidative membrane breakdown, autophagic digestion of the organelles and decline in the production of ATP and of energy-dependent protein synthesis.

In further work from our laboratory (Fleming et al., 1982) our hypothesis of *intrinsic mitochondrial mutagenesis* by oxy-radicals was made more specific as regards the probable genetic damage sustained by the organelles: "The simultaneous presence of malonaldehyde and DNA on or near the inner mitochondrial membrane could result in cross-linking reactions, which are likely to cause deletion errors in transcription of the mitochondrial genome because of polymerization of the mtDNA". In that article we also discussed the bioenergetic implications of the primary injury to the mitochondrial genome: "Differentiated cells must budget their ATP production between the support of their specialized physiological function and their own maintenance needs. Therefore, if enough mitochondria are injured as a consequence of mitochondrial genome inactivation, the remaining (intact) mitochondria may be unable to divert a sizable amount of their energy production for the formation of new mitochondria at the expense of their own needs and of the support of whole-body homeostasis. Thus, a vicious circle of energy loss triggering further energy loss may result". Further, a theoretical article by Miquel and Fleming (1984) comments on the role of both cell differentiation and genomic instability in setting up the stage for cellular and organismic aging. "We propose that, as already suggested by Minot in 1907, senescence is the result of cell differentiation. We further maintain that *the fundamental cause of cell aging is an instability of the mitochondrial genome* because of a lack of balance between mitochondrial repair and the disorganizing effect of oxygen radicals which arise in the respiring mitochondria of terminally differentiated cells. This probably results in intrinsic mitochondrial mutagenesis which may be followed by endonuclease degradation of the altered mitochondrial DNA".

Theoretical and experimental basis of the hypothesis

A fundamental fact in support of the above concept is that, as previously reviewed (Harman, 1986; Miquel and Fleming, 1986; Fridovich, 1988), the mitochondrial respiratory chain causes a steady flow of oxy-radicals. These include the highly toxic superoxide radical (O_2^-) because, although two electrons are transferred from the substrate to the respiratory chain, the reduction of ubiquinone (and probably flavin) occurs by a single electron transfer (Boveris et al., 1979). The ubisemiquinone thus generated is, in large part, reoxidized by the cytochrome b complex, but may also react with O_2 to generate O_2^-.

As pointed out by Richter (1988), about 1% of the oxygen used in the mitochondria is transformed into superoxide, with a daily output as high as 10^7 molecules of O_2^-/mitochondrion/day. Most intramitochondrial superoxide is destroyed by Mn-superoxide dismutase. However, the detoxifying effects of this mitochondrial enzyme (Fridovich, 1988) are far from perfect since, as the result of enzyme-mediated dismutation, H_2O_2 is produced and accumulates in the mitochondria. In turn, H_2O_2 (which is not totally eliminated by intramitochondrial catalase and peroxidase) reacts with superoxide to produce the extremely active hydroxyl radical (HO·). This oxygen species, in addition to singlet molecular oxygen (also released in respiring cells), may lead to radical and lipid peroxidation chain reactions and widespread cellular injury (Tappel, 1975; Pryor, 1978; Lippman, 1983; Miquel et al., 1989), including damage to the mitochondrial genome.

Metals with catalytic activity such as iron and copper (which are also present in the respiratory chain) play a key role in promoting free radical release and lipid peroxidation. Thus, according to Valdimirov et al. (1981), the peroxidative disorganization of lipid biomembranes is activated by the ferrous ions arising from iron reduction in the respiratory chain. He concludes that the participation of iron in lipid peroxidation reactions may be considered as an alternative metabolic pathway of ferrous ions which normally are used in heme synthesis.

Since the special vulnerability of mtDNA to oxy-radical attack has been discussed in detail in our previous articles on the mitochondrial DNA mutation hypothesis of aging (Miquel et al., 1980; Fleming et al., 1982; Miquel and Fleming, 1984; Miquel, 1991), only the most salient data that back our hypothesis will be summarized here.

Our hypothesis is in agreement with the fact that H_2O_2, and the superoxide and hydroxyl radical (all of which are present in the mitochondria) are able to react with all kinds of biomolecules, including mtDNA (Greenstock, 1986; Emerit, 1986; Cochrane et al., 1988).

It is also important as regards our proposed mechanism of cell aging that mitochondria possess their own system of gene expression including ribosomal RNA, transfer RNA and messenger RNA. The mitochondrial DNA and ribosomes are involved in the synthesis of the hydrophobic protein subunits of at least three enzyme complexes of the inner membrane which play an essential role in mitochondrial respiration and energy transduction.

The above synthesis also requires the intervention of the nuclear genome since the mtDNA and RNA polymerases, aminoacyl tRNA synthetases, tRNA transformylase, and the initiation, elongation and termination factors are presumably coded by the nucleus. Thus, it is apparent that the biogenesis and preservation of mitochondrial structure and function are dependent on a complicated interplay of nuclear and mitochondrial products. Although, theoretically, intrinsic aging of the nuclear genome could result in age-related mitochondrial changes, it is more likely, for the reasons discussed below, that mtDNA is the initial target in a cascade of free radical peroxidative or cross-linking injuries that may eventually reach the nuclear genome.

It seems that mtDNA is quite sensitive to damage by toxic chemical agents. The formation of complexes between these substances and mtDNA may be favored by the closed circular structure of the mitochondrial genome (Wunderlich et al., 1970), and probably involves intercalation of the foreign chemical in the DNA. As pointed out by Linnane et al. (1989), human tRNA$_s$ and the protein coding genes are arranged in the mtDNA without non-coding sequences in between except for a small region involved in the replication and transcription of the mtDNA. Therefore, a mutation of the nuclear DNA may be harmless if it affects a non-expressed region of its genome, while most mutational changes in mtDNA must have injurious effects.

As noted by Richter (1988), 8-hydroxydeoxyguanosine is present in mtDNA from rat liver mitochondria at 16 times the level found in nuclear DNA. That chemical is one of the products of base oxidation resulting from OH˙ attack to DNA. Therefore, it seems that oxidative damage to mtDNA may be quite high.

Theoretically, the damage caused to mtDNA by mutagenic substances could be repaired by mechanisms similar to those protecting the identity of the nuclear genome. However, recent research suggests that mtDNA may be lacking adequate repair mechanisms (Fukanaga and Yielding, 1979). Specifically, the mitochondrial genome seems to be unable to repair strand breaks. Since it has been shown that hydrogen peroxide (which is normally present in the mitochondria) induces single-strand scission and cross-linking of DNA (Samis et al., 1972), the potential for intrinsic mitochondrial mutagenesis clearly exists.

Experimental evidence for the oxygen radical-mitochondrial damage hypothesis of aging

As pointed out in the reviews by Miquel et al. (1980), and Hansford (1983), the subject of age-dependent changes in respiration and energy metabolism abounds in conflicting data, probably because of technical pitfalls, including the fact that usually the mitochondria are obtained from tissue samples that contain, side by side, aging differentiated cells and relatively non-aging dividing cells. Nevertheless, research on insects and mammals suggests that differentiated cells such as those of *Drosophila* muscle and the mammalian hepatocytes and neurons suffer a senescent change in the amount and/or structrure of their mitochondrial populations (Miquel et al., 1980, 1984; Miquel and Fleming, 1986). It also seems well established that there is a senescent diminution in the ability to phosphorylate ADP with 3-hydroxybutyrate as substrate. As pointed out by Hansford (1983), this age-related change may be the result of a decrease in the activity of several enzymes responsible for fatty acid oxidation.

The activity of cytochrome oxidase, which is often used as a marker enzyme for the mitochondrial inner membrane, shows a senescent decline in mitochondria isolated from liver as well as in homogenates of striated muscle (see review by Miquel and Fleming, 1986).

Unfortunately, there are very few reports on the effects of aging on ATP content and synthesis in fixed postmitotic cells (Foris and Leovey, 1980). This is a very important area in which more work is needed to confirm the preliminary data on senescent impairment of energy production, which suggests that "mitochondrial dysfunction may play a role in the immunodeficiency of aging" (Verity et al., 1983) and in other age-related functional losses.

It is very interesting as regards the central tenet of our hypothesis, i.e. that there is a senescent impairment of the mitochondrial genome, that *Drosophila* aging may be accompanied by a total loss of mtDNA while the amount of nuclear DNA does not show any change (Massie et al., 1975). A senescent decrease in extractable mtDNA has also been observed in the liver of Fisher rats (Stocco and Hutson, 1978). This was interpreted as evidence of an age-related decrease in the number of mitochondria, since enzyme and morphological studies rule out any mitochondrial loss during the isolation procedure.

Senescent changes in the structure and replication of the mitochondrial genome have been found. Thus, Murray and Balcavage (1982) showed an increase in dimeric catenated forms of mtDNA and the appearance of DNA with altered ethidium bromide binding properties in aging mice and rats. Further research by Piko et al. (1984) on mtDNA of brain, heart, kidney and liver of rats has shown that the frequency of circular dimers is strikingly increased with age in mouse brain and rat kidney and moderately increased in heart mtDNA of both species.

The effect of aging on mitochondrial replication and function is a problem area that has been almost totally neglected by experimental gerontologists. In 1973, Kiessling et al. published data showing that training has a very different effect on the volume fraction of the mitochondria in two age groups of humans. In young men, physical training resulted in an increase of the total mitochondrial volume of about 100%, while the increase was only 20% in older men. The large mitochondrial volume found in the young group was caused by an increase in the number of mitochondria without an enlargement of their individual dimensions. By contrast, in older men the larger volume was the result of an increase in the size of the individual mitochondria without an increase in their amount. This suggests that, in agreement with our hypothesis, aging results in an impairment in the replicative ability of the mitochondria.

Conclusions: Implications of the mitochondrial damage hypothesis for an integration of stochastic and programmed theories of aging

There is considerable evidence in support of the view that in irreversibly differentiated cells (such as the neuron) the tug of war between peroxidative damage to the mitochondria and the antioxidant and regenerative mechanisms that protect these organelles is decided in favor of a progressive structural and biochemical disorganization. Thus, though senescence does not result in striking changes in mitochondrial respiration or coupling, it seems well established that there is an age-related decrease in key structural and functional components of the inner membrane, such as cardiolipin, ubiquinone, and the cytochromes. These changes (which probably impair mitochondrial synthesis of ATP) are the result of the high levels of oxygen radical release in the respiratory chain of differentiated cells.

We hypothesize that there is no mitochondrial aging in fast-replicating cells because their organelles are protected by a most effective rejuvenation mechanism, i.e. frequent renewal of their macromolecules (including the structural lipids and the hydrophobic proteins of the inner membrane) through the process of mitochondrial division at the time of mitosis (Fig. 2). By contrast, though mitochondria also divide in irreversibly differentiated cells, the turnover of the above-mentioned components of the inner membrane is much slower than in replicating cells. Therefore, the mitochondria of this cell type may accumulate peroxidative damage in their structural components, with resulting impairment of their enzyme biochemistry and energy production. If, in addition, the mitochondrial genome is mutated or inactivated by oxygen radicals or their products, the mitochondria will lose their replicative ability and thus the stage will be set for a higher degree of peroxidative damage and eventual autophagic breakdown of the altered mitochondria.

In our opinion, it is evident that, though the free radical theory of aging provides a reasonable explanation for the senescent changes occurring at the molecular level, a much more complete understanding of the aging process is gained by integrating the free radical concepts with the classic views of Minot (1907) and Pearl (1928) on the role of cell differentiation (as the foundation of aging) and metabolic rate (as the main determinant of rate of aging and life span). We feel that this cluster of (metabolic) wear-and-tear theories of aging offers a logical and parsimonious explanation of the main features of senescence, from the molecular to the systematic levels.

An advantage of this integrated theory is that it bridges the gap between molecular and systemic concepts of aging that, in Sacher's (1968) view, has hindered gerontological progress. Further, as we have pointed out before (Miquel, 1991), "this assemblage of matching pieces of the puzzle reconciles programmed and stochastic concepts of aging,

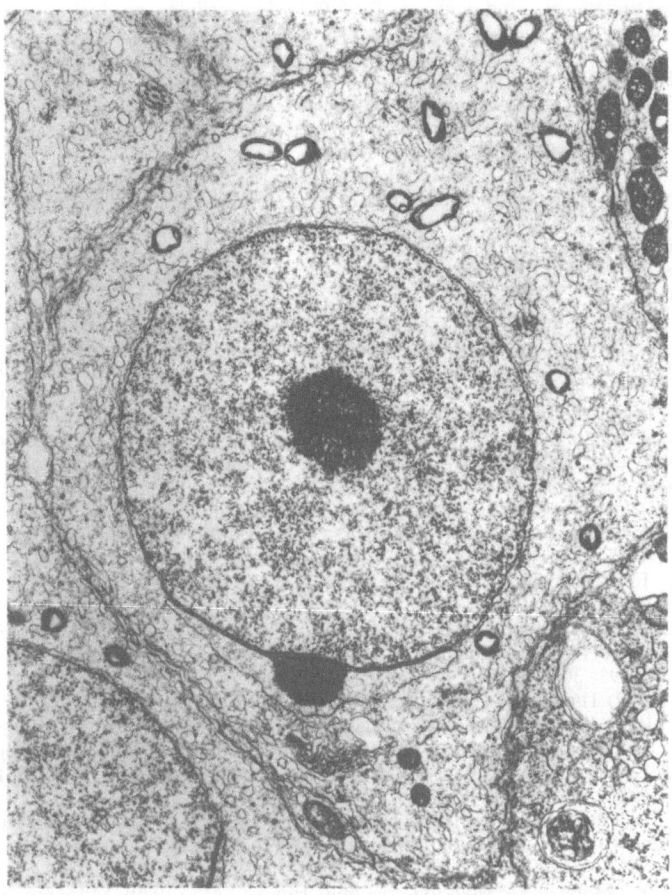

Figure. 2. This electron micrograph from the testis of a young mouse shows a spermatogonion, whose nucleus carries a copy of the genes to be passed to future generations. In agreement with the classic views of Weissman and Minot and with our own hypothesis, this immortal cell line does not show any age-related fine structural alterations. Note the scarcity of mitochondria and of mitochondrial cristae, that reveals a very low respiration rate and, therefore, minimum release of injurious oxy-radicals. This is in striking contrast to the example of the somatic cell (Sertoli cell, shown on the upper right corner), with its numerous mitochondria packed with cristae, which are exposed to a high degree of oxidative stress with resulting mitochondrial breakdown, bioenergetic decline and cell death. ×18,000 (From Miquel, 1989).

since both are consistent with the facts. Cells are first programmed to differentiate and, then, they fall pray to (stochastic) wear-and-tear chemical insults as the side effect of mitochondrial energy production".

Acknowledgments. The authors thank Ms. Margarita Blasco for her help in the preparation of the manuscript. This work was supported in part by grants from the CICYT: ESP-88/0351, and FISS 84/1994 and 89/0877 (Spain).

56

Boveris, A., Oshimo, N., and Chance, B. (1972) The cellular production of hydrogen peroxide. Biochem. J. 128: 617–630.

Chance, B., Sies, H., and Boveris, A. (1979) Hydroperoxide metabolism in mammalian organs. Physiol. Rev. 59: 527–605.

Cochrane, C. G., Schraufstatter, I. U., Hyslop, P., and Jackson, J. (1988) Cellular and biochemical events in oxidant injury, in: Oxygen Radicals and Tissue Injury. B. Halliwell, ed. Upjohn/FASEB, Bethesda, Md, pp. 49–54.

Emerit, I. (1986) Oxygen-derived free radicals and the DNA damage in autoimmune disease, in: Free Radicals, Aging and Degenerative Disease. J. Johnson Jr, R. Walford, D. Harman and J. Miquel, eds. Alan R. Liss Inc., New York, pp. 307–324.

Fleming, J. E., Miquel, J., Cottrell, S. F., Yengoyan, L. S., and Economos, A. C. (1982) Is cell aging caused by respiration-dependent injury to the mitochondrial genome? Gerontology 28: 44–53.

Flohe, L. (1982) Free Radicals in Aging, Vol. 5. Academic Press, New York, pp. 223–254.

Foris, G., and Leovey, A. (1980) Age related changes in cAMP and cGMP levels during phagocytosis in human polymorphonuclear leukocytes. Mech. Ageing Dev. 27: 233–237.

Fridovich, I. (1988) The biology of oxygen radicals: general concepts, in: Oxygen Radicals and Tissue Injury. B. Halliwell, ed. Upjohn/FASEB, Bethesda, Md, pp. 1–5.

Fukanaga, M., and Yielding, K. L. (1979) Fate during cell growth of yeast mitochondria and nuclear DNA after photolytic attachment of the monazide analog of ethidium. Biochem. Biophys. Res. Commun. 90: 582–586.

Gerschman, R. (1962) Man's dependence on the earthly atmosphere, in: Proc. 1st Symp. Submarine and Space Medicine. K. S. Schaeffer, ed. MacMillan, New York, p. 475.

Greenstock, C. L. (1986) Radiation induced aging and induction and promotion of biological damage, in: Free Radicals, Aging and Degenerative Diseases. J. E. Johnson Jr, R. Walford, D. Harman and J. Miquel, eds. Alan R. Liss, New York, pp. 197–219.

Hansford, R. G. (1983) Bioenergetics in aging. Biochim. Biophys. Acta 726: 41–80.

Harman, D. (1956) Aging: a theory based on free radical and radiation chemistry. J. Gerontol. 11: 298–300.

Harman, D. (1972) The biologic clock: the mitochondria? J. Amer. Geriatr. Soc. 20: 145–147.

Harman, D. (1986) Free radical theory of aging: role of free radicals on the origination and evolution of life, aging and disease processes, in: Free Radicals, Aging and Degenerative Diseases. J. E. Johnson Jr, R. Walford, D. Harman and J. Miquel, eds. Alan R. Liss, New York, pp. 3–49.

Kiesling, K. H., Pilstrom, I., Karlsson, J., and Piehl, K. (1973) Mitochondrial volume in skeletal muscle from young and old physically untrained healthy men and from alcoholics. Clin Sci. 44: 547–554.

Linnane, A. W., Ozawa, T., Marzuki, S., and Tanaka, M. (1989) Mitochondrial DNA mutations as an important contributor to ageing and degenerative diseases. Lancet 1: 642–645.

Lippman, R. D. (1983) Chemiluminescence measurement of free radicals and antioxidant molecular protection inside living rat mitochondria. Exp. Gerontol. 15: 339–351.

Massie, H. R., Baird, M. B., and McMahon M. M. (1975) Loss of mitochondrial DNA with aging. Gerontology 21: 231–237.

Minot, C. S. (1907) The problem of age, growth and death. Pop. Sci. Mon. 71: 496.

Miquel, J. (1971) Aging of male *Drosophila melanogaster*, histological, histochemical and ultrastructural observations, in: Advances in Gerontological Research, Vol. 3. B. L. Strehler, ed. Academic Press, London, New York, pp. 39–71.

Miquel, J. (1991) An integrated theory of aging as the result of mitochondrial DNA mutation in differentiated cells. Arch. Gerontol. Geriatr. 12: 99–177.

Miquel, J., and Fleming, J. E. (1984) A two-step hypothesis on the mechanism of *in vitro* cell aging: cell differentiation followed by intrinsic mitochondrial mutagenesis. Exp. Gerontol. 19: 31–36.

Miquel, J., and Fleming, J. E. (1986) Theoretical and experimental support for an "oxygen radical-mitochondrial injury" hypothesis of cell aging, in: Free Radicals, Aging and Degenerative Diseases. J. E. Johnson Jr, R. Walford, D. Harman and J. Miquel, eds. Alan R. Liss, New York, pp. 51–74.

Miquel, J., Tappel, A. L., Dillard, C. J., Herman, M. M., and Bensch, K. G. (1974) Fluorescent products and lysosomal components in aging *Drosophila melanogaster*. J. Gerontol. 29: 622–637.

Miquel, J., Oro, J., Bensch, K. G., and Johnson, J. E. Jr (1977) Lipofuscin: fine structural and biochemical studies, in: Free Radicals in Biology. W. Pryor, ed. Academic Press, New York, pp. 133–182.

Miquel, J., Lundgren, P. R., and Johnson, J. E. (1978) Spectrophotometric and electron microscopic study of lipofuscin accumulation in the testis of aging mice. J. Gerontol. 33: 5–19.

Miquel, J., Economos, A. C., Fleming, J., and Johnson, J. E. Jr. (1980) Mitochondrial role in cell ageing. Exp. Gerontol. 15: 575–591.

Miquel, J., Binnard, R., and Fleming, J. E. (1983) Role of metabolic rate and DNA repair in *Drosophila* ageing: Implications for the mitochondrial mutation theory of cell aging. Exp. Gerontol. 18: 161–171.

Miquel, J., Quintanilha, A. T., and Weber, H. (1989) CRC Handbook of Free Radicals and Antioxidants in Biomedicine, 3 Vols. Boca Raton, Fl, pp. 997.

Murray, M. A., and Balcavage, W. X. (1982) Changes in mitochondrial DNA during aging. Mech. Ageing Dev. 20. 233–241.

Pearl, R. (1928) The Rate of Living. University of London Press.

Piko, L., Bulpitt, K. J., and Meyer, R. (1984) Structural and replicating forms of mitochondrial DNA in tissues from adult and senescent BALB/c mice and Fisher 344 rats. Mech. Ageing Dev. 26: 113–131.

Pryor, W. A. (1978) Formation of free radicals and the consequences of their reaction *in vivo.* Pathochem. Photobiol. 28: 287–307.

Richter, C. (1988) Do mitochondrial DNA fragments promote cancer and aging? FEBS lett. 241: 1–5.

Sacher, G. A. (1968) Molecular versus systemic theories on the genesis of aging. Exp. Gerontol. 3: 265–271.

Samis, H. V., Baird, M. B., and Massie, H. R. (1972) Molecular genetic mechanisms in development and aging. M. Rockstein and G. Baker, eds. Academic Press, New York, pp. 113–143.

Schapira, A. H. V., Cooper, J. M., Dexter, D., Jenner, P., Clark, J. B., and Marsden, C. D. (1989) Deficiencia del complejo I mitocondrial en la enfermedad de Parkinson. Lancet 15: 247–275 (Spanish edition).

Stocco, D. M., and Hutson, J. C. (1978) Quantitation of mitochondrial DNA and protein in the liver of Fisher 344 rats during aging. J. Gerontol. 33: 802–810.

Takahashi, A., Philpott, D. E., and Miquel, J. (1970) Electron microscope studies on aging *Drosophila melanogaster.* I. Dense bodies. J. Gerontol. 25: 210–217.

Tappel, A. L. (1975) Pathology of cell membranes, Vol. 1. B. F. Trump and A. W. Arstila, eds. Academic Press, New York, pp. 145–172.

Verity, M. A., Tam, C. F., Chezig, M. K., Mock, D. C., and Walford, D. C. (1983) Delayed phytohemagglutinin-stimulated production of adenosine triphosphate by aged human lymphocytes: Possible relation to mitochondrial disfunction. Mech. Ageing Dev. 23: 53–65.

Vladimirov, Yu. A., Olenev, V. I., Suslova, T. B., and Cheremisina, Z. P. (1981) Lipid peroxidation in mitochondrial membranes. Adv. Lipis Res. 17: 173–249.

Wunderlich, V., Schutt, M., Bottgerd, M., and Graffi, A. (1970) Preferential alkylation of mitochondria deoxyribonucleic acid by N-nitrosourea. Biochem. J. 118: 99–109.

Free Radicals and Aging
ed. by I. Emerit & B. Chance
© 1992 Birkhäuser Verlag Basel/Switzerland

Instabilities of metabolic regulations in aging

Britton Chance[a], William W. Bank[b] and Cheng-dou Zhang[b]

[a]*Department of Biochemistry and Biophysics, Johnson Foundation, University of Pennsylvania, Philadelphia, PA 19104, USA, and* [b]*Department of Neurology, Hospital of the University of Pennsylvania*

Summary. 1) That non-invasive NMR and optical methods can a) quantify the work stress on mitochondria for ATP production, and b) indicate the tissue O_2 tension in the capillary bed that is responsible for the rate of radical generation. 2) That free radical damage to mitochondrial function càn be quantified by reciprocal plots of inverse slope giving the extrapolated V_m of mitochondria. 3) That a particular genetically deficient individual requiring high dosages of menadione has survived over 9 years. 4) That mitochondrial deficiency leads to an exercise hyperoxia.

Introduction

Basic to any consideration of free radicals in aging are the underlying features of metabolic control of energy metabolism and oxygen delivery to the tissue. The oxygen demand of tissues is set by their metabolic rate, high for the heart and brain, and low for the skeletal muscle with liver being intermediate between the two. It is characteristic of regulatory systems to "abhor V_{max}" (as nature abhors a vacuum), V_{max} being the maximal rate capability of oxidative metabolism. It is by virtue of the setpoint of metabolism that the steady state of Claude Bernard (1878) might be obtained in *le milieu fixe* and is furthermore the hallmark of a stable metabolic system that V/V_{max} be in the 10–30% region, as indeed it is for reliable control of metabolic activity. It is implicit in this concept that indeed such systems could accommodate by virtue of small V/V_m's, significant deterioration of V_m due to free radical insults such as measured by chemiluminescence, which would simply drop V_m due to the smaller number of competent mitochondria.

The second topic, the delivery of oxygen to tissue, is of importance not only to be adequate for mitochondrial function at a low V/V_m, i.e., an oxygen sufficiency, but also the oxygen level in the tissue should be capable of being "down regulated" to the point where the oxygen concentration is sufficiently low that free radical generation, which in the end, all originates from oxygen in tissue, be minimized. Thus, tissue O_2 must be measured.

Lastly, I will comment briefly upon a third noninvasive technique where the V of noninvasive readout of cortical function becomes possible from a simple optical method for measuring blood concentration.

NMR as a non-invasive criterion of the V/V_m

In the steady state function of tissues, the controllers of ATP synthesis, principally ADP but in some cases P_i, substrate, NADH or indeed oxygen, are set at such levels so that at least one is rate limiting and may be varied over a significant interval of concentration to afford essential metabolic regulation (Chance et al., 1985; Chance et al., 1986).

ADP control. The theory of ADP control metabolism came naturally from the experiment of Chance and Williams (1955) which showed unprecedentedly low concentrations of ADP control of respiration and ATP synthesis of isolated mitochondria in a hyperbolic fashion with a K_m of 20 mM. The realization of this experiment in vivo awaited the development of NMR for the study of muscle and the understanding of equations appropriate to the calculation of ADP from the creatine kinase equilibrium (Chance et al., 1985; Chance et al., 1986). Suffice to say that V/V_m both theoretically and experimentally is obtained by the simple equation

$$\frac{V}{V_m} = \frac{1}{1 + \dfrac{0.6}{\dfrac{PCr}{P_i}}}$$

where PCr is phosphocreatine, P_i is inorganic phosphate and 0.6 is the effect of Michaelis-Menten constant for ADP control, and ADP is calculated from P_i at pH 7; ADP = 3×3 P_i/PCr (K_m for ADP = 20 μM (Veech et al., 1979)).

The Hanes plot is most appropriate to demonstrating this relationship where the hyperbolas is converted into a linear plot from the equation

$$S/V = [S + K_m]/V_m.$$

In a plot of S/V, V on the ordinate and S on the abscissa, a straight line is expected of slope $1/V_m$ and an intercept K_m where S = ADP and $K_m = 0.6(3)$.

Figure 1 demonstrates how this relationship can be used in working skeletal tissue to obtain V_m with a very high value for a rower and an intermediate value for a yachting winch grinder, and a much lower value for a chemical/biochemical bypass of a genetic defect in a human subject. Each point in this graph corresponds to work at an increasing level, as S increases with corresponding increases of V, the rate of mitochondrial ATP synthesis must be sufficient to match the rate of ATP utilization for the steady state condition. It is clear that this method affords a non-invasive in vivo method of determining the oxidative capacity of mitochondria which should furthermore be used

60

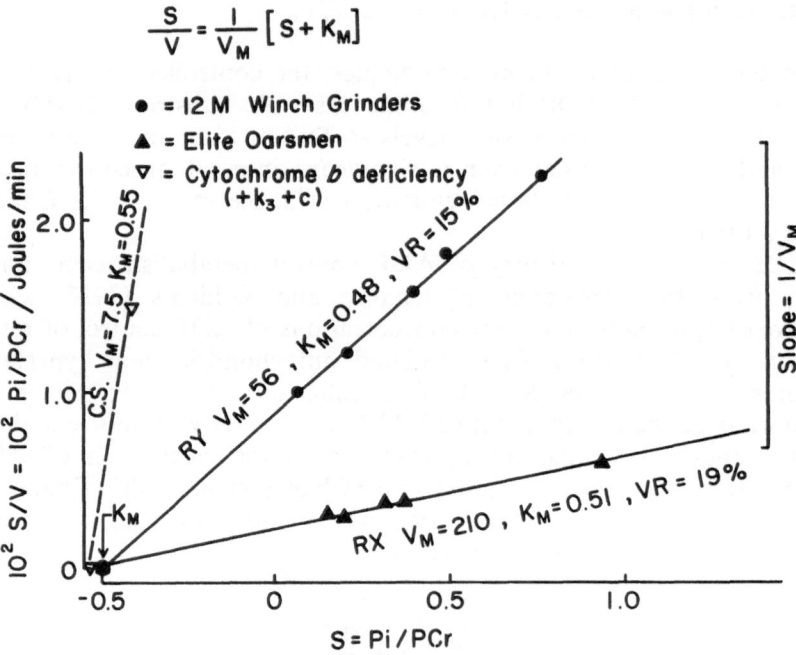

Figure 1. An illustration of the determination of V_m for individuals of differing mitochondrial function.

by experimenters in this field who have postulated in vivo damage to mitochondria due to the free radicals.

Oxygen radical exposures of mitochondrial disease. Genetic deficiencies that may lead to special susceptibilities to oxygen free radicals. In order to maximize, or to restore, a modest level of phosphorylation in these limbs, I recommended that the in vitro work of Nosoh et al., (1968) and Sanadi (1964) be applied in vivo, namely that menadione Vitamin K_3 or K_1 should be employed as a biochemical bypass, a radical generating system. However, an excess of ascorbic acid has minimized radical production, but never-theless the danger remains after the nine years of injesting high levels of Vitamin K. Figure 2 shows that the V_m of a biochemical bypass of a genetic deficiency remains low compared to normal and indicates how these tissues are stressed to their utmost with even mild exercise so that the subject is capable of climbing a few stairs or walking a block or two only. The dose response profile is shown in Figure 2 showing that even 80 mg/day does not evoke full V_{max} of the mitochondria, but does impose a heavy radical load due to the menadione radicals.

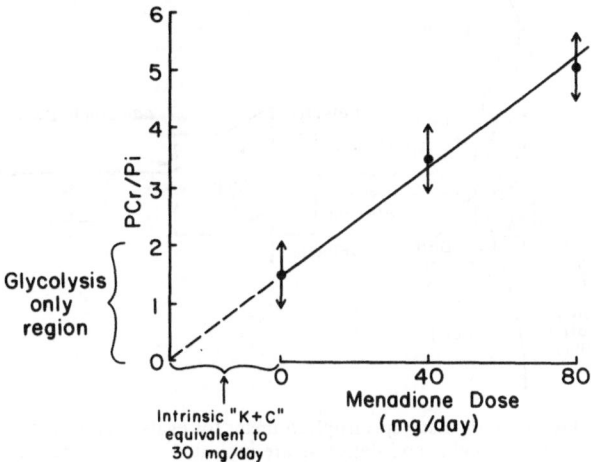

Figure 2. An illustration of NMR determined dose-response profile of a genetic diseased skeletal muscle.

Optical spectroscopic methods

Tissue oxygen levels

Of most importance in radical generation is the oxygen concentration itself since low values of tissue oxygen would afford some protection against excess oxygen radical generation. In order to be able to parallel the noninvasive measurements of mitochondrial function by NMR, we have developed a simple dual wavelength apparatus to monitor the tissue oxygen tension in the exercising muscle. Hemoglobin provides a well-calibrated oxygen indicator that can be used in vivo, together with myoglobin, to indicate whether the tissue oxygen tension can be deleterious to health. The method depends upon the diffusion of light through the skin, a fat layer of fascia into the muscle and exiting on the reverse path to a pick up point several centimeters from the input. This input/output separation affords a satisfactory depth of penetration of the light as shown by Figure 3.

For these studies, a simple unit using continuous light provides a sufficiently selective path in and out of the skeletal muscle for recording changes of saturation accompanying exercise. Whilst in normal subjects characteristic deoxygenation occurs, due to massive utilization of oxygen in the muscle mitochondria, recent studies show that metabolic disease (ragged-red syndrome) causes a most remarkable

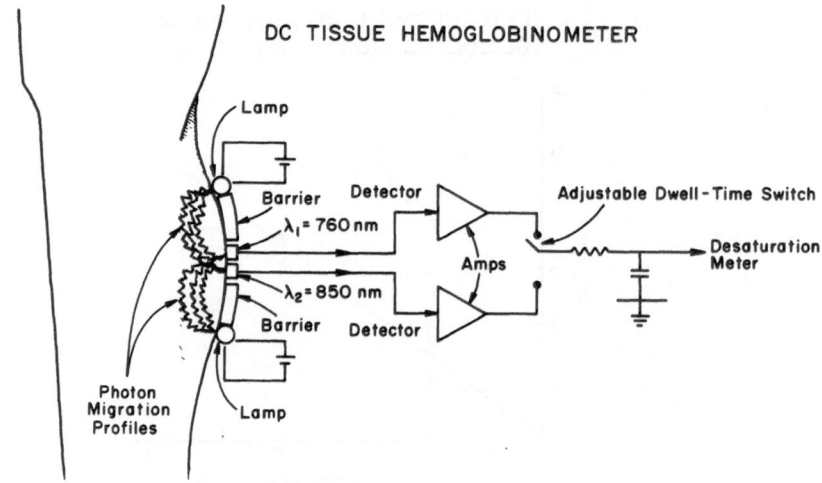

DC TISSUE HEMOGLOBINOMETER

Figure 3. A dual wavelength tissue spectrophotometer for evaluation of muscle deoxygenation during exercise. The input lights and detectors are indicated as are the scattered light paths in the tissue.

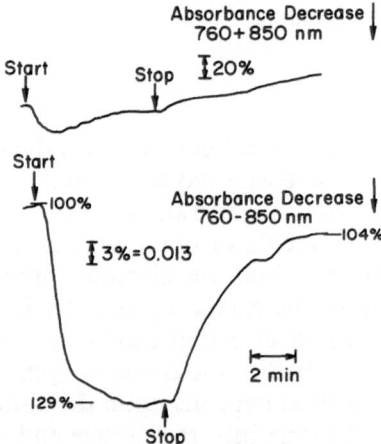

PATIENT "L.C." TREADMILL EXERCISE
AT 2 MPH LEFT GASTROCNEMIUS

Figure 4. Patient with myopathy in treadmill exercise at 2 mph (left gastrocnemius). The optical device of Fig. 3 was used.

response of the muscle to increased exercise as shown in Figure 4. Here the direction of increased oxygenation of hemoglobin and myoglobin is downwards and the very start of exercise is marked with such a deflection corresponding to an oxygenation of the entire content of myoglobin and hemoglobin of the muscle; a paradoxal result during increased O_2 demand in the exercising tissue.

Figure 4 shows a much increased oxygenation of the muscle tissue, even though the oxygen demand of the diseased mitochondria is much less than that of normal tissue. Such a muscle is in a much more "endangered" oxygenation status than the normal muscle which is often oxygen-deficient during function. Thus mitochondrial deficiencies can lead to a pathological tissue hypoxia.

In summary, mitochondria, while radical generators themselves, do down regulate tissue pO_2's and act as a safeguard against tissue hypoxia.

Acknowledgments. Supported in part by National Institutes of Health Grants HL18708 and AA07186.

Bernard, C., Les Phenomenes de la Vie, Vols 1, 2. J. B. Balliere et fils, Paris, 1878.

Chance, B., and Williams, G. (1955) Respiratory enzymes in oxidative phosphorylation. I. Kinetics of oxygen utilization. J. Biol. Chem. 217(1): 383–393.

Chance. B., Leigh, Jr. J. S., Clark, B. J., Maris, J., Kent, J., and Smith, D. (1985) Control of oxidative metabolism and oxygen delivery in human skeletal muscle: A steady-state analysis of the work/energy cost transfer function. Proc. Natl. Acad. Sci. USA 82: 8384–8388.

Chance, B., Leigh, Jr. J. S., Kent, J., McCully, K., Nioka, S., Clark, B. J., Maris, J. M., and Graham, T. (1986) Multiple controls of oxidative metabolism of living tissues as studied by ^{31}P MRS. Proc. Natl. Acad. Sci. USA 83: 9458–9462.

Eleff, S., Kennaway, N. G., Buist, N. R. M., Darley-Usmar, V. M., Capaldi, R. A., Bank, W. J., and Chance, B. (1984) ^{31}P NMR Study of improvement in oxidative phosphorylation by Vitamins K_3 and C in a Patient with a defect in electron transport at complex III in skeletal muscle. Proc. Natl. Acad. Sci. USA 81: 3529–3533.

Nosoh, Y., Kajioka, J., and Itoh, M. (1968) Effect of menadione on the electron transport pathway of yeast mitochondria. Arch. Biochem. Biophys. 127: 1–6.

Sanadi, D. R. (1964) Energy-requiring reduction of pyridine nucleotide by ascorbate in the presence of coenzyme Q or menadione. J. Biol. Chem. 238: PC482–3.

Veech, R. L., Lawson, J. W. R., Cornell, N. W., and Krebs, H. A. (1979) Cytosolic phosphorylation potential. J. Biol. Chem. 254: 6538–6547.

Free Radicals and Aging
ed. by I. Emerit & B. Chance

Protein modification in aging

E. R. Stadtman*[1], P. E. Starke-Reed[§], C. N. Oliver[‡], J. M. Carney[‖]
and R. A. Floyd[¶]

*Laboratory of Biochemistry, National Heart, Lung, and Blood Institute, National Institutes
of Health, Bethesda, Maryland 20892 USA, [§]Geriatrics Program, National Institute on
Aging, National Institutes of Health, Bethesda, Maryland 20892, USA, [‡]Merck Sharp &
Dohme Research Laboratories, WP16100, West Point, Pennsylvania 19486, USA,
[‖]Department of Pharmacology, Chandler Medical Center, University of Kentucky,
MS 305, Lexington, Kentucky 40536, USA, [¶]Department of Molecular Toxicology,
Oklahoma Medical Research Foundation, Oklahoma City, Oklahoma 73104, USA, and
[1]To whom correspondence should be addressed: National Institutes of Health,
9000 Rockville Pike, Building 3, Room 222, Bethesda, Maryland 20892, USA

Summary. During aging a number of enzymes accumulate as catalytically inactive or less
active forms. The age-related changes in catalytic activity are due in part to reactions of the
protein with "active" oxygen species such as ozone, singlet oxygen, or with oxygen free
radicals as are produced during exposure to ionizing radiation or to metal ion catalyzed
oxidation (MCO) systems. The levels of oxidized proteins in cultured human fibroblasts from
individuals of various ages and in liver and brain extracts of rats of different ages increase
progressively with age, and in old rats can represent 30–50% of the total cellular protein. The
age-related increase in oxidized protein in rat liver and brain tissue is accompanied by a loss
of glutamine synthetase (GS) and glucose-6-P dehydrogenase (G-6-PDH) activities, and to a
decrease in the level of cytosolic neutral protease activity which is responsible for the
degradation of oxidized (denatured) protein. Of particular significance are the results of
experiments showing that similar age-related changes occur in the gerbil brain and that these
changes are accompanied by a loss of short-term memory as measured by the radial arm maze
technique. Chronic treatment (intraperitoneal injections) of old animals with the free radical
spin-trap reagent, N-tert-butyl-α-phenylnitrone (PBN) resulted in normalization of the several
biochemical parameters to those characteristic of the young animals; coincidentally, the
short-term memory index was restored to the young animal values. These results provide the
first evidence that there is likely a linkage between the age-dependent accumulation of
oxidized enzymes and the loss of physiological function.

Introduction

A number of enzymic and nonenzymic systems, variously referred to
as mixed-function oxidation (MFO) or metal catalyzed oxidation
(MCO) (Stadtman, 1986; Stadtman, 1990) systems are able to catalyze
the oxidative modification of proteins (Oliver et al., 1981, Fucci et al.,
1983; Stadtman, 1990). The composition of some of these systems is
given in Table 1. The most physiologically relevant systems are the
NADH and NADPH oxidase, xanthine oxidase and cytochrome P_{450}
reductases, which in the presence of Fe(III), O_2 and their specific
electron donor substrates (NADH, NADPH, xanthine) are able to

catalyze the oxidation of the side chains of certain amino acid residues in proteins. In addition, nonenzymic systems in which ascorbate or mercaptans serve as auto-oxidizable electron donors are able to catalyze the O_2 dependent oxidation of proteins in the presence of Fe(III), and these have been widely used in mechanistic studies (Levine, 1983; Amici et al., 1989; Climent and Levine, 1991).

Role of H_2O_2 and Fe(II)

All of the MFO systems listed in Table I have been shown to catalyze the reduction of O_2 to H_2O_2, and of Fe(III) to Fe(II), via either O_2^- independent or dependent pathways (Fucci et al., 1983) (Fig. 1). The production of H_2O_2 and Fe(II) is the only function that the MFO systems have in common, and is all that is required for protein oxidation. The importance of H_2O_2 and Fe(II) in the oxidation of proteins is confirmed by the observations: (a) that a mixture of H_2O_2 and Fe(II) (but not Fe(III)) catalyzes rapid oxidation of proteins, and (b) that the

Table 1. Mixed-function oxidation systems that catalyze the modification of enzymes

Enzyme systems
  NAD(P)H oxidases/NAD(P)H/Fe(III)/O_2
  Xanthine oxidase/hypoxanthine/Fe(III)/O_2
  Nicotinate hydroxylase/NADPH/Fe(III)/O_2
  Cytochrome P_{450} reductase/cytochrome P_{450}/NADPH/Fe(III)/O_2
  Redoxin reductase/redoxin/cytochrome P_{450}/NADH/Fe(III)/O_2

Nonenzymic systems
  Ascorbate/Fe(III)/O_2
  RSH/Fe(III)/O_2
  Fe(II)/O_2
  Fe(II)/H_2O_2

For references see Stadtman (1990).

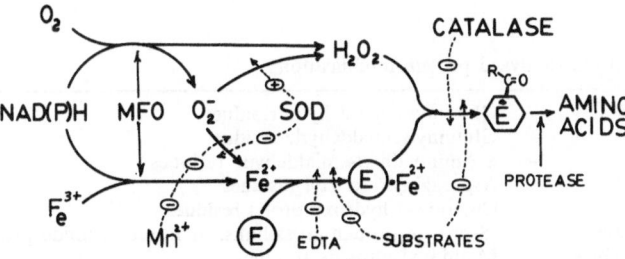

Figure 1. Mechanism of enzyme oxidation and degradation. SOD, superoxide dismutase, E, enzyme, MFO, mixed-function oxidation system. The symbols: ⊖, inhibition; ⊕ activation. From Stadtman & Oliver (1991).

oxidation of proteins by all of the MCO systems listed in Table 1 is inhibited by catalase (Fucci et al., 1983).

The MFO-catalyzed oxidation of enzymes is generally accompanied by a loss of catalytic activity and by conversion of the enzyme to a form that is highly susceptible to proteolytic degradation, especially to degradation by the neutral/alkaline cytosolic proteinases, which exhibit high specificity for the "oxidized" as opposed to the native form of enzymes (Rivett, 1985a; Rivett, 1985b; Roseman and Levine, 1987).

Site-specific nature of MCO systems

When proteins are exposed to ozone or to free radicals generated by ionizing radiation, most of the cysteine, histidine, methionine, tyrosine, phenylalanine and tryptophan residues are modified in an almost random manner. In contrast, histidine, arginine, lysine, proline, and cysteine are the most common targets for MCO systems (Amici et al., 1989). As shown in Table 2, in the oxidation of proteins by MCO systems, histidine residues are converted to aspartate/asparagine residues, arginine residues are converted to glutamic semialdehyde residues, lysine residues are converted to 2-aminoadipic semialdehyde derivatives, proline residues are converted to glutamate/pyroglutamate, and glutamic semialdehyde derivatives and cysteine residues are converted to disulfide derivatives. In addition, the nonheme iron centers of proteins are highly sensitive to oxidation by MCO systems (Turnbough and Switzer, 1975).

The site-specific nature of the metal ion catalyzed oxidation reactions is indicated by the facts: (a) The reactions are only slightly sensitive to inhibition by radical scavengers (formate, mannitol, SOD), suggesting that oxygen radial intermediates are not formed in free solution. (b) Only one or a few amino acid residues in a given protein are modified by MCO systems. (c) Only amino acid residues at metal binding sites of the protein are readily oxidized; for example, the oxidation of GS leads

Table 2. Metal ion catalyzed protein modification

Arginyl residues	→	Glutamylsemialdehyde residues
Prolyl residues	→	Glutamylsemialdehyde residues
Lysyl residues	→	α-Aminoadipylsemialdehyde residues
Histidyl residues	→	Asparaginyl/aspartyl residues
Prolyl residues	→	Cis/trans-4-hydroxy prolyl residues
Cysteinyl residues	→	-S-protein-protein cross-links; or mixed-disulfide protein S-S-R
Methionyl residues	→	Methionylsulfoxide residues
Tyrosyl residues	→	Tyrosyl-tyrosyl cross-links

For references see Stadtman (1990).

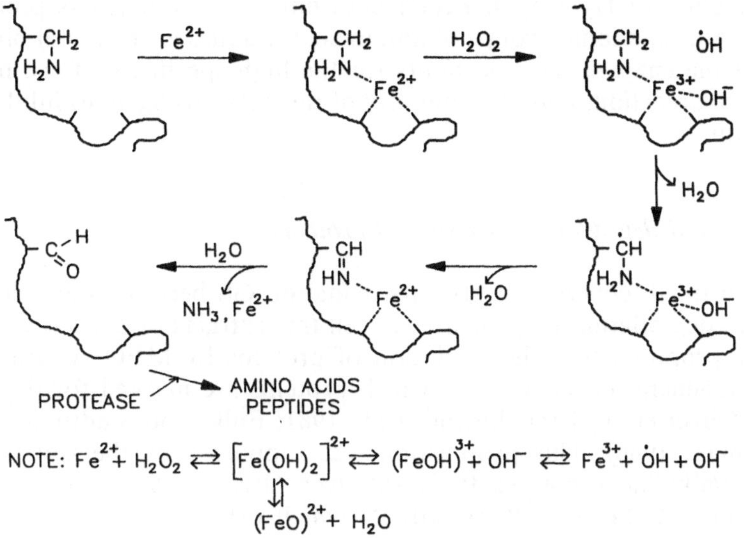

Figure 2. Site specific oxidation of enzymes. From Stadtman (1990).

to the modification of two histidine and one arginine residues, all of which are situated at metal binding sites on the enzyme (Farber and Levine, 1986; Ciment et al., 1989).

A site-specific mechanism by which the lysyl residue at a metal binding site of an enzyme could be oxidized to an aldehyde derivative by the combined action of Fe(II) and H_2O_2 is illustrated in Figure 2. In this mechanism, it is assumed that the ϵ-amino group of the lysyl residue ($-CN_2NH_2$) is one of several ligands to which Fe(II) can bind. Reaction of the Fe(II)-protein complex with H_2O_2 would lead to the production of Fe(III), ȮH and OH^- (Fenton chemistry). Because it is so reactive, the ȮH produced would not diffuse into the surrounding media, but would preferentially abstract a H atom from the carbon atom bearing the ε-amino group to form an alkyl radical. Subsequent transfer of the lone electron of the alkyl group to Fe(III) in the complex would lead to regeneration of Fe(II), and coincidentally, to formation of the imine derivative ($-CH=NH$), which upon spontaneous hydrolysis, would give rise to an aldehyde derivative and NH_3. This leads overall to destruction of the metal binding site, release of the Fe(II), conversion of the protein to a form which is highly susceptible to proteolytic degradation. In Figure 2, the reaction of Fe(II) with H_2O_2 is assumed to yield ȮH as a primary product; other mechanisms in which $(FeO)^{2+}$ or $(Fe(OH_2)_2)^{2+}$, or $(FeOH)^{3+}$ is the reactive intermediate are also possible. The important point is that by whatever mechanism oxidation is likely a "caged" reaction in which the active oxygen species produced in the reaction of H_2O_2 with Fe(II) does not escape but reacts preferentially with functional groups of amino acid residues at the metal binding site of the enzyme. This accounts for the high specificity of the metal-catalyzed reaction, and also inability of radical scavengers to inhibit the reaction.

Preferential degradation of oxidized proteins

With the discovery that after its oxidation GS becomes more susceptible to degradation by proteases in cell-free extracts of *Escherichia coli*, it was proposed that the oxidation of proteins by MFO systems provides a general mechanism by which proteins are marked for degradation (Oliver et al., 1981; Levine et al., 1981; Fulks and Stadtman, 1985; Stadtman, 1986). This proposition was supported by the isolation of neutral/alkaline proteases from rat liver (Rivett, 1985a) and *E. coli* (Roseman & Levine, 1987) extracts which exhibited almost complete specificity for the oxidized form of GS; neither protease could degrade the native form of GS. In the meantime, this concept was supported also by the extensive studies of Davies and coworkers (Lee et al., 1988; Pacifici et al., 1989; Davies et al., 1989; Davies and Lin, 1988; Marcillat et al., 1988) showing that the degradation of endogenous proteins in *E. coli*, red blood cells, and liver and heart mitochondria is greatly enhanced following exposure of the cells and organelles to oxygen-free radical generating systems.

Accumulation of oxidized proteins during aging

At least two-thirds of over 30 enzymes examined were found to be converted to catalytically inactive forms upon exposure to one or more of the MFO systems shown in Table 1 (Oliver et al., 1987). Many of the enzymes that are readily oxidized by MFO systems had been shown by other workers to accumulate as inactive or less-active, more thermal labile forms during aging. This suggested that some of the age-related changes in protein are the result of free-radical-mediated modification reactions. To test this possibility, advantage was taken of the fact that the metal ion catalyzed oxidation of proteins involves the conversion of the side chains of some amino acid residues (viz, arginine, lysine, proline) to carbonyl derivatives (cf, Table 2, Fig. 2). Through the use of highly sensitive techniques for measurement of the carbonyl content of proteins, it was found that there is in fact an age-dependent increase in the level of oxidized proteins. The relationship between age and the level of oxidized protein was examined in five different aging systems; namely, *the human erythrocyte system* (Oliver et al., 1987), the *cultured human fibroblast system* (Oliver et al., 1987), the *rat hepatocyte system* (Starke-Reed & Oliver, 1989), the *gerbil brain system* (Carney et al., 1991), and the *human brain model* (Smith et al., 1991). In all four systems, it was found that the level of oxidized protein (carbonyl content) increased with animal age, and, in all but the fibroblast model which was not tested, there was a concomitant decrease in the catalytic activities of GS and G-6-PDH. Moreover, in the hepatocyte system, it was established that the age-related decrease in the catalytic activities of GS and G-6-PDH was associated with the accumulation of catalytically inactive protein, as measured by immunotitration techniques. With reasonable assumptions, it is estimated that the level of oxidized protein represents about 10 and 25–40% of the total cellular protein in young and old animals, respectively. This is clearly an underestimate of the amount of damaged protein since the oxidation of some amino acid residues does not yield carbonyl derivatives; e.g., the oxidation of histidine to aspartate, of proline to glutamate, and of cysteinyl residues to disulfide derivatives.

Why do oxidized proteins accumulate?

The steady-state level of oxidized protein in a cell is determined by the relative rates of protein oxidation and the degradation of oxidized protein. Both rates are complex functions of numerous parameters. The rate of oxidation is determined among other factors by: the availability of O_2, Fe(III), electron donors (NAD(P)H); the activities of NAD(P)H oxidases and the cytochrome P_{450} enzymes; the concentrations of metal ion chelators, and free radical scavengers (SOD, uric acid, α-tocopherol, sulfhydryl compounds); the concentrations of enzymes that consume H_2O_2 (catalase, peroxidases); and the efficiency of the terminal electron transport system (cytochrome oxidase). The degradation of oxidized protein is a function of the levels of the proteases, the concentration of substrates which protect enzymes from oxidative modification (Fucci et al., 1981; Levine, 1983; Stadtman and Wittenberger, 1985), and the concentration of factors (peptides) that inhibit the activity of the proteases. In any case, it is likely that the age-related accumulation of oxidized proteins is due in part to the loss of protease activity since the level of neutral protease activity in hepatocyte and the brain models has been shown to decline with animal age (Starke-Reed and Oliver, 1989; Carney et al., 1991). In fact there is a linear inverse relationship between the amounts of altered forms of G-6-PDH that accumulate with age and the level of the neutral protease activity which is responsible for degrading oxidized protein (Fig. 3).

70

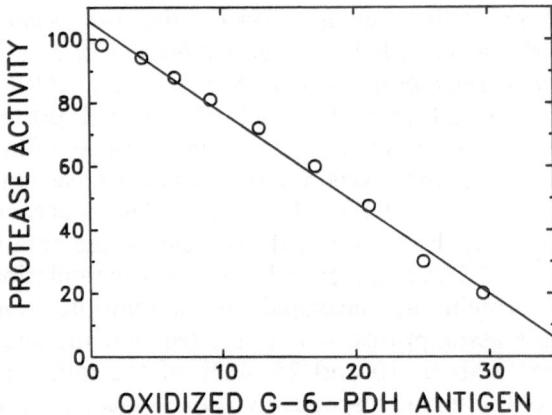

Figure 3. Age-dependent relationship between neutral protease activity and accumulation of inactive (oxidized) 6-6-PDH antigen. 100% protease activity refers to activity in rat hepato- cyte extracts of young (3-month-old) animals. Values are calculated from data in Starke-Reed and Oliver (1989) as described in Stadtman and Oliver (1991).

Physiological consequences of oxidized protein accumulation

It stands to reason that age-related decreases of 40–50% in the levels of key enzymes, and the accumulation of up to 40% of the total cellular protein as catalytically inactive, oxidized protein, would seriously com- promise some physiological functions. Especially, in view of the fact that the solvent capacity of cell water is finite, and that during evolution a number of highly sophisticated mechanisms have been developed for the rigorous control of both the intracellular concentrations and the catalytic activities of key enzymes. Indeed, possible linkage between the accumu- lation of oxidized protein and the manifestation of age-related physiolog- ical dysfunction is indicated by the recent studies of Carney et al. (1991). In these studies, it was shown that the levels of oxidized protein, GS and neutral protease activity in old gerbil brain cortex were 180, 63, and 36%, respectively, of the levels found in young gerbil brains (Table 3). More- over, the spatial/temporal memory acuity of young (3-month-old) gerbils was much better than that of old (15–18-month-old) animals. Thus, when placed in an octagonal radial arm maze chamber, the young animals made only 4 errors (as measured by the number of times they reentered an arm they had already explored, before exploring all 8 arms of the maze); whereas the old animals made 9–10 errors (Table 2). But of particular interest was the finding that when the old animals were treated (twice daily injections of 32 mg/kg of body weight) with the free-radical spin trap, N-tert-butyl-α-phenylnitrone (PBN), there was a progressive decrease in the level of oxidized protein and an increase in the levels of GS and neutral protease activity. After 14 days of treatment, the

Table 3. Effect of aging and PBN treatment on gerbil brain function[§]

Gerbils	Protein carbonyl	GS	Protease relative activity	Maze behavior index
	nmol/mg	units/mg	%	errors
Young	4.3	2.35	100	4.1
Old	8.1	1.55	36	9.5
Old + PBN*	4.1	2.13	69	4.2

[§]Calculated from data of Carney et al. (1991) Proc. Natl. Acad. Sci. USA 88: 3633–3636.
*After 14 days of PBN treatment.

levels of all three biochemical parameters approached the levels found in young gerbils (Table II). Even more exciting, however, was the observation that after 14 days of PBN treatment, the memory acuity of the old animals was restored to the young gerbil level. Taken together, the results of these studies show: (a) that the increase in the level of oxidized proteins is likely due in part to a decrease in the neutral protease activity that degrades oxidized protein; (b) that the age-related changes can be reversed by chronic chemotherapeutic measures (viz, by PBN treatment); and (c) that there may be a linkage between the age-related loss of a physiological function (i.e., memory) and the accumulation of oxidized proteins.

Amici, A., and Levine, R. L., Tsai, L., and Stadtman, E. R. (1989) Conversion of amino acid residues in proteins and amino acid homopolymers to carbonyl derivatives by metal-catalyzed oxidation reactions. J. biol. Chem. 264: 3341–3346.

Carney, J. M., Starke-Reed, P. E., Oliver, C. N., Landum, R. W., Cheng, M. S., Wu, J. F., and Floyd, R. A. (1991) Reversal of age-related increase in brain protein oxidation, decrease in enzyme activity, and loss in temporal and spatial memory by chronic administration of the spin-trapping compound N-tert-butyl-α-phenylnitrone. Proc. Natl. Acad. Sci. USA 88: 3633–3636.

Climent, I., and Levine, R. L. (1991) Oxidation of the active site of glutamine synthetase: Conversion of arginine-344 to γ-glutamylsemialdehyde. Archs Biochem. Biophys. 289: 371–375.

Climent, I., Tsai, L., and Levine, R. L. Derivatization of γ-glutamyl semialdehyde residues in oxidized proteins by fluoresceinamine. Analyt. Biochem. 182: 226–232.

Davies, K. J. A., and Lin, S. W. (1988) Degradation of oxidatively denatured proteins in Escherichia coli. Free Radical Biol. Med. 5: 215–223.

Davies, K. J. A., Lin, S. W., and Pacifici, R. E. (1987) Protein damage and degradation by oxygen radicals. IV. Degradation of denatured protein. J. biol. Chem. 262: 9914–9920.

Farber, J. M., and Levine, R. L. (1986) Sequence of a peptide susceptible to mixed-function oxidation: Probable cation binding site in glutamine synthetase. J. biol. Chem. 261: 4574–4578.

Fucci, L., Oliver, C. N., Coon, M. J., and Stadtman, E. R. (1983) Inactivation of key metabolic enzymes by mixed-function oxidation reactions: Possible implication in protein turnover and ageing. Proc. Natl. Acad. Sci. USA 80: 1521–1525.

Fulks, R. M., and Stadtman, E. R. (1985) Regulation of glutamine synthetase aspartokinase, and total protein turnover in Klebsiella aerogenes. Biochim. Biophys. Acta 843: 214–229.

Lee, Y. S., Park, S. C., Goldberg, A. L., and Chung, C. H. (1988) Protease So from Escherchia coli preferentially degrades oxidatively damaged glutamine synthetase. J. biol. Chem. 263: 6643–6646.

72

Levine, R. L. (1983) Oxidative modification of glutamine synthetase. II. Characterization of the ascorbate model system. J. Biol. Chem. 258: 11828–11833.

Levine, R. L., Oliver, C. N., Fulks, R. M., and Stadtman, E. R. (1981) Turnover of bacterial glutamine synthetase: Oxidative inactivation precedes proteolysis. Proc. Natl. Acad. Sci. USA 78: 2120–2124.

Marcillat, O., Zhang, Y., Lin, S. W., and Davies, K. J. A. (1988) Mitochondria contain a proteolytic system which can recognize and degrade oxidatively denatured proteins. Biochem. J. 254: 677–683.

Oliver, C. N., Ahn, B.-W., Moreman, E. J., Goldstein, S., and Stadtman, E. R. (1987) Age-related changes in oxidized proteins. J. biol. Chem. 262: 5488–5491.

Oliver, C. N., Levine, R. L., and Stadtman, E. R. Regulation of glutamine synthetase degradation; in: (Holzer, H., ed.) Metabolic Interconversion of Enzymes, pp. 259–268. Ed. H. Holzer. Springer-Verlag, Berlin 1981.

Oliver, C. N., Levine, R. L., and Stadtman, E. R. (1987) A role of mixed-function oxidation of altered enzyme forms during aging. J. Am. Geriat. Soc. 35: 947–956.

Pacifici, R. E., Salo, D. C., and Davies, K. J. A. (1989) Macroxyproteinase (M.O.P.): A 670 kDa proteinase complex that degrades oxidatively denatured proteins in red blood cells. Free Radical Biol. Med. 7: 521–536.

Rivett, A. J. (1989) Preferential degradation of the oxidatively modified form of glutamine synthease by intracellular proteases. J. biol. Chem. 260: 300–305.

Rivett, A. J. (1985b) Purification of a liver alkaline protease which degrades oxidatively modified glutamine synthetase: Characterization as a high molecular weight cysteine proteinase. J. biol. Chem. 260: 12600–12606.

Roseman, J. E., and Levine, R. L. (1987) Purification of a protease from *Escherichia coli* with specificity for oxidized glutamine synthetase. J. biol. Chem. 262: 2101–2110.

Smith, C. D., Carney, J. M., Starke-Reed, P. E., Oliver, C. N., Stadtman, E. R., Floyd, R. A., and Merkesbery, W. R. (1991) Excess brain protein oxidation and enzyme dysfunction in normal aging and in Alzheimer disease. Proc. Natl. Acad. Sci. USA 88: 10540–10543.

Stadtman, E. R. (1986) Oxidation of proteins by mixed-function oxidation systems: Implication in protein turnover, ageing and neutrophil function. Trends Biochem. Sci. 11: 11–12.

Stadtman, E. R. (1990) Metal ion-catalyzed oxidation of proteins: Biochemical mechanism and biological consequences. Free Radical Biol. Med. 9: 315–325.

Stadtman, E. R., and Oliver, C. N. (1991) Metal-catalyzed oxidation of proteins. J. biol. Chem. 266: 2005–2008.

Stadtman, E. R., and Wittenberger, M. E. (1985) Inactivation of *Escherchia coli* glutamine synthetase by xanthine oxidase, nicotinate hydroxylate, horseradish peroxidase, or glucose oxidase: Effects of ferredoxin, putaredoxin, and menadione. Archs Biochem. Biophys. 239: 379–387.

Starke-Red, P. E., and Oliver, C. N. (1989) Protein oxidation and proteolysis during aging and oxidative stress. Archs Biochem. Biophys. 275: 559–567.

Turnbough, L. L., and Switzer, R. L. (1975) Oxygen-dependent inactivation of glutamine phosphoribosylpyrophosphate amido transferase *in vitro*: Model for *in vivo* inactivation. J. Bacteriol. 121: 115–120.

Free Radicals and Aging
ed. by I. Emerit & B. Chance
© 1992 Birkhäuser Verlag Basel/Switzerland

Mitochondrial production of oxygen free radicals in the heart muscle during the life span of the rat: Peak at middle age

Carlo Guarnieri, Claudio Muscari and Claudio M. Caldarera

Department of Biochemistry, Centre of Research on Heart Metabolism, University of Bologna, v. Irnerio 48, I-40126 Bologna, Italy

Summary. Mitochondria extracted from Wistar rat hearts at 3, 14–18 and 24 months of age showed no change in state 3-mitochondrial respiration measured in the presence of glutamate or succinate.

Again no changes were found in the SMP–O_2^- production at the level of the rotenone-inhibited region, whilst at the level of the antimycin-inhibited region there was a marked increase in O_2^- production in the group of 14–18-month-old rats.

In the same age period, the production of mitochondrial H_2O_2 supported by glutamate or succinate and the level of GSSG increased in comparison to the young group, accompanied by a decrease in the GSH level.

Mitochondrial TBARS levels did not change during a life span, while a progressive accumulation in the mitochondrial lipofuscin content with age was measured.

Mitochondria have been recognized as a major physiological source of reactive oxygen species which can be generated during mitochondrial respiration (Turrens and Boveris, 1980). Superoxide radicals (O_2^-) formed from minor side-reactions of the mitochondrial electron-transport chain or by a NADH-independent enzyme located in the inner mitochondrial membranes (Nohl, 1987) can originate H_2O_2 and the powerful oxidant hydroxyl radical.

For these reasons, mitochondria are also considered the most important sites where peroxidative processes can arise, causing damage to electron transfer activities (Guarnieri et al., 1985), decreased membrane potential and altered permeability to cations (Bindoli, 1988).

Cardiac mitochondrial membranes are rich in polyunsaturated fatty acids susceptible to oxidation in vitro (Muscari et al., 1990) or in vivo in different pathological conditions such as ischemia and reperfusion or severe hypertrophy (Guarnieri et al., 1985), causing an impairment of mitochondrial function.

On the other hand, although the results are still controversial, the mitochondrial dysfunction is often considered the most important cause of functional modifications occurring in the aging heart.

In fact, cardiac mitochondria from senescent animals have specific reductions in the activities and contents of electron transport components (Chen et al., 1972), decreases in ADP/ATP exchange rates (Kim et al., 1988) and age-linked modifications in the activity of enzymes of the tricarboxylate cycle (Hansford and Castro, 1982).

Following the oxygen radicals theory of aging, we decided to study whether aging causes changes in cardiac mitochondrial function and whether there is a possible correlation between these changes; we also analyzed the rate of superoxide formation and the entity of lipid peroxidation at the mitochondrial level.

Results and discussion

ADP-induced mitochondrial respiration which is representative of cardiac mitochondrial function in vivo, did not change during the life span of the rat heart muscle (Table 1). The same table shows that in the 14–18-month-old group, the O_2^- production at the level of the antimycin-inhibited region of the respiratory chain was enhanced in comparison to the 3-month-old group, whereas the O_2^- production in the 24-month-old group remained unchanged. On the contrary, no differences were observed in the rate of O_2^- production at the level of the rotenone-inhibited region during aging. The production of mitochondria H_2O_2 increased at 14–18-months of age compared to the young group, when either glutamate or succinate was used as a substrate (Table 2). In the same age period, the levels of GSH significantly decreased while GSSG levels increased. At 24 months of age, only H_2O_2 production induced by succinate remained elevated, while that induced

Table 1. ADP-induced state-3 mitochondrial respiration and superoxide (O_2^-) formation by cardiac mitochondria from rats at different ages

	Months		
	3	14–18	24
State-3 (Glutamate)	113 ± 7	107 ± 8	99 ± 10
State-3 (Succinate)	153 ± 7	140 ± 12	160 ± 25
O_2^- (Rotenone-inhibited region)	2.5 ± 0.6	2.6 ± 0.5	2.4 ± 0.6
O_2^- (Antimycin-inhibited region)	0.9 ± 0.2	$2.3 \pm 0.9^*$	1.1 ± 0.8

The mitochondrial function and the rate of O_2^- formation were evaluated as previously described (Muscari et al., 1990).
State 3, natoms oxygen uptake in presence of ADP/mg mitochondrial protein/min.
O_2^-, nmol O_2^-/mg protein/min.
The values represent the average $\pm$SEM of 10 mitochondrial samples of 5 different animals.
$^*p < 0.05$, significantly different from the 3-month-old group.

Table 2. Mitochondrial H_2O_2 production and levels of glutathione from rats at different ages

	Months		
	3	14–18	24
H_2O_2 (glutamate)	0.28 ± 0.01	0.38 ± 0.04*	0.30 ± 0.03
H_2O_2 (succinate)	0.40 ± 0.02	0.58 ± 0.06*	0.49 ± 0.05*
GSH	8.2 ± 1.1	6.2 ± 4*	7.9 ± 3
GSSG	2.8 ± 1.1	8.1 ± 1.2*	3.3 ± 1.2

The rate of H_2O_2 formation was evaluated by the fluorimetric method that employs scopoletine as a reagent (Guarnieri et al., 1985). The levels of glutathione were determined as previously described (Muscari et al., 1990).
H_2O_2, nmol/mg protein/min.
GSH, GSSG, nmol/mg protein
The values represent the average ±SEM of 10 mitochondrial samples of 5 different animals.
*$p < 0.05$, significantly different from the 3-month-old group.

Table 3. Mitochondrial levels of peroxidative-derived metabolites in rats at different ages

	Months		
	3	14–18	24
TBARS	0.34 ± 0.01	0.39 ± 0.04	0.32 ± 0.04
Lipofuscin	48 ± 8	59 ± 10	99 ± 4

Lipid peroxidation parameters were determined as previously described (Muscari et al., 1990).
TBARS, thiobarbituric reactive substance (nmol/mg protein).
Lipofuscin (% fluorescence/mg protein).
The values represent the average ±SEM of 10 mitochondrial samples of 5 different animals.

by glutamate decreased to the value of the younger group. In the oldest rats (24 months) the levels of glutathione metabolites also decreased to those of 3 months of age. Mitochondrial TBARS levels did not change during a life span, while a progressive accumulation with age was observed in the cardiac mitochondria of fluorescent products such as lipofuscin (Table 3).

All these observations indicate that the cardiac muscle retains an efficient mitochondrial function during aging. Moreover, an increased mitochondrial production of O_2^- and H_2O_2 accompanied by a decrease of the GSH/GSSG ratio in the 14–18-month-old group suggests that at this age the heart muscle supports a kind of oxidative stress. On the contrary, we found no difference in mitochondrial O_2^- and H_2O_2 generation with respect to the younger animals in the group of rats that survived 24 months.

In accordance with the hypothesis that aging is a continuation of development influenced by a genetically programmed phenomena (Sohal and Allen, 1990), our results indicate that the appearance of oxidative stress at middle age might represent a signal for the heart muscle to adapt against senescence-induced modifications. For example, an increased oxygen radical production could be a signal to accentuate mitochondrial biogenesis (Davies et al., 1982) and protein turnover (Davies, 1987), or to influence gene expression at the mitochondrial level (Sohal and Allen, 1990). All these events can lead to a condition of muscle hypertrophy that is particularly frequent in the cardiac muscle of middle-aged animals (Jullien et al., 1989). It is interesting to note that in a 20-month-old myocardium Anversa et al. (1990) found a process of myocyte cellular hyperplasia which tends to regenerate the ventricular mass during aging, providing evidence that middle age represents a critical period for heart development.

The elevated flux of oxygen radicals at middle age is probably not thoroughly tolerated by rats since it corresponds to a marked decline in the survival curve of our colony of animals, suggesting that the 24-month-old rats represent a "survivor" group able to support the oxidative stress at middle age. Nevertheless, in the older group of animals there are also signs of oxidative damage at the myocardial level, evidenced by an accumulation of lipofuscin. Therefore, these data seem to indicate that the oxidative damage accumulated during aging is a secondary effect rather than a direct cause of senescence; they also suggest that a protective action against lipofuscin accumulation evident at 24 months of age, can preserve the mitochondrial function in the cardiac muscle for the rest of the life cycle.

Acknowledgments. The authors wish to thank Maddalena Zini and Massimo Sgarbi for their excellent technical assistance. We also thank Giovanna Grandi for her careful secretarial support. The research was supported by grants from C.N.R. (Project on Aging) and M.U.R.S.T.

Anversa, P., Palackal, T., Sonnenblick, E. H., Olivetti, G., Meggs, L. G., and Capasso, J. M. (1990) Myocyte cell loss and myocyte cellular hyperplasia in the hypertrophied aging rat heart. Cir. Res. 67: 871–885.

Bindoli, A. (1988) Lipid peroxidation in mitochondria. Free Rad. Biol. Med. 5: 247–261.

Chen, J. C., Warshaw, J. B., and Sanadi, D. R. (1972) Regulation of mitochondrial respiration in senescence. J. Cell. Physiol. 80: 141–148.

Davies, K. J. A., Quintanilha, A. T., Brooks, G. A., and Parker L. (1982) Free radicals and tissue damage produced by exercise. Biochem. Biophys. Res. Commun. 107: 1198–1205.

Davies, K. J. A. (1987) Protein damage and degradation by oxygen radicals. J. Biol. Chem. 262: 9895–9901.

Guarnieri, C., Muscari, C., and Caldarera, C. M. (1985) Oxygen radicals and tissue damage in heart hypertrophy. Advances in Myocardiology, Plenum Publishing, vol. 5, pp. 191–199.

Hansford, R. G., and Castro, F. (1982) Age-linked changes in the activity of enzymes of the tricarboxylate cycle and lipid oxidation and of carnitine content, in muscles of the rat. Mech. Aging Dev. 19: 191–201.

Jullien, T., Cand, F., Fargier, C., and Verdetti, J. (1989) Age-dependent differences in energetic status, electrical and mechanical performance of rat myocardium. Mech. Aging Dev. 48: 243–254.

Kim, J. H., Shrago, E., and Elson, C. E. (1988) Age-related changes in respiration coupled to phosphorylation. Cardiac mitochondria. Mech. Aging Dev. 46: 279–290.

Muscari, C., Frascaro, M., Guarnieri, C., and Caldarera, C. M. (1990) Mitochondrial function and superoxide generation from submitochondrial particles of aged rat hearts. Biochim. Biophys. Acta 1015: 200–204.

Nohl, H. (1987) A novel superoxide radical generation in heart mitochondria. FEBS Lett. 214: 269–273.

Sohal, R. S., and Allen, R. G. (1990) Oxidative stress as a casual factor in differentiation and aging: a unifying hypothesis. Exp. Geront. 25: 499–522.

Turrens, J. F., and Boveris, A. (1980) Generation of superoxide anion by the NADH dehydrogenase of bovine heart mitochondria. Biochem. J. 191: 421–427.

Free Radicals and Aging
ed. by I. Emerit & B. Chance
© 1992 Birkhäuser Verlag Basel/Switzerland

Lipofuscinogenesis in a model system of cultured cardiac myocytes

Massoud R. Marzabadi, Dazhong Yin and Ulf T. Brunk

Department of Pathology, University of Linköping, S-581 85 Linköping, Sweden

Summary. Lipofuscin, or age pigment, is a yellowish-brown, autofluorescent, protein, and lipid containing pigment that accumulates in the lysosomal vacuome of a variety of post-mitotic cells in man and animals during aging. Lipofuscin seems to be an end product of peroxidation, fragmentation and polymerization of proteins and lipids. Protein and lipid containing materials are regularly sequestered to the lysosomal system by means of endocytosis and autophagocytosis.

Lysosomes and acidified endosomes constitute cellular compartments where iron for short periods of time may exist in reactive form allowing for peroxidative processes. The importance was demonstrated in model systems of cultured cells subjected to oxidative stress and augmented amounts of $FeCl_3$ in the culture medium. This combination was considered to result in increased intracellular, mainly mitochondrial, production of hydrogen peroxide as well as increased intralysosomal concentration of iron. One of the results of this situation was dramatically increased lipofuscinogenesis.

Introduction

Lipofuscin, or age pigment, accumulates with age in the lysosomal vacuome of a variety of post-mitotic cell types in animals belonging to widely divergent phylogenetic groups (Strehler, 1964; Porta and Hartroft, 1969; Sohal, 1981; Donato, 1982). Although these cellular deposits have been widely regarded by the classical cytologists as an index of aging, their significance and relationship to aging processes has remained somewhat ambiguous because the chemical composition of the pigment is heterogeneous and its identification cannot be based on any single or distinct compound. However, lipofuscin invariably contains polymerized residues of peroxidized lipids and proteins and seems to be incorporated into the secondary lysosomes by autophagocytosis of cellular constituents undergoing peroxidation events within this vacuome (Brunk and Ericsson, 1972; Collins et al., 1980; Brunk and Collins, 1981; Sohal and Wolfe, 1986). A variety of experimental and comparative studies have been conducted to identify the factors that influence lipofuscinogenesis and may link it to the aging process. Although we are still in the midst of this quest, a considerable amount of knowledge has been gained to lead to the postulation of two hypotheses regrading the causes of lipofuscin accumulation and its role in the aging process. One hypothesis proposes oxidative stress, i.e. the ratio of pro-oxidants to antioxidants as the main factor (Tappel, 1975; Donato, 1982; Sohal and Wolfe, 1986; Brunk and Sohal, 1991), while the other hypothesis regards a decline in the degradative ability of cells to be primarily responsible for lipofuscin formation (Elleder, 1981; Wolfe et al., 1988; Ivy and Gurd, 1988; Marzabadi et al., 1991). The rationale for understanding the nature and mechanism of lipofuscinogenesis remains compelling, namely, that processes associated with its formation, or cellular effects, may be directly or indirectly related to the mechanism of cellular aging.

Characteristics of lipofuscin

Lipofuscin is a yellowish-brown pigment which is sequestered within cytoplasmic granules and can occupy up to 20% of the cytoplasmic volume in old cells (Brizzee et al., 1969) as granules up to 5 μm in diameter. Electron micrographs demonstrate lipofuscin-containing granules, or residual bodies, as having a single limiting membrane enclosing variable amounts of materials with vacuolar, granular, or lamellar organization (Wolman, 1980).

Lipofuscin has been histochemically characterized as a heterogeneous compound with some unifying characteristics as follows: basophilic due to the presence of acidic groups, with a pK_a of about 3 to 4; periodic acid-Schiff positive, probably due to the presence of carbonyl lipids or carbon chains with adjacent aldehyde groups; stainable by lipid stains such as Sudan black B and by reagents specific for proteins; and exhibiting positive localization of lysosomal enzymes (Strehler, 1964; Nandy, 1971). Lipofuscin granules emit intense yellowish auto-fluorescence when excited with ultraviolet and blue light (Fig. 1).

Chemical analyses of isolated lipofuscin have indicated the presence of three major components: 19 to 51% lipids, 30 to 50% proteins, and 9 to 30% acid hydrolysis-resistant residues. Extracts of isolated lipofuscin in chloroform contain a complex mixture of fluorescent substances (Björkerud, 1964; Jolly et al., 1987).

Mechanisms of lipofuscin formation

The lysosomal membrane-bound system of hydrolytic enzymes controls a major part of intracellular catabolism of macromolecules sequestered during autophagy in autophagosomes through fusion between lysosomes and phagosomes and encloses some 40 enzymes, such as proteases, nucleases, glycosidases, lipases, phospholipases, phosphatases, and sulfatases. These enzymes, being acid hydrolases, function in the pH 4.5–6 acidity maintained within the lysosomal

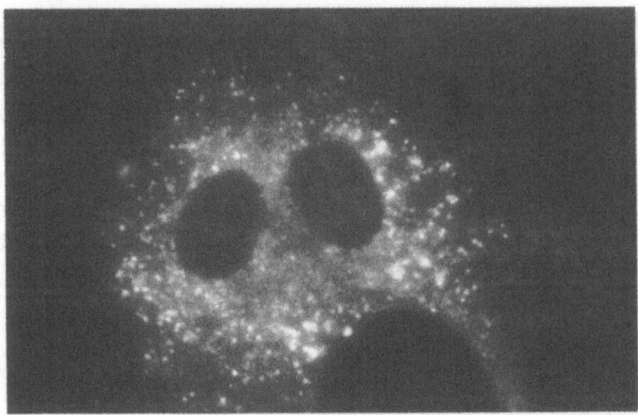

Figure 1. Photomicrograph of cardiac myocytes from a culture kept in 40% oxygen, 55% nitrogen and 5% carbon dioxide for 12 days. Cells show numerous autofluorescent yellow granules in the perinuclear area when excited with blue light (Reproduced from Sohal et al., 1989).

system by the ATP-dependent H^+ pump of the lysosomal membrane (Ohkuma et al., 1982). Lipofuscin granules, or residual bodies, are special types of secondary lysosomes with characteristic ultrastructure and special cytochemical features (Strehler, 1964) formed by incomplete degradation of material taken up during endocytosis and/or incomplete decomposition following autophagocytosis. Residual bodies accumulate as a result of continuous autophagocytosis in combination with inability of the cells to rid themselves of the residues of digestion to a sufficient degree (Collins et al., 1980; Brunk and Collins, 1981).

Studies on cultured human glial cells with electron dense marker particles added to the culture medium have shown residual bodies (lipofuscin granules) to form part of a vacuome system which by fusion and fission allows material within membrane-limited vacuoles to spread throughout the system although at any one moment it is discontinuous (Brunk, 1973). Lipofuscin is the end product of molecular damage to cell organelles by oxygen free radicals (Tappel, 1975; Sohal et al., 1989). Studies supporting this view have attempted to identify molecular events which could inflict oxidative damage to cell organelles, leading to their autophagocytosis and subsequent partial degradation or, alternatively, to intralysosomal peroxidation occurring after cellular material becomes sequestered within the lysosomal vacuome during normal autophago-cytosis. Following peroxidation of degration fragments, polymerization of these fragments would result in products resistent to normal enzymatic degradation. Most experimental studies have focused on peroxidation of lipids, the subsequent reactions of the lipid breakdown products, and the reactions resulting in the production of the characteristic fluorophores

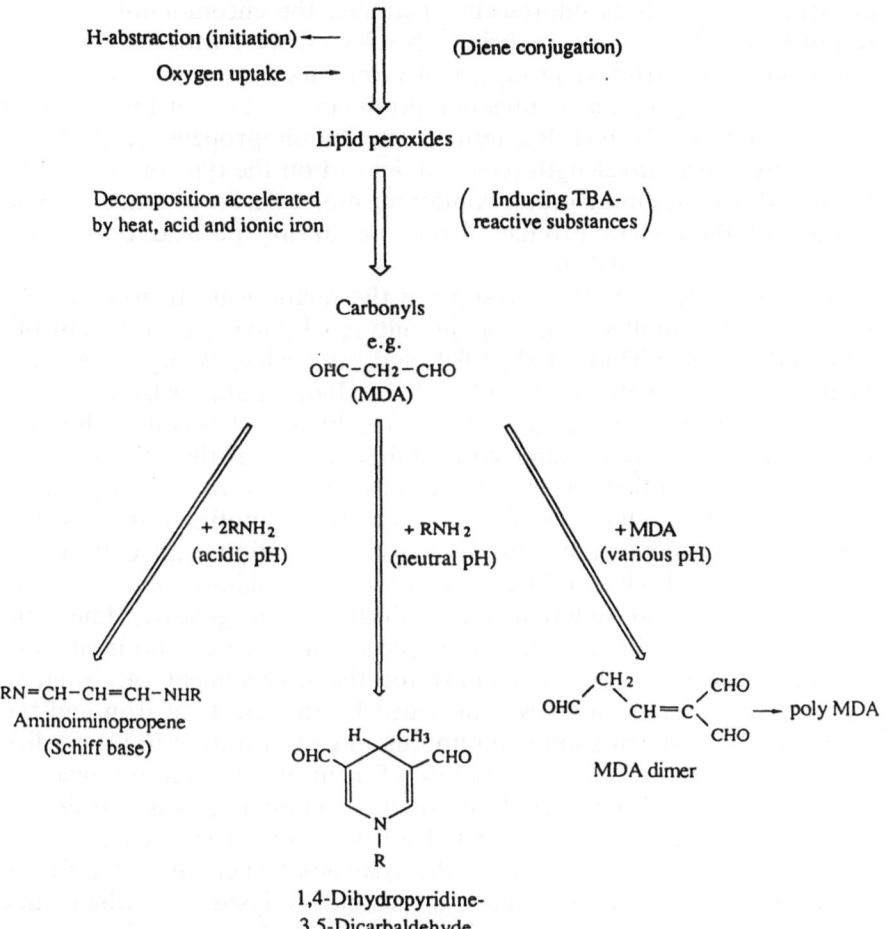

Figure 2. Fluorescent materials produced from iron catalyzed peroxidation of membrane lipids. (Modified from Halliwell and Gutteridge, Eds. "Free radicals in biology and medicine", Clarendon Press, Oxford 1989).

within lipofuscin (Tappel, 1975). The biochemical concept of fluorophore formation, developed mainly by Tappel and coworkers (Tappel, 1975), considers lipid autooxidative reactions to play a crucial role. Figure 2 illustrates the formation of fluorescent products during lipid peroxidation events. Following the free radical initiation and propagation steps, lipid hydroperoxides, cyclic peroxides, and cyclic endoperoxides are formed, which further decompose to carbonyl compounds, among them malondialdehyde. The reaction of carbonyl compounds with side-chain amino groups present in proteins, free amino acids, amino phospholipids, and nucleic acids produces conjugated Schiff bases that have strong optical absorptions as well as fluorescence features, the chromophoric system responsible for fluroescence being $-N=C-C=C-N-$. Malondialdehyde, possessing two carbonyl groups, can cross-link two amino-containing compounds to produce fluorescent products of the general structure $R_1-N=CH-CH=CH-NH-R_2$, termed aminoiminopropene Schiff bases. The fluorescence wavelength seems to depend on the type of side chains (R_1 and R_2) in the amino group-containing molecule. An alternative route leading to fluorescent product formation during peroxidation is the polymerization of aldehydes.

We have experimentally investigated the relationship between oxidative stress and lipofuscinogenesis in cultured human glial cells (Brunk and Collins, 1981; Thaw et al., 1984) and rat cardiac myocytes (Marzabadi et al., 1988; Sohal et al., 1989; Marzabadi et al., 1990). Effects of various ambient oxygen concentrations (5, 20, and 40%) on the lipofuscin concentration were examined at different ages of the culture. Lipofuscin was quantified by microspectrofluorometry of individual cells (see below). The yellow autofluorescence due to lipofuscin increased in direct relationship to ambient oxygen concentration with culture age (Figs. 1 and 3). Light- and transmission electron microscopic examination of the cells at different ages indicated a progressive time- and oxygen-dependent increase in the frequency and size of lipofuscin-containing organelles. Further support for the involvement of oxidative stress in lipofuscinogenesis was provided by the effects of iron and the iron chelator desferrioxamine on lipofuscin concentration in rat cardiac myocytes in culture. Augmentation of iron in the culture medium markedly increased the level of lipofuscin accumulation, whereas desferrioxamine had the opposite effect. Both iron and desferrioxamine are endocytosed and will end up in the lysosomal vacuome after fusion between endosomes and primary or secondary lysosomes where they influence the intralysosomal milieu and the condition for lipofuscinogenesis. These findings can be understood in terms of iron potentiation of oxidative stress, probably by means of a superoxide (O_2^-) or other reducing agent such as glutathion, ascorbic acid or cysteine-driven Fenton reaction (Halliwell and Gutteridge, 1985) reactions [1 and 2].

$$X-Fe^{3+} + O_2^- \longrightarrow X-Fe^{2+} + O_2 \qquad [1]$$

$$X-Fe^{2+} + H_2O_2 \longrightarrow X-Fe^{3+} + OH^- + OH^· \qquad [2]$$

Hydroxyl radicals ($OH^·$) are very potent radicals abstracting hydrogen from all available molecules and initiating peroxidation (Fig. 2). It is believed that lipofuscinogenesis is mainly an intralysosomal process and that iron added to the culture medium enters the lysosomal vacuome following endocytotic uptake. The effects of 30 μM ferric iron

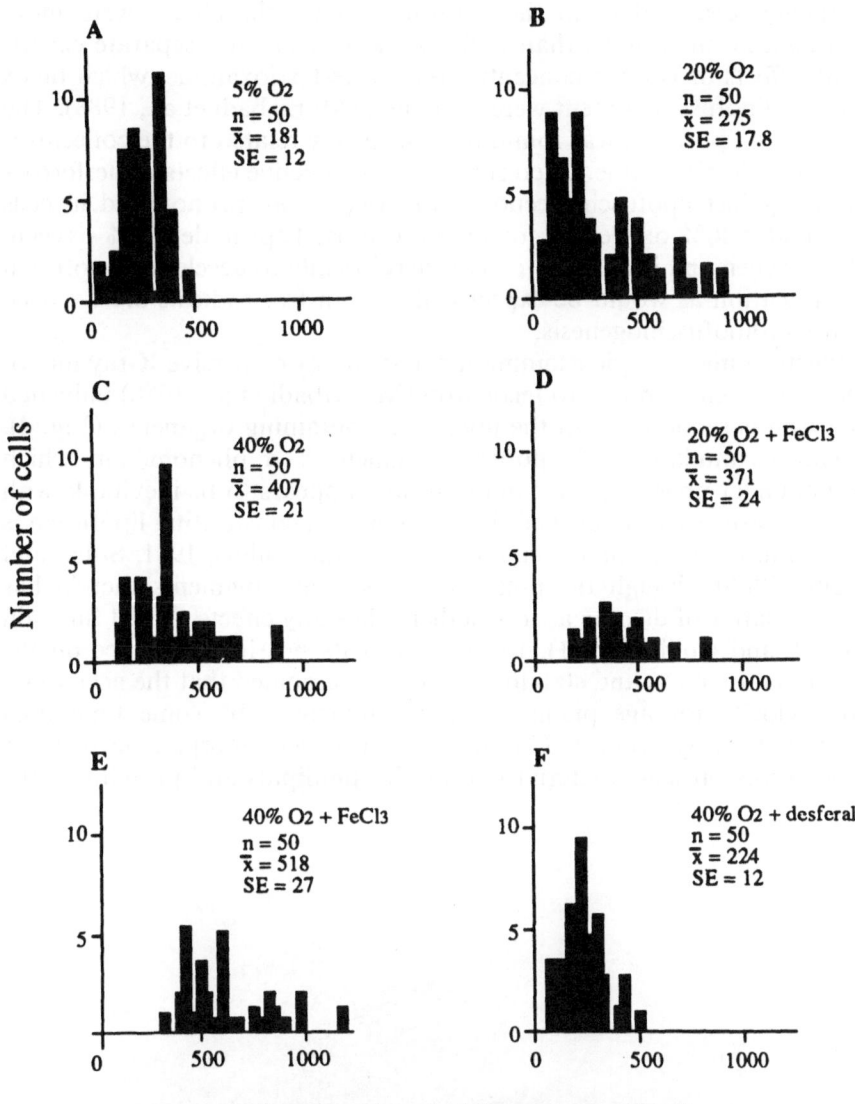

Figure 3. Computer-generated histograms of autofluorescence exhibited by cardiac myocytes kept in 5%, 20% or 40% oxygen for 12 days. The additional effect of FeCl₃ and desferrioxamine is also shown. Abscissa: Autofluorescence/cell (arbitrary units). Ordinate: Number of cells. A total of 50 cells were examined in each group.

added to the culture medium on lipofuscin levels were mainly examined on neonatal cardiac myocytes kept under 20% and 40% oxygen (Marzabadi et al., 1988). At both 6 and 12 days of culture age, the amount of lipofuscin found in myocytes exposed to 30 μM ferric iron was

84

markedly greater than in the controls. Again, the effects were more pronounced under 40% than under 20% oxygen. In a separate experiment, effects of various concentrations of desferrioxamine, which binds iron, on lipofuscin content were examined (Marzabadi et al., 1988). The amount of lipofuscin was found to be inversely related to the concentration of desferrioxamine. Interestingly, the protective effects of desferroxamine against lipofuscin accumulation were more pronounced in cells kept under 40% oxygen as compared to those kept under 20% oxygen. Thus oxygen and iron seem to act synergistically to accelerate lipofuscin accumulation as would be expected if oxygen free radicals are a causal factor in lipofuscinogenesis.

Electron microscopic examination and energy dispersive X-ray microanalysis of cells exposed to ferric iron (Marzabadi et al., 1988) indicated that iron is sequestered in the lipofuscin-containing organelles (Fig. 4).

The accumulation of lipofuscin pigment is a phenomenon which occurs under physiological conditions and appears to be inevitable with age. It is now well established that the material constituting lipofuscin is located in the lysosomal vacuome (Brunk and Collins, 1981; Sohal and Wolfe, 1986), though the presence of these age pigments therein has been a matter of dispute as to whether it has any effects on cell function (Brunk and Collins, 1981). Regardless of its possible influence on the lysosomal function and stability, it could be assumed that the accumulation velocity for age pigment will be influenced by some functional properties of lysosomes: (1) the rate at which autophagocytosis or endocytosis presents material rich in phospholipids and proteins to the

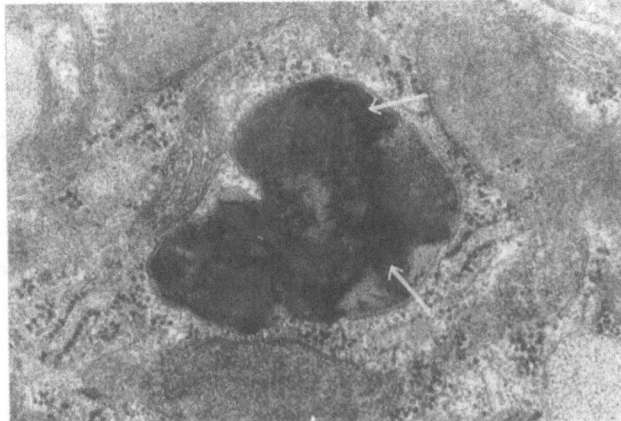

Figure 4. Electron micrograph of a lipofuscin-containing organelle in a myocyte cultured for 8 days in a medium containing 30 μM FeCl$_3$ showing dense needle-shaped material (arrow) shown to be iron when analyzed with energy dispersive X-ray analysis (Reproduced from Marzabadi et al., 1988).

lysosomal system, (2) the efficiency of the lysosomal enzymes in degrading this material, (3) the degree of peroxidation inside the lysosomal vacuome, and (4) the cellular capacity to exocytose undegradable residues. Peroxidation will also be influenced by cytosolic formation and degradation of O_2^- and H_2O_2 as well as the amount of reactive iron outside and inside the lysosomal vacuome. Peroxidation initiated in the cytosol may influence factors leading to autophagocytosis. Age pigment, being located in lysosomes and therefore mixed with lytic enzymes which are capable of degrading almost all types of organic material, is essentially a nondegradable end product. Possibly this nondigestible lipofuscin may cause cellular impairment by forcing the cell to supply large amounts of the multiple degradative enzymes to a system containing material which cannot be lysed. Also, lipofuscin-overloaded lysosomes might be unable to further handle peroxidized material formed during oxidative stress, which would increase intracellular concentration of products of lipid peroxidation, such as malondialdehyde (MDA), impairing several critical cellular targets. Under normal conditions then, the intralysosomal localization of lipofuscin could be rationalized as a secondary antioxidant defense in as far as this complex polymerized compound does not interfere with cell functions, at least not until loading is very substantial (Brunk and Cadenas, 1988).

Quantification of lipofuscin

Lipofuscin is identified and measured by four different methods:

(1) The most reliable method measures the autofluorescence of lipofuscin granules and employs microspectrofluorometry of cells in situ.

(2) Another method uses cytochemical staining and morphometry of sectioned material and can only provide semi-quantitiative measurements. The drawback of this method is that it cannot determine if all the granules under study exhibit fluorescence; and if they do, of what intensity.

(3) Another morphological method employs electron microscopy and morphometry of sectioned material. It suffers from the handicap that it cannot specifically identify the relative fluorescent content of the secondary lysosomes.

(4) One method that has been frequently used in recent studies and the one which has contributed the most to the current confusion in terminology is based on the measurement of blue-emitting fluorescence in the organic solvent extracts of tissues (Csallany and Ayaz, 1976). Studies using this method have generated much controversy because the results are often at odds with those obtained by microspectrofluorometry. We therefore suggest that the term "lipofuscin" be used only for the material that specifically exhibits in situ the characteristics of lipofuscin,

namely greenish-yellowish autofluorescence eminating from the lysoso-mal vacuome. In our opinion, microspectrofluorometry of the in situ graunles is currently the most reliable method for the quantification of lipofuscin. The discrepancies between emission values for fluorometric and microfluorometric measurements on tissue extracts and intact cells, respectively, may be due to the methodological differences. The basic difference in methodological approach is that pathologists and histolo-gists examine intact lipofuscin granules in sectioned cells by light microscopy and collect data using microscopic fluorometers, whereas biochemists measure extractable lipofuscin material in diluted solutions by spectrofluorometry. It must be stressed that the lipofuscin granules are highly condensed when studied in a microscope with an attached fluorometer, and the fluorophores' concentration in accumulated pig-ment granules may be thousands of times higher than in extracts. For instance, if we suppose that the extractable fluorophores amount to 2% of a certain sample, the extraction introduced dilution is 1000 times using the usual chloroform-methanol extraction method (sol-vent:tissue = 20:1). However, blue in situ lipofuscin fluorescence has been reported (Barden, 1980), but a careful study of that report shows the barrier filters used were chosen on the assumption that the maxi-mum emission peak of lipofuscin was between 400 and 500 nm.

In contrast to absorbance spectroscopy, fluorescence spectroscopy values do not follow the Lambert-Beer Law at high fluorophore concen-tration (Bashford, 1987). A main cause is known as the inner filter effect, which arises because the excitation or emission light may be strongly reabsorbed by the fluorophore. Such a loss of quantum yield is often accompanied by spectral shifts (Pesce et al., 1971). Several mech-anisms are known for these alterations: (a) re-absorption of emitted light, (b) formation of fluorophore dimers or polymers, and (c) forma-tion of fluorophore eximers. In many cases the fluorescence spectra of various compounds will be shifted against longer wavelengths at in-creasing concentrations (Pesce et al., 1971). As reported in a recent study (Yin and Brunk, 1991), the blue → yellow shift may occur within a range as narrow as one or two orders of magnitude with respect to concentration.

In histotechnology and histochemistry such concentration-dependent color changes (or fluorescence shifts) are well known and denoted as metachromasia (Sculthorpe, 1978). Many dyes may accumulate differ-ently within various cell compartments, thus resulting in different col-oration by the same stain. For example, when acridine orange and 9-aminoacridine are accumulated by acidic vacuoles within living cells, it results in quenching of their blue fluorescence, so that these vesicles appear orange-red in the fluorescent microscope. This orange fluores-cence is the characteristic emission of very concentrated solutions of these dyes (Bashford, 1987). These exact mechanisms of such metachromatic effects are quite complicated and many theories have been proposed (Sculthorpe, 1978); (a) polymerization of dye molecules, (b) combination of the chromotrope with resonance forms of the dye, (c) orderly dye aggregation, characterized by the formation of new intermolecular bonds between adjacent dye molecules. In a recent study from this laboratory (Yin and Brunk, 1991), these characteristics are shown to be valid for three types of compounds all supposed to be important part of lipofuscin fluorophores, namely, 1,4-dihydropyridines, Schiff bases and MDA polymers. Hence, the blue → yellow fluorescence shift of the lipofuscin-associated fluorophores may be explained as resulting from the concentration-dependent metachromatic effect.

Conjoined shift of excitation with emission maxima were observed (Yin and Brunk, 1991). This effect matches well with the phenomenon recognized by microscopists that the blue light excitation can often result in a stronger yellow emission than UV light excitation. Katz (1984), Dowson (1981) and their colleagues reported that a 390–490/515 nm-filter combination produced the greatest intensity of lipofuscin autofluorescence (yellowish) under microspectral fluorometers, when compared with using excitation filters in UV area.

The correction of instrumental sensitivity for different wavelengths supported the existence of a concentration–dependent fluorescence shift of lipofuscin-like fluorophores (Yin and Brunk, 1991). The concentration-dependent metachromasia, however, suggests a direct explanation for the discrepancies in the literature on lipofuscin fluorescence emission spectra.

Barden, H. (1980) Interference filter microfluorometry of neuromelanin and lipofuscin in human brain. J. Neuropathol. Exp. Neur. 39: 329.

Bashford, C. L. (1987) Spectrophotometry and fluorimetry of cellular compartments, in: Spectrophotometry & Spectrofluorimetry: A Practical Approach. Bashford, C. L., and Harris, D. A., eds. IRL Press, Oxford, p. 115.

Björkerud, S. (1964) Isolated lipofuscin granules – a survey of a new field. Adv. Gerontol. Res. 1: 257.

Brizzee, K. R., Cancilla, P. A., Sherwood, N., and Timmiras, P. S. (1969). The amount and distribution of pigment in neurons and glia of the cerebral cortex. J Gerontol. 24: 127.

Brunk, U., and Cadenas, E. (1988) The potential intermediate role of lysosomes in oxygen free radical pathology. APMIS 96: 3.

Brunk, U. T., and Collins, V. P. Lysosomes and age pigment in cultured cells, in: Age Pigments. Sohal R. S., ed. Elsevier, Amsterdam, p. 243.

Brunk, U. T. (1973) Distribution and shifts of ingested marker particles in residual bodies and other lysosomes. Exp. Cell Res. 79: 1.

Brunk, U. T., and Ericsson, J. L. E. (1972) Electron microscopical studies on rat brain neurons. Localization of acid phosphatase and mode of formation of lipofuscin. J. Ultrastruct. Res. 38: 1.

Brunk, U. T., and Sohal, R. S. (1991) Mechanisms of lipofuscin formation, in: Membrane Lipid Oxidation, Vol. II. Vigo-Pelfrey C, ed. Boston, Boca Raton CRC Press, p. 191.

Collins, V. P., Arborgh, B., Brunk, U. T., and Schellens, S. P. M. (1980) Phagocytosis and degradiation of rat liver mitochondria by human glia cells. Lab. Invest. 42: 209.

Csallany, A. S., and Ayaz, K. L. (1976) Quantitative determination of organic solvent soluble lipofuscin pigments in tissues. Lipids 11: 412.

88

Donato, H. Jr, and Sohal, R. S. (1982) Lipofuscin, in: Handbook of Biochemistry in Aging. Florini J, ed. Cleveland, CRC Press, p. 221.

Dowson, J. H., and Harris, S. J. (1981) Quantitative studies of the autofluorescence derived from neuronal lipofuscin. J. Microsc. 123: 249.

Elleder, M. (1981) Chemical characterization of age pigments, in: Age Pigments. Sohal R. S., ed. Elsevier, Amsterdam, p. 203.

Halliwell, B., and Gutteridge, J. M. C. (1985) The importance of free radicals and catalytic metals in human diseases. Mol. Aspects Med. 8: 89.

Ivy, G. O., and Gurd, J. W. (1987) A proteinase inhibitor model of lipofuscin formation, in Lipofuscin – 1987: State of the Art. Zs.-Nagy I, ed. Akademiai Kiado, Budapest and Elsevier, Amsterdam, p. 83.

Jolly, R. D., Barns, A. B., and Palmer, D. N. (1987) Ovine ceroid-lipofuscinogenesis: chemical constituents of the lipopigment, their pathogenic significance and similarities to age pigment. Adv. Biol. Sci. 64: 197.

Katz, M. L., Robison, W. G. Jr, Herrmann, R. K., Groome, A. B., and Bieri, J. G. (1984) Lipofuscin accumulation resulting from senescence and vitamin E deficiency: Spectral properties and tissue distribution. Mech. Ageing Dev. 25: 149.

Marzabadi, M. R., Sohal, R. S., and Brunk, U. T. (1990) Effect of vitamin E and some metal chelators on lipofuscin accumulation in cultured neonatal rat myocytes. Anal. cell. pathol. 2: 333.

Marzabadi, M. R., Sohal, R. S., and Brunk, U. T. (1988) Effect of ferric iron and desferrioxamine on lipofuscin accumulation in cultures rat heart myocytes. Mech. Ageing Dev. 46: 145.

Marzabadi, M. R., Sohal, R. S., and Brunk, U. T. (1991) Mechanisms of lipofuscinogenesis: Effect of the inhibition of lysosomal proteases and lipases under varying concentrations of ambient oxygen in cultured rat neonatal myocardial cells. APMIS 99: 416.

Nandy, K. (1971) Properties of neuronal lipofuscin pigment in mice. Acta Neuropathol. 19: 25.

Ohkuma, S., Moriyama, Y., and Takano, T. (1982) Identification and characterization of a proton pump on lysosomes by fluorescein isothiocyanate-dextran fluorescence. Proc. Natl. Acad. Sci. USA 79: 2758.

Pesce, A. J., Rosen, C. G., and Pasby, T. L., eds. (1971) Fluroescence Spectroscopy; an Introduction for Biology and Medicine. Marcel Dekker Inc., New York, p. 37.

Porta, E., and Hartroft, W. S. Lipid pigments in relation to aging and dietary factors (lipofuscin), in: Pigments in Pathology. Wolman, M., ed. Academic Press, New York, p. 191.

Sculthorpe, H. H. (1978) Metachromasia, Med. Lab. Sci. 35: 365.

Sohal, R. S., ed. (1981) Age Pigments. Elsevier, Amsterdam.

Sohal, R. S., Marzabadi, M. R., Galaris, D., and Brunk, U. T. (1989) Effect of ambient oxygen concentration on lipofuscin accumulation in cultured rat heart myocytes – a novel in vitro model of lipofuscinogenesis. Free Radic. Biol. Med. 6: 23.

Sohal, R. S., and Wolfe, L. S. (1986) Lipofuscin: characteristics and significance, in: Progress in Brain Research, Vol. 70. Swaab, D. F., Fliers, E., Mirmiran, M., Van Gool, W. A., and Van Haaren, F., ed. Elsevier Biomedical, New York, p. 171.

Strehler, B. L. (1964) On the histochemistry and ultrastructure of age pigment in: Advances in Gerological Research, Vol. 1. Strehler, B. L., ed. Academic Press, New York, p. 343.

Tappel, A. L. (1970) Lipid peroxidation and fluorescent molecular damage to membranes, in: Pathobiology of Cell Membranes, Vol. 1. Trump B, Arstila A, eds. Academic Press, New York, p. 145.

Thaw, H. H., Brunk, U. T., and Collins, V. P. Influence of oxygen tension, prooxidants and antioxants on the formation of lipid peroxidation products (lipofuscin) in individual cultivated human glial cells. Mech. Ageing Dev. 24: 211.

Wolfe, l. S. Gauthier, S., and Durham, H. D. (1988) Dolichols and phosphorylated dolichols in the neuronal ceroid lipofuscinosis, other lysosomal storage diseases and Alzheimer's disease. Induction of autolysosomes in fibroblasts, in: Lipofuscin–1987: State of the Art, Zs-Nagy, I., ed. Akademiai Kiado, Budapest and Elsevier, Amsterdam, p. 389.

Wolman, M. (1980) Lipid pigments (chromolipids): their origin, nature and significance, in: Pathobiology Annual, Vol. 10. Joachim, H. L., ed. Raven Press, New York, p. 253.

Yin, D., and Brunk, U. T. (1991) Microfluorometric and fluorometric lipofuscin spectral discrepancies: A concentration dependent metachromatic effect? Mech. Ageing Dev. 59: 95.

Free Radicals and Aging
ed. by I. Emerit & B. Chance
© 1992 Birkhäuser Verlag Basel/Switzerland

Cellular clones and transgenic mice overexpressing copper–zinc superoxide dismutase: Models for the study of free radical metabolism and aging

Irène Ceballos-Picot, Annie Nicole and Pierre-Marie Sinet

CNRS URA 1335, Laboratoire de Biochimie Génétique, Hôpital Necker-Enfants Malades, 149 rue de Sèvres, F-75015 Paris, France

Summary. Down's Syndrome (DS), the most frequent of congenital birth defects, results from the trisomy of the chromosome numbered 21 in all cells of affected patients. This disease is characterized by developmental anomalies, mental retardation and features of rapid aging, particularly in the brain where the occurrence of Alzheimer's disease (AD) is observed in all trisomy 21 patients over the age of 35. Elucidation of the biological mechanisms leading to brain aging in DS might provide new insight into the understanding of brain aging and AD in normal people.

Copper-zinc superoxide dismutase (CuZnSOD) is one of the genes encoded by chromosome 21. As a consequence of gene dosage excess, CuZnSOD activity and protein are increased by 50% in all DS tissues. The level of CuZnSOD protein and mRNA is particularly high in hippocampal pyramidal neurons susceptible to degenerative processes in AD and in dopaminergic melanized-neurons vulnerable in Parkinson's disease. Increased CuZnSOD activity in these age-related neurodegenerative disorders might result in H_2O_2 overproduction and subsequently promote peroxidative damages within cells. Increase of seleno-dependent glutathione peroxidase (Se-GPx) in DS cells supports this concept.

In order to test this hypothesis, cell and animal models of CuZnSOD overexpression have been designed. In cells transfected with the human CuZnSOD gene, and increased Se-GPx activity is observed, a situation which mimics DS. In mice transgenic for the human CuZnSOD, the expression pattern of the transgene in the brain is similar to that in humans, and we can observe an increased peroxidation in this tissue. These data, like others in the literature, support the hypothesis that excess CuZnSOD induces an imbalance in the regulation of oxygen-derived free radical production which might result in peroxidative brain damage and possibly contribute to accelerated aging and age-related neuropathology.

Introduction

Down's syndrome (DS) or trisomy 21 is the most frequent birth defect in humans, afflicting about one newborn in 1000. In most of the cases, DS results from the presence of an extra copy of the entire chromosome 21 in all cells. DS is characterized by morphological abnormalities, frequent visceral malformations, high incidence of leukemia and phenotypic and biological signs of accelerated aging. Dysfunction of the central nervous system is manifested by two pathological features: mental retardation and Alzheimer-like neuropathology.

Neuropathology is encountered in potentially all DS patients over the age of 35, with senile plaques and neurofibrillary tangles distributed in the same brain territories as in Alzheimer's disease (AD). Assessment of the AD lesions in DS patients at various ages compared to the age-related occurrence of AD lesions in the general population leads to the conclusion that the neuropathological processes might start in a DS brain some 50 years earlier than in a normal brain (Rumble et al., 1989). Thus, DS can be regarded as a disease model for identifying biological mechanisms leading to accelerated aging and AD-like neuropathology. We can infer from this model that excess of one or several genes on chromosome 21 is responsible for the pathogenesis of these features.

Human copper-zinc superoxide dismutase (CuZnSOD; superoxide:superoxide oxidoreductase, EC 1.15.1.1), a key enzyme in the metabolism of oxygen free radicals (Mc Cord and Fridovich, 1969), is encoded by a gene residing on the 21q22.1 band of chromosome 21 (Sinet et al., 1976). The CuZnSOD gene is expressed in all tissues. In the human brain, its level of expression is particularly high in neurons (Ceballos et al., 1989; Ceballos et al., 1990a). Immunohistochemistry (Delacourte et al., 1988) and cDNA-mRNA in situ hybridization (Ceballos et al., 1989; 1991a; Somerville et al., 1991) studies on human hippocampus have shown that the amounts of CuZnSOD protein and mRNA are high in large pyramidal neurons and granular cells, suggesting that the production of superoxide radicals within these cells might be particularly elevated. Overexpression of the CuZnSOD gene, due to gene dosage, has been recorded in all tissues of trisomic 21 patients and in a variety of trisomy 21 cells in culture (Sinet et al., 1982) as well as in the brains of DS patients (Brooksbank and Balazs, 1984). Moreover, elevated CuZnSOD activity has been reported in two other neurodegenerative disorders. First, in brains of patients with AD, increased CuZnSOD levels were found in certain regions, for instance the nucleus caudatus (Marklund et al., 1985). Second, increased SOD activities have also been found in the brains of patients who suffer from Parkinson's disease (Marttila et al., 1988). This enzyme is particularly well expressed in the age-vulnerable dopaminergic melanized neurons (Ceballos et al., 1990a). Since CuZnSOD catalyzes the dismutation of superoxide radicals $O_2°-$ to oxygen and hydrogen peroxide (H_2O_2), it has been hypothesized that increased CuZnSOD in trisomy 21 might induce overproduction of H_2O_2 within cells and accelerate oxidative damage to cell components (Sinet, 1982). Supporting this hypothesis, the selenodependent glutathione peroxidase (Se-GPx) activity, which catalyzes the reduction of H_2O_2, has been found to be increased in trisomy 21 tissues like red cells and platelets (Sinet et al., 1975; Crosti et al., 1989). In fibroblasts, normal (Feaster et al., 1977; Crosti et al. 1989) or increased (Sinet et al., 1979; Anneren and Epstein, 1987) Se-GPx activity has been reported whereas in the fetal brain Se-GPx was found to be normal (Brooksbank et al., 1984). The increase in Se-GPx activity in trisomy 21 tissues does not result from a gene dosage effect since the gene encoding Se-GPx is on chromosome 3. This increase is actually linked to CuZn-SOD excess (Ceballos et al., 1988). Recent reports have shown that CuZnSOD is able to catalyze hydroxyl radical ($OH^.$) production from hydrogen peroxide (Yim et al., 1990) and that elevated level of SOD is cytotoxic (Norris and Hornsby, 1990) and enhance the cytotoxicity of active oxygen species (Scott et al., 1987; 1989). Such a mechanism may be in part responsible for certain clinical symptoms of DS patients and for the accelerated aging and AD pathology associated with the Down's phenotype (Sinet et al., 1982); it could also contribute to the neuropathology associated with AD and Parkinson's disease (Ceballos et al., 1990b).

To investigate the possible involvement of CuZnSOD overproduction in the etiology of these neurodegenerative diseases, the cloned gene and cDNA could be introduced into cells as part of recombinant plasmids. Work with expression vectors containing CuZnSOD indicates that an oversupply of this enzyme is not necessarily beneficial to a mammalian cell and can even cause increased lipid peroxidation (Elroy-Stein, 1986). To further delineate this relationship, we (Ceballos-Picot et al., 1991b) and other groups (Epstein et al., 1987) have created transgenic mice for the human CuZnSOD gene. Is CuZnSOD overexpression toxic or beneficial for the cells? Could CuZnSOD overexpression induce modifications in free radical metabolism and aging-related biological perturbations? A review of the literature on cellular clones and transgenic mice overexpressing CuZnSOD will help us answer these questions.

Cellular clones overexpressing CuZnSOD: Biological consequences

The human CuZnSOD gene and cDNA were introduced into human HeLa cells, mouse L cells and rat PC12 cells (Elroy-Stein et al., 1986; 1988) and in mouse L cells and mouse neuroblastoma cells (Ceballos et al., 1988). Cell clones expressing elevated levels of enzymatically active human CuZnSOD showed increased lipid peroxidation which may result in the alteration of the structure and function of cellular membranes (Elroy-Stein et al., 1986) and higher resistance to the toxic effect of paraquat (Elroy-Stein et al., 1986; Krall et al., 1988; Kelner and Bagnell, 1990). In stable clones producing increased amounts of CuZn-SOD a marked increase in endogenous Se-GPx was observed (Ceballos et al., 1988; Kelner and Bagnell, 1990), and a smaller increase in glutathione transferase activity also occurred. Manganese superoxide dismutase (MnSOD) activity was decreased in all clones, whereas catalase and NADPH reductase were not affected (Kelner and Bagnell,

1990). Alteration in Se-GPx and MnSOD correlated with increases in CuZnSOD activity. Whereas all clones were resistant to paraquat, a direct correlation between CuZnSOD activity and resistance to paraquat did not exist. Clones expressing the highest CuZnSOD activity did not display the highest resistance to paraquat. However, there was a direct correlation between the increase in Se-GPx activity and paraquat resistance (Kelner and Bagnell, 1990). An explanation for the decrease in MnSOD activity is that the presence of increased CuZnSOD activity leads to a reduction in the intracellular steady-state superoxide concentrations, and MnSOD is not produced when not required by the cell. Alternatively, the overproduction of CuZnSOD may be harmful to the cell by producing increased intracellular concentrations of hydrogen peroxide, and the response of endogenous glutathione peroxidase supports this hypothesis. The cell may try to respond by decreasing endogenous superoxide dismutase production such as MnSOD or endogenous CuZnSOD as a regulatory defense. Further studies are required to determine how these antioxidant enzymes are controlled intracellularly. The ways in which stable adaptations might occur are through heritable changes in the cell's ability to transcribe certain genes selectively, to diminish cellular turnover of specific mRNAs, or to prolong the intracellular life span of the enzymes themselves. These results support the previous hypothesis that balanced increments of antioxidant enzymes are necessary for cell integrity and growth and that a deregulation of this complex antioxidant system may disturb the steady state equilibrium of oxy-radicals within cells, resulting in oxidative damage to biologically important molecules. Such a mechanism may in part be responsible for certain clinical symptoms of DS patients such as accelerated aging and for the AD pathology associated with the DS phenotype. Experiments support this interpretation: the transfected PC12 cells overexpressing CuZnSOD had impaired uptake of neurotransmitters such as dopamine and norepinephrine due to a dimunution in the PH gradient across their chromaffin granule membrane (Elroy-Stein and Groner, 1988). The authors attributed the pertubation of PH gradient to the uncoupling of ATPase or increased permeability of the granule membranes to protons, likely secondary to lipid peroxidation. It was described that peroxidative damage resulting from excess CuZnSOD expression in transfected adrenocortical cells causes cell fusion, nuclear fragmentation and subsequent propagation of cell death (Norris and Hornsby, 1990). Because the metabolism of oxygen free radicals and in particular the process of lipid peroxidation are tightly connected to the biosynthesis of prostaglandins, the synthesis of prostaglandin E2 (PGE2) was diminished in transfected PC12 cells, as well as in fibroblasts from DS patients (Minc-Golomb et al., 1991).

Increased SOD activity in *Drosophila* did not confer greater resistance to paraquat-generated superoxide radicals. On the contrary, in some transformants, increased activity resulted in a greater sensitivity to paraquat (Seto et al., 1990). It may be argued that with elevated levels of SOD activity, the $O_2^{\div}$ generated by paraquat would be dismutated to H_2O_2 as well as the other active form of oxygen (OH$^{\cdot}$, $^1O_2$), which may be more toxic than $O_2^{\div}$ itself. However, it was reported by another group that expression of enzymatically active bovine SOD in *Drosophila* confers resistance to paraquat (Reveillaud et al., 1991). This is consistent with data on adult mortality, because there was a slight but significant increase in the mean life span of several of the transgenic lines with a 1.6 increase in SOD activity. Higher levels may have led to the formation of toxic levels of H_2O_2 during development, since flies that died during the process of eclosion showed an unusual accumulation of lipofuscin (age pigment) in some of their cells (Reveillaud et al., 1991). Alternatively, Seto et al. (1990) have reported that additional SOD activity in *Drosophila* transformants (2- to 3-fold) does not affect adult life span.

Likewise, *Escherichia coli* that overexpress FeSOD or MnSOD (Scott et al., 1987; Liochev and Fridovich, 1991) was more sensitive to paraquat. Thus, elevated levels of SOD activity alone are not necessarily advantageous. In this way, the radiation sensitivity of *E. coli* increases as bacterial SOD activity increases. Elevated intracellular SOD activity sensitizes *E. coli* to radiation-induced mortality, whereas SOD-deficient bacteria show normal or decreased radiosensitivity (Scott et al., 1989). Therefore, radiation sensitivity of Down's syndrome fibroblasts might be due to overexpressed CuZnSOD (Schwaiger et al., 1989). *E. coli* double mutants completely lacking superoxide dismutase greatly enhanced mutation rates during aerobic growth (Farr et al., 1986).

Transgenic mice overexpressing CuZnSOD: Biological consequences

In order to further explore the deleterious effect of excess CuZnSOD, transgenic mice for the human CuZnSOD were obtained by us (Ceballos-Picot et al., 1991b) and others (Epstein et al., 1987). In our transgenic strain, one copy was integrated in the mouse genome and CuZnSOD activity was higher than in control mice in almost all tissues (Ceballos et al., 1991c). In the brain, CuZnSOD activity was increased by a factor of 2 and the cell-specificity of the transgene expression was similar to that observed in humans, i.e. in the hippocampus large pyramidal neurons and granular cells (Ceballos-Picot et al., 1991b). Transgenic mice carrying the human CuZnSOD gene are not phenotypically abnormal. If CuZnSOD excess is indeed involved in the pathogenesis of DS, we can expect its effects to be on more subtle aspects of tissue function, particularly in tissues that might be affected by altered metabolism of free radicals over long periods of time. The nervous system was the prime candidate for our investigations. Looking for biochemical modifications which could support this hypothesis, we studied the activities of enzymes involved in oxygen radical detoxification in the brain of transgenic mice. Despite the 2-fold increase of CuZnSOD in the brains of transgenic mice, Se-GPx, glutathione reductase and glutathione S-transferase activities were not modified (Ceballos et al., 1991c). It appears that, at least in young mice and in a normal oxygen environment, an increase in CuZnSOD activity similar to that found in Down's syndrome does not by itself produce alterations of the enzymes associated with oxygen radical detoxification. Thus, unlike what has been observed in DS erythrocytes (Anneren et al., 1985; Sinet et al., 1975), in lymphoid cells and fibroblasts (Anneren et al., 1985) and in transfected cells overexpressing CuZnSOD (Ceballos et al., 1988; Kelner and Bagnell, 1990), similar to that found in the DS fetal brain (Brooksbank and Balazs, 1984), Se-GPx activity was not increased in the brain of transgenic mice. The reason for this discrepancy between the results in the brains of transgenic mice and in the transfected cells with the same human CuZnSOD gene (Ceballos et al., 1988) is not clear. One possibility is that Se-GPx is not inducible in the brain by increased CuZnSOD activity due to a differential cellular localization of these two enzymes. To further delineate the relationship between peroxidative stress and increased CuZnSOD activity without an adaptative rise in Se-GPx activity in the brain of transgenic mice, the basal lipid peroxidation (as measured by thiobarbituric acid (TBA)-reactive material) was determined and found to be significantly higher in transgenic brains compared to controls (Ceballos-Picot et al., 1991b). This result was consistent with that reported for the human fetal DS brain (Brooksbank and Balazs, 1984) and for human CuZnSOD transfected cells (Elroy-Stein et al., 1986), as well as in blood platelets from human CuZnSOD transgenic mice (Schickler et al., 1989). Such reports are evidence in favor of the hypothesis that oxidant damage due to CuZnSOD excess contributes to the aging process, providing at least partial support for the free-radical theory of aging (Harman, 1981). Moreover, platelets of transgenic CuZnSOD animals, which overexpress the transgene, contain lower levels of serotonin than non-transgenic littermate mice, due to a reduced rate of uptake of neurotransmitters by the dense granules of the platelets (Schickler et al., 1989). Furthermore, a significantly lower than normal serotonin accumulation rate was also detected in dense granules isolated from blood platelets of Down's syndrome individuals. These findings suggest that CuZnSOD gene dosage is affecting the dense granule transport system and is thereby involved in the depressed level of blood serotonin found in patients with Down's syndrome. Other interesting features have been observed in these transgenic mice, such as ultrastructural changes in the neuromuscular junction of the tongue muscle similar to those seen in the muscles of aging mice and rats as well as in the tongue muscles of Down's syndrome patients (Avraham et al., 1988). Whether or not these anomalies are linked to increased peroxidative damages remains to be proved. Nevertheless, these observations suggest that CuZnSOD excess could have deleterious effects on the nervous system. However, no electrophysiological differences were found in mouse dorsal root ganglion neurons transgenic for human CuZnSOD (Ault et al., 1989), indicating that increased dosage of the CuZnSOD gene alone is not causal to action potential dysfunction found in trisomy 21 (Nieminen et al., 1988).

The development of these transgenic mice has opened the way for the evaluation of a number of problems related to the involvement of increased CuZnSOD gene dosage in Down's syndrome. In addition, these transgenic mice will also be of importance in the investigation of the role of CuZnSOD in the protection against the adverse effects of a variety of exogeneous agents and pathophysiological situations implicating oxygen free radicals such as hyperoxia, ischemia, radiation, chemical agents, i.e. paraquat and MPTP.

Transgenic mice with the human CuZnSOD gene exhibit several differences from control mice when exposed to oxidative stress. After a 30-second cold injury to the cerebral cortex, transgenic mice with approximately 3-times the normal CuZnSOD activity had less cerebral edema and extravasation of Evans blue (a measure of increased vascular permeability) and a smaller area of infarction than the control animals (Chan et al., 1991). It was concluded that superoxide radicals play an important role in the pathogenesis of these lesions in the cold-traumatized brain. Transgenic mice were also protected to a significant degree against the effects of transient focal cerebral ischemia of the middle cerebral artery, and had less edema and smaller infarcts, with penumbra region surrounding the ischemic brain tissue being particularly spared (Kinouchi et al., 1991). Moreover, cultured primary cortical neurons from transgenic fetuses demonstrated a significant reduction in the toxic effects of exposure to H_2O_2 or 0.5 mM glutamate, as measured by intracellular 3-O-methyl glucose space, efflux of lactate dehydrogenase, and determination of cell viability (Chan et al., 1990). Transgenic mice had an increased development of tolerance to hyperoxia associated with decreased histopathologic evidence of lung damage and mortality (White et al., 1991). These results suggest that excessive superoxide, or one of its toxic products such as OH·, causes or contributes to pulmonary oxygen toxicity.

Conclusion

The use of genetic manipulation to construct cells or transgenic mice that specifically lack or overexpress a single antioxidant enzyme represent a mean for understanding the role of individual antioxidant enzymes in cellular aging and degenerative diseases. Extended biochemical and ultrastructural studies of the brains of the transgenic mice described here should explain whether CuZnSOD excess participates in the accelerated aging process in Down's syndrome, especially with regard to Alzheimer-like neuropathology. Moreover, the study of these animal models and of clones overexpressing CuZnSOD has general implications, particularly with respect to the potential role of free radicals in aging and age-related diseases.

Anneren, G., Edqvist, L. E., Gebre-Medhin, M., and Gustavson, K. H. (1985) Glutathione peroxidase activity in erythrocytes in Down's syndrome. Abnormal variation in relation to age and sex through childhood and adolescence. Trisomy 21. 1: 9–17.

Anneren, G., and Epstein, C. J. (1987) Lipid peroxidation and superoxide dismutase-1 and glutathione peroxidase activities in trisomy 16 fetal mice and human trisomy 21 fibroblasts. Pediat. Res. 21: 88–92.

Ault, B., Caviedes, P., Hidalgo, J., Epstein, C. J., and Rapoport, S. I. (1989) Electrophysiological analysis of cultured fetal mouse dorsal root ganglion neurons transgenic for human superoxide dismutase-1, a gene in the Down syndrome region of chromosome 21. Brain Res. 497: 191–194.

Avraham, K. B., Schickler, M., Sapoznokow, D., Yarom, R., and Groner, Y. (1988) Down's syndrome: abnormal neuromuscular junction in tongue of transgenic mice with elevated levels of human CuZn-superoxide dismutase. Cell 54: 823–829.

Brooksbank, B. W. L., and Balazs, R. (1984) Superoxide dismutase, glutathione peroxidase and lipoperoxidation in Down syndrome fetal brain. Dev. Brain Res. 16: 37–44.

Ceballos, I., Delabar, J. M., Nicole, A., Lynch, R. E., Hallewel, R. A., Kamoun, P., and Sinet, P. M. (1988) Expression of transfected human CuZn superoxide dismutase gene in mouse L cells and NS20Y neuroblastoma cells induces enhancement of glutathione peroxidase activity. Biochim. Biophys. Acta 949: 58–64.

Ceballos, I., Javoy-Agid, F., Hirsch, E. C., Dumas, S., Kamoun, P., Sinet, P. M., and Agid, Y. (1989) Localization of copper-zinc superoxide dismutase mRNA in human hippocampus by in situ hybridization. Neurosci. Lett. 105: 41–46.

Ceballos, I., Lafon, M., Javoy-Agid, F., Hirsch, E., Nicole, A., Sinet, P. M., and Agid, Y. (1990a) Superoxide dismutase and Parkinson's disease. The Lancet, 335: 1035–1036.

Ceballos, I., Javoy-Agid, F., Delacourte, A., Defossez, A., Nicole, A., and Sinet, P. M. (1990b) Parkinson's disease and Alzheimer's disease: neurodegenerative disorders due to brain antioxidant system deficiency? In: Antioxidants in Therapy and Preventive Medicine, Emerit et al. (Eds), Plenum Press, N.Y., p. 493–498.

Ceballos, I., Javoy-Agid, F., Delacourte, A., Defossez, A., Lafon, M., Hirsch, E., Nicole, A., Sinet, P. M., and Agid, Y. (1991a) Neuronal localization of copper-zinc superoxide dismutase protein ans mRNA within the human hippocampus from control and Alzheimer's disease brains. Free Rad. Res. Comms. 12–13: 571–580.

Ceballos-Picot, I., Nicole, A., Briand, P., Grimber, G., Delacourte, A., Defossez, A., Javoy-Agid, F., Lafon, M., Blouin, J. L., and Sinet, P. M. (1991b) Neuronal-specific expression of human copper-zinc superoxide dismutase gene in transgenic mice: animal model of gene dosage effect in Down's syndrome. Brain Res. 552: 198–214.

Ceballos, I., Nicole, A., Briand, P., Grimber, G., Delacourte, A., Flament, S., Thevenin, J. M., Kamoun, P., and Sinet, P. M. (1991c) Expression of human CuZn superoxide dismutase gene in transgenic mice: model for gene dosage effect in Down's syndrome. Free Rad. Res. Comms. 12–13: 581–589.

Chan, P. H., Chu, L., Chen, S., Carlson, E. J., and Epstein, C. J. (1990) Reduced neurotoxicity in transgenic mice overexpressing copper-zinc superoxide dismutase. Stroke Suppl. III, 21: 80–82.

Chan, P. H., Yang, G. Y., Chen, S. F., Carlson, E., and Epstein, C. J. (1991) Cold-induced brain edema and infarction are reduced in transgenic mice overexpressing CuZn-superoxide dismutase. Ann. Neurol. 29: 482–486.

Crosti, N., Bajer, J., Gentile, M., Resta, G., and Serra, A. (1989) Catalase and glutathione peroxidase activity in cells with trisomy 21. Clin. Genet. 36: 107–116.

Delacourte, A., Defossez, A., Ceballos, I., Nicole, A., and Sinet, P. M. (1988) Preferential expression of copper–zinc superoxide dismutase in the vulnerable cortical neurons in Alzheimer's disease. Neurosci. Lett. 92: 247–253.

Elroy-Stein, O., Bernstein, Y., and Groner, Y. (1986) Overproduction of human CuZn superoxide dismutase in transfected cells: extenuation of paraquat-mediated cytotoxicity and enhancement of lipid peroxidation. EMBO J. 5: 615–622.

Elroy-Stein, O., and Groner, Y. (1988) Impaired neurotransmitter uptake in PC12 cells overexpressing human Cu-Zn superoxide dismutase: implications for gene dosage effect in Down's syndrome. Cell 52: 259–267.

Epstein, C. J., Avraham, K. B., Lovett, M., Smith, S., Elroy-Stein, O., Rotman, G., Bry, C., and Groner, Y. (1987) Transgenic mice with increased Cu/Zn-superoxide dismutase activity: Animal model of gene dosage effects in Down syndrome. Proc. Natl Acad. Sci. 84: 8044–8048.

Farr, S. B., D'Ari, R., and Touati, D. (1986) Oxygen-dependent mutagenesis in Escherichia coli lacking superoxide dismutase. Proc. Natl. Acad. Sci. 83: 8268–8272.

Feaster, W. W., Kwok, L. W., and Epstein, C. J. (1977). Dosage effects for superoxide dismutase-1 in nucleated cells aneuploid for chromosome 21. Am. J. Hum. Genet. 29: 563–570.

Harman, D. (1981) The aging process. Proc. Natl Acad. Sci. 78: 7124–7128.

Kelner, M. J., and Bagnell, R. (1990) Alteration of endogenous glutathione peroxidase, manganese superoxide dismutase, and glutathione transferase activity in cells transfected with a copper-zinc superoxide dismutase expression vector. J. Biol. Chem. 265: 10872–10875.

Kinouchi, H., Mizui, T., Carlson, E., and Epstein, C. J. (1991) Focal cerebral ischemia and the antioxidant system in transgenic mice overexpressing CuZn-superoxide dismutase. J. Cereb. Blood Flow Metab. 11: S423.

Krall, J., Bagley, A. C., Mullenbach, G. T., Hallewell, R. A., and Lynch, R. E. (1988) Superoxide mediates the toxicity of Paraquat for cultured mammalian cells. J. Biol. Chem. 263: 1910–1914.

Liochev, S. I., and Fridovich, I. (1991). Effects of overproduction of superoxide dismutase on the toxicity of paraquat toward Escherichia coli. J. Biol. Chem. 266: 8747–8750.

Marklund, S. L., Adolfsson, R., Gottfries, C. G., and Winblad, B. (1985) Superoxide dismutase isoenzymes in normal brains and in brains from patients with dementia of Alzheimer type. J. Neurol. Sci., 67: 319–325.

Marttila, R. J., Lorentz, H., and Rinne, U. K. (1988) Oxygen toxicity protecting enzymes in Parkinson's disease. Increase of superoxide dismutase-like activity in the substantia nigra and basal nucleus. J. Neurol. Sci. 86: 321–331.

McCord, J. M., and Fridovich, I. (1969) Superoxide dismutase, an enzymatic function for erythrocuprein (hemocuprein). J. Biol. Chem. 244: 6049–6055.

Minc-Colomb, D., Knobler, H., and Groner, Y. (1991) Gene dosage of CuZnSOD and Down's syndrome: diminished prostaglandin synthesis in human trisomy 21, transfected cells and transgenic mice. EMBO 10: 2119–2124.

Nieminen, K., Suarez-Isla, B. A., and Rapoport, S. I. (1988) Electrical properties of cultured dorsal root ganglion neurons from normal and trisomy 21 human fetal tissue. Brain Res. 474: 246–254.

Norris, K. H., and Hornsby, P. J. (1990) Cytotoxic effects of expression of human superoxide dismutase in bovine adrenocortical cells. Mut. Res. 237: 95–106.

Reveillaud, I., Niedzwieki, A., Bensh, K. G., and Fleming, J. E. (1991) Expression of bovine superoxide dismutase in Drosophila melanogaster augments resistance to oxidative stress. Mol. Cell. Biol. 11: 632–640.

Rumble, B., Tetallack, R., and Hilbich, C. (1989) Amyloid A4 protein and its precursor in Down's syndrome and Alzheimer's disease. New England J. Med. 22: 1446–1452.

Schickler, M., Knobler, H., Avraham, K. B., Elroy-Stein, O., and Groner, Y. (1989) Diminished serotonin uptake in platelets of transgenic mice with increased Cu/Zn-superoxide dismutase activity. EMBO 8: 1385–1392.

Schwaiger, H., Weirich, H. G., Brunner, P., Rass, C., Hirsch-Kauffman, M., Groner, Y., and Schweiger, M. (1989) Radiation sensitivity of Down's syndrome fibroblasts might be due to overexpressed C/Zn-superoxide dismutase (EC 1.15.11). Eur. J. Cell Biol. 48: 79–87.

98

Scott, M. D., Meshnick, S. R., and Eaton, J. W. (1987) Superoxide dismutase rich bacteria. Paradoxal increase in oxidant toxicity. J. Biol. Chem. 262: 3640–3645.

Scott, M. D., Meshnick, S. R., and Eaton, J. W. (1989) Superoxide dismutase amplifies organismal sensitivity to ionizing radiation. J. Biol. Chem. 264: 2498–2501.

Seto, N. O. L., Hayashi, S., and Tener, G. M. (1990) Overexpression of Cu-Zn superoxide dismutase in Drosophila does not affect life span. Proc. Natl Acad. Sci. 87: 4270–4274.

Sinet, P. M., Michelson, A. M., Bazin, A., Lejunne, J. and Jerome, H. (1975) Increase in glutathione peroxidase activity in erythrocytes from trisomy 21 subjects. Biochem. Biophys. Res. Comm. 67: 910–915.

Sinet, P. M., Couturier, J., Dutrillaux, A., and Jerome, H. (1976). Trisomie 21 et superoxide dismutase-1 (IPO-A). Tentative de localisation sur la sous-bande 21q22.1. Exp. Cell Res. 97: 47–55.

Sinet, P. M., Lejeune, J., and Jerome, H. (1979) Trisomy 21 (Down's syndrome), glutathione peroxidase, hexose monophosphate shunt and IQ. Life Sci. 24: 29–34.

Sinet, P. M. (1982) Metabolism of oxygen derivatives in Down's syndrome. Ann. N.Y. Acad. Sci. 386: 82–94.

Somerville, M. J., Percy, M. E., Bergeron, C., Yoong, L. K. K., Grima, E. A. and McLachlan, D. R. C. (1991) Localization and quantitation of 68 kda neurofilament and superoxide dismutase-1 mRNA in Alzheimer brain. Molec. Brain Res. 9: 1–8.

White, C. W., Avraham, K. B., Shanley, P. F., and Groner, Y. (1991) Transgenic mice with expression of elevated levels of copper-zinc superoxide dismutase in the lungs are resistant to pulmonary oxygen toxicity. J. Clin. Invest. 2162–2168.

Yim, M. B., Chock, P. B., and Stadtman, E. R. (1990). Copper, zinc superoxide dismutase catalyses hydroxyl radical production from hydrogen peroxide. Proc. Natl Acad. Sci. 87: 5006–5010.

Free Radicals and Aging
ed. by I. Emerit & B. Chance
© 1992 Birkhäuser Verlag Basel/Switzerland

The importance of antioxidant enzymes in cellular aging and degeneration

José Remacle, Carine Michiels and Martine Raes

Laboratoire de Biochimie Cellulaire, Facultés Universitaires, 61 Rue de Bruxelles, B-5000 Namur, Belgium

Summary. Aerobic cells contain various amounts of the three main antioxidant enzymes: superoxide dismutase (SOD), catalase and GSH peroxidase. These three enzymes are necessary for cell survival since inhibition of their activity leads to the arrest of cell mitosis and to cell death. Amongst them, GSH peroxidase was shown to be more efficient than catalase and much more than SOD. This result was obtained by comparing the cell protection against oxidative stress after their microinjection in the cytoplasm. With age, the level of these antioxidant enzymes does not change in several experimental models, so that it is not possible to explain the aging process by a lack of protection due to a decrease in the activity of these three enzymes. However, tissues and cells are more susceptible to free radical attacks with age. In order to understand the importance of free radicals in this process, we have to distinguish between their respective effects on cell mitosis, cell death and cell aging. The effects on mitosis and cell death are well described, and the results clearly show a threshold of response which is determined by the antioxidant content of the cell. There is now evidence that short free radical stresses can also speed up the aging of in vitro cultured human fibroblasts. However, such effects are not typical of free radicals but are also obtained with many other deleterious substances so that free radicals have to be considered as one amongst other factors responsible for influencing the evolution of a cell to an older stage or to cell death. The lowering of the general metabolism and of the free energy in old cells are probably the main factors responsible for the increased susceptibility of these cells to stresses such as oxidative stresses.

Introduction

Free radicals are ubiquitous reactive molecules mostly derived from oxygen. They are generated naturally in living cells, but their production can be increased by the presence of xenobiotic molecules or in pathological conditions (Freeman and Crapo, 1982). Because of their high reactivity, they can affect most of the biological molecules containing unsaturated bonds such as polyunsaturated lipids but also DNA and proteins. However, cells are well protected by a series of antioxidants and compounds such as vitamin E and C, quinones, glutathione. . . and also by different antioxidant enzymes. Superoxide dismutases (SOD) are probably the most abundant ones. The cytoplasmic SOD is a Cu/Zn-containing enzyme also found in the mitochondrial intermembrane space while the mitochondrial matrix contains a Mn SOD. They accelerate the dismutation of the superoxide anion into oxygen and hydrogen peroxide. This molecule is then decomposed into oxygen and water by catalase which is mostly located in the peroxisomes. Glutathione (GSH) peroxidase is in the cytoplasm and mitochondrial matrix where it reduces hydroperoxides into alcohols using the reductive potential of GSH. The main activity is ascribed to a Se-containing enzyme but there is also a Se-independent activity and recently a phospholipid hydroperoxide GSH peroxidase active near the plasma membrane has been reported (Ursini et al., 1985).

Because of the damaging effects of free radicals and the accumulation of lipofuscin which is one of the compounds resulting from the free radical peroxidation of lipids, free radicals have been proposed as possible candidates for explaining the aging process. In this review, we will examine the respective role and the importance of the three antioxidant enzymes and show that the effects of free radicals have to be evaluated taking into account the reactivity of these toxic molecules but also the potential of cell defense and repair. We have also to discriminate between a toxic effect leading either to the arrest of mitosis and cell death or a real pro-aging effect considered as the speeding up of the normal time evolutive process. In order to understand the relevance of these results for aging, we will first briefly examine the relative efficiency of the three antioxidant enzymes for cell protection.

The efficiency of the antioxidant enzymes

The question of the importance of the defense system against free radicals during aging is directly related to their possible involvement in some aging processes or in cell death. Indeed, it is the balance between the free radical production with the damages they cause on cellular constituents and the capacity of the cell to destroy these radicals or to repair the induced alterations, which is the decisive criterion for the biological consequences of free radicals on cell degeneration. Examples for the existence of such a balance are numerous, and we know that an overproduction of free radicals by enzymes like xanthine oxidase, by molecules like paraquat, nitrofurantoin, daunomycin or by radiations, will first stop cell division if the production is low and kill the cells at higher concentrations. Comparison of the sensitivity of different cell types having different levels of antioxidant enzymes already gives a better indication of the relative importance of the free radical damaging concentration compared to the cell potential to defense. This type of comparison can also be very useful in order to understand the evolution of tissues in some chronic diseases like arthritis or lung emphysema. One good example is given by the comparison of human fibroblasts and endothelial cells in culture (Michiels et al., 1990b). The levels of GSH peroxidase, reductase and catalase were significantly lower in endothelial cells and represented respectively 73, 53 and 45% of the corresponding activity in the fibroblasts; the activity of SOD was identical in the two cell types. When exposed to 1 or 2 atm of oxygen both cells died within one week but there was at least a two-day difference in the mortality curves, the endothelial cells being more sensitive to the oxygen stress.

In such experiments, the susceptibility to free radicals could be explained by the presence of various antioxidant molecules or by the resistance of lipids to peroxidation but other factors than the levels of the antioxidant enzymes themselves can also probably influence the cell behavior.

The best approach in order to avoid such interferences with other factors is to use the same cell but to vary the level of the antioxidant enzymes. Experiments with SOD-transfected cells were rather disappointing since increased levels of SOD did not give any supplement of protection or even increased the susceptibility to free radicals (Elroy-Stein et al., 1986; Seto et al., 1990). Another more limited but more refined approach is to inject the different antioxidant enzymes into the cell using the microinjection technique and to test for the induced resistance given by the additional enzyme on cell resistance to free radical attacks (Raes et al., 1987). The three antioxidant enzymes were injected in increasing amounts into human fibroblasts which were then incubated under one or two atmospheres of 95% oxygen. The number of surviving cells were counted every day and the mortality curves were compared to the control cells injected with buffer only. The percentage of protection calculated as the integration of the enzyme-injected cell survival curve during the first 4 days compared to the control increased in each case with increasing amounts of the injected antioxidant enzymes. However, comparison of the concentrations needed for a 20% protection shows that the solution of SOD had to contain 1.1 million U/ml while only 41 U/ml was necessary for catalase and 5.3 U/ml for glutathione peroxidase. The same experiment was performed in different conditions using for example nitrofurantoin as a free radical generating system (Michiels et al., 1988c). In all cases, glutathione peroxidase was the most potent protector with an efficiency about 2,000-fold higher than for SOD and 7- to 23-fold higher than for catalase.

The opposite approach was also undertaken, which is to lower the activity of these enzymes and to study the effects of this inhibition on cell survival. Enzyme inhibition could be obtained either by chemical inhibitors or after injection of specific antibodies (Michiels et al., 1988a, 1988b). In all cases, inhibition of the three enzymes induced a higher sensitivity of the cells to the presence of free radicals. When the enzymes were highly inhibited, cell death could be obtained even in normal 20% oxygen incubation conditions. Here also, GSH peroxidase was found to be the most essential enzyme since only a 21% inhibition was sufficient to reduce cell survival to 50% in normal conditions while a 55% inhibition was needed for catalase to achieve the same effect.

The balance between the free radical production and the level of antioxidant activities was further demonstrated in experiments where cells were injected with increasing amounts of antioxidant enzymes and then tested for the maximum time they could sustain incubation with 1 atm O_2 without losing their mitotic capacity (Michiels et al., 1990a). Clearly, the higher the concentration of the injected enzymes, the longer the cells could stay under oxygen without affecting their capability to divide. The level of antioxidant enzymes determines the level of oxidative stress and radical production that cells can sustain before the mitotic activity is inhibited. The opposite approach leads to the same conclusion. If antioxidant enzymes are inactivated either by chemical inhibition or after injection of specific antibodies, the mitotic activity of the cells is blocked even in a normal atmosphere. In these experiments, the inhibition of cell division was taken as the criterion to test the negative effect of free radicals because of the higher sensitivity of the mitotic capacity to such a stress. In fact, when increasing the free radical concentration, the mitotic activity is first blocked, and only at a higher concentration does the cell enter a progressive degeneration leading to cell death. Both effects are modulated by the concentrations of the antioxidant enzymes and other protective molecules which will determine the threshold of damage that cells can sustain before their survival is irreversibly affected. When considering the free radical deleterious effects, the composition of the cell, meaning its differentiation or specialized state which also determine the level of the antioxidant protection, has also to be taken into account. So, the effects of free radicals on living organisms have to be seen as the result of multiple effects on the various cells and tissues with some of them being more susceptible to alterations than others.

Antioxidant and life span potential

One way to determine a possible role of these antioxidant enzymes in aging is to see if all the long-living species that have been selected along the natural evolution process, also display higher antioxidant enzyme contents. Such a study has been extensively conducted by Cutler who tried to correlate the level of antioxidant enzymes and molecules with the maximum life span of various species (Cutler, 1991). He found that the activity of SOD and the concentration of antioxidant molecules like urate and GSH are much higher in the long-living species than in the short-living ones. A linear correlation was obtained between the plasma level of urate and the maximum life span potential of different primates. For SOD in the liver, the same correlation was obtained if the activity was corrected according to the specific metabolic rate of the animals, meaning the calories produced per year and per gram of tissue. This positive correlation indicates that long-living species have relatively higher SOD contents for an identical metabolic activity. This is especially true for human tissues which contain large amounts of SOD. However, if the latter correlations are very good, other antioxidants do not show any correlation or even show an inverse relation between their tissue antioxidant concentrations and the maximum life span of the species. This is the case of catalase and GSH peroxidase which are the key enzymes in the cell defense. The reasons for this discrepancy are not clear. Ames et al. (1981) and Cutler (1982) suggest that in the case of vitamin C an optimized adaptation could have occurred in favor of urate. They found that the ascorbate level does not vary much in species with a maximum life span higher than 30 years while the plasma level of urate increases sharply in long-living animals including the human species. The reason for this shift in favor of some antioxidants could be the negative effect of further increasing the vitamin C concentration which can be converted into a toxic free radical species. In fact, when used for protection in models of free radical toxicity, vitamin C showed very slight or no protective effect and the protective concentration is very close to the toxic range (Remacle et al., 1984; Michiels et al., 1988c). Such an hypothesis is very appealing but can not hold for SOD which is already found in the primitive aerobic species as a Fe, Mn or later as a Cu/Zn enzyme and is the most important defense against high levels of superoxide anion. It would have been less costly for the cell to increase the production of catalase and GSH peroxidase which are so much more efficient than SOD. Whatever the reason for such an increased activity in long-living species, the data reported by Cutler show that long-living species have increased levels of many antioxidant molecules so that they are better protected. This was clearly illustrated by estimating the rate of autooxidation in brain homogenates assayed by the level of malonaldehyde accumulation (Cutler, 1985). The rate of oxidation was strongly correlated with the life span of the species from which the brain was obtained, with the lowest aldehyde formation in humans and a very high one in short-living species like mice. If there is a correlation between the free radical protection and the life span of the species and tissues, it is only one of the many protective factors which make human cells so much more resistant than murine tissues for example, as far as aging is concerned. These factors have been extensively described by Cutler (1982) as longevity determinant processes, and we will just mention some of them here: the better DNA protection, the redundancy of genes, the detoxification processes, the guanylate cyclase activity, the level of dehydroepiandrosterone, some vitamins, the protein turnover rate and the life span of red blood cells.

Antioxidant enxymes in aging tissues

The evolution of antioxidant enzyme activities, especially of catalase and GSH peroxidase, could give us information on the possibility of deleterious effects of free radicals on aging if this protection decreases with age. Unfortunately, results are not always concordant depending on the animal or cellular model considered (Balin, 1982). However in many cases, the activity of these enzymes does not change with age. In two reports for example, GSH peroxidase was assayed during the growth and aging of rats and there was no decrease of its activity with age (Pinto and Bartley, 1967; Santa Maria and Machado, 1987). Also in the model of aging cells in culture, no decrease of GSH peroxidase was obtained if sufficient amounts of selenomethionine were added to the medium (Pigeolet et al., 1991).

However, the concentration of GSH seems to be lower in old tissues while the GSSG concentration does not change. For example, Hazelton and Lang (1980) found a 30, 34 and 20% decrease in the GSH concentration, respectively in 31 months old rat liver, kidney and heart compared to mature animals.

In the model of in vitro aging fibroblasts, the GSH content was stable along with the serial passages, then increased for two passages, and suddenly dropped in cells of the last two cultures. The level of GSSG did not change (Mbemba et al., 1985). GSH reductase was also shown to decrease in these aging cells so that the regeneration of GSH is not at its maximal capacity, and this could hamper cell defense if high concentrations of hydroperoxides are formed.

In the same culture model, the activities of catalase and SOD were not changed in old cells compared to young cells (Remacle et al., 1980). In aged rats, Reiss and Gershon (1976) found a decreased activity of SOD in the liver while Kellogg and Fridovich (1976) reported a constant specific activity in brain and liver.

The fact that there is no correlation between the catalase and GSH peroxidase activities with the maximum life span of various species and that at least in some aging models, there is also no decreased activity, strongly indicates that aging is not associated with any shortage in antioxidant enzyme protection at least in these models.

How is it then possible to explain the higher susceptibility of old tissues and cells to free radicals and their higher rate of peroxidation when cell homogenates are left for spontaneous oxidation for example. One possible explanation could be that the enzymes are modified in old tissues and that their efficiency is lowered. In fact, the antioxidant enzymes were found to be gradually modified with age. One way to detect such altered enzymes is to demonstrate the presence of immunologically cross-reacting materials which are enzyme molecules able to react with specific antibodies but which have no catalytic activity any more. Such inactive SOD has been observed in old cells (Somville et al., 1985). The presence of thermolabile enzymes is another expression of enzyme modification. When heated at 70°C, the rate of inactivation of cytoplasmic SOD follows a first order kinetic characterized by an exponential decay of the activity in young cells. In old cells, however, two SOD populations are observed, one with a normal denaturation rate and another one which inactivates much faster. The proportion of this heat-labile SOD increases in the cells during the late passages in culture. The molecular mechanism beyond this enzyme modification has been shown to be an association of the normal dimers into active but more heat-sensitive tetramers (Somville et al., 1985). This association is dependent on the cytoplasmic composition and especially on its reducing capacity. GSH peroxidase heat-denaturation rate also strongly increases with age but, in this case, the overall enzyme population seems to be more affected (Pigeolet et al., 1991). In this case, the proposed mechanism involves a transformation of the enzyme into an intermediate active enzyme, which can then be converted into an inactive protein. All these changes are posttranslational modifications detected only at high temperatures; the enzyme activity is normal at 37°C so that these modifications should not affect the enzyme's protective capacity in aging cells. Another possibility would be that other antioxidant molecules are lacking in aged cells. The amounts of vitamin E and C change with age but not enough to account for the higher sensitivity of old cells. We would rather prefer to link the increased susceptibility of old cells to a general lowering of cell metabolism and perhaps to the presence of already modified cellular components rather than to the lowering of the antioxidant enzyme protective system. Indeed, old cells are more susceptible than young cells whatever kind of stress is applied, free radicals being only one type. The threshold of free radical production leading the old cells to death would then be lowered not necessarily because of a lowering in the antioxidant molecules per se, but because of the resulting altered molecules and functions and of a lower capacity of the cells to resist to any disturbance which requires a supplement of energy, as for instance for the regeneration of GSH or the replacement of oxidized cellular components like the lipid hydroperoxides. In this respect, the general lowering of the cell metabolism and of the free energy production would be the main factor explaining the higher susceptibility of old cells to all types of stress.

Is it possible that free radicals have some influence not only on the cell mitotic capacity or cell death but on the rate of aging itself? In other words, does repeated exposure of cells to free radicals increase their rate of aging? In vitro aging human fibroblasts are a good model for testing such a hypothesis. Cells were cultivated under various concentrations of oxygen ranging from 5 to 95% (Balin et al., 1976). At very low and very high oxygen concentrations, the life span was shortened while for a very large range of concentrations, there was no influence on the maximum population doubling level. This experiment indicates that cells are well protected against oxygen free radicals. However, when this production reaches a too high level, a threshold is reached which affects cell cultures. We do not know, however, if high oxygen concentrations speed up the aging process of cells or rather leads to cell toxicity so that the global cell population and its mitotic capacity are reduced. In order to answer such a question, a more detailed observation of the stress effect on individual cells was performed. When serially cultivated, human fibroblasts go through seven morphological stages or cell types, with the last one leading to cell degeneration (Bayreuther et al., 1988). The first three types are mitotic cells and the four last ones postmitotic cells mostly obtained in the last passages. Individual cells always to through these seven stages successively, and this evolution can be defined as an aging process. When short oxidative stress is applied, as for example the present of t-butylhydroperoxide at 10^{-4} M for a few minutes in the culture medium, it can be observed that within a few days after the stress, individual cells speed up their transformation from one stage to the next. Such an evolution is strongly dependent on the duration and intensity of the stress (Toussaint et al., unpublished). This result clearly indicates that in this model, free radical stress can speed up cellular aging defined as an accelerated evolution through the seven successive cell types. If too strong, the stress will lead the cells directly into a postmitotic state and when even stronger to cell death.

The situation is however more complex since many other forms of stress will provoke the same shift in the cell types so that again free radicals have to be considered as one of the deleterious processes affecting aging and cell death. The free radical theory has thus to be replaced by a more general approach to aging. It considers the cell as a global system which optimizes its energy production in order to maintain its structure and its functions according to its differentiation and its specific role in the organism. The synthesis of the free radical protecting enzymes and the maintenance of a high concentration of reduced GSH are energy-consuming processes which have to be regarded in this global optimization of cell behavior. With age or under stress conditions, cells will defend themselves but if they cannot cope with such a detrimental situation, they will have to readapt to a lower level of free energy production. This gradual decrease of energy in the cell explains why young cells can better resist to any stress, like free radicals, while old cells are more vulnerable even when considering the level of antioxidant enzymes as constant. We propose that the free radical influence on aging has to be seen in comparison with an antioxidant defense capacity of the cell and to be replaced in a more general view of aging considering the cell as a global thermodynamical open system evolving with time through

several stages characterized by a lowering of the entropy production
(Toussaint et al., 1991).

Acknowledgments. C. M. is a Senior Research Assistant, and M. R. a Research Associate of
the F.N.R.S. (Bruxelles, Belgium). This work was supported by the F.N.R.S.

Ames, B. N., Cathart, R., Schwiers, E., and Cochstein, P. (1981) Uric acid: an antioxidant
defence in human against oxidant and radical-caused aging and cancer. A hypothesis. Proc.
Natl. Acad. Sci. USA 78: 6858–6862.

Balin, A. K., Goodman, D. B. P., Rasmussen, H., and Cristofalo, V. J. (1976) The effect of
oxygen tension on the growth and the metabolism of WI-38 cells. J. Cell. Physiol. 89:
235–250.

Balin, A. K. (1982) Testing the free radical theory of aging, in: Testing the Theories of
Ageing. R. Adelman and G. Roth, eds. Boca Raton, FL., CRC Press, pp. 137–182.

Bayreuther, K., Rodemann, H. P., Hommel, R., Dittmann, K., Albiez, M., and Francz, P. I.
(1988) Human skin fibroblasts in vitro differentiate along a terminal cell lineage. Proc. Natl.
Acad. Sci. USA 85: 5112.

Cutler, R. G. (1982) Longevity is determined by specific genes: testing the hypothesis, in:
Testing the Theories of Ageing. R. Adelman and G. Roth, eds. Boca Raton, FL., CRC
Press, pp. 25–114.

Cutler, R. G. (1985) Peroxide-producing potential of tissue: correlation with the longevity of
mammalian species. Proc. Natl. Acad. Sci. USA 82: 4798–4802.

Cutler, R. G. (1991) Antioxidants and ageing. Am. J. Clin. Nutr. 53: 373S–379S.

Elroy-Stein, O., Brenstein, Y., and Groner, Y. (1986) Overproduction of human Cu/Zn
superoxide dismutase in transfected cells: extenuation of paraquat-mediated cytotoxicity
and enhancement of lipid peroxidation. EMBO J. 5: 615–622.

Freeman, B. A., and Crapo, J. D. (1982) Biology of disease. Free radicals and tissue injury.
Lab. Invest. 47: 417–426.

Hazelton, G., and Lang, C. (1980) Glutathione contents of tissues in the aging mouse.
Biochem. J. 188: 25.

Kellogg, E. W., and Fridovich, I. (1976) Superoxide dismutase in the rat and mouse as a
function of age and longevity. J. Gerontol. 31: 405.

Mbemba, F., Houbion, A., Raes, M., and Remacle, J. (1985) Subcellular localization and
modification with ageing of glutathione, glutathione peroxidase and glutathione reductase
activities in human fibroblasts. Biochim. Biophys. Acta 838: 211–220.

Michiels, C., and Remacle, J. (1988a) Use of the inhibition of enzymatic antioxidant systems
in order to evaluate their physiological importance. Eur. J. Biochem. 177: 435–441.

Michiels, C., Raes, M., Zachary, M.-D., Delaive, E., and Remacle, J. (1988b) Microinjection
of antibodies against superoxide dismutase and glutathione peroxidase. Exp. Cell Res. 179:
581–589.

Michiels, C., and Remacle, J. (1988c) Quantitative study of natural antioxidant systems for
cellular nitrofurantoin toxicity. Biochim. Biophys. Acta 967: 341–347.

Michiels, C., Raes, M., Pigeolet, E., Corbisier, P., Lambert, D., and Remacle, J. (1990a)
Importance of a threshold for error accumulation in cell degenerative processes. I. Modula-
tion of the threshold in a model of free radical-induced cell degeneration. Mech. Ageing
Dev. 51: 41–54.

Michiels, C., Toussaint, O., and Remacle, J. (1990b) Comparative study of oxygen toxicity in
fibroblasts and endothelial cells. J. Cell. Physiol. 144: 295–302.

Pigeolet, E., and Remacle, J. (1991) Alteration of enzymes in ageing human fibroblasts in
culture. V. Mechanisms of glutathione peroxidase modification. Mech. Ageing Dev. 58:
93–109.

Pinto, R. E., and Bartley, W. (1969) The effect of age and sex on GSSG reductase and
glutathione peroxidase activities and on aerobic GSH oxidation in rat liver homogenates.
Biochem. J. 112: 109–115.

Raes, M., Michiels, C., and Remacle, J. (1987) Comparative study of the enzymatic defense
systems against oxygen-derived free radicals: the key role of glutathione peroxidase. Free
Rad. Biol. Med. 3: 3–7.

Reis, U., and Gershon, D. (1976) Rat liver superoxide dismutase. Purification and age-related modifications. Eur. J. Biochem. 63: 617–62.

Remacle, J., Houbion, A., and Houben, A. (1980) Subcellular fractionation of WI-38 fibroblasts comparison between young and old cells. Biochem. Biophys. Acta 630: 57–70.

Remacle, J., Lenoir, G., Mbemba, F., Houben, A., Raes, M., Houbion, A., and Delaive, E. (1984) Free radicals in cell degenerescence. Example of cellular ageing, in: New Trends on Atherosclerosis. C. L. Malmendier and J. Polanovski, eds. Fondations de recherche sur l'athérosclérose Belgium, pp. 23–29.

Santa Maria, C., and Machado, A. (1987) Effect of development and ageing on pulmonary NADPH-cyt C reductase, glutathione peroxidase, glutathione reductase and thioredoxin reductase activities in male and female rats. Mech. Ageing Dev. 37: 183–195.

Seto, N. O. L., Hayashi, S., and Tener, G. M. (1990) Overexpression of Cu-Zn superoxide dismutase in Drosophila does not affect lifespan. Proc. Natl. Acad. Sci. USA 87: 4270–4274.

Somville, M., Houben, A., Raes, M., Houbion, A., Henin, V., and Remacle, J. (1985) Alteration of enzymes in ageing human fibroblasts in culture. III. Modification of superoxide dismutase as an environmental and reversible process. Mech. Ageing Dev. 29: 35–51.

Toussaint, O., Raes, M., and Remacle, J. (1991) Aging as a multi-step process leading the cell from stage to stage characterized by a lowering of entropy production. Mech. Ageing Dev. 61: 45–64.

Ursini, F., Maiorino, M., and Gregolin, C. (1985) The selenoenzyme phospholipid hydroperoxide glutathione peroxidase. Biochim. Biophys. Acta 839: 62–70.

Free Radicals and Aging
ed. by I. Emerit & B. Chance

Relationship between antioxidants, lipid peroxidation and aging

G. Barja de Quiroga, M. López-Torres and R. Pérez-Campo

Departamento de Biología Animal-II (Fisiología Animal), Facultad de Biología, Universidad Complutense, Madrid 28040, Spain

Summary. Experiments performed on species as different as flies, rats and frogs are not conclusive about the possibility that antioxidant defenses decrease in old animals. Even when these decreases are found, their physiological meaning is far from clear. Furthermore, a constancy of antioxidant capacity in old age is consistent with the fact that aging is a progressive phenomenon which occurs at a rather constant rate from the mature young to the very old animal, without showing a great acceleration rate in the aged. Nevertheless, experimental results strongly suggest that the maintenance of an appropriate antioxidant/prooxidant balance does have an important role in maintaining health in the aging animal. It is possible that the continuous presence of small amounts of free radicals in the adult tissues of both mature adults and old animals is an important factor in aging (a progressive phenomenon) and susceptibility to disease. Since, similarly to what occurs in procariota, the whole antioxidant system seems to be under homeostatic control in vertebrates, it is imperative to perform comprehensive and detailed studies on the effects of carefully controlled doses of antioxidants on biomarkers of health as well as on the different endogenous cellular antioxidant and prooxidant systems. These studies should have as a final goal the knowledge of which doses of antioxidants are high enough to increase antioxidant protection but low enough to avoid feedback depression of other endogenous antioxidants; this could further improve the health state of humans situated in the middle and last phases of their life span.

Introduction

Among the theories of aging, the free radical hypothesis (Harman, 1956) is currently receiving strong attention (Sohal et al., 1990a,b,c; Fraga et al., 1990; Cutler, 1991; Miquel, 1991). Much work has focused on the variations of antioxidants in old age. Nevertheless, we think that two different possibilities must be considered regarding the theory:

(1) Cellular scavenging of active oxygen species can not be 100% effective. Therefore, small amounts of free radicals, present in the tissues of adult animals (both young and old mature adults), can be causative agents of the progressive decline of functionality observed during aging.

(2) The total antioxidant capacity of the tissues decreases in old age.

We think that only the first possibility is fundamental for the acceptance of the free radical theory of aging. The second one depends upon the acceptance of the first, and only implies that aging rate is accelerated in old age. On the contrary, maintenance of antioxidant capacity throughout a life span could lead (if the first possibility is true) to a continous decline of maximum physiological capacities from the young mature adults to the old ones. This would be concordant with the progressive character of these decreases observed in mammals during aging (Cutler, 1984).

Antioxidants during aging

Musca domestica is perhaps the animal for which previous data is better correlated with the second possibility mentioned above. In whole body homogenates of this animal species, clear decreases of the principal antioxidants including superoxide dismutase (SOD), catalase (CAT) and GSH, together with sharp increases of GSSG and in vivo and in vitro lipid peroxidation (TBA-RS), have been described in old age (Sohal et al., 1983; 1985). This is in contrast to the contradictory results reported for rodents (as will be shown below). One might want to consider the fundamental differences between insects and mammals, in resolving those contradictions. Nevertheless, a recent report from the same laboratory has shown that the pattern of variations of free radical related parameters is not fully consistent with the "second possibility" in another insect species. In *Drosophila melanogaster*, even though CAT, GSH-reductase (GR) and GSH decreased and GSH/GSSG and NADPH/NADP$^+$ decreased in old age, SOD and NADH/NAD$^+$ increased, steady state H_2O_2 concentration did not change and in vivo TBA-RS decreased during aging (Sohal et al., 1990a).

In aging tissues of rats and mice (such as brain and liver) many contradictory results have been reported concerning the second possibility. Thus, the brain of old rodents have been reported to show similar (Kellog and Fridovich, 1976; Mizuno and Ohta, 1986; Cand and Verdetti, 1989; Ansari et al., 1989; Kurobe, 1990; Semsei, 1991), decreased (Benzi et al., 1988a; Vanella et al., 1989; Rao et al., 1990a), or even increased (Dahn et al., 1983; Sohal et al., 1990b) SOD activity in relation to that of fully mature young adults; CAT to be unchanged (Ansari et al., 1989; Tayarami et al., 1989), decreased (Cand and Verdetti, 1989; Semsei et al., 1989; Rao et al., 1990a), or increased (Sohal et al., 1990b); GSH-peroxidase (GPx) to be unchanged (Mizuno and Ohta, 1986; Ansari et al., 1989; Rao et al., 1990a; Sohal et al., 1990b) or increased (Vitorica et al., 1984; Benzi et al., 1989); GR to be unchanged (Mizuno and Ohta, 1986), decreased (Benzi et al., 1989) or increased (Sohal et al., 1990b); GSH to be unchanged (Rikans and Moore, 1988; Benzi et al., 1990) or decreased (Benzi et al., 1988b).

On the other hand, rat liver SOD was found to be unchanged (Lammi-Keefe et al., 1984; Kurobe et al., 1990; Sohal et al., 1990b) or decreased (Kellog and Fridovich, 1976; Reiss and Gershon, 1976; Cand and Verdetti, 1989; Semsei et al., 1989; Rao et al., 1990b; Semsei et al.,

1991) during aging; CAT to be unchanged (Koizumi et al., 1987; Sohal et al., 1990b) or decreased (Cand and Verdetti, 1989; Laganiere and Yu, 1989; Semsei et al., 1989; Rao et al., 1990b); GPx to be increased (Rao et al., 1990b; Sohal et al., 1990b) or unchanged (Laganiere and Yu, 1989); GR to be increased (Sohal et al., 1990b) or unchanged (Langaniere and Yu, 1989), and GSH to be decreased (Hazelton and Lang, 1980; Stohs and Lawson, 1986; Laganiere and Yu, 1989) or unchanged (Rikans and Moore, 1988; Carrillo et al., 1989). A very recent report suggests that the effect of aging is sex-dependent in the case of CAT liver (Rikans and Moore, 1991), males showing a decrease and females an increase during aging. But a close inspection of the previously cited literature shows that sex can not explain the contradictory character of the results since both decreases (Laganiere and Yu, 1989; Semsei et al., 1989; Rao et al., 1990b) and absence of changes (Sohal et al., 1990b) have been described for male rat liver CAT, whereas Koizumi et al. (1987) showed an absence of change in the same enzyme and tissue for female rats during aging.

It is difficult to clarify these highly contradictory results. Nevertheless, some points are worth mentioning:

1) GPx is the only enzyme for which no decreases during aging have been reported in liver or brain; in some reports it has even shown a tendency to increase.

2) Many reports use a limited number of ages. The most extensive report in this respect is that of Kurobe et al. (1990) which included 7 different ages throughout the rat life span and did not find any changes during aging for SOD activities or immunoreactive SOD in three brain areas, liver or kidney.

3) Apart from differences in methodology, some decreases can be due to suboptimal conditions in the long-term maintenance of animals. This is difficult to survey, however, since few details about this aspect are usually given in most of the reports. The absence of simultaneous measurement of other biochemical markers of a non-antioxidant type (in practically all the reports) hampers the clarification of this subject.

4) When decreases of a particular antioxidant are found, they are usually of small magnitude and their physiological meaning is not evident. The observation of such decreases can not be readily interpreted as a higher susceptibility to oxidative stress in the absence of information concerning changes during aging for in vivo rates of generation of free radicals. Decreases of oxygen radical generation in vitro have been repeatedly observed in aged rats (Floyd et al., 1984; Muscari et al., 1990; LeBel and Bondy, 1991) except for an ESR study in which strong increases were found (Sawada and Carlsson, 1987). If in vivo oxygen radical generation is in fact depressed in old animals, the decreases of the antioxidant defenses occasionally observed should be

112

interpreted as a physiological compensatory down regulation instead as a deleterious change leading to additional oxidative damage.

5) Finally, most of the reports concentrate on a single or a few antioxidants. This can complicate the interpretation of the results since it is known that cellular antioxidants are under homeostatic control (Sohal et al., 1984; Cutler, 1984; frog results after CAT inhibition in this work). Thus, a decrease in a particular antioxidant can be compensated by an increase in a different one not included in the study. This is why comprehensive studies of the whole antioxidant system during aging are needed.

Thus, we studied in a comprehensive way antioxidant and prooxidant relating parameters in the brain, liver and lung of old and young adult Wistar rats, specially raised under barrier conditions at CERJ (France). Figure 1 shows the results obtained for brain tissue. No significant changes were found for SOD, CAT, Se and non-Se GPx, GR, GSH, GSSG, GSSG/GSH (an indicator of oxidative stress) or in vivo lipid peroxidation (TBA-RS). It must be mentioned that in the same tissue COX activity (a marker of maximum mitochondrial capacity) was found to be depressed. As a result, when the activities of the antioxidant enzymes were referred to COX activity instead of to protein, even significant increases were found for both kinds of GPx. Results of similar character were found in the liver and lung of the same animals (Barja de Quiroga et al., 1990a; Pérez-Campo et al., 1991).

A further concern arises with regard to the oxygen radical generation rates. The concept that aging can be due to more than a single kind of cause is widespread among gerontologists. Since mitochondria are a constant and important source of oxygen radicals in almost every metabolically active tissue, it is logical to assume that the higher the metabolic rate, the higher the rate of oxygen radical production at mitochondria. Indeed, this has been shown to be true for various mammalian species (Sohal et al. 1989, 1990b, 1990c). Thus, as it can be seen in Table 1, the housefly shows an O_2 consumption rate 82 times higher than that of the cow and its rates of submitochondrial O_2^- and mitochondrial H_2O_2 generation rates are 25 and 167 times higher than those of the cow. If free radicals explain only a part of the aging phenomenon, it is then theoretically possible that they are the main cause of aging in houseflies but that they are of much smaller relative importance in the cow. Testing the free radical theory in animals with moderate metabolic rates (such as that of humans) is then a relevant subject.

Taking into account this problem, we decided to study the variations of SOD, CAT, Se and non-Se GPx, GR, COX, GSH, GSSG, GSSG/GSH, ascorbate (ASC; HPLC), and MDA (HPLC) in the brain, liver and lung of the frog *R. perezi*, an animal showing a moderate metabolic rate (even somewhat lower than that of humans; Table 1) and a

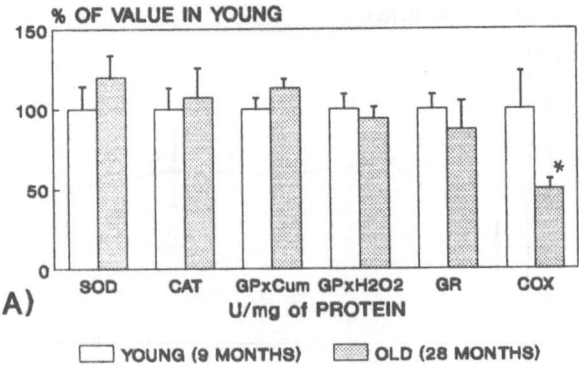

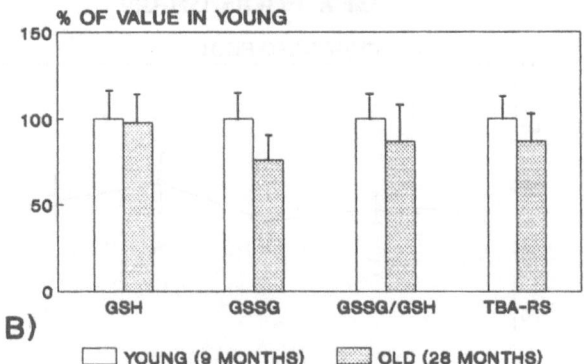

Figure 1. Changes as a function of age in the cerebral cortex of male Wistar rats; *A*: SOD, CAT, TOTAL (Cumene-OOH) or Se-GPx, GR and Cytochrome oxydase; *B*: GSH, GSSG, GSSG/GSH and in vivo TBA-RS. Data from Barja de Quiroga et al., 1990a.

Table 1. Relationship between oxygen consumption and oxygen radical production in different species

	VO_2	$O_2^{\cdot-}$	H_2O_2
Housefly	14,000	3.3	1000
Mouse	2,500	0.66	58
Human	200	-	-
Cow	170	0.13	6
Frog	65	-	-

$O_2^{\cdot-}$ in nmol and H_2O_2 in pmol (/min/mg prot.); VO_2 in μl O_2/hr/g body weight. $O_2^{\cdot-}$ and H_2O_2 from Sohal et al. 1989, 1990b and c.

114

SOD FROG LIVER

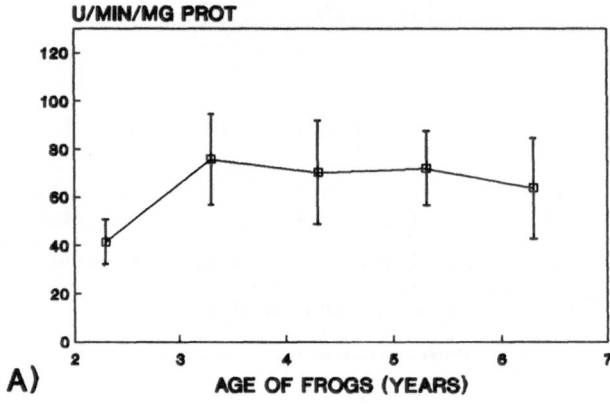

A)

GPx FROG BRAIN

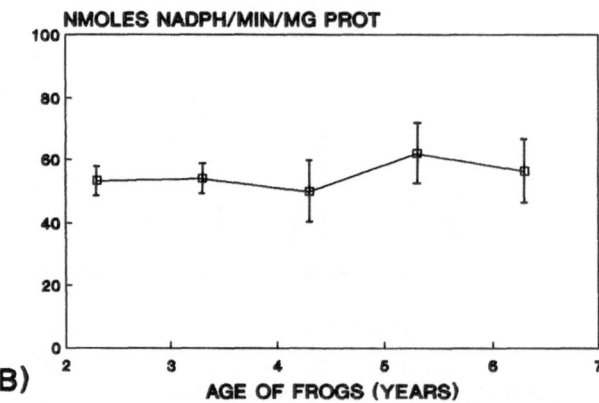

B)

Figure 2. Values of liver SOD (*A*) and brain GPx (Cumene-OOH) (*B*) throughout the whole adult life of *R. perezi* frogs.

maximum life span of 7 years. The general pattern observed in all the possible comparisons was that of an absence of age-related changes. As an example, Figure 2 shows the constancy of liver SOD and brain GPx throughout the adult life span of this frog. These results confirm those previously reported in cross-sectional studies performed with the same frog species (Barja de Quiroga et al., 1990b; Pérez-Campo et al., 1990; López-Torres et al., 1991). In the case of humans, it is difficult to obtain reliable data from tissues. Nevertheless, it can be mentioned that the measurement of antioxidant enzymes in extracellular fluids or blood cells usually shows an absence of age-related changes. A particularly large study (2397 normal individuals) showed no changes in erythrocyte

SOD activity among five different age groups ranging from 18 to 65 years of age (De la Torre et al., 1990). Thus, it can be summarized that the possibility that aging is accelerated in old age due to a decrease in antioxidant capacity can not be accepted on the basis of the information available.

Lipid peroxidation

Similarly to what occurs for antioxidants, increases (Devasagayam, 1986; Mizuno and Ohta, 1986; Sawada and Carlson, 1987) or decreases (Devasagayam and Pushpendran, 1987; Cand and Verdetti, 1989; Levitsky et al., 1989) of in vivo TBA-RS values (direct application of the TBA test to biological samples) have been described in rodents. However, one must remember that the TBA test is very sensitive but shows a very poor specificity. We have observed (unpublished results) that the amount of "malondialdehyde" (MDA) found in unstimulated tissue samples is five to ten times higher when measured by the TBA test than when measured in the same samples by HPLC (true MDA). This means that as much as 80–90% of the TBA-RS in vivo values is not MDA and corresponds to other compounds related (hydroperoxides) or not (as it happens for glucose) to lipid peroxidation (Sevanian and Hochstein, 1985). Nevertheless, if the sample is subjected to a strong peroxidative stress (in vitro TBA-RS) before the TBA assay, the TBA values can be multiplied by an order of magnitude (this is what happens in rat liver) and the interference from substances non-related to lipid peroxidation can be minimized to 8–9%. Unfortunately, to our knowledge a kinetic study on the rate of production of these more reliable in vitro TBA-RS values has been reported only two times. One comes from our laboratory and the results obtained in ther rat liver can be observed in Figure 3. It is clear from this figure that sensitivity to lipid peroxidation (stimulated with ascorbate-iron) was not higher in old rats than in young rats. In fact the opposite was the case, with old animals showing the smaller rates of in vitro peroxidation. The other report is fully coincident with our data since it shows a progressive decline in peroxidation rate of rat liver microsomes (NADPH-induced) from young adult to middle-aged and from middle-aged to old animals (Fig. 4; Devasagayam, 1986). The same result has been reported by the same authors for ascorbate-induced peroxidation of liver microsomes. On the other hand, a decrease of in vitro TBA-RS (measured at only one time-point) has also been described by Ansari et al. (1989) in rat brain, whereas our results show an absence of changes for in vitro TBA-RS in the cerebral cortex of the rat (Fig. 3b).

What is the meaning of these results? We do not think that they suggest a smaller degree of oxidative stress in old animals. In vitro

116

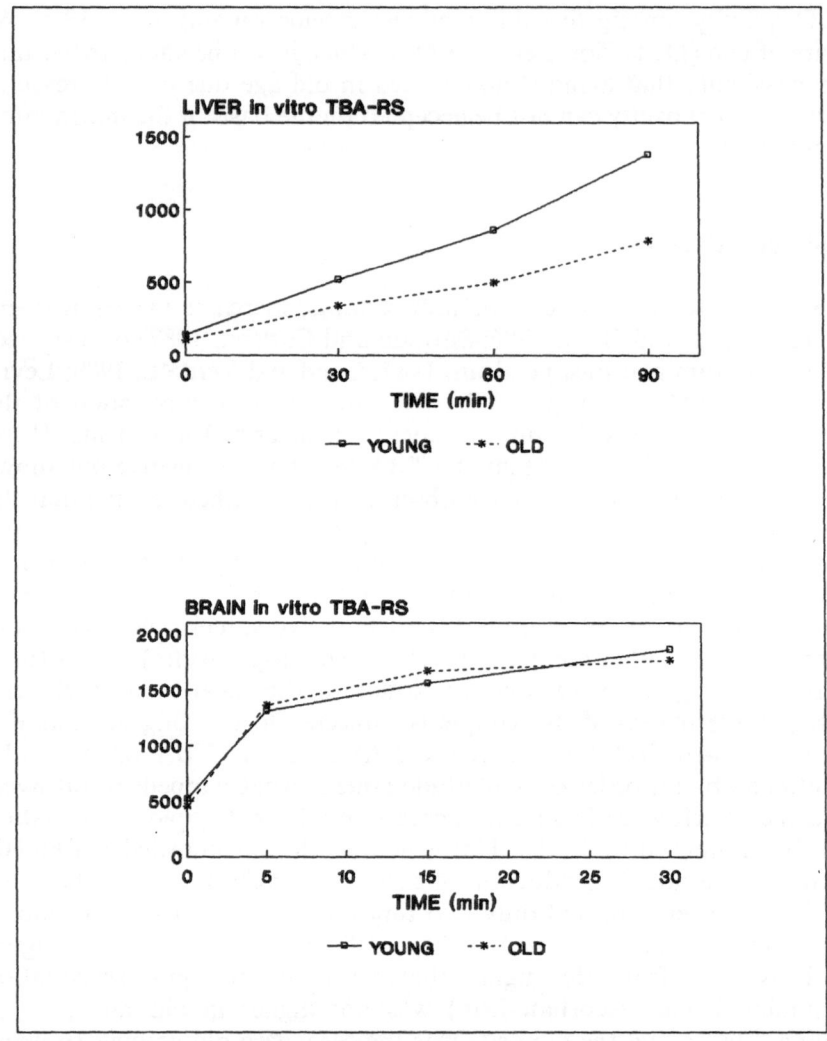

Figure 3. Production of TBA-RS in liver and cerebral cortex homogenates from young (9 months) or old (28 months) Wistar rats in the presence of 50 μM FeSO$_4$ plus 0.4 mM ascorbate.

TBA-RS can be influenced by a number of factors including the differential amounts of peroxidizable substrates and lipid-soluble antioxidants initially present in the samples obtained from young and old animals. Thus, relative decreases of polyunsaturated fatty acids (Bordoni et al., 1988; the preferred substrates of lipid peroxidation) accompanied by corresponding increases in the microviscosity of liver membranes (Nukobo, 1985) and strong increases of lipid-soluble an-

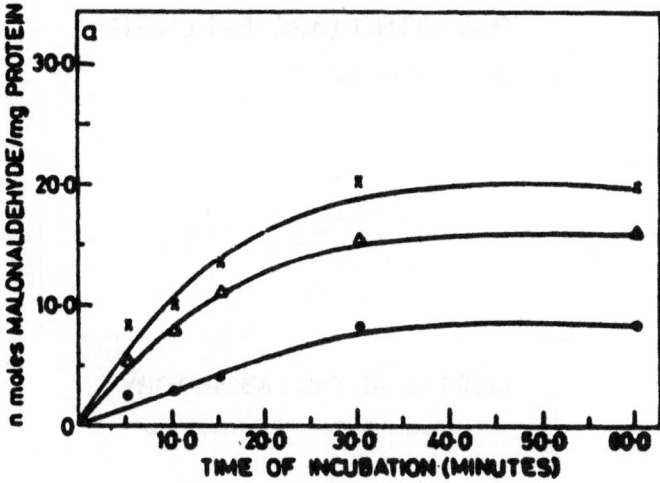

Figure 4. Production of TBA-RS in liver microsomes from young (×), middle aged (△), or old (○) Wistar rats. Data from Devasagayam, 1986.

tioxidants such as α-tocopherol, have been described in tissues from old animals (Vatassery et al., 1984; Rikans and Moore, 1991) and could be responsible for the observed decrease of in vitro TBA-RS in old age. Furthermore, it should be signaled that these studies coincidentally show that vitamin E progressively accumulates in the liver (but not in the cerebral cortex) of male rats from 3–4 to 29–30 months of age, reaching values almost two times higher in old animals in relation to young animals. This could explain why the rate of production of in vitro TBA-RS is smaller in the liver of the old rat but not in the cerebral cortex (Fig. 3). It has been shown recently in the liver of male rats (Rikans and Moore, 1991) that the increase in vitamin E as a function of age is not accompanied by reductions in TBA-RS, but these results corresponded to in vivo, not to in vitro measurements of TBA-RS. The decrease of in vitro TBA-RS observed in old animals is most probably related to "time" itself rather than to aging. This is supported by the fact that, when adult frogs are captured in the field, old animals show levels of in vitro TBA-RS that are similar to those of young ones, but these levels decrease acutely (both in initially young and old animals) as a result of the years of captivity at the laboratory, where they continuously receive a mineral and vitamin mix in appropriate amounts for normal maintenance (Fig. 5). Thus, the decrease observed in TBA-RS in old animals could be a "time" (nutrition) dependent change instead of an age-related change. If this is true, the utility of this test in relation to aging studies is limited.

118

TBA-VITRO (ASC-Fe) LIVER

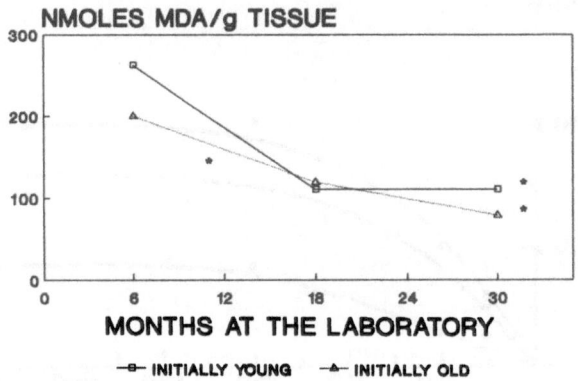

TBA-VITRO (ASC-Fe) LUNG

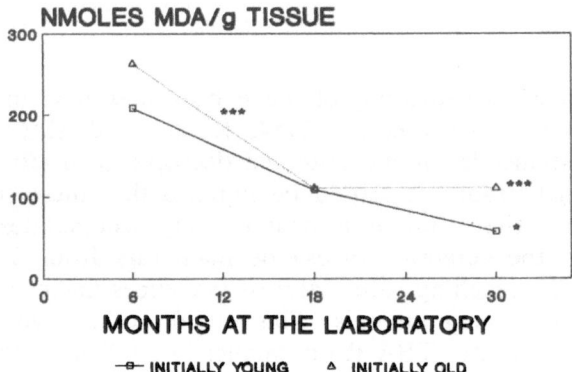

Figure 5. Effect of time of stay at the laboratory on the levels of in vitro TBA-RS in liver and brain of initially young (2.3 years) or old (4.3 years) *Rana perezi* frogs in the presence of 50 μM $FeSO_4$ plus 0.4 mM ascorbate.

Effect of modulation of endogenous antioxidants on survival

Several approaches have been used in trying to ascertain if the modification of the levels of cellular antioxidants can alter the life span of animals. One approach has been to study the effects of antioxidant supplementation in the diet. Feeding animals with relatively high amounts of antioxidants has resulted sometimes in moderate increases in mean (Harman, 1968) but not in maximum life span (Balin, 1982; Cutler, 1984). It has even been shown that the effect of antioxidants on the mean life span of mice disappeared when the experiment was performed under optimal conditions for survival (Kohn, 1971). Negative results have been repeatedly attributed to compensatory reactions among the different antioxidants whenever an appropriate comprehensive approach was chosen to design the experiments. Thus, supplementing the life-long diets of flies with ascorbate, β-carotene, or α-tocopherol did not change or even reduce the mean life span, probably because of the compensatory decreases in SOD, GSH, or CAT (Sohal et al., 1985). Similarly, continous inhibition of SOD (with diethyldithiocarbamate) or CAT depletion (with aminotriazole) did not affect the maximum life span of *Musca domestica* flies but resulted in a decrease of metabolic rate and induction of GSH (Sohal et al., 1984) or in a decrease in metabolic rate and induction of SOD plus GSH (Allen et al., 1983). A decrease in the metabolic rate of flies after antioxidant supplementation has been also described in *D. melanogaster* (Miquel and Lindset, 1983).

For the first time in a vertebrate animal, we studied the effects of continuous catalase depletion (greater than 95% in every organ) by aminotriazole treatment (i.p. injection every two weeks) during three years in liver, kidney, brain and lung; at the same time we followed longitudinally the survival rate of 220 animals. Biochemical measurements were taken after 2.5, 14.5 and 26.5 months of continuous CAT depletion. At the first time point (2.5 months), survival was 100% in all groups and the only change observed was a 100% induction of SOD and GR in CAT-depleted animals (López-Torres et al., 1991; Pérez-Campo et al., 1990; Barja de Quiroga et al., 1990b). As shown in Table 2, after 14.5 months of treatment the animals reacted with important inductions of GR (which reached 900% of control values), SOD, GSH, and ASC (measured by HPLC) in kidney, liver, brain or lung. To our knowledge, this is the first time that quantitatively important inductions of four different antioxidants simultaneously occurred during long periods of time in many vital organs of a vertebrate. These animals showed smaller amounts of lipofuscin accumulation in tissues (electron microscopy) and a survival rate two times higher than that of controls. It must be signaled that neither the oxygen consumption of whole animals at rest or tissue cytochrome oxidase activities were altered by the treatment. On the contrary, at 26.5 months of experimentation the above-mentioned inductions were partially (GR-liver and kidney, SOD-liver) or totally (GSH-liver and kidney, ASC-kidney, GR-brain and lung) lost in a spontaneous manner together with the accumulation of high amounts of tissue lipofuscin. This was followed by a high death rate during the next two months. In other words, a clear correlation was observed between a high antioxidant/prooxidant balance in the tissue, a small accumulation of lipofuscin, and a high probability of survival of the individual. The reason why such consistent results were not obtained in the experiments of antioxidant supplementation performed in other species and mentioned before can be that only in the "frog experiment"

Table 2. Correlation between endogenous induction of antioxidants and survival in cat-depleted frogs throughout the life span

Months	Antioxidant	% Induction	Survival (treated/control)
14.5	GR-KIDNEY	773%	2
	GSH-KIDNEY	80%	
	ASC-KIDNEY	65%	
	SOD-LIVER	126%	
	GR-LIVER	800%	
	GSH-LIVER	114%	
	ASC-LIVER	120%	
	GR-BRAIN	77%	
	GR-LUNG	48%	
26.5	GR-KIDNEY	495%	0.4
	GSH-KIDNEY	0%	
	ASC-KIDNEY	0%	
	SOD-LIVER	61%	
	GR-LIVER	564%	
	GSH-LIVER	0%	
	ASC-LIVER	220%	
	GR-BRAIN	0%	
	GR-LUNG	0%	

Treated frogs were continuously depleted (95%) in every organ during 3 years by i.p. injection of aminotriazole (1 mg/g); ASC = ascorbate.

four endogenous antioxidants were induced simultaneously. Furthermore, these were endogenous inductions instead of single doses chosen by the particular researcher (it is known that the effect of many antioxidants can be beneficial or detrimental depending on the dose chosen). Thus, on the basis of the present information, it is logical to expect that the supplementation of a particular antioxidant to the diet should result in compensatory decreases in other endogenous antioxidants. Detailed studies about the effects of different doses of antioxidants on the different endogenous and exogenous antioxidants, peroxidative damage and free radical generation rates are clearly needed.

It can be summarized that the information available to date can not tell us if the modulation of antioxidant capacity of the tissues can increase the maximum life span of animals. Methods of increasing more than one antioxidant defense simultaneously (obtained without using oxidative stress as a stimulus) are needed in order to clarify the issue. Nevertheless it appears that, when increased in appropriate proportions, antioxidants such as ASC and vitamin E can increase the general health state of the animals. This concept is consistent with their general stimulatory action upon the immune system (Tengerdy, 1989; Meydani et al., 1989; 1990) and with their preventive value against cardiovascular diseases (Gey, 1990) and cancer (Buiatti et al., 1990; Comstock et al., 1991). This should encourage further studies leading to a true knowledge of safe and beneficial doses of antioxidant supplementation in animals and humans. These doses should be capable of decreasing the levels of oxidative damage in the tissues without acutely depressing the rest of the endogenous antioxidants so that they can help improve the health status of the aging organism.

Acknowledgments. This work was supported by funds from the FISss (National Health Research Foundation of the Spanish Ministry of Health), grant no. 89/0074 and 90/0013.

Allen, R. G., Farmer, K. J., and Sohal, R. S. (1983) Effect of catalase inactivation on levels of inorganic peroxides, superoxide dismutase, glutathione, oxygen consumption and life span in adult houseflies (*Musca domestica*). Biochem. J. 216: 503–506.

Ansari, K. A., Kaplan, E., and Shoeman, D. (1989) Age-related changes in lipid peroxidation and protective enzymes in the central nervous system. Growth Dev. Aging 53: 117–121.

Balin, A. K. (1982) Testing the free radical theory of aging, in: Free Radicals in Biology, vol. V, pp. 255–275. Ed. W. A. Pryor. Academic Press, New York.

Barja de Quiroga, G., Pérez-Campo, R., and López-Torres, M. (1990A) Antioxidant defenses and peroxidation in liver and brain of aged rats. Biochem. J. 272: 247–250.

Barja de Quiroga, G., Pérez-Campo, R., and López-Torres, M. (1990b) Changes on cerebral antioxidant enzymes, peroxidation, and the glutathione system of frogs after aging and catalase inhibition. J. Neurosci. Res. 26: 370–376.

Benzi, G., Pastoris, O., and Villa, R. F. (1988a) Changes induced by aging and drug treatment on cerebral enzymatic antioxidant system. Neurochem. Res. 13: 467–478.

Benzi, G., Pastoris, O., Marzatico, F., and Villa, R. F. (1988b) Influence of aging and drug treatment on the cerbral glutathione system. Neurobiol. Aging 9: 371–375.

Benzi, G., Pastoris, O., Marzatico, F., and Villa, R. F. (1989) Cerebral enzyme antioxidant system. Influence of aging and phosphatidylcholine. J. Cer. Blood Flow and Metabolism 9: 373–380.

Benzi, G., Marzatico, F., Pastoris, O., and Villa, R. F. (1990) Influence of oxidative stress on the age-linked alterations of the cerebral glutathione system. J. Neurosci. Res. 26: 120–128.

Bordoni, A., Biagi, P. L., Turchetto, E., and Hreila, S. (1988) Aging influence on delta-6-desturase activity and fatty acid composition of rat liver microsomes. Biochem. Int. 17: 1001–1009.

Buiatti, E., Palli, D., Decarli, A., Amadori, D., Avellini, C., Bianchi, S., Bonaguri, C., Cipriani, F., Cocco, P., Giacosa, A., Marubini, E., Minacci, C., Puntoni, R., Russo, A., Vindigni, C., Fraumeni, J. F., and Blot, W. J. (1990) A case-control study of gastric cancer and diet in Italy: II. Association with nutrients. Int. J. Cancer 45: 896–901.

Cand, F., and Verdetti, J. (1989) Superoxide dismutase, glutathione peroxidase, catalase, and lipid peroxidation in the major organs of the aging rats. Free. Rad. Biol. Med. 7: 59–63.

Carrillo, M. C., Kitani, K., Kanai, S., Sato, Y., Nokubo, M., Ohta, M., and Otsubo, K. (1989) Differences in the influence of diet on the hepatic glutathione S-transferase activity and gluthathione content between young and old C57 black female mice. Mech. Ageing Dev. 47: 1–15.

Comstock, G. W., Helzlsouer, K. J., and Bush, T. L. (1991) Prediagnostic serum levels of carotenoids and vitamin E as related to subsequent cancer in Washington county, Maryland. Am. J. Clin. Nutr. 53: 260S–264S.

Cutler, R. G. (1984) Antioxidants, aging, and longevity, in: Free Radicals in Biology, vol. VI, pp. 379–428. Ed. W. A. Pryor. Academic Press, New York.

Cutler R. G. (1991) Recent progress in testing the longevity determinant and dysdifferentiation hypotheses of aging. Arch. Gerontol. Geritatr. 12: 75–98.

Danh, H. C., Benedetti, M. S., and Destert, P. (1983) Differential changes in superoxide dismutase activity in brain and liver of old rats and mice. J. Neurochem. 40: 1003–1007.

Devasagayam, T. P. A. (1986) Senescence-associated decrease of NADPH-induced lipid peroxidation in rat liver microsomes. FEBS Lett. 205: 246–250.

Devasagayam, T. P. A., and Pushpendran, C. K. (1986) Changes in ascorbate-induced lipid peroxidation of hepatic rough and smooth microsomes during postnatal development and ageing of rats. Mech. Ageing Dev. 34: 13–21.

Floyd, R. A., Zaleska, M. M., and Harmon, H. (1984) Possible involvement of iron and oxygen free radicals in aspects of aging brain, in: Free Radicals in Molecular Biology, Aging, and Disease, pp. 143–161. Ed. D. Armstrong. Raven Press, New York.

Fraga, C. G., Shinegaga, M. K., Park, J.-W., Degan, P., and Ames, B. (1990) Oxidative damage to DNA during aging: 8-Hydroxy-2'-deoxyguanosine in rat organ DNA and urine. Proc. Natl. Acad. Sci. 87: 4533–4537.

Gey, K. F. (1990) The antioxidant hypothesis of cardiovascular disease: epidemiology and mechanisms. Biochem. Soc. Transact. 18: 1041–1045.

Harman, D. (1956) Aging: a theory based on free radical and radiation chemistry. J. Gerontol. 11: 298–300.

Harman, D. (1968) Free radical theory of aging: Effects of free radical reaction inhibitors on the mortality rate of male LAF, mice. J. Gerontol. 23: 475–482.

Hazelton, A., and Lang, C. A. (1980) Glutathione contents of tissues in the aging mouse. Biochem. J. 188: 25–30.

Kellogg, E. W. III, and Fridovich, I. (1976) Superoxide dismutase in the rat and mouse as a function of age and longevity. J. Gerontol. 31: 405–408.

Khon, R. R. (1971) Effect of antioxidants on life-span of C57BL mice. J. Gerontol. 26: 378–380.

Koizumi, A., Weindruch, R., and Walford, R. L. (1987) Influences of dietary restriction and age on liver enzyme activities and lipid peroxidation in mice. J. Nutr. 117: 361–367.

Kurobe, N., Suzuki, F., Kato, K., and Sato, T. (1990) Sensitive immunoassay of rat Cu/Zn superoxide dismutase: concentrations in the brain, liver, and kidney are not affected by aging. Biomed. Res. 11: 187–194.

Laganiere, S., and Yu, B. P. (1989) Effect of chronic food restriction in aging rats II. Liver cytosolic antioxidants and related enzymes. Mech. Ageing Dev. 48: 221–230.

Lammi-Keefe, C. J., Swan, P. B., Hegarty, P. V. J. (1984) Copper-zinc and manganese superoxide dismutase activities in cardiac and skeletal muscles during aging in male rats. Gerontol. 30: 153–158.

LeBel, C. P., and Bondy, S. C. (1991) Persistent protein damage despite reduced oxygen radical formation in the aging rat brain. Int. J. Dev. Neurosci. 9: 139–146.

Levitsky, E. L., Gubsky, Y. I., Goldstein, N. B., and Litoshenko, A. Y. (1989) Lipid peroxidation and polymerase activities of chromatin fractions of the rat liver in aging. Bull. Exp. Biol. Med. 6: 693–694.

López-Torres, M., Pérez-Campo, R., and Barja de Quiroga, G. (1991) Effect of natural ageing and antioxidant inhibition on liver antioxidant enzymes, glutathione system, and peroxidation, and oxygen consumption, in *Rana perezi*. J. Comp. Physiol. B. 160: 655–661.

Meydani, S. N., Mehdani, M., Barklund, P. M., Liu, S., Miller, R. A., Cannon, R. A., Rocklin, R., and Blumberg, J. B. (1989) Effect of vitamin E supplementation on immune responsiveness of the aged. Ann. N. Y. Acad. Sci. 570: 283–290.

Meydani, S. N., Barklund, M. P., Liu, S., Meydani, M., Cannon, J. G., Morrow, F. D., Rockling, R., and Blumberg, J. B. (1990) Vitamin E supplementation enhances cell-mediated immunity in healthy elderly subjects. Am. J. Clin. Nutr. 52: 557–563.

Miquel, J., and Lindseth, K. (1983) Determination of biological age in antioxidant-treated Drosophila and mice, in: Intervention in the Aging Process, pp. 317–358. Alan R. Liss, New York.

Miquel, J. (1991) An integrated theory of aging as the result of mitochondrial-DNA mutation in differentiated cells. Arch. Gerontol. Geriatr. 12: 99–117.

Mizuno, Y., and Ohta, K. (1986) Regional distributions of thiobarbituric acid-reactive products, activities of enzymes regulating the metabolism of oxygen free radicals, and some of the related enzymes in adult and aged rat brains. J. Neurochem. 46: 1344–1352.

Muscari, C., Frascaro, M., Guarneri, C., and Caldarera, C. M. (1990) Biochim. Biophys. Acta 1015: 200–204.

Nukobo, M. (1985) Physical-chemical and biochemical differences in liver plasma membranes in aging F-344 rats. J. Gerontol. 40: 409–414.

Pérez-Campo, R., López-Torres, M., Patón, D., Sequeros E., and Barja de Quiroga, G. (1990) Lung antioxidant enzymes, peroxidation, glutathione system and oxygen consumption in catalase inactivated young and old *Rana perezi* frogs. Mech. Ageing Dev. 56: 281–292.

Pérez-Campo, R., López-Torres, M., and Barja de Quiroga, G. (1991) Aging and lung antioxidant enzymes, glutathione, and lipid peroxidation in the rat. Free Rad. Biol. Med. 10: 35–39.

Rao, G., Xia, E., and Richardson, A. (1990a) Effect of age on the expression of antioxidant enzymes in male Fischer F344 rats. Mech. Ageing Dev. 53: 49–60.

Rao, G., Xia, E., Nadakavukaren, M. J., and Richardson, A. (1990b) Effect of dietary restriction on the age-dependent changes in the expression of antioxidant enzymes in rat liver. J. Nutr. 120: 602–609.

Reiss, U., and Gershon, D. (1976) Comparison of cytoplasmic superoxide dismutase in liver, heart, and brain of aging rats and mice. Biochem. Biophys. Res. Commun. 73: 255–262.

Rikans, L. E., and Moore, R. (1988) Effect of aging on aqueous phase antioxidants in tissues of male Fisher rats. Biochim. Biophys. Acta 966: 269–275.

Rikans, L. E., Moore, D. R., and Snowden, C. D. (1991) Sex-dependent differences in the effects of aging on antioxidant defense mechanisms of rat liver. Biochim. Biophys. Acta 1074: 195–200.

Sawada, M., and Carlson, J. C. (1987) Changes in superoxide radical and lipid peroxide formation in the brain, heart and liver during the lifetime of the rat. Mech. Ageing Dev. 41: 125–137.

Semsei, I., Rao, G., and Richardson, A. (1989) Changes in the expression of superoxide dismutase and catalase as a function of age and dietary restriction. Biochem. Biophys. Res. Commun. 164: 620–625.

Semsei, I., Rao, G., and Richardson, A. (1991) Expression of superoxide dismutase and catalase in rat brain as a function of age. Mech. Ageing Dev. 58: 13–19.

Sevanian, A., and Hochstein, P. (1985) Mechanisms and consequences of lipid peroxidation in biological systems. Ann. Rev. Nutr. 5: 365–390.

Sohal, R. S., Farmer, K. J., Allen, R. G., and Cohen, N. R. (1983) Effect of age on oxygen consumption, superoxide dismutase, catalase, glutathione, inorganic peroxides and chloroform-soluble antioxidants in the adult male housefly, *Musca domestica*. Mech. Ageing Dev. 24: 185–195.

Sohal, R. S., Farmer, K. J., Allen, R. G., and Ragland, S. S. (1984) Effects of diethyldithiocarbamate on lifespan, metabolic rate, superoxide dismutase, catalase, inorganic peroxides and glutathione in the adult male housefly, *Musca domestica*. Mech Ageing Dev. 24: 175–183.

Sohal R. S., Müller, A., Koletzko, B., and Sies, H. (1985) Effect of age and ambient temperature on n-pentane production in adult housefly, *Musca domestica*. Mech. Ageing Dev. 29: 317–326.

Sohal R. S., Svensson, I., Sohal., B. H., and Brunk, U. T. (1989) Superoxide anion radical production in different animal species. Mech. Ageing Develop. 49: 129–135.

Sohal, R. S., Arnold, L., and Orr, W. C. (1990a) Effect of age on superoxide dismutase, catalase, glutathione reductase, inorganic peroxides, TBA-reactive material, GSH/GSSG, NADPH/NADP$^+$ and NADH/NAD$^+$ in *Drosophila melanogaster*. Mech. Ageing Dev. 56: 223–235.

Sohal, R. S., Arnold, L. A., and Sohal, B. H. (1990b) Age-related changes in antioxidant enzymes and prooxidant generation in tissues of the rat with special reference to parameters in two insect species. Free Rad. Biol. Med. 9: 495–500.

Sohal, R. S., Svensson, I., and Brunk, U. T. (1990c) Hydrogen peroxide production by liver mitochondria in different species. Mech. Ageing Dev. 53: 209–215.

Stohs, S. J., and Lawson, T. (1986) The role of glutathione and its metabolism in aging, in: Liver and Brain, pp. 59–70. Ed. K. Kitani. Elsevier, Amsterdam.

Tayarami, I., Cloez, I., Clément, M., and Bourre, J. M. (1989) Antioxidant enzymes and related trace elements in aging brain capillaries and choroid plexus. J. Neurochem. 53: 817–824.

Tengerdy, R. P. (1989) Vitamin E immune response and disease resistance. Ann. N. Y. Acad. Sci. 570: 335–344.

Vatassery, G. T., Angerhofer, C. K., and Knox, C. A. J. (1984) Effect of age on vitamin E concentrations in various regions of the brain and a few selected peripheral tissues of the rat, and on the uptake of radioactive vitamin E by various regions of rat brain. J. Neurochem. 43: 409–412.

Vanella, A., Villa, R. F., Gorini, A., Campisi, A., and Giuffrida-Stella, A. M. (1989) Superoxide dismutase and cytochrome oxidase activities in light and heavy synaptic mitochondria from rat cerebral cortex during aging. J. Neurosci. Res. 22: 351–355.

Vitorica, J., Machado, A., and Satrustegni, J. (1984) Age-dependent variations in peroxide-utilizing enzymes from rat brain mitochondria and cytoplasm. J. Neurochem. 42: 351–356.

Free Radicals and Aging
ed. by I. Emerit & B. Chance
© 1992 Birkhäuser Verlag Basel/Switzerland

The metabolism of 4-hydroxynonenal, a lipid peroxidation product, is dependent on tumor age in Ehrlich mouse ascites cells

Werner G. Siems[a], Tilman Grune[a], Beatrix Beierl[a], Helmward Zollner[b] and Hermann Esterbauer[b]

[a]*Institute of Biochemistry, Medical Faculty (Charité), Humboldt University of Berlin, Hessische Strasse 3-4, D-(0)-1040 Berlin, Federal Republic of Germany, and* [b]*Institute of Biochemistry, University of Graz, Schubertstrasse 1, A-8010 Graz, Austria*

Summary. 4-Hydroxynonenal is a major product formed by lipid peroxidation from omega 6-polyunsaturated fatty acids as linoleic acid and arachidonic acid. This aldehyde is cytotoxic at high concentrations (in the range of 100 μM), disturbs cell proliferation at low concentrations and exhibits genotoxic effects. Furthermore, in the submicromolar range 4-hydroxynonenal is chemotactic and stimulates phospholipase C. 4-Hydroxynonenal is rapidly metabolized in eucaryotic cells. Here the metabolism of 4-hydroxynonenal was studied in suspensions of Ehrlich mouse ascites cells at different periods of the tumor age. The Ehrlich ascites tumor is a convenient biological model for the investigation of tumor cells in different age and proliferation phases of the tumor. The main products of 4-hydroxynonenal which were identified in the Ehrlich ascites cells were glutathione-HNE-conjugate, hydroxynonenoic acid and 1,4-dihydroxynonene. The formation of glutathione conjugates following the addition of 4-hydroxynonenal was higher in cells of the early phase in comparison with cells of the late phase of tumor growth. That was in accordance with the increased consumption of the reduced form of glutathione during 4-hydroxynonenal utilization. The degradation of 4-hydroxynonenal and other aldehydic products of lipid peroxidation is postulated to be an important part of the intracellular antioxidative defense system.

Abbreviations. EATC – Ehrlich ascites tumor cells; GSH – glutathione (reduced form); HNE – 4-hydroxynonenal; DHN – 1,4-dihydroxynonene; HNA – 4-hydroxy-2-nonenoic acid; PUFA – polyunsaturated fatty acids.

Introduction

Aldehydes are generated during degradation of lipid hydroperoxides, the primary products of free radical-initiated lipid peroxidation (Esterbauer, 1982; Esterbauer et al., 1989; Schaur et al., 1991). Several aldehydes react quickly with cellular constituents, and it is assumed that the free radical-medicated functional and morphological disturbances of cells are in part induced by such aldehydic lipid peroxidation products (Esterbauer et al., 1988; Cerutti, 1985). Therefore, the aldehydes should be seen as "second toxic messengers" (Esterbauer et al., 1989) for the primary free radicals which initiated the lipid peroxidation processes. 4-Hydroxynonenal (HNE) is a degradation product of omega 6-PUFA.

It exhibits cytotoxic, hepatotoxic, mutagenic, and genotoxic properties (Esterbauer, 1982; Esterbauer et al., 1988). It was discussed as an important component of clastogenic or chromosome breakage factors (Emerit, 1987a,b; Emerit et al., 1991). Increased levels of HNE were found in plasma and various tissues under conditions of oxidative stress (Esterbauer et al., 1989). In our laboratory an increase of HNE was observed during the early phase of postischemic reperfusion of small intestine (Siems et al., 1991; Kowalewski et al., 1991).

HNE appears to be not only an intermediate generated at situations of imbalances between prooxidants and antioxidant defenses, but also a normal constituent of plasma, blood cells and parenchymal organs in mammals, including man (Schaur et al., 1991; Van Kuijk et al., 1986). Its concentration in tissues and body fluids depends on the relationship between the rate of HNE formation and the rates of HNE-consuming reactions.

Metabolic changes in Ehrlich ascites tumor cells during the transition from the proliferating into the resting phase of the in vivo growth of the tumor

During the growth of Ehrlich ascites tumor cells (EATC) a proliferating phase exists, during which the number of cells increases quasi-exponentially. This period is followed by a resting phase, where the number of cells remains practically constant. The growth retardation is not due to an increased rate of cell death, but is caused by the prolongation of all phases of the cell cycle (G_1, S, G_2, M) and the increased fraction of cells in G_0, whereas the percentage of S-phase cells is decreased (see Schmidt et al., 1991). The mechanisms of these changes have been obscure until now. The effects of several growth stimulating and inhibiting factors have been discussed. During the transition of ascites tumor cells from the proliferating into the plateau period of growth a number of changes occur, i.e., structural deterioration and decrease in number of mitochondria (Siems et al., 1989), decrease in DNA-, RNA-, and protein synthesis, increase in protein degradation (Tessitore et al., 1987), and changes of the protein pattern (Benndorf et al., 1988). Furthermore, a tremendous loss of intracellular purine and pyrimidine nucleotides, nucleosides, and bases was measured (Siems et al., 1989) with increasing age of the tumor following its transplantation into the animal. These changes correlate to a decline in the incorporation rates of adenine, hypoxanthine and adenosine (Siems et al., 1989) and a decrease in oxygen consumption and ATP turnover (Müller et al., 1986; Schmidt et al., 1989; Schmidt et al., 1991). The aim of this study was to investigate whether proliferating EATC (early phase of the tumor growth/5th day after inoculation of the tumor) and resting EATC (late phase/12th day) also differ in their metabolism of lipid peroxidation products. As a model compound we selected 4-hydroxynonenal, since the metabolism of this lipid peroxidation product has been studied in detail already in other cell lines.

Endogeneous lipid peroxidation and sensitivity of tumor cells against oxidative stress in relation to tumor age and proliferation stage

One of the characteristics of the oxidative metabolism of tumor cells is their high resistance to inducers of free radical oxidative stress resulting in low steady-state concentrations of endogeneous lipid peroxidation products (Galeotti et al., 1984; Cheeseman et al., 1986; Masotti et al., 1988; Kagan et al., 1991). Similarly, in proliferating and nondifferentiated embryonic tissues, the endogeneous lipid peroxidation rate and the accessibility to lipid peroxidation inducing agents are decreased (Devasagayam, 1986; Kagan et al., 1991). It was reported that the steady-state concentrations of lipid hydroperoxides in tumor cells are approximately five to seven times lower as compared to normal tissues (Kagan et al., 1991). The production of superoxide radicals is also decreased in tumor cells, which was measured in hepatomas in comparison with the normal rat liver (Peskin et al., 1977; Bize et al., 1980). Tumor cells are characterized by a low lipid peroxidation rate in vitro, too (Galeotti et al., 1984). The high concentrations of antioxidants and particularly of vitamin E are considered to be an important factor for the resistance of tumor cells against an oxidative stress (Kagan et al., 1991; Cheeseman et al., 1984). Additionally, increased amounts of saturated and monoenoic fatty acids and decreased amounts of PUFA, which are the main substrates for lipid peroxidation processes, were found in cancer cells (Hostetler et al., 1979; Borrello et al., 1985; Poli et al., 1986). Therefore, it was concluded that the α-tocopherol/polyenoic fatty acid ratio is drastically increased in tumor cells as compared to the corresponding normal tissue. Kagan et al. (1991) suggested that protein factors, too, contribute to the high resistance of tumor cells against oxidative stress. In addition, malignant cells possess low activities of antioxidant enzymes like glutathione peroxidase, catalase and superoxide dismutase (Peskin et al., 1977).

A further aspect of the sensitivity of tumor cells against an oxidative stress is the inhibition of cell proliferation by products of lipid peroxidation. It has been suggested that 4-hydroxyalkenals play a role in the regulation of cell division (Burdon and Rice-Evans, 1989). HNE added to cultures of dividing cells blocks proliferation and leads to an arrest of most cells in the G_1 and G_2 phases. Hydroxyalkenals are suggested to exert carcinostatic activity in tumor-bearing animals (Hauptlorenz et al., 1985).

Metabolic pathways of 4-hydroxynonenal in Ehrlich ascites tumor cells in different periods of tumor growth

Ehrlich ascites tumor cells at different growth periods of the tumor rapidly degrade exogenously added HNE. Figure 1 shows the rates of HNE disappearance in suspensions of cells of the 5th and of the 12th day after inoculation of the tumor within 30 minutes of incubation at 37°C, pH 7.4 and a cytocrit of 1% following the addition of 100 μM HNE. The initial concentration of HNE added in all experiments with EMAT cells was 100 μM. There was no difference in the capacity of HNE metabolism between the cells of the early and late phases of tumor growth.

Current knowledge on the in vitro metabolism of HNE in hepato-cytes (Esterbauer et al., 1985; Schaur et al., 1991; Fauler, 1987; Siems et

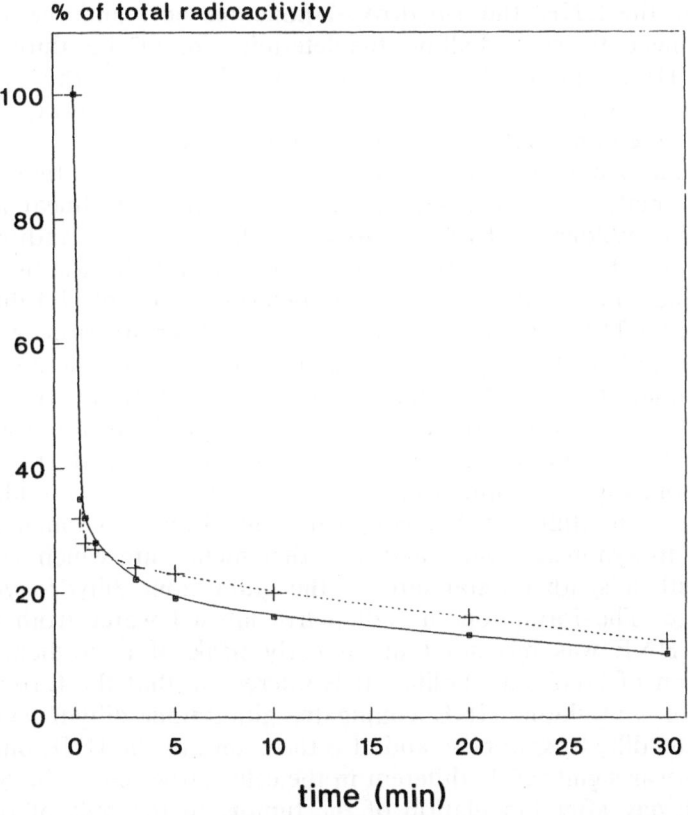

Figure 1. Rate of HNE degradation in Ehrlich ascites tumor cells at the 5th and the 12th day after inoculation of the tumor. The initial HNE concentration was 100 M. The cytocrit during the incubation experiments was about 1%. The values are given as % of the initial concentration which is identical to nmoles/ml suspension. Solid line: EATC of the 5th day; dotted line: EATC of the 12th day after innoculation of the tumor.

al., 1990), hepatoma cells (Ferro et al., 1988) and enterocytes of the small intestine (Grune et al., 1991) indicates that the main enzymes involved in the metabolism are glutathione transferases, aldehyde dehydrogenases, and alcohol dehydrogenases. The main primary metabolites of HNE are therefore the glutathione-HNE conjugate, the 4-hydroxy-2-nonenoic acid and 1,4-dihydroxynonene. To verify if these metabolites were in fact formed, and if so in which proportion, sensitive analytical procedures which were developed in studies with hepatocytes (Fauler, 1987) were used for the qualitative and the quantitative determination of these compounds. These methods include HPLC separations with precolumn derivatization, HPLC with UV detection, HPLC of the ^{3}H-labeled compounds, TLC of the ^{3}H-labeled compounds. The use of tritium-labeled (2-^{3}H) HNE enabled us to quantify the formation of metabolites by TLC without prior derivatization. Two different solvent systems were used for the TLC separation of HNE metabolites: one separates the HNE, the 4-hydroxy-2-nonenoic acid and the 1,4-dihydroxynonene, the second allows the determination of glutathione conjugates of HNE and of 1,4-dihydroxynonene. The position of a particular metabolite on the TLC-plate was determined by separating the standards on the same plate. The concentration was determined by measuring the radioactivity associated with the TLC-spots. For that purpose the TLC-plates were scanned with an automatic TLC linear analyzer (Berthold, Wildbach, FRG). Based on the recovered radioactivity, HPLC and TLC separations gave concurrent results for the three HNE metabolites mentioned above. The determination of 1,4-dihydroxynonene by TLC received preference over its determination by HPLC, since the TLC method is less tedious and much faster. Therefore, most of the quantitative data result from TLC separations. It could be demonstrated that the three primary HNE products are formed in Ehrlich ascites tumor cells after the addition of HNE to the cell suspension. Figure 2 summarizes the metabolites which were identified. These are glutathione-HNE conjugate, the 4-hydroxynonenoic acid, 1,4-dihydroxynonene, water and a further metabolite which was identified with a synthetic standard as the glutathione-dihydroxynonene-conjugate. The formation of radioactive labeled water from tritium-labeled HNE was measured in an early peak of a gradient HPLC separation of HNE metabolites. It is interesting that the formation of water, of glutathione-HNE conjugate, glutathione-dihydroxynonene-conjugate, dihydroxynonene and also the extent of the HNE binding to proteins was significantly different in the cells harvested at the 5th or at the 12th day after inoculation of the tumor. In the cells of the early phase of tumor growth the accumulation of the corresponding alcohol of the HNE, the 1,4-dihydroxynonene, is increased – beginning at the third minute of cell incubation (Fig. 3), whereas in the cells of the late phase of tumor growth, even under conditions of a decreased mitochon-

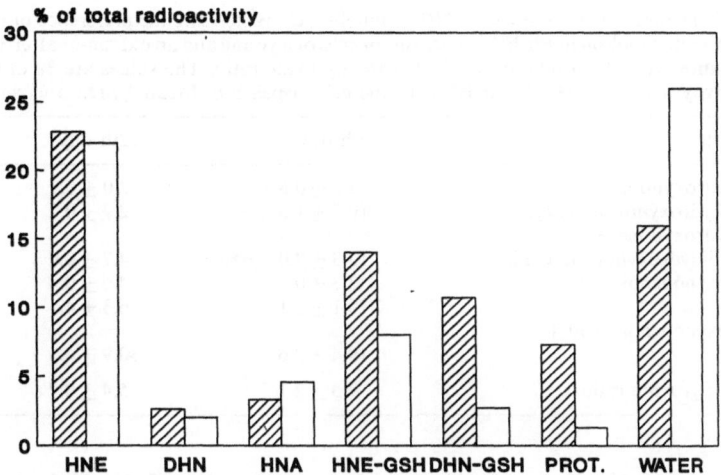

Figure 2. Distribution of the radioactivity in different products of HNE-metabolism after the addition of 100 μM tritium-labeled HNE to EATC suspensions. Values are given as % of the total radioactivity. Symbols: (▨) EATC of the 5th day, (□) EATC of the 12th day after innoculation of the tumor.

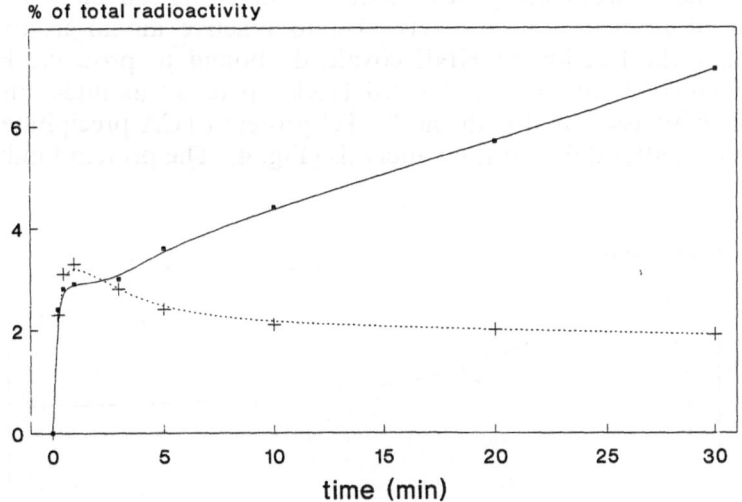

Figure 3. Formation of the 1,4-dihydroxy-nonene from exogeneously added HNE in EATC. Symbols as in Fig. 1.

drial volume in the cells, the oxidative pathway with the formation of the 4-hydroxynonenoic acid and water (due to beta-oxidation of the acid) is significantly higher. Table 1 shows the values of the accumulation of the alcohol plus the glutathione-alcohol conjugate dominating in the cells of the 5th day and of the 4-hydroxynonenoic acid plus water

130

Table 1. The accumulation of four HNE metabolites characterizing the preference of reductive or oxidative metabolism in Ehrlich ascites tumor cells of a young and an old tumor after 3 minutes of incubation at 37°C in relation to the lactate: pyruvate ratio. The values are % of the total radioactivity added as ^{3}H-labeled HNE to the cell suspension. Mean ± SD.; 5 experiments

Metabolite	5th day	12th day
1,4-Dihydroxynonene	2.6 ± 0.5	2.0 ± 0.2
GSH-dihydroxynonene conj.	10.7 ± 1.5	2.7 ± 0.7
1,4-Dihydroxynonene + GSH-dihydroxynonene conj.	13.3 ± 2.0	4.7 ± 0.9
4-Hydroxynonenoic acid	3.3 ± 0.5	4.6 ± 0.4
Water	16.1 ± 2.1	26.3 ± 4.1
4-Hydroxynonenoic acid + water	19.4 ± 2.6	30.9 ± 4.5
Lactate: pyruvate ratio	9.3 ± 2.1	5.4 ± 0.7

formed from HNE dominating in the ascites cells of the 12th day after tumor inoculation. The preference of the reductive or oxidative metabolism of the HNE is in accordance with the cytosolic NADH:NAD$^+$ ratio which can be directly derived from the lactate:pyruvate ratio in cell suspensions.

It is known from many other studies that HNE can bind to thiol-groups of proteins and is likely also to reactive amino-groups. To determine the fraction of HNE covalently bound to proteins, EATC were incubated with tritium-labeled HNE up to 30 minutes, and the radioactivity associated with the EATC protein (TCA precipitate) was determined after different time intervals (Fig. 4). The protein binding of

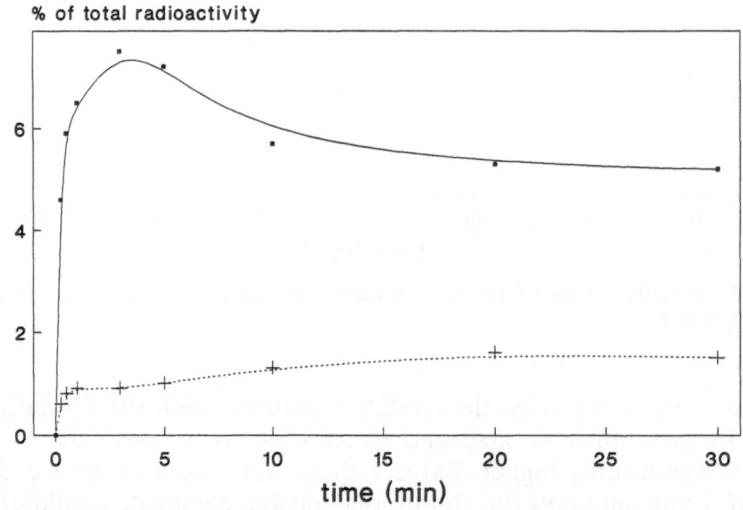

Figure 4. Binding of HNE to proteins in EATC. Symbols as in Fig. 1.

Table 2. Concentration of reduced glutathione, activity of glutathione-S-transferases and formation of glutathione-conjugates in Ehrlich mouse ascites cells from different growth phases of the tumor. The incubations were carried out at pH 7.4, 37°C at an initial HNE concentration of $100 \mu M$. The concentration changes were monitored within 30 min of incubation at 37°C. Values are given as mean ± S.D.; 5 independent experiments

	Dimension	5th day	12th day
Initial GSH concentration	mmol/1 cells	5.04 ± 0.16	3.55 ± 0.89
Activities of:			
glutathione-S-transferases	nmol/min/mg protein	575 ± 105	574 ± 170
glutathione peroxidase	nmol/min/mg protein	520 ± 174	233 ± 120
GSH loss within 30 min	mmol/1 cells	1.37 ± 0.31	0.61 ± 0.08
Accumulation of GSH-HNE and			
GSH-DHN conjugates in 30 min	mmol/1 cells	1.22 ± 0.33	0.53 ± 0.15

HNE was more than doubled in the cells of the early phase of tumor growth compared with the cells of the stationary phase. The fraction of HNE binding to proteins accounted only for 6% (proliferating phase) or 2% (resting phase of tumor growth) of the total HNE degradation. Nevertheless, even such a quantitatively low fraction is expected to be of great functional importance for cellular damage by HNE in situations of increased HNE formation.

Table 2 shows data on the formation of glutathione conjugates: (i) the GSH-HNE conjugate and (ii) the GSH-DHN conjugate. The formation of the GSH-HNE conjugate occurs mainly via enzymatic conversion. But HNE readily reacts with glutathione in a nonenzymatic reaction as well. The GSH transferase-catalyzed conversion can, however, proceed about 500 times faster than the nonenzymatic reaction. The GSH depletion in the EATC both in the cells of the 5th day and of the 12th day after inoculation of the tumor is in good agreement with the rate of the GSH-conjugate formation (GSH-HNE + GSH-DHN). Furthermore, the increased GSH concentration of the cells of the early phase of tumor growth at conditions of unchanged glutathione-S-transferases activity corresponds with the increased generation of GSH-conjugates in these cells. It is assumed that the GSH-dihydroxynonene conjugate in the Ehrlich ascites tumor cells is generated rather by enzymatic conversion of the GSH-HNE conjugate than by the enzymatically catalyzed direct reaction between GSH and 1,4-dihydroxynonene.

Several HPLC peaks which amount to about 20 to 30% of the total radioactivity of HNE added to the cell suspensions could not be identified. It is assumed that the unidentified substances include prod-

132

ucts of the beta-oxydation of the 4-hydroxynonenoic acid as acetyl-CoA or compounds of the tricarboxylic acid cycle.

Comparison of the flux rates of the 4-hydroxynonenal metabolism in Ehrlich ascites tumor cells and other eucaryotic cell types

The metabolism of HNE and also of 4-hydroxypentenal was studied in vitro with hepatocytes (Esterbauer et al., 1985; Fauler, 1987; Siems et al., 1990; Schaur et al., 1991), hepatoma cells (Ferro et al., 1988), isolated perfused rat hearts (Ishikawa et al., 1986), in enterocytes of the rat small intestine (Grune et al., 1991), in thymocytes (unpublished results from Grune and Siems) and in this study in Ehrlich ascites tumor cells. Furthermore, experiments with subcellular fractions of rat liver cytosol, microsomes, and mitochondria as well as with purified enzymes of the HNE metabolizing pathways (aldehyde dehydrogenases, alcohol dehydrogenase, and glutathione transferase) were carried out. The latter enzymes were found to catalyze the main reaction steps of the metabolism of HNE, i.e. the steps of the formation of primary products of HNE. Figure 5 represents a scheme of the formation of the main metabolites of HNE in Ehrlich ascites tumor cells, which is similar for the different cell types which have been investigated as yet. There is a very rapid degradation of exogeneous HNE in all of investigated cell types. Table 3 gives a comparison of the HNE degradation rates in different eucaryotic cell types. One can find the highest HNE degradation rate in hepatocytes and enterocytes. Esterbauer et al. (1985) reported that, in rats, liver has the highest capacity to metabolize HNE. Rat hepatocytes with a cell concentration of 10^6 cells/ml suspension were able to metabolize about 90% of 100 μM HNE within the first three minutes of normothermic incubation (Esterbauer et al., 1985; Fauler, 1987; Siems et al., 1990). The contribution of the single path-

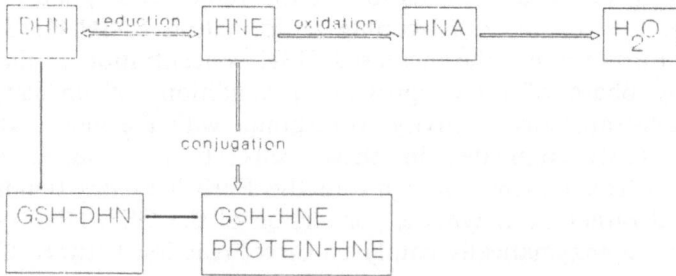

Figure 5. Scheme of the main pathways of HNE metabolism in Ehrlich ascites tumor cells. The changed quality of arrows directed to the GSH-DHN pool should demonstrate that it is not yet clear via which reaction GSH-DHN is formed in EATC, either by direct reaction between GSH and DHN or by the conversion of GSH-HNE.

Table 3. 4-Hydroxynonenal utilization rates in different eucaryotic cell types. The incubations were carried out at 37°C, pH 7.4 and an initial HNE concentration of 100 μM. Values are given as nmoles/mg w.w./min

Cell type	Species	HNE utilization rate	n
Enterocytes	Rat	25.2 ± 2.7	15
Hepatocytes	Rat	28.4 ± 1.8	20
Ehrlich ascites tumor cells	Mouse	8.7 ± 1.5	10

ways to the formation of HNE intermediates, however, depends on the capacity of the HNE metabolizing enzymes in the different cell types. HNE metabolizing aldehyde dehydrogenase isoenzymes are present in liver cytosol, mitochondria, and also in microsomes (for review see Esterbauer et al., 1991). The HNE-metabolizing NADH-dependent alcohol dehydrogenase is localized in hepatocytic cytosol (Esterbauer et al., 1985). Glutathione transferases are ubiquitous enzymes. The capacity of the EATC to metabolize HNE was in the range of such organs as kidney or heart, but higher than in the skeletal muscle or the spleen (Esterbauer et al., 1985).

The main difference between tumor cells including EATC and liver cells consists of the different activities of HNE metabolizing enzymes. Alcohol dehydrogenase activity is low or totally absent, and the isoenzyme pattern of aldehyde dehydrogenases is altered. HNE metabolism by tumor cells has already been studied in the hepatoma MH_1C_1 cell line (Ferro et al., 1988). The studies described here clearly show that EATC have a significantly lower reductive metabolism of HNE leading to 1,4-dihydroxynonene than hepatocytes. On the other hand, the oxidative pathway leading to 4-hydroxynonenoic acid is about equal to that in hepatocytes, and the 4-hydroxynonenoic acid is rapidly further metabolized to water and carbon dioxide by beta-oxidation.

The major difference between proliferating and resting EATC is that the young cells mainly from the glutathione conjugate(s) of the HNE, whereas the old non-proliferating cells mainly oxidize HNE to 4-hydroxynonenoic acid and subsequently to water and carbon dioxide.

Biological importance of HNE metabolism and its age dependency

We postulate that the degradation of 4-hydroxynonenal and other aldehydic lipid peroxidation products is an important part of the intracellular antioxidative defense system. Up to now nothing was known on the toxicity of the primary metabolites of the HNE described above. One can assume that the formation of 4-hydroxynonenoic acid

134

leads to an effective detoxification because of the beta-oxidation of the acid to carbon dioxide and water. HNE toxicity is obviously correlated to HNE-protein binding. We could demonstrate that the flux rates through the pathways of HNE metabolism change as the tumor ages. The age-dependency of the metabolic pathways of the cytotoxic lipid peroxidation product 4-hydroxynonenal – which seem to exist in all eucaryotic cells – underlines the data on the changes of the balance between prooxidant and antioxidant mechanisms during aging.

Acknowledgments. The authors' work has been supported by the Deutsche Forschungsgemeinschaft, F.R.G. (Ge 669/1-1-284/91), by the University Agreement between the University of Graz/Austria and the Humboldt University of Berlin/F.R.G., and by the Association for International Cancer Research (AICR), U.K.

Benndorf, R., Nürnberg, P., and Bielka, H. (1988) Growth phase-dependent proteins of the Ehrlich ascites tumor analyzed by one- and two-dimensional electrophoresis. Exp. Cell Res. 174: 130–138.

Bize, J. B., Overley, L. W., and Morris, H. P. (1980) Superoxide dismutase and superoxide radical in Morris hepatoma. Cancer Res. 40: 3686–3693.

Borrello, S., Minotti, G., Palombini, G., Grattagliano, A., and Galeotti, T. (1985) Superoxide-dependent lipid peroxidation and vitamin E content of microsomes from hepatomas with different growth rates. Archs Biochem. Biophys. 238: 588–595.

Burdon, R. H., and Rice-Evans, C. (1989) Free radicals and the regulation of mammalian cell proliferation. Free Radical Res. Commun. 6: 345–358.

Cerutti, P. A. (1985) Prooxidant states and tumor promotion. Science 227: 375–381.

Cheeseman, K. H., Burton, G. W., Ingold, K. U., and Slater, T. F. (1984) Lipid peroxidation and lipid antioxidants in normal and tumor cells. Toxicol. Pathol. 12: 552–557.

Cheeseman, K. H., Collins, M., Proudfoot, K., Slater, T. F., Burton, G. W., Webb, A. C., and Ingold, K. U. (1986) Studies on lipid peroxidation in normal and tumor tissues. Biochem. J. 235: 507–514.

Devasagayam, T. P. A. (1986) Low level of lipid peroxidation in new born rats. FEBS Lett. 199: 203–207.

Emerit, I. (1987a) Icosanoids, chromosome damage and cancer, in: Prostaglandins in Cancer Research, pp. 29–34. Eds E. Garaci, R. Paoletti and M. G. Santoro. Springer-Verlag, Berlin, Heidelberg.

Emerit, I. (1987b) Clastogenic factors, a link between chronic inflammation and carcinogenesis, in: Anticarcinogenesis and Radiation Protection, pp. 59–62. Eds P. A. Cerutti, O. F. Nygaard and M. G. Simic. Plenum Press, New York.

Emerit, I., Khan, S. H., and Esterbauer, H. (1991) Hydroxynonenal, a component of clastogenic factors? Free Radical Biol. Med. 10: 371–377.

Esterbauer, H. (1982) Aldehydic products of lipid peroxidation, in: Free Radicals, Lipid Peroxidation and Cancer, pp. 101–128. Eds D. C. H. McBrien and T. F. Slater. Academic Press, New York.

Esterbauer, H., Zollner, H., and Lang, J. (1985) Metabolism of the lipid peroxidation product 4-hydroxynonenal by isolated hepatocytes and by liver cytosolic fractions. Biochem. J. 228: 363–373.

Esterbauer, H., Zollner, H., and Schaur, R. J. (1988) Hydroxyalkenals: cytotoxic products of lipid peroxidation. ISI Atlas Sci. Biochem. 1: 311–317.

Esterbauer, H., Zollner, H., and Schaur, R. J. (1989) Aldehydes formed by lipid peroxidation: mechanisms of formation, occurrence, and determination, in: Membrane Lipid Oxidation, vol. 1, pp. 239–283. Ed. C. Vigo-Pelfrey. CRC Press, Boca Raton, FL.

Fauler, G. (1987) Investigations on the metabolism of 4-hydroxy-alkenals. Austria: Univ. Graz; Ph.D. Thesis.

Ferro, M., Marinari, U. M., Poli, G., Dianzani, M. U., Fauler, G., Zollner, H., and Esterbauer, H. (1988) Metabolism of 4-hydroxynonenal by the rat hepatoma cell line MH1C1. Cell Biochem. Funct. 6: 245–250.

Galeotti, T., Borrello, S., Minotti, G., Palombini, G., Masotti, L., Sartor, G., Cavatorta, P., Arcioni, A., and Zannoni, C. (1984) Lipid composition, physical state, and lipid peroxidation of tumor membranes. Toxicol. Pathol. 12: 324–330.

Grune, T., Siems, W., Kowalewski, J., Zollner, H., and Esterbauer, H. (1991) Identification of metabolic pathways of the lipid peroxidation product 4-hydroxynonenal by enterocytes of rat small intestine. Biochem. Int. 25: 963–971.

Hauptlorenz, S., Esterbauer, H., Moll, W., Pümpel, R., Schauenstein, E., and Puschendorf, B. (1985) Effects of the lipid peroxidation product 4-hydroxynonenal and related aldehydes on proliferation and viability of cultured Ehrlich ascites tumor cells. Biochem. Pharmacol. 34: 3803–3809.

Hostetler, K. Y., Zenner, B. D., and Morris, H. P. (1979) Phospholipid content of mitochondrial and microsomal membranes from Morris hepatomas of varying growth rates. Cancer Res. 39: 2978–2983.

Ishikawa, T., Esterbauer, H., and Sies, H. (1986) Role of cardiac glutathione transferase and of the glutathione S-conjugate export system in biotransformation of 4-hydroxynonenal in the heart. J. Biol. Chem. 261: 1576–1586.

Kagan, V. E., Bakalova, R. A., and Karakashev, P. H. (1991) Lipid peroxidation in the tumor cells and tissues of tumor-bearing animals, in: Membrane Lipid Oxidation, vol. 3, pp. 191–208. Ed. C. Vigo-Pelfrey. CRC Press, Boca Raton, FL.

Kowalewski, J., Siems, W., Grune, T., Werner, A., Esterbauer, H., and Gerber, G. (1991) Nucleotide degradation and oxygen radical formation during ischemia and reperfusion of the rat small intestine, Z. Klin. Med. 46: 143–146.

Masotti, L., Gasali, E., and Galeotti, T. (1988) Lipid peroxidation in tumor cells. Free Rad. Biol. Med. 4: 377–386.

Müller, M., Siems, W., Buttgereit, F., Dumdey, R., and Rapoport, S. M. (1986) Quantification of ATP-producing and consuming processes of Ehrlich ascites tumour cells. Eur. J. Biochem 161: 701–705.

Peskin, A. V., Koen, Y. M., and Zbarsky, J. B. (1977) Superoxide dismutase and glutathione peroxidase activities in tumors. FEBS Lett. 78: 41–45.

Poli, G., Cecchini, G., Biasi, F., Chiarpotto, E., Canuto, R. A., Biocca, M. E., Muzio, G., Esterbauer, H., and Dianzani, M. U. (1986) Resistance to oxidative stress by hyperplastic and neoplastic rat liver tissue monitored in terms of production of unpolar and medium polar carbonyls. Biochim. Biophys. Acta 883: 207–214.

Schaur, R. J., Zollner, H., and Esterbauer, H. (1991) Biological effects of aldehydes with particular attention to 4-hydroxynonenal and malonaldehyde, in: Membrane Lipid Oxidation, vol. 3, pp. 141–163. Ed. C. Vigo-Pelfrey. CRC Press, Boca Raton, FL.

Schmidt, H., Siems, W., Müller, M., Dumdey, R., Jakstadt, M., and Rapoport, S. (1989) Balancing of mitochondrial and glycolytic ATP production and of the ATP-consuming processes of Ehrlich ascites tumour cells in a high phosphate medium. Biochem. Int. 19: 985–992.

Schmidt, H., Siems, W., Müller, M., Dumdey, R., and Rapoport, S. (1991) ATP-producing and consuming processes of Ehrlich mouse activities tumor cells in proliferating and resting phases. Exp. Cell Res. 194: 122–127.

Siems, W. G., Zollner, H., and Esterbauer, H. (1990) Metabolic pathways of the lipid peroxidation product 4-hydroxynonenal in hepatocytes – quantitative assessment of an antioxidant defense system. Free Radic. Biol. Med. 9: 110.

Siems, W., Schmidt, H., Werner, A., Uerlings, I., David, H., and Gerber, G. (1989) Changes in the nucleotide metabolism of Ehrlich ascites tumour cells during their growth in vivo. Cell. Molec. Biol. 35: 255–262.

Siems, W., Kowalewski, J., David, H., Grune, T., and Bimmler, M. (1991) Discrepancy between biochemical normalisation and morphological recovery of jejunal mucosa during postischemic reperfusion in presence of the xanthine oxidase inhibitor oxypurinol. Cell. Molec. Biol. 37: 213–226.

Tessitore, L., Bonelli, G., Cecchini, G., Amenta, J. S., and Baccino, F. M. (1987) Regulation of protein turnover versus growth state: Ascites hepatoma as a model for studies both in the animal and in vitro. Archs Biochem. Biophys. 255: 372–384.

Van Kuijk, F. J. G. M., Thomas, D. W., Stephens, R. J., and Dratz, E. A. (1986) Occurrence of 4-hydroxyalkenals in rat tissues determined as pentafluorobenzyl oxime derivatives by gas chromatography-mass spectrometry. Biochem. Biophys. Res. Comm. 139: 144–149.

Free Radicals and Aging
ed. by I. Emerit & B. Chance

Effect of aging on glutathione metabolism. Protection by antioxidants

J. Viña, J. Sastre, V. Anton, L. Bruseghini*, A. Esteras*
and M. Asensi

*Departamento de Fisiología, Facultad de Medicina, Universidad de Valencia, and
Zambon Laboratories, Santa Perpetua (Barcelona) Spain

Summary. The free radical theory of aging suggests that oxygen free radicals may be involved in the aging process. Thus, changes in antioxidant mechanisms may occur with aging. Since glutathione is one of the most effective antioxidant systems in the cell, its metabolism may change with aging. In this chapter we describe experiments which show the involvement of glutathione in the aging process and which provide a rationale for the administration of antioxidants to old organisms to protect them against some of the changes that occur with aging.

Introduction

One of the most prominent theories used to explain aging-associated phenomena is the free radical theory of aging, first proposed by Harman (for review see Harman, 1981). This theory suggests that free radicals play a prominent role in the pathogenesis of aging. Glutathione protects against the effects of free radicals (for review see Viña, 1990). Thus, it was likely that glutathione metabolism would be involved in aging. Since the pioneering work of Pinto and Bartley (1969) in the sixties, many authors have paid attention to the relationship between glutathione metabolism and aging. However, some relevant questions remain open. For instance, is age-associated glutathione oxidation caused by a change in enzymes involved in the glutathione redox cycle? Is cysteine and glutathione synthesis from methionine adequate in tissues from old animals? Can changes in glutathione status that occur with aging be prevented by administration of antioxidants? Is the administration of antioxidants able to increase the life span and the vitality of animals? Recent work from several laboratories, including our own, will be reviewed in this chapter.

Glutathione redox status in the aged cell

If free radicals are associated with the phenomenon of aging, one could expect that glutathione would be oxidized in the tissues of old

Table 1. Effect of age on GSH, GSSG and GSSG/GSH ratio in various organs of mice

	Brain	Testes	Liver
GSH			
Young	$1.4 \pm 0.6(12)$	$3.2 \pm 1.0(6)$	$5.5 \pm 0.9(9)$
Old	$1.1 \pm 0.2(13)**$	$2.4 \pm 0.2(12)*$	$4.6 \pm 0.7(13)*$
GSSG			
Young	$32 \pm 13(9)$	$130 \pm 43(6)$	$209 \pm 47(12)$
Old	$57 \pm 18(13)**$ $\pm 109(13)*$	$131 \pm 49(12)$	283
GSSG/GSH ($\times 10^{-3}$)			
Young	22.8	40	38
Old	51.8	54	61

Values are means $\pm$S.D. for the number of observations in parentheses. GSH concentrations are μmol/g of fresh tissue. GSSG concentrations are nmol/g of fresh tissue. $*p < 0.05$; $**p < 0.005$

animals. This is indeed the case and we show in Table 1 that the GSH/GSSG ratio falls with age in several organs of mice of the C57BJ strain. This finding is in agreement with those of previous authors who studied the effect of aging on glutathione metabolism in tissues of both animals (Hazelton and Lang, 1980; Lauterburg, 1980; Allen and Sohal, 1986) and humans (Goldschmidt, 1970).

However, the age-associated oxidation of glutathione has to be explained; is it due to an increase in oxidative challenge, to a decrease in reductive capacity or to both? The following section attempts to provide an answer to these questions.

The effect of aging on enzymes related to the glutatione redox cycle

Changes in glutathione redox status reflect the equilibrium between the generation of oxidative challenge and the production of reductive equivalents to counteract such a challenge. Helmut Sies coined the term oxidative stress to denote an imbalance between the former and the latter in favor of oxidation (Sies, 1986). In Figure 1 we show the main enzymes involved in the oxidation and in the reduction of glutathione. The effect of aging on the activity of glutathione-related enzymes has been studied in several laboratories, using various cell types (Pinto and Bartley, 1969; Santa María and Machado, 1987; Al-Turk et al., 1987; Yen et al., 1989; Benzi et al., 1989). We have measured the activities of glutathione reductase, glutathione peroxidase, glutathione transferase and glucose 6 phosphate dehydrogenase in various organs of the mouse such as liver, brain and testes. Figure 2 shows the effect of age on glutathione-related enzymes in mouse brain, testes

138

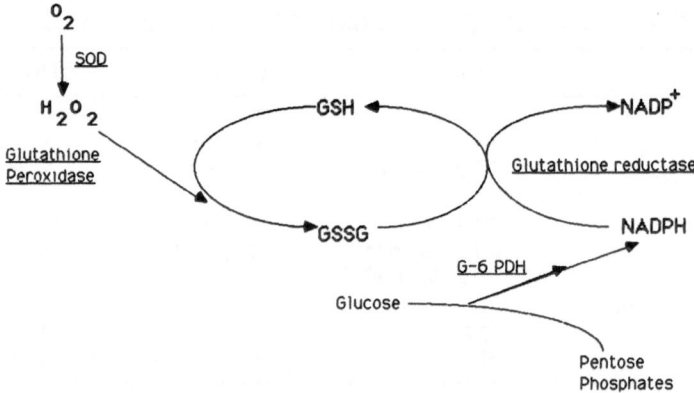

Figure 1. Glutathione red-ox cycle.

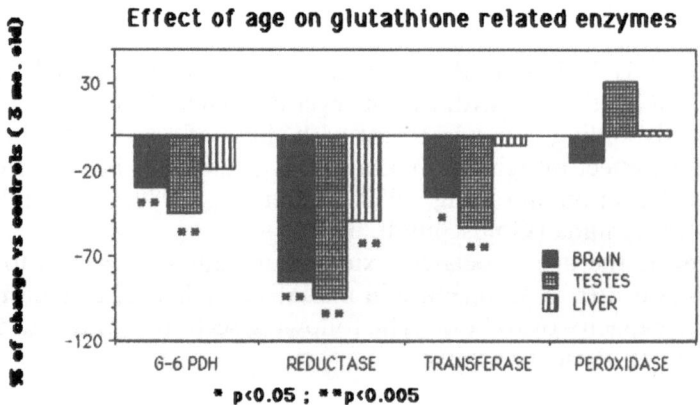

Figure 2. Effect of age on glutathione related enzymes.

and kidney. We found that aging is associated with a decrease in the activity of enzymes that catalyze reactions that tend to reduce GSSG, i.e. G-6P DH or glutathione reductase, rather than with an increase in the activity of those that tend towards oxidation of glutathione, i.e. glutathione peroxidase or transferase.

Effect of antioxidants on changes in glutathione redox status associated with aging

Various authors have emphasized the importance of maintaining an effective antioxidant system during aging (Frei et al., 1990; Cutler, 1991; Barja de Quiroga et al., 1991). Others, suggested that administration of antioxidants might help to prevent changes in cell function (Miquel and Economos, 1979) or in blood parameters (Tolonen et al., 1988) that occur with aging. The fact that aging is associated with an oxidation of glutathione in various organs of the mouse (see Table 1) prompted us to see if administration of antioxidants could prevent, at least in part, this shift towards oxidation that occurs in various organs of the mouse.

In Figure 3 we show the effect of oral administration of thiol-containing antioxidants on the gutathione redox status of mouse brain. Antioxidants were administered mixed with food for six months (from the age of 12 mo. until the age of 18 mo., when the animals were killed). We found that administration of glutathione significantly increased GSH levels in mouse brain and partially prevented the increase in GSSG/GSH ratio that occurs with aging.

We also studied the effect of oral administration of antioxidants on the glutathione redox status of testes and found (Fig. 4) that both oral

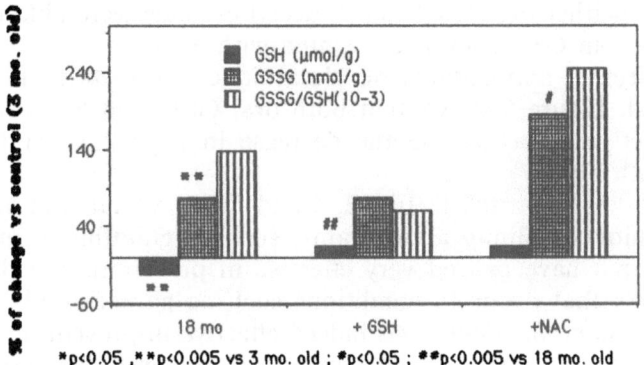

Figure 3. Effect of antioxidants on glutathione status of brain.

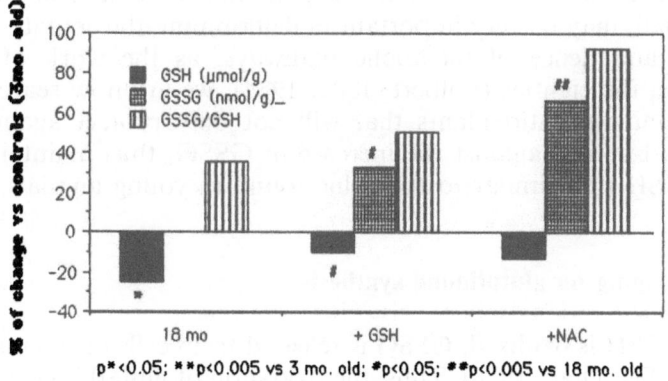

Figure 4. Effect of antioxidants on GSH status in testes.

140

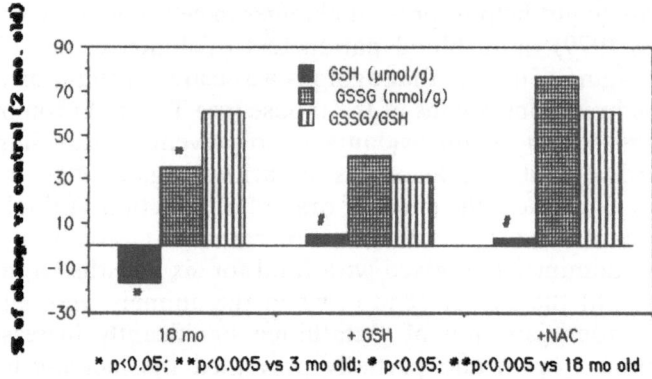

Figure 5. Effect of antioxidants on GSH status in liver.

GSH or N-acetyl cysteine increase the GSH levels in these organs. However, neither oral GSH nor N-acetyl cysteine were able to prevent the increase in GSSG levels associated with aging.

The effect of antioxidants on glutathione redox status in liver was also tested. Figure 5 shows that both oral GSH and N-acetyl cysteine were effective in preventing the decrease in hepatic GSH levels that occurs with aging.

Our studies (reported in Fig. 3, 4 and 5) show that oral administration of thiol-containing antioxidants, such as glutathione or N-acetyl cysteine, that have proved very effective in preventing the decrease in GSH levels that occur in conditions such as paracetamol toxicity or fasting (Viña et al., 1989), was indeed effective in preventing the fall in GSH that occurs with aging. However, neither oral GSH nor N-acetyl cysteine were effective in preventing the age-associated increase in GSSG. Thus, these antioxidants were not effective in preventing the oxidation of the GSSG/GSH redox pair. Since changes in this ratio, albeit small, may be very important in determining the activity of many enzymes and hence of metabolic pathways, as the work of Gilbert showed in the eighties (Gilbert, 1982, 1984), we are now searching for thiol-containing antioxidants that will not only protect against GSH depletion but also against the increase in GSSG, thus maintaining the GSSG/GSH pair similar to the value found in young animals.

Effect of aging on glutathione synthesis

When GSH is oxidized, GSSG is released from cells of organs such as the liver (Sies et al., 1978). Thus, the oxidation of glutathione associated with aging may increase the rate of glutathione released from cells. This, together with the fact that Ritchie and Lang (1988) showed that a decrease in cysteine level causes a glutathione deficiency in aging mosquitos, prompted us to study the effect of aging on glutathione synthesis in various organs of mice. We did not find any significant changes in the rate of GSH synthesis from precursors in cells such as hepatocytes or erythrocytes. However we did find a qualitative difference in GSH synthesis in the eye lenses of old animals when compared with young ones. We found (Ferrer et al., 1990) that methionine, which is an excellent precursor of glutathione in several cell types such as hepatocytes (Viña et al., 1978), is indeed a good glutathione precursor in lenses of young animals but not in those of old animals. However, N-acetyl cysteine is an excellent precursor of glutathione in lenses of both young and old animals. Since glutathione synthesis from methionine requires the activity of the cysthathionase pathway (Reed and Orrenius, 1977) we measured the activity of cysthathionase in lenses of both young and old animals and found that it is absent in lenses of old animals. Thus it is the low cysteine synthesis from methionine that is responsible for the impaired glutathione synthesis from methionine, when compared with N-acetyl cysteine. Since methionine is a major glutathione precursor in vivo, this low ability to synthesize glutathione may aggravate the oxidative stress that occurs in the old lens, and that is, at least in part responsible for the occurrence of cataracts (Ferrer et al., 1991).

Effect of oral administration of antioxidants on life span and vitality of animals

Several authors have pointed to the fact that since the maximum life span of animals of a given species cannot be increased, our efforts should concentrate on improving the vitality of old animals and in increasing the proportion of animals that reach old age. Since our studies, and those of other authors, have shown that there is an oxidative stress associated with aging (see previous sections of this chapter), we tested whether administration of thiol-containing antioxidants was effective in increasing both the life span and the vitality of animals. Table 2 shows the effect of antioxidants on the motor coordination ability of mice. We measured this parameter using a method described by Miquel and Economos (1979). Each animal passed the test on one occassion each month from the age of 12 to 18 months. Table 2 shows that at the age of 15 months the animals actually improved their performance (probably because they learnt how to perform the test). However, from this age onwards, they lost their capacity to pass the test. This was greatly diminished when they reached the age of 17 and 18 months. However, oral administration of thiol-containing antioxidants, such as GSH or N-acetyl cysteine improved the performance of

142

Table 2. Effect of oral antioxidant administration of motor coordination of mice

Months	Controls	+GSH	+N-acetyl cysteine
12	100	100	100
15	108	50	133
16	82	50	100
17	67	67	66
18	50	67	100

Mice were submitted to a motor coordination test (see the text) at the indicated ages starting at month 12, which was considered 100% in all cases. Values are means of the mice that passed the test at each give age.

mice, and indeed in the case of N-acetyl cysteine the rats did not lose their ability to pass the motor coordination test.

We also studied the effect of oral administration of thiol-containing antioxidants in increasing the life span of *Drosophila*. Figure 6 shows that both N-acetyl cysteine and GSH induced a shift of the survival curve to the right. This means that the administration of antioxidants did indeed improve the survival rate of mice; for instance, 20% survival occurred at day 49 in controls but at day 56 in *Drosphila* treated with GSH and at day 61 in those treated with N-acetyl cysteine. These results are in keeping with those of Miquel and Economos (1979) who showed that administration of thiazolidine carboxylate increases the life span and the vitality of *Drosophila*.

Concluding remarks

Present evidence suggests that free radicals play a major role in the pathogenesis of aging (Harman, 1991). Since oxygen free radicals cause

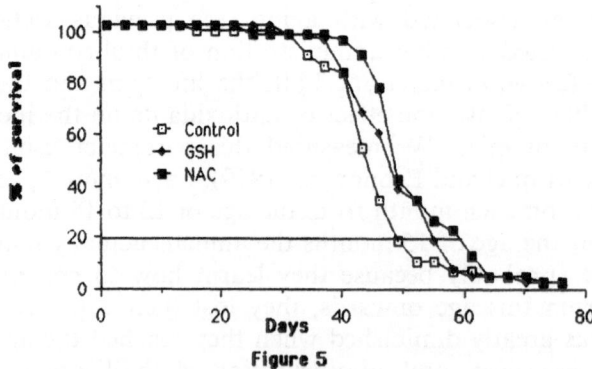

Figure 6. Effect of antioxidants on survival curves of *Drosophila*.

oxidative stress, it is reasonable to suggest that antioxidants may have favorable effects on life span, vitality and metabolic performance of cells of old animals. Furthermore, it must be pointed out that the synthesis of metabolites that act as antioxidants can be impaired in old cells, as we have shown in the crystalline lens (Ferrer et al., 1991). Our results, as well as those of other authors show that indeed oral administration of antioxidants may be beneficial to those animals.

Al-Turk, W., Stohs, S. J., El-Rashidy, F. H., and Othman, S. (1987) Changes in glutathione and its metabolizing enzymes in human erithrocytes and lymphocytes with age. J. Pharm. Pharmacol. 39: 13–16.

Allen, R. G., and Sohal, R. S. (1986) Role of glutathione in the aging and development of insects, in: Insect Aging. Collatz, K. G. and Sohal, R. S. Eds. Springer Verlag, Berlin/ Heidelberg. pp. 168–181.

Barja de Quiroga, G., Perez-Campo, R., and Lopez Torres, M. (1990) Anti-oxidant defences and peroxidation in liver and brain of aged rats. Biochem. J. 272: 247–250.

Benzi, G., Marzatico, F., Pastoris, O., and Villa, R. F. (1989) Relationship between aging, drug treatment and the cerebral enzymatic antioxidant system. Exp. Geront. 24: 137–148.

Cutler, R. G. (1991) Antioxidants and aging. Amer. J. Clin. Nutr. 53: S373–S379.

Ferrer, J. V., Gascó, E., Sastre, J., Pallardó, F. V., Asensi, M., and Viña, J. (1990) Age-related changes in glutathione synthesis in the eye lens. Biochem. J. 269: 531–534.

Ferrer, J. V., Sastre, J., Pallardó, F. V., Asensi, M., Antón, V., Estrela, J. M., Viña, J., and Miquel, J. (1991) Senile cataract: a review on free radical related pathogenesis and antioxidant prevention. Archs Geront. Geriat. 13: 51–59.

Frei, B., Stocker, R., and Ames, B. (1990) Antioxidant defenses and lipid peroxidation in human blood plasma. Proc. Natl. Acad. Sci USA 85: 9748–9752.

Gilbert, H. (1984) Redox control of enzyme activities by thiol/disulfide exchange. Meth. Enzymol. 107: 330–335.

Gilbert, H. F. (1982) Biological disulfides: The third messenger? J. Biol. Chem. 257: 12086–12091.

Goldschmidt, L. (1970) Seasonal variations in red cell glutathione levels with aging in mental patients and normal controls. Proc. Soc. Exp. Biol. Med. 133: 555–559.

Harman, D. (1981) The free radical theory of aging. Proc. Natl. Acad. Sci. USA 78: 7124–7128.

Harman, D. (1991) The aging process – major risk factor for disease and death. Proc. Natl. Acad. Sci. USA 88: 5360–5363.

Hazelton, G. A., and Lang, C. A. (1980) Glutathione contents of tissues in the aging mouse. Biochem. J. 188: 25–30.

Lauterburg, B. H., Vaishnav, Y., Stillwell, W. G., and Mitchell, J. (1980) The effects of age and glutathione depletion on hepatic glutathione turnover in vivo determined by acetaminophen probe analysis. J. Pharmacol. Exp. Ther. 213: 54–58.

Miquel, J., and Economos, A. C. (1979) Favorable effects of the antioxidants sodium and magnesium thiaxolidine carboxylate on the vitality and life span of Drosophila and mice. Exp. Geront. 14: 279–285.

Pinto, R. E., and Bartley, W. (1969) A negative correlation between oxygen uptake and glutathione oxidation in rat liver homogenates. Biochem. J. 114: 5–9.

Reed, D. J., and Orrenius, S. (1977) The role of methionine in glutathione biosynthesis by isolated hepatocytes. Biochem. Biophys. Res. Commun. 77: 1257–1264.

Richie, J. P., and Lang, C. A. (1988) A decrease in cysteine levels causes the glutathione deficiency of the aging mosquito. Proc. Soc. Exp. Biol. Med. 187: 235–240.

Santa María, C., and Machado, A., (1987) Effects of development and ageing on pulmonary NADP-cytochrome c reductase, glutathione peroxidase, glutathione reductase and thioredoxin reductase activities in male and female rats. Mech. Ageing Dev. 37: 183–195.

Sies, H. (1986) Biochemistry of oxidative stress. Angew. Chem. 25: 1058–1071.

Sies, H., Bartoli, G. M., Burk, R. F., and Waydhas, Ch. (1978) Glutathione efflux from perfused rat liver after phenobarbital treatment, during drug oxidations, and in selenium deficiency. Eur. J. Biochem. 89: 113–118.

144

Tolonen, M., Sarna, S., Halme, M., Tuominen, S., Westermarck, T., Nordberg, U., Keinonen, M., and Schrijver, J. (1988) Antioxidant supplementation decreases TBA reactants in serum of elderly. Biol. Tract Elements 17: 221–228.

Viña, J., (Ed.) (1990). Glutathione: Metabolism and Physiological Functions. CRC Press, Boston.

Viña, J., Hems, R., and Kreds, H. A. (1978) Maintenance of glutathione content in isolated hepatocytes. Biochem. J. 170: 627–630.

Viña, J., Perez, C., Furukawa, T., Palacin, M., and Viña, J. R. (1989) Effect of oral glutathione on hepatic glutathione levels. Br. J. Nutr. 62: 663–691.

Yen, T. C., Chen, Y. S., King, K. L., Yeh, S. H., and Wei, Y. H. (1989) Liver mitochondrial respiratory functions decline with age. Biochem. Biophys. Res. Commun. 165: 994–1003.

Free Radicals and Aging
ed. by I. Emerit & B. Chance
© 1992 Birkhäuser Verlag Basel/Switzerland

Inhibition of LDL oxidation by antioxidants

Hermann Esterbauer, Georg Waeg, Herbert Puhl,
M. Dieber-Rotheneder and Franz Tatzber

Institute of Biochemistry, University of Graz, A-8010 Graz, Austria

Summary. Low density lipoprotein (LDL) consists of about 3000 fatty acids (50% polyunsaturated) and a single molecule apolipoprotein B (500 kDa). The endogenous antioxidants of LDL consist mainly of tocopherols and few carotenoids, which protect the PUFAS against oxidation. That native LDL contains traces of oxidation products has not been proved yet.

Oxidatively modified LDL (oLDL) exhibits cytotoxic and chemotactic activities, furthermore it leads to foam cell formation, a critical step in atherogenesis. The oxidation of LDL is a free radical process and leads to various aldehydic products. The oxidation of LDL is initiated by cells as well as by transition metals like Cu^{2+}. In both cases the oxidation goes through three consecutive phases. The lag-phase is characterized by minimal degradation of PUFAs but a loss of the antioxidants. Thereafter the PUFAs are oxidized to lipid hydroperoxides, which are only intermediates (propagation-phase). These intermediates will decompose to aldehydic products, accompanied by several additional changes in the LDL particle (decomposition-phase). For increased macrophage uptake oLDL must reach the late decomposition-phase; the presence of lipid hydroperoxides in LDL is not sufficient. It is suggested that binding of aldehydes to free amino groups of Apo B is the reason for macrophage uptake. This is supported by the finding that antibodies against aldehyde-modified LDL are able to recognize oxidized LDL in atherosclerotic lesions. Antioxidants like α-tocopherol are able to protect LDL against oxidation. The duration of the lag-phase shows a linear relationship with the content of α-tocopherol in LDL. Yet the efficiency of α-tocopherol to protect LDL shows strong individual variation.

Introduction

Oxidatively modified low density lipoprotein (oLDL) exhibits in vitro a number of properties which could be of potential importance in atherogenesis. Most important are its uncontrolled uptake by macrophages, its cytotoxicity and its chemotactic activity (for review see Steinberg et al., 1989; Steinbrecher et al., 1990; Esterbauer et al., 1990a; Esterbauer et al., 1991c). Oxidation of LDL by cells or in cell-free systems is principally a free radical process (Esterbauer et al., 1990a; Juergens et al., 1987; Steinbrecher, 1987; Bedwell et al., 1989; Jessup et al., 1990), in which the polyunsaturated fatty acids (PUFAS) contained in LDL are converted by lipid peroxidation to lipid hydroperoxides, which are then further degraded to a variety of products including reactive aldehydes such as malonaldehyde (MDA), 4-hydroxynonenal (HNE), hexanal and others. Various lines of research suggest that some of these lipid peroxidation products generated within the LDL particle covalently bind to amino acid residues of apolipoprotein B (apo B) as

146

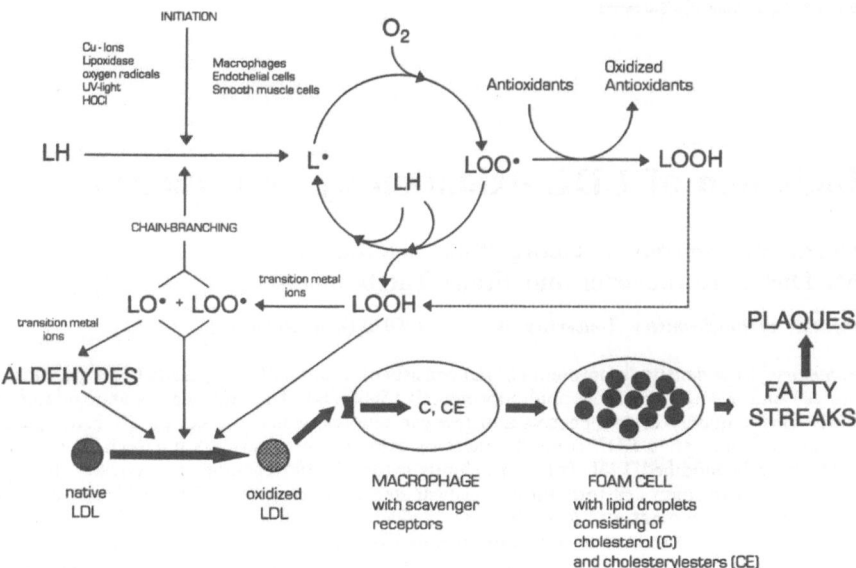

Figure 1. Scheme showing the major events of LDL oxidation and formation of foam cells. LH = an LDL lipid containing PUFAS; LOO' = lipid peroxyl radical; LO' = lipid alkoxyl radical.

for example ϵ-amino groups of lysine residues, and thereby produce modifications in apo B which have a high affinity to the macrophage scavenger receptors (Juergens et al., 1987; Steinbrecher, 1987; Hoff et al., 1988, 1989) and lead to foam cell formation (Fig. 1). The cytotoxicity of oLDL may also be mediated, at least to some extent, by aldehydic lipid peroxidation products (Juergens et al., 1987; Hoff et al., 1989). HNE, 2-alkenals and 2,4-alkadienals are cytotoxic at micromolar concentrations towards endothelial cells (Kaneko et al., 1988). Aldehydes emanating from oLDL present in the intima could therefore lead to damage of the endothelial layer of the vascular system. The chemotactic activity of oLDL towards monocytes has been ascribed to lysophosphatidyl choline formed from peroxidized phosphatidyl choline by the action of phospholipase A_2 (Quinn et al., 1987). An additional chemotactic factor might be HNE, which shows chemotactic properties towards neutrophils at nanomolar concentrations (Curzio et al., 1986). These, together with the findings (Haberland et al., 1988; Rosenfeld et al., 1990; Palinski et al., 1989) that oxidized LDL and MDA and HNE modified LDL are indeed present in atherosclerotic lesions point out that lipid peroxidation and aldehydic lipid peroxidation products play an important role in the initial processes of atherogenesis. Furthermore, if this is so, antioxidants should play a crucial role in preventing oxidation of LDL and its conversion into an atherogenic form. In this article we give a brief review of published work on the mechanism of LDL oxidation and the inhibition by antioxidants, particularly vitamin E.

Composition of LDL

LDL is a large spherical particle with a mean diameter of 22 nm and a molecular weight of about 2.5 million Da. According to our analysis (Esterbauer et al., 1990a; Esterbauer et al., 1991c), it consists on average of 1600 molecules of cholesteryl esters, 600 molecules of free cholesterol, 700 molecules of phospholipids and 170 molecules of triglycerides. The phospholipids and free cholesterol are arranged in monolayers surrounding the lipophilic core of cholesteryl esters and triglycerides. The mean number of fatty acids contained in the various lipids of an LDL particle is about 3000, and half of them are PUFAS, mainly linoleic acid with minor amounts of arachidonic and docosahexaenoic acid (Table 1). Linoleic acid is mainly bound in the cholesteryl esters whereas arachidonic and docosahexaenoic acid are mainly present in the phospholipids.

Embedded in the outer monolayer is apo B, a large protein with a molecular weight of about 500,000 Da. It contains a total of 356 lysine residues, and those of them with positively charged ϵ-amino groups are essential for the recognition of the apo B by the LDL receptor. Modification of these lysine amino groups (e.g. acetylation or conjugation with MDA or other aldehydes) attenuates binding to the B/E receptor and strongly increases binding and uptake by the scavenger receptor of macrophages.

Table 1. Polyunsaturated fatty acids and antioxidants in LDL. Values are in mol/mol LDL. Given is also the oxidation resistance of LDL as measured by the lag-phase in Cu^{++} stimulated oxidation.

	n	Mean	$\pm$SD	Range
linoleic acid	31	1101	$\pm$298	680–1832
arachidonic acid	31	153	$\pm$55	48–250
docosahexaenoic acid	15	29	$\pm$17	15–62
alpha-tocopherol	87	6.37	$\pm$1.84	2.9–14.9
gamma-tocopherol	88	0.51	$\pm$0.20	0.18–1.26
beta-carotene	122	0.29	$\pm$0.26	0.03–1.87
alpha-carotene	28	0.12	$\pm$0.14	0.02–0.52
lycopene	136	0.16	$\pm$0.11	0.03–0.70
cryptoxanthin	114	0.14	$\pm$0.13	0.03–0.70
cantaxanthin	53	0.02	$\pm$0.04	0.01–0.24
lutein + zeaxanthin	113	0.04	$\pm$0.03	0.01–0.16
phytofluene	10	0.05	$\pm$0.03	0.02–0.11
ubiquinol-10	7	0.10	$\pm$0.10	0.03–0.35
lag-phase (min)	72	68	$\pm$15	34–114

The PUFAS in LDL are protected against free radical attack and lipid peroxidation by a variety of endogenous antioxidants (Table 1). On a molar basis the most relevant is α-tocopherol; on average 6 such molecules are present in a LDL particle. Gamma-tocopherol and carotenoids are present in much lower quantities. The PUFA content and also the antioxidant content of LDL varies strongly from donor to donor, which is probably important for the differences in the oxidation resistance of LDL in vitro as well as in vivo (Esterbauer et al., 1990a,b; 1991a,c).

That native LDL contains traces of lipid peroxides or lipid peroxidation products has yet to be proved. The iodometric and the TBA assay give a weak positive result, but the limitation of these assays are known. With more specific methods, e.g. HPLC with post column derivation and chemiluminescence detection, no lipid hydroperoxides could be detected in freshly prepared LDL (Stocker et al., 1991). On the other hand the work of Avogaro's group (Avogaro et al., 1988; Cazzolato et al., 1991) suggests that LDL from healthy donors contains a small subfraction which has properties similar to a minimally oxidized LDL. We also find that some but not all of the LDL samples contain traces of HNE or some other aldehydes which might be indicative of minimal oxidation.

Mechanism of LDL oxidation, formation of aldehydes

In the experiments (Steinbrecher et al., 1984), which led to the discovery that cells can oxidize LDL, LDL was incubated with cultured rabbit aortic endothelial cells in F10 medium for 24 h; after that time the medium contained TBARS and the LDL was more rapidly degraded by macrophages than native LDL. Principally the same procedure with a 24- or 48-h end point was used later in many experiments with other cell lines and also in oxidation experiments in cell-free systems. We have done intensive studies on the time course of the change of the chemical composition of LDL when it is exposed to copper ions as a pro-oxidant (Esterbauer et al., 1990a). The whole process can be divided roughly into three consecutive phases, i.e. a lag-phase, a propagation-phase and a decomposition-phase.

Immediately after addition of 1.66 μM CuCl$_2$ to an LDL solution (0.25 mg/ml) in oxygen-saturated phosphate-buffered saline, the endogenous antioxidants of LDL start to disappear, α-tocopherol being the first to go and β-carotene the last. During this period only minimal lipid peroxidation occurs as evidenced by analysis of fatty acids, TBARS, lipid hydroperoxides or conjugated dienes. This suggests that during the lag-phase the PUFAS are protected against lipid peroxidation by the endogenous antioxidants. The first defence line appears to be vitamin E, the second line consists of carotenoids and probably other not yet identified components. The length of the lag-phase is on average 68 min, but strong individual variations occur in the range of 34–114 min (Table 1). If the antioxidants are exhausted, the LDL is unprotected and the PUFAS are oxidized to lipid hydroperoxides in the propagation phase within 1 to 2 h. Shortly after the onset of lipid peroxidation the concentration of aldehydic lipid peroxidation products starts to rise, and in parallel many alterations occur in LDL such as an increase in fluorescence at 430 nm (excitation 360 nm), increase in relative electrophoretic mobility, fragmentation of apo B, increase in dienes and lysophosphatides. The lipid hydroperoxides are only transient intermediates, their concentration increases as long as the rate of their formation exceeds the decomposition rate. When most of the PUFAS (about 70 to 80%) are oxidized the decomposition becomes prevalent and the lipid hydroperoxides concentration starts to decrease again. We call the period following the lipid hydroperoxides' peak (which coincides with the peak of conjugated dienes) the decomposition-phase, although one shluld keep in mind that it temporarily overlaps with the propagation phase. The decomposition is a long-lasting process, which after about 15 h approaches a stage, where no further major changes occur in LDL. The nature of the chemical reactions during the decomposition are complex. The aldehydes continue to increase to a plateau, e.g. HNE from 13.8 to 63 mol/mol LDL, which indicates that decomposition of hydroperoxides by β-cleavage reactions keep going on. After passing a minimum, the 234 nm diene absorption increases again. But this second increase is not due to newly formed dienes but to other compounds with absorption around 230 nm e.g. α,β-unsaturated carbonyls. The 430 nm fluorescence of oLDL increases, mainly due to new chromophores formed on apo B. That the structure of apo B is strongly altered during the decomposition-phase is also evident from the relative electrophoretic mobility of LDL. The relative electrophoretic mobility increases during the propagation-phase from 1.00 to about 1.5 and during the decomposition-phase to about 3.5. It should be noted that the uptake of oLDL by macrophages correlates with its relative electrophoretic mobility (Esterbauer et al., 1990a). It has been shown experimentally (Bedwell et al., 1989) that oxidation of PUFAS to the corresponding lipid hydroperoxides is, per se, not sufficient for accelerated macrophage uptake. The decomposition of the lipid hydroperoxides and probably the formation of aldehydes are necessary requirements for the formation of an LDL recognized by the scavenger receptor.

The three consecutive phases described above for Cu^{++}-stimulated LDL oxidation are also characteristic for the oxidation of LDL by cultured macrophages (Jessup et al., 1990), and it is reasonable to assume that they occur regardless of how LDL oxidation is initiated.

Jessup et al. (1990) have measured the kinetics of modification of LDL by macrophages and compared it to oxidation by copper ions. Macrophages first deplete LDL from vitamin E, then the rate of lipid peroxidation rapidly accelerates, and when the lipid hydroperoxides start to decrease again, LDL is converted into a form showing a high uptake rate in other macrophages. Incubation of LDL with Fe^{++} in the absence of cells also leads to lipid peroxidation but not to the decomposition of lipid hydroperoxides and also not to the formation of a high-uptake LDL. This clearly indicates that the formation of lipid hydroperoxides is not sufficient for the formation of a high-uptake LDL.

From the temporal change of the chemical composition of LDL it is clear that an oLDL with defined chemical properties does not exist; its compositional and functional properties change continuously during the lag-, propagation- and decomposition-phases. Only a few laboratories have done kinetic experiments as part of their LDL oxidation studies; however, from the available data, in particular from the long incubation time of about 24 to 48 h, and the reported TBARS values, one can assume that the LDL used in most biological studies (macrophage uptake, cytotoxicity, chemotaxis, raising of antibodies) was an oLDL at the end of the decomposition-phase.

The decomposition of lipid hydroperoxides to aldehydes is a general phenomenon for lipid peroxidation in biological systems and many studies ascribe the ultimate damage to the action of such reactive aldehydes, e.g. MDA, HNE and others (for review see Esterbauer et al., 1989b; 1991b; Esterbauer et al., 1987). A part of the LDL oxidation hypothesis implies that aldehydes generated in the LDL particle by decomposition of lipid hydroperoxides react with free amino groups in apo B. It is usually assumed that MDA reacts to amino-iminopropene cross-links, aminopropenals or dihydropyridine derivatives, whereas saturated aldehydes can form Schiff's basis and HNE can react to Michael adducts, to Schiff's bases or both. In our opinion, the reaction of aldehydes with proteins is much more complex than normally assumed and more work should be done to clarify the structure of aldehyde apo B reaction products.

The potential importance of the aldehydes for LDL modification is supported by several findings. Firstly, treatment of not previously oxidized LDL with MDA or HNE leads to modifications which are more rapidly degraded by macrophages than native LDL (Hoff et al., 1989; Haberland et al., 1982). Secondly, treatment of LDL with certain aldehydes (e.g. HNE) strongly increases the relative electrophoretic mobility and generates an apo B with the same fluorescence as found in oLDL (Juergens et al., 1987). Thirdly, HNE treated LDL exhibits cytotoxic properties very similar to oLDL (Hoff et al., 1988). Finally the most convincing evidence comes from studies with polyclonal and monoclonal antibodies raised against oLDL or LDL modified by aldehydes (Haberland et al., 1988; Palinski et al., 1989; Rosenfeld et al., 1990; Juergens et al., 1990). Antibodies against MDA or HNE-conjugated LDL do react with oLDL, this indicates that MDA and HNE conjugates with apo B are formed when LDL is oxidized. Even more importantly, it was shown by immunohistochemical methods that MDA and HNE modified proteins occur extracellularly and within macrophages in atherosclerotic lesions of Watanabe heritable rabbits. Autoantibodies directed against MDA or HNE modified LDL are also present in serum of rabbits and humans. All this shows that aldehyde-modified LDL is indeed formed in vivo.

Role of vitamin E and other antioxidants in preventing the oxidation of LDL

As discussed above, when the LDL is depleted from all antioxidants, lipid peroxidation rapidly accelerates and all the PUFAs are quickly oxidized. This sequence of events suggest that the lag-phase and hence the oxidation resistance of LDL is mainly determined by its antioxidant content. Based on the changes in the 234 nm absorption (diene absorption), which occurs during LDL oxidation, we have developed (Esterbauer et al., 1989a) a standardized routine assay for the measurement of the diene vs time profile, from which the length of the lag-phase can easily be deduced (Fig. 2).

To dissociate the effect of vitamin E from the protective effect of the other antioxidants, LDL samples from given donors were loaded with α-tocopherol, by incubating plasma samples with increasing (100–1000 μM) concentrations of α-tocopherol prior to the isolation of LDL. By that, the α-tocopherol of a given LDL could be increased several-fold, whilst all other properties of the LDL were probably not affected (Esterbauer et al., 1990b; Esterbauer et al., 1991a). The oxidation resistance of such α-tocopherol-loaded LDL samples always increased linearly ($r^2 = 0.96$) with the α-tocopherol content according to the equation $y = kx + a$, where y is the lag-phase in minutes and x is mol α-tocopherol/mol LDL, k is an efficiency constant of α-tocopherol and a is a vitamin E independent variable (Fig. 2).

The determination of the values for the constants k and a in different subjects revealed a very strong individual variation with a mean value $\pm$ SD for $k = 4.39 \pm 3.05$ (range 0.7 to 17), and a mean $\pm$ SD value for $a = 35.99 \pm 35.86$ min (range minus 68.6 to plus 108.6 min) (Esterbauer et al., 1991c and unpublished data). This suggests strong individual differences in the protective effects of vitamin E. Persons with large k values are good responders to vitamin E, whereas those with a low k-value respond only marginally with an increase of LDL oxidation resistance.

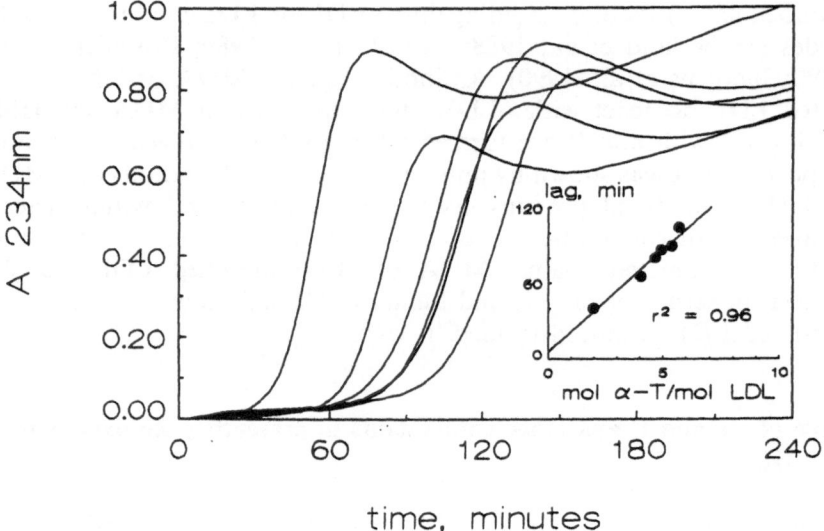

Figure 2. Determination of the efficacy of α-tocopherol in increasing the oxidation resistance of LDL. The LDL samples with different α-tocopherol contents were prepared by adding increasing concentrations of α-tocopherol to plasma of a donor prior to isolation of LDL. The α-tocopherol content of isolated LDL was determined by HPLC. LDL (0.25 mg/ml equal 0.1 μM) in oxygen saturated PBS was oxidized with 1.67 μM Cu^{++}; the curves show the diene vs time profile. The relationship between lag-phase (y) and α-tocopherol (x) for this donor is: $y = 5.41x + 6.48$ ($r^2 = 96$). From Esterbauer et al. (1991c), with permission.

This relationship found for LDL loaded in vitro with vitamin E is also valid for oral vitamin E intake (Dieber-Rotheneder et al., 1991). Daily doses of 150, 225, 800 or 1200 iu RRR-α-tocopherol over three weeks increased the LDL α-tocopherol content to 138 ± 12, 158 ± 32, 144 ± 12 and $215 \pm 47\%$ compared to the baseline values ($= 100\%$) and in parallel the oxidation resistance ($=$ lag-phase) increased to 118 ± 17, 156 ± 22, 135 ± 23 and $175 \pm 21\%$ compared to the baseline ($= 100\%$). The increase of the lag-phase was linearly correlated ($y = kx + a$) with the α-tocopherol content with r^2 0.56 to 0.95. Figure 3 shows the response of a subject receiving 1200 iu RRR-α-tocopherol for 3 weeks. Mino et al. (1989) found a strict linear correlation between the lag-phase (T inhibition) and the α-tocopherol content of LDL and HDL with AAPH-induced oxidation.

Due to the strong individual variation of k and a, the α-tocopherol content of a given LDL is, per se, not predictive for the oxidation resistance. However, based on a large number of different LDL samples with different α-tocopherol contents (baseline values, supplemented in vitro or by oral intake) for which the oxidation resistance was determined, a statistical prediction is possible (Esterbauer et al., 1991c). We now find a statistical correlation between lag-phase (y) and LDL

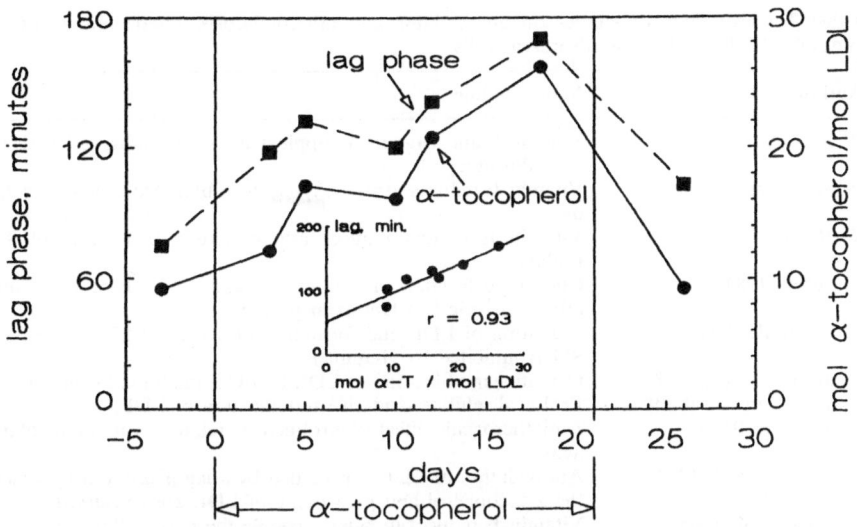

Figure 3. Plot showing the effect of oral intake of vitamin E on α-tocopherol content and oxidation resistance (=lag-phase) of LDL. The subject recieved 1200 iu RRR-α-tocopherol daily for 3 weeks, baseline values were determined 2 days prior (day −2) and 5 days after termination of supplementation (day 26). The correlation in the insert is: $y = 4.4x + 52.4$; $r^2 = 0.868$. From Esterbauer et al., 1991c, with permission.

α-tocopherol content in mol/mol LDL (x), the equation being: $y = 2.94x + 52.4$, $r^2 = 0.46$, $p < 0.001$, $n = 206$. In the statistical average with 6.88 mol vitamin E/mol LDL (Table 1) the lag-phase should therefore be 72 min; this agrees very well with the experimentally determined value (Table 1) of 68 ± 15 min ($n = 76$). Statistically, vitamin E (α- + gamma-tocopherol) contributes about 30% to the lag-phase, whereas 70% is due to the vitamin E independent variable. However, it should be emphasized again that LDLs of some subjects (with low or negative a value) solely depend on protection by α-tocopherol.

A chronological list of studies related to the prevention of atherosclerosis by antioxidants and inhibition of LDL oxidation by vitamin E is given in Table 2. The therapeutic potential of vitamin E in the treatment of atherosclerosis was recently extensively reviewed by Janero (1991). Epidemiological studies in the European population revealed a highly significant inverse correlation between the incidence of ischemic heart disease mortality and the level of plasma α-tocopherol (Gey et al., 1991; Gey et al., 1989); low levels of α-tocopherol and ascorbate appear to be risk factors in early angina pectoris (Riemersma et al., 1991). Gaziano et al. (1990) presented preliminary findings from a study (USA physician study) in which a group of patients receiving 50 mg β-carotene every second day showed significant reduction of major

154

Table 2. Chronology of studies on antioxidants and atherosclerosis. abbreviations: EC= endothelial cells, SM = smooth muscle cells

Author(s)	Major finding
Weitzel et al. 1956	Vitamin E and vitamin A supplements reduce atherosclerosis in hens during egg-laying period.
Wilson et al. 1978	Vitamin E reduces atherosclerosis in rabbits fed atherosclerotic diet.
Westrope et al. 1982	Vitamin E in diet reduces atherosclerosis in cholesterol fed rabbits.
Liu et al. 1984	Chronic deficiency of vitamin E and C is associated with atherosclerotic like lesions in primate.
Morel et al. 1984	Oxidation of LDL and formation of cytotoxic LDL by EC or SM is inhibited by vitamin E.
Steinbrecher et al. 1984	Oxidative modification of LDL by EC is inhibited by vitamin E.
Parthasarathy et al. 1986	Probucol inhibits oxidative modification of LDL.
Carew et al. 1987	Antiatherogenic effect of probucol is related to its antioxidant activity.
Esterbauer et al. 1987	Autoxidation of LDL is preceded by a lag-phase, during which the α-tocopherol and β-carotene of LDL are consumed.
Tolonen et al. 1988	Vitamin E reduces in elderly people the plasma TBARS.
Harrats et al. 1989	Short term dietary vitamin E and C protects LDL against in vivo oxidation by acute cigarette smoking.
Esterbauer et al. 1989a	Determination of lag-phase by the diene vs. time profile, kinetics of disappearance of antioxidants in LDL, ascorbate has a sparing effect on vitamin E.
Mino et al. 1989	Lag-phase (T-inhibition) in LDL oxidation with AAPH increases with LDL α-tocopherol content.
Smith et al. 1989	Vitamin E reduces atherosclerosis in restricted ovulatory chicken.
Gey et al. 1989	Inverse correlation between plasma vitamin E and mortality from IHD in the European population.
Esterbauer et al. 1990a	Review on oxidation of LDL and effect of endogenous antioxidants in LDL.
Gaziano et al. 1990	50 mg β-carotene supplementation every second day significantly reduces major coronary and vascular events in men.
Boscoboinic et al. 1991	Vitamin E inhibits SM proliferation.
Esterbauer et al. 1991a	In vitro loading of LDL with vitamin E leads to a linear increase of lag-phase.
Esterbauer et al. 1991c	Review of the effects of vitamin E on the oxidation resistance of LDL, efficiency of vitamin E shows strong individual variations.
Verlangieri et al. 1991	Vitamin E reduces atherosclerosis in cholesterol fed primate.
Janero et al. 1991	Review of therapeutic potential of vitamin E in pathogenesis of atherosclerosis.
Dieber-Rotheneder et al. 1991	Dietary vitamin E (125–1200 iu/day) increases in a concentration. Dependent manner resistance of LDL against copper oxidation.
Riemersma et al. 1991	Risk of angina pectoris increases with low plasma concentrations of vitamin E, A and C.

coronary and vascular events. A number of animal studies were performed until 1982 to test the antiatherosclerotic effect of dietary vitamin E (reviewed in Janero, 1991); some reported that vitamin E did not reduce diet-induced lesions, whereas others reported a protective effect.

Very recently, Verlangiery et al. (1991, in press) reported that vitamin E significantly reduces symptoms of atherosclerosis in primates fed an atherogenic cholesterol diet. Chronic deficiency of vitamin E and C is associated with atherosclerosis-like lesions in rodents, pigs and primates (reviewed in Gey et al., 1991; Gey et al., 1989). Vitamin E supplementation of diabetic rats prevents oxidation of LDL in vivo and its conversion to a cytotoxic form (Morel et al., 1989). A number of studies deal with probucol, a widely used drug for the treatment of hypercholesterolemia in men. Animal experiments with WHHL rabbits showed that probucol reduces the extent of lesions to a much larger extent than expected from its cholesterol-lowering effect; this is most likely due to its antioxidant activity (Kita et al., 1988). LDL isolated from probucol-treated rabbits or humans is highly resistant to in vitro oxidation by macrophages or transition metal ions (Parthasarathy et al., 1986). In vitro experiments show that vitamin E and possibly other natural or synthetic antioxidants might prevent initiation and progression of atherosclerosis also by effects not related to their chainbreaking antioxidant activity in LDL. For example, physiological concentrations of vitamin E inhibit the proliferation of smooth muscle cells (Boscoboinik et al., 1991), provide general protection to endothelial cells and inhibit lipoxygenase activities (reviewed in Janero, 1991).

Many more basic biochemical studies are required to understand the potential benefits of antioxidants in the prevention or treatment of atherosclerosis and other diseases.

Acknowledgment. The authors' work has in part been supported by the Association for International Cancer Research, AICR, U.K.

Avogaro, P., Bittolo-Bon, G., and Cazzolato, G. (1988) Presence of a modified low density lipoprotein in humans. Arteriosclerosis 8: 79–87.

Bedwell, S., Dean, R. T., and Jessup, W. (1989) The action of defined oxygen-centred free radical on human low-density lipoprotein. Biochem. J. 262: 707–712.

Boscoboinik, D., Szewczyk, A., Hensey, C., and Azzi, A. (1991) Inhibition of cell proliferation by α-tocopherol. J. Biol. Chem. 266: 6188–6194.

Carew, T. E., Schwenke, D. C., and Steinberg, D. (1987) Antiatherogenic effect of probucal unrelated to its hypocholesterolemic effect: evidence that antioxidants in vivo can selectively inhibit low density lipoprotein degradation in macrophage-rich fatty streaks and slow the progression of atherosclerosis in the Watanabe heritable hyperlipidemic rabbit. Proc. Natl. Acad. Sci. USA 84: 7725–7729.

Cazzolato, G., Avogaro, P., and Bittolo-Bon, G. (1991) Characterization of a more electronegatively charged LDL subfraction by ion exchange HPLC. Free Rad. Biol. Med. 11: 247–253.

Curzio, M., Esterbauer, H., Di Mauro, C., Cecchini, G., and Dianzani, M. U. (1986) Chemotactic activity of the lipid peroxidation product 4-hydroxynonenal and homologous hydroxyalkenals. Biol. Chem. Hoppe-Seyler 367: 321–329.

Dieber-Rotheneder, M., Puhl, H., Wäg, G., Striegl, G., and Esterbauer, H. (1991) Effect of oral supplementation with d-alpha-tocopherol on the vitamin E content of human low density lipoproteins and its oxidation resistance. J. Lipid Res. 8: 1325–1332.

Esterbauer, H., Juergens, G., Quehenberger, O., and Koller, E. (1987) Autoxidation of human low density lipoprotein. Loss of polyunsaturated fatty acids and vitamin E and generation of aldehydes. J. Lipid Res. 28: 495–509.

Esterbauer, H., Quehenberger, O., and Juergens, G. (1988) Oxidation of human low density lipoprotein with special attention to aldehydic lipid peroxidation products, in: Free Radi-

156

cals: Methodology and Concepts, pp. 243–268. Eds C. Rice-Evans and B. Halliwell. Richelieu Press, London.

Esterbauer, H., Striegl, G., Puhl, H., and Rotheneder, M. (1989a) Continous monitoring of in vitro oxidation of human low density lipoprotein. Free Rad. Res. Commun. 6: 67–75.

Esterbauer, H., Zollner, H., and Schaur, R. J. (1989b) Aldehydes formed by lipid peroxidation: Mechanism of formation, occurrence and determination, in: Membrane Lipid Oxidation, vol. 1, pp. 239–268. Ed. C. Vigo-Pelfrey. CRC Press, Boca Raton.

Esterbauer, H., Rotheneder, M., Waeg, G., Striedl, G., and Juergens, G. (1990a) Biochemical, structural, and functional properties of oxidized low-density lipoproteins. Chem. Res. Toxicol. 3: 77–92.

Esterbauer, H., Dieber-Rotheneder, M., Waeg, G., Puhl, H., and Tatzber, F. (1990b) Endogenous antioxidants and lipoprotein oxidation. Biochem. Soc. Trans. 18: 1059–1061.

Esterbauer, H., Dieber-Rotheneder, M., Striegl, G., and Wäg, G. (1991a) Role of vitamin E in preventing the oxidation of low density lipoprotein. Am. J. Clin. Nutr. 53: 314S–321S.

Esterbauer, H., Schaur, R. J., and Zollner, H. (1991b) Chemistry and biochemistry of 4-hydroxynonenal, malonaldehyde and related aldehydes. Free Rad. Biol. Med. 11: 81–128.

Esterbauer, H., Puhl, H., Dieber-Rotheneder, M., Waeg, G., and Rabl, H. (1991c) Effect of antioxidants on oxidative modification of LDL. Ann. Med. 23: 574–581.

Gaziano, J. M., Manson, J. E., Ridker, P. M., Buring, J. E., and Hennekens, C. H. (1990) Beta carotene for chronic stable angina. Circulation 82 (4 Suppl. III): 202.

Gey, K. F., and Puska, P. (1989) Plasma vitamins E and A inversely related to mortality from ischemic heart disease in cross-cultural epidemiology. Ann. N.Y. Acad. Sci. 570: 268–282.

Gey, F., Puska, P., Jordan, P., and Moser, U. K. (1991) Inverse correlation between plasma vitamin E and mortality from ischemic heart disease in cross-cultural epidemiology. Am. J. Clin. Nutr. 53: 326S–334S.

Haberland, M. E., Fogelman, A. M., and Edwards, P. A. (1982) Specificity of receptor-mediated recognition of malondialdehyde-modified low density lipoprotein. Proc. Natl. Acad. Sci. USA 79: 1712–1716.

Haberland, M. E., Fong, D., and Cheng, L. (1988) Malondialdehyde-altered protein occurs in atheroma of Watanabe heritable hyperlipidemic rabbits. Science 241: 215–218.

Harats, D., Ben-Naim, M., Dabach, Y., Hollander, G., Stein, O., and Stein, Y. (1989) Cigarette smoking renders LDL susceptible to peroxidative modification and enhanced metabolism by macrophages. Atherosclerosis 79: 245–252.

Hoff, H. F., Chisolm, G. M. III, Morel, D. W., Juergens, G., and Esterbauer, H. (1988) Chemical and functional changes in LDL following modification by 4-hydroxynonenal, in: oxy-Radicals Molecular Biology and Pathology, pp. 459–472. Eds P. A. Cerutti, J. M. McCord and T. Fridovich. Alan R. Liss, New York.

Hoff, H. F., O'Neil, J., Chisolm, G. M., Cole, T. B., Quehenberger, O., Esterbauer, H., and Juergens, G. (1989) Modification of low-density lipoprotein with 4-hydroxynonenal induces uptake by macrophages. Arteriosclerosis 9: 538–549.

Janero, D. R. (1991) Therapeutic potential of vitamin E in the pathogenesis of spontaneous athersclerosis. Free Rad. Biol. Med. 11: 129–144.

Jessup, W., Rankin, S. M., De Whalley, C. V., Hoult, J. R. S., Scott, J., and Leake, D. S. (1990) Alpha-tocopherol consumption during low-density lipoprotein oxidation. Biochem. J. 265: 399–405.

Juergens, G., Hoff, H. F., Chisolm, G. M., and Esterbauer, H. (1987) Modification of human serum low density lipoprotein by oxidation-characterization and pathophysiological implications. Chem. Phys. Lipids. 45: 315–336.

Juergens, G., Ashy, A., and Esterbauer, H. (1990) Detection of new epitopes formed upon oxidation of low-density lipoprotein, lipoprotein (a) and very-low-density lipoprotein. Use of an antiserum against 4-hydroxynonenal-modified low-density lipoprotein. Biochem. J. 265: 605–608.

Kaneko, T., Kaji, K., and Matsuo, M. (1988) Cytotoxicities of a linoleic acid hydroperoxide and its related aliphatic aldehydes toward cultured human umbilical vein endothelial cells. Chem.-Biol. Interact. 67: 295–304.

Kita, T., Nagano, Y., Yokode, M., Ishii, K., Kume, N., Narumiya, S., and Kawai, C. (1988) Prevention of atherosclerotic progession in Watanabe rabbits by probucol. Am. J. Cardiol. 62: 13B–19B.

Liu, S. K., Dolensek, E. P., and Tappe, J. P. (1984) Cardiomyopathy associated with vitamin E deficiency in seven gelada baboons. J. Am. Vet. Med. Assoc. 185: 1347–1350.

Mino, M., Miki, M., Miyake, M., and Ogihara, T. (1989) Nutritional assessment of vitamin E in oxidative stress. Ann. N.Y. Acad. Sci. 570: 296–310.

Morel, D. W., DiCorleto, P. E., and Chisolm, G. M. (1984) Endothelial and smooth muscle cells alter low density lipoprotein in vitro by free radical oxidation. Arteriosclerosis 4: 357–364.

Morel, D. W., and Chisolm, G. M. (1989) Antioxidant treatment of diabetic rats inhibits lipoprotein oxidation and cytotoxicity. J. Lipid Res. 30: 1827–1834.

Palinski, W., Rosenfeld, M. E., Ylä-Herttuala, S., Gurtner, G. C., Socher, S. S., Butler, S. W., Parthasarathy, S., Carew, T. E., Steinberg, D., and Witztum, J. L. (1989) Low density lipoprotein undergoes oxidative modification in vivo. Proc. Natl. Acad. Sci. USA 86: 1372–1376.

Parthasarathy, S., Young, S. G., Witztum, J. L., Pittman, R. C., and Steinberg, D. (1986) Probucol inhibits oxidative modification of low density lipoprotein. J. Clin. Invest. 7: 641–644.

Quinn, M. T., Parthasarathy, S., Fong, L. G., and Steinberg, D. (1987) Oxidatively modified low density lipoproteins: A potential role in recruitment and retention of monocyte/macrophages during atherogenesis. Proc. Natl. Acad. Sci. USA 84: 2995–2998.

Riemersma, R. A., Wood, D. A., Macintyre, C. C. A., Elton, R. A., Gey, K. F., and Oliver, M. F. (1991) Risk of angina pectoris and plasma concentrations of vitamins A, C and E and carotene. Lancet 337: 1–5.

Rosenfeld, M. E., Palinski, W., Ylä-Herttuala, S., Butler, S., and Witztum, J. L. (1990) Distribution of oxidation specific lipid-protein adducts and apolipoprotein B in atherosclerotic lesions of varying severity from WHHL rabbits. Arteriosclerosis 10: 336–349.

Smith, T. L., and Kummerow, F. A. (1989) Effect of dietary vitamin E on plasma lipids and atherogenesis in restricted ovulatory chickens. Atherosclerosis 75: 105–109.

Steinberg, D., Parthasarathy, S., Carew, T. E., Khoo, J. C., and Witztum, J. L. (1989) Beyond cholesterol. Modifications of low-density lipoprotein that increase its atherogenicity. J. Engl. J. Med. 302: 915–924.

Steinbrecher, U. P., Parathasarathy, S., Leake, D. S., Witztum J. L., and Steinberg, D. (1984) Modification of low density lipoprotein by endothelial cells involves lipid peroxidation and degradation of low density lipoprotein phospholipids. Proc. Natl. Acad. Sci. USA 81: 3883–3887.

Steinbrecher, U. P. (1987) Oxidation of human low density lipoprotein results in derivatisation of lysine residues of apolipoprotein B by lipid peroxide decomposition products. J. Biol. Chem. 262: 3603–3608.

Steinbrecher, U. P., Zhang, H., and Lougheed, M. (1990) Role of oxidatively modified LDL in atherosclerosis. Free Rad. Biol. Med. 9: 155–168.

Stocker, R., Bowry, V. W., and Frei, B. (1991) Ubiquinol-10 protects human low density lipoprotein more efficiently against lipid peroxidation than does α-tocopherol. Proc. Natl. Acad. Sci. USA 88: 1646–1650.

Tolonen, M., Sarna, S., Halme, M., Tuominen, S. E. J., Westermarck, T., Nordberg, U., Keinonen, M., and Schrijer, J. (1988) Antioxidant supplementation decreases TBA reactants in serum of elderly. Biol. Trace Element Res. 17: 221–228.

Verlangieri, A. J., and Bush, M. J. (1992) Effects of d-alpha-tocopherol supplementation on experimentally induced primate atherosclerosis. J. Am. Coll. Nutr. 11: 131–138.

Weitzel, G., Schön, H., Gey, K. F., and Buddecke, E. (1956) Lipid-soluble vitamins and atherosclerosis. Hoppe Seylers Z. Physiol. Chem. 304: 247–72

Westrope, K. L., Miller, R. A., and Wilson, R. B. (1982) Vitamin E in a rabbit model of endogenous hypercholesterolemia and atherosclerosis. Nutr. Reports Int. 25: 83–88.

Wilson, R. B., Middleton, C. C., and Sun, G. Y. (1978) Vitamin E antioxidants and lipid peroxidation in experimental atherosclerosis of rabbits. J. Nutr. 108: 1858–67.

Free Radicals and Aging
ed. by I. Emerit & B. Chance
© 1992 Birkhäuser Verlag Basel/Switzerland

Desialylated low density lipoproteins and atherosclerosis

Nicole Dousset[a], Jean Claude Dousset[b], Marie Laure Soléra[a] and Pierre Valdiguié[a]

[a]*Department of Biochemistry, Rangueil University Hospital, 31054 Toulouse Cedex, France, and* [b]*Department of Biochemistry, Faculty of Pharmaceutical Sciences, 31062 Toulouse Cedex, France*

Summary. Oxidative modification of LDL is accompanied by a number of compositional and structural changes, now well known. In addition, other atherogenic modifications of LDL exist, such as desialylation. The present article summarizes the recent data related to desialylated LDL and to the presence of these LDL in blood plasma of patients with coronary atherosclerosis. In addition, this review examines the sensitivity of these LDL to peroxidative stress.

Introduction

Hypercholesterolemia is associated with premature atherosclerosis, but the precise mechanisms involved in Low Density Lipoproteins (LDL)-related atherogenesis remain unclear. The physiopathology of atherosclerosis is complex, and implicates multiple interacting mechanisms.

LDL can undergo free radical oxidation, due to prolonged contact between LDL and arterial endothelial cells that can be reproduced by copper-dependent oxidation in the absence of cells. The biochemical, structural and functional changes accompanying LDL oxidation are presently well known and reported (Esterbauer et al., 1990; Steinbrecher et al., 1990).

LDL attack by free radicals induces several modifications of the apo B (modifications of charge and molecular weight) and of the lipids: decrease of polyunsaturated fatty acids (PUFA), increase of aldehyde compounds and thiobarbituric reactive substances (Esterbauer et al., 1987; Jurgens et al., 1987; Steinbrecher et al., 1990), hydrolysis of phosphatidylcholine (Steinbrecher et al., 1984) and also, as demonstrated in our group, formation of dicarboxylic phosphatidylcholines (Dousset et al., 1969; Dousset et al., 1981).

Modifications of LDL (glycation, desialylation)

These modifications of LDL are observed in diabetes mellitus, via the ability of glucose to glycate proteins. They are responsible for the increased vulnerability to oxidation of lipoproteins in diabetic patients (Mullarkey et al., 1990; Wolff et al., 1991).

LDL desialylation is an atherogenic modification of LDL leading to the accumulation of intracellular fat. Sialic acid is a component of native LDL (Swaminathan and Aladjem, 1976) which plays an important role in lipoprotein function (Filipovic and Buddecke, 1979; Filipovic et al., 1979; Malmendier et al., 1979; Malmendier et al., 1980; Camejo et al., 1985) and controls receptor-mediated binding and uptake of LDL by cultured cells (Filipovic et al., 1979). Desialylation of LDL with neuraminidase leads to increased binding and uptake of LDL by cultured fibroblasts and smooth muscle cells (Filipovic et al., 1979).

Sialic acid, exposed at the LDL surface, plays a determinent role in the in vitro association of LDL with proteoglycans (Camejo et al., 1985). After neuraminidase treatment the reduction of the sialic acid content induces an increase in the affinity for proteoglycans. The in vitro interaction between LDL and aortic proteoglycans has been studied and it is clearly demonstrated that proteoglycans from aortas form complexes (soluble and insoluble) with LDL (Steele and Wagner, 1987). *The binding of proteoglycans to LDL may be an early and important step in atherogenesis.* Sambandam et al. (1991) have demonstrated the specificity of the LDL-glycosaminoglycan interaction, because the degree of sulfation of glycosaminoglycan influences the binding to LDL.

Furthermore, patients with coronary atherosclerosis present modified LDL which are desialylated (Orekhov et al., 1989). These LDL possess the ability to induce cholesterol accumulation in cultured human aortic smooth muscle cells, while LDL from healthy donors have no such atherogenic properties. LDL of atherosclerotic patients have a 2- to 5-fold lower level of sialic acid as compared with non-atherogenic LDL of healthy donors (Orekhov et al., 1989). Atherogenic properties of patients LDL are explained by a low degree of sialylation. This fact is proved by experiments where non-atherogenic LDL are partially desialylated by neuraminidase treatment inducing a substantial increase of cellular cholesterol in cultured cells. Thus, these experiments clearly demonstrated that *LDL desialylation is an atherogenic modification of lipoprotein leading to accumulation of intracellular lipids.* On the other hand, Orekhov et al. (1991) have detected immunoglobulins with a higher affinity for desialylated LDL and for LDL obtained from patients with coronary artery disease than for native LDL. This result is a very good argument in favor of the presence of desialylated LDL in these patients. These data from the Russian group show that two factors are present in the blood plasma of atherosclerotic patients. The first one concerns the modified LDL (desialylated in this case, or possibly glycosylated or other type of modification) and the second is represented by autoantibodies produced in response to the emergence of modified LDL. The desialylated LDL have less sialic acid, triglycerides and cholestrol, smaller particle size, greater density and negative charge, higher aggregative activity. Recently, Orekhov et al. (1991) have demonstrated that desialylated LDL are senescent LDL with altered physicochemical properties and represent a part of circulating LDL. It was shown (Tertov et al., 1989) that modification of LDL by oxidation, glycosylation or desialylation was accompanied by aggregate formation and intracellular lipid accumulation. Similar results reported by Naruszewicz et al. (1991) are relative to the modification of lipoprotein (a) [Lp (a)] through copper ion treatment and show protein fragmentation, the tendency to aggregate and a higher density. *Aggregates of LDL in the artery wall may induce foam cell formation and favor the formation of fatty streak* (Khoo et al., 1988). The works of Khoo et al. suggest the possibility that denaturation of LDL and self-aggregation could play a role in atherogenesis. Aggregation of LDL can be inhibited by the addition of HDL or simply by the addition of apo A_I (Khoo et al., 1990).

Relationships between oxidation and sialic acid content

LDL particles constitute a heterogenous group and it was demonstrated that some subgroups were associated with an atherogenic potential (Chapman et al., 1988). Differences in carbohydrate content of LDL (La Belle and Krauss, 1990) have been reported: predominance of small, dense LDL with significantly less sialic acid content is associated with a greater affinity for proteoglycans. Interaction of LDL with human arterial proteoglycans, probably as a consequence of desialylation, can stimulate its uptake by human monocyte – derived macrophages and may contribute to the formation of foam cells (Hurt et al., 1990).

In trying to explain the differences in response to oxidation among LDL subpopulations, De Graaf et al. (1991) have compared the susceptibility of LDL subpopulations to in vitro lipid peroxidation and shown that the dense LDL subfractions are the most sensitive to peroxidative stress. The differences in response to oxidation can be explained by the large variability in fatty acid and antioxidant contents in the LDL subfractions and also in the subjects studied. Therefore, the more dense LDL subfractions may have a larger contribution to atherosclerosis than the less dense LDL subfraction. Jiahal et al. (1991) have shown that each patient's LDL have a particular "oxidation profile" but they are not precise on what factor is responsible for the wide variability of susceptibility to oxidation (sialic acid content for example).

The correlation of proteoglycan content and atherosclerotic status can be explained by the ability of proteoglycans to form complexes with lipoproteins. Several studies have implicated the carbohydrate moieties of glycoproteins in preventing protein component from proteolysis. Görög and Pearson (1985) indicated that removal of surface sialic acid renders the intact endothelium less resistant to proteolysis. It seems therefore interesting to understand the effects of desialylation of LDL on their susceptibility to peroxidative stress induced by copper ions.

In our laboratory, we are investigating the modifications of LDL, consecutive to the desialylation (by neuraminidase) before or after peroxidation by copper ions. Our experimental results demonstrate that the desialylation of LDL alone is not followed by any change in chemical composition of LDL, except the evident decrease of sialic acid content. A second point concerns the observation that peroxidative stress is not accompanied by any loss of sialic acid content. This result permits us to make the hypothesis that *desialylation of LDL, observed in atherosclerotic patients, is not consecutive to peroxidation.*

Polyacrylamide gel electrophoresis pattern of apo B100 showed no modification of the molecular weight of desialylated LDL, whereas the apparent molecular weight of apo B100 was dramatically altered in all peroxidized LDL. A loss of Coomassie blue staining in the apo B100 region was observed, together with the formation of low molecular weight peptides, suggesting that a fragmentation of apo B100 occurs. On the contrary the absence of modification of apo B100 and TNBS reactivity observed in desialylated LDL is in agreement with previous reports on the uptake of desialylated LDL by smooth muscle cells (Filipovic and Buddecke, 1979; Filipovic et al., 1979; Orekhov et al., 1989). However, when desialylated LDL are peroxidized, interaction of LDL with human arterial proteoglycans induced by desialylation may contribute to the formation of foam cells, as reported by Hurt et al. (1990) and Steinbrecher et al. (1990).

In fact, the irreversible interaction of LDL with arterial proteoglycans leads to the formation of insoluble aggregates which induce lipid accumulation. In addition, LDL with a lower sialic acid content should have a stronger binding that LDL presenting a higher sialic acid content (La Belle and Krauss, 1990).

Furthermore, desialylation of LDL is not followed by precipitation or self-aggregate of LDL, as was observed in all the peroxidized LDL, under our experimental conditions.

In addition, we tested the effect of desialylation on peroxidized LDL. This treatment was followed by a loss of total fatty acid content, but the ratio polyunsaturated fatty acids/total fatty acids remains unchanged (Dousset et al., 1991). The data concerning the treatment of LDL by neuraminidase are in agreement with previous studies (Sattler et al., 1991) concerning the comparison of the oxidation of Lp (a) and low density lipoproteins. The authors have shown that Lp (a) was more resistant against oxidative stress than LDL; this might be due to the higher content of N-acetylneuraminic acid in Lp (a).

162

If desialylation of LDL determines their capacity of increasing the cellular lipids, the mechanism of the desialylation is not yet known (perhaps sialidase action after viral infection).

Acknowledgment. The authors wish to gratefully thank Mrs M. A. Delpech for her excellent technical assistance.

Camejo, G., Lopez, A., Lopez, F., and Quinones, J. (1985) Interaction of Low Density Lipoproteins with arterial proteoglycans. The role of charge and sialic acid content. Atherosclerosis 55: 93–105.

Chapman, M. J., Laplaud, P. M., Luc, G., Forgez, P., Bruckert, E., Goulinet, S., and Lagrange, D. (1988) Further resolution of the LDL spectrum in normal human plasma: physicochemical characteristics of discrete subspecies separated by density gradient ultracentrifugation. J. Lipid Res. 29: 442–458.

De Graaf, J., Hak-Lemmers, H. L. M., Hectors, M. P. C., Demacker, P. N. M., Hendriks, J. C. M., and Stalenhoef, A. F. H. (1991) Enhanced susceptibility to in vitro oxidation of the dense low density lipoprotein subfraction in healthy subjects. Arteriosclerosis and Thrombosis 11: 298–306.

Dousset, N., and Douste-Blazy, L. (1969) Etude chromatographique des lécithines plasmatiques irradiées (1969) Bull. Soc. Chim. Biol. 51: 1013–1020.

Dousset, N., Dousset, J. C., and Douste-Blazy, L. (1981) Influence of dicarboxylic phosphatidylcholines on the stability and phase transition of phosphatidylcholine liposomes. Biochim. Biophys. Acta 641: 1–10.

Dousset, N., Dousset, J. C., Soléra, M. L., Salvayre, R., and Valdiguié, P (1991) Biochemical modifications of desialylated and peroxidized LDL. 57th EAS Meeting, Lisbon, p. 12.

Esterbauer, H., Jurgens, G., Qhehendergen, O., and Koller, E. (1987) Autoxidation of human LDL: loss of polyunsaturated acids and vitamin E and generation of aldehydes. J. Lipid Res. 28: 495–509.

Esterbauer, H., Dieber-Rotheneder, M., Waeg, G., Striegl, G., and Jürgens, G. (1990) Biochemical, structural, and functional properties of oxidized low density lipoprotein. Chem. Res. Toxicol. 3: 77–92.

Filipovic, I., and Buddecke, E. (1979) Role of net charge of low density lipoproteins in high affinity binding and uptake by cultured cells. Biochem. Biophys. Res. Com 38: 485–490.

Filipovic, I., Schwarzmann, G., Mraz, W., Wiegandt, H., and Buddecke, E. (1979) Sialic acid content of low density lipoproteins controls their binding and uptake by cultured cells. Eur. J. Biochem. 93: 51–55.

Görög, P., and Pearson, J. D. (1985) Sialic acid moieties on surface glycoproteins protect endothelial cells from proteolytic damage. J. Pathol. 146: 205–212.

Hurt, E., Bondjers, G., and Camejo, G. (1990) Interaction of LDL with human arterial proteoglycans stimulates its uptake by human monocyte-derived macrophages. J. Lipid Res 31: 443–454.

Jialal, I., Freeman, D. A., and Grundy, S. M. (1991) Varying susceptibility of different low density lipoproteins to oxidative modification. Arteriosclerosis and Thrombosis 11: 482–488.

Jurgens, G., Hoff, H. F., Chilsom, G. M., and Esterbauer, H. (1987) Modification of human serum low density lipoprotein by oxidation. Characterization and pathophysiological implications. Chem. Phys. Lipids 45: 315–336.

Khoo, J. C., Miller, E., McLoughlin, P., and Steinberg, D. (1988) Enhanced macrophage uptake of low density lipoprotein after self-aggregation. Arteriosclerosis 8: 348–358.

Khoo, J. C., Miller, E., McLoughlin, P., and Steinberg, D. (1990) Prevention of low density lipoprotein aggregation by high density lipoprotein or apolipoprotein A-1. J. Lipid Res. 31: 645–652.

La Belle, M., and Krauss, R. M. (1990) Differences in carbohydrate content of low density lipoproteins associated with low density lipoprotein subclass patterns. J. Lipid Res. 31: 1577–1588.

Malmendier, C., Feremans, W. W., and Fontaine M. (1979) The effect of sialic acid removal on very low density lipoprotein. Artery 6: 144–156.

Malmendier, C., Delcroix, C., and Fontaine, M. (1980) Effect of sialic acid removal on human low density lipoprotein catabolism in vivo. Atherosclerosis 37: 277–284.

Mullarkey, C. J., Edelstein, D., and Brownlee, M. (1990) Free radical generation by early glycation products: a mechanism of accelerated atherogenesis in diabetes. Biochem. Biophys. Res. Commun. 173: 932–939.

Naruszewicz, M., Selinger, E., Dufour, R., and Davignon, J. (1991) Modification of Lp(a) and the effect of antioxidants. 57th EAS Meeting, Lisbon, p. 5.

Orekhov, A. N., Tertov, V. V., Mukhin, D. N., and Mikhailenko, I. A. (1989) Modification of low density lipoprotein by desialylation causes lipid accumulation in cultured cells: Discovery of desialylated lipoprotein with altered cellular metabolism in the blood of atherosclerotic patients. Biochem. Biophys. Res. Commun. 162: 206–211.

Orekhov, A. N. (1991) Cellular mechanisms of atherosclerosis. 57th EAS Meeting, Lisbon, pp. 64–65.

Orekhov, A. N., Tertov, V. V., Kabakav, A. E., Adamova, I. Y., Pokrovsky, S. N., and Smirnov, V. N. (1991) Autoantibodies against modified low density lipoprotein. Non lipid factor of blood plasma that stimulates foam cell formation. Arteriosclerosis and Thrombosis 11: 316–326.

Sambandam, T., Baker, J. R., Christner, J. E., and Ekborg, S. L. (1991) Specificity of the low density lipoprotein-glycosaminoglycan interaction. Arteriosclerosis and Thrombosis 11: 561–568.

Sattler, W., Kostner, G. M., Waeg, G., and Esterbauer, H. (1991) Oxidation of lipoprotein Lp(a). A comparison with low density lipoproteins. Biochim. Biophys. Acta 1081: 65–74.

Steele, R. H., and Wagner, W. D. (1987) Lipoprotein interaction with artery wall derived proteoglycan: Comparisons between atherosclerosis-susceptible WC-2 and resistant Show Racer pigeons. Atherosclerosis 65: 63–73.

Steinbrecher, U. P., Parthasarathy, S., Leake, D. S., Witztum, J. L., and Steinberg, D. (1984) Modification of low density lipoprotein by endothelial cells involves lipid peroxidation and degradation of low density lipoprotein phospholipids. Proc. Natl. Acad. Sci. USA 81: 3883–3887.

Steinbrecher, U. P., Zhang, H., and Lougheed, M. (1990) Role of oxidatively modified LDL in atherosclerosis. Free Radical Biol. Med. 9: 155–168.

Swaminathan, N., and Aladjem, F. (1976) The monosaccharide composition and sequence of the carbohydrate moiety of human serum low density lipoproteins. Biochemistry 15: 1516–1522.

Tertov, V. V., Sobenin, I. A., Gabbasov, Z. A., Popov, E. G., And Orekhov, A. N. (1989) Lipoprotein aggregation as an essential condition of intracellular lipid accumulation caused by modified low density lipoproteins. Biochem. Biophys. Res. Commun. 163: 489–494.

Wolff, S. P., Jiang, Z. Y., and Hunt, J. V. (1991) Protein Glycation and oxidative stress in diabetes mellitus and ageing. Free Radical Biol. Med. 10: 339–352.

Free Radicals and Aging
ed. by I. Emerit & B. Chance
© 1992 Birkhäuser Verlag Basel/Switzerland

Molecular basis of α-tocopherol inhibition of smooth muscle cell proliferation in vitro

Daniel Boscoboinik, Eric Chatelain, Gianna-M. Bartoli and Angelo Azzi

Institut für Biochemie und Molekularbiologie, Universität Bern, CH-3012 Bern, Switzerland

Summary. The molecular events responsible for the inhibition of cell proliferation by α-tocopherol have been investigated. Smooth muscle cells in vitro have been shown to be specifically inhibited by α-tocopherol with a concomitant inhibition of protein kinase C activity. β-Tocopherol was inactive, despite its similar radical scavenging activity. The point of inhibition of α-tocopherol relative to the cell cycle was localized in the late G_1 phase. A second effect of α-tocopherol observed with smooth muscle cells was the stimulation of protein kinase C biosynthesis in both the S and G_2 phases of the cell cycle. The implications of these findings for the onset of arteriosclerosis are discussed.

Abbreviations: FCS, fetal calf serum; HU, hydroxyurea; PDBu, phorbol dibutyrate; PDGF, platelet derived growth factor; PKC, protein kinase C; VSMC, vascular smooth muscle cells.

Introduction

Proliferation of vascular smooth muscle cells (VSMC) is a key event in the development of several pathological conditions including arteriosclerotic lesions (Ross, 1986; Munro and Cochran, 1988) and hypertension (Owens and Reidy, 1985; Schwartz et al., 1986).

Many agents have been identified which induce smooth muscle cell proliferation, including serum, platelet-derived growth factor[1] (PDGF), epidermal growth factor, angiotensin, interleukin-1 and endothelin (Ross et al., 1974; Gospodarowicz et al., 1977; Geisterfer et al., 1988; Raines et al., 1989; Ku et al., 1988; Bobik et al., 1990). After binding to the cell surface of cultured VSMC, these agonists trigger a sequence of events (increase in cytosolic calcium and pH, diacylglycerol dependent-activation of protein kinase C, calmodulin, *c-fos* mRNA and phosphatidylinositol hydrolysis) that leads to DNA synthesis and cell division (Rozengurt, 1986). On the other hand, several molecules have been described that can inhibit the growth of cultured smooth muscle cells, including heparin (Castellot et al., 1989), prostaglandins (Loesberg et al., 1985), staurosporine (Matsumoto and Sasaki, 1989), nisoldipin (Thyberg and Palmberg, 1987), and TMB-8 (Desgranges et al.,

1991). We have recently found that α-tocopherol (the most active form of vitamin E) at physiological concentrations inhibits proliferation of smooth muscle cells. Moreover, this effect was correlated with the inhibition by α-tocopherol of protein kinase C activity (Boscoboinik et al., 1991).

Vitamin E and esterified forms of vitamin E have multiple roles in cellular functions (Packer and Landvik, 1989). Besides their well-known biological antioxidant activity (Tappel, 1972), these lipid-soluble tocopherols stabilize membranes (Urano et al., 1988), inhibit growth in several types of cells (Kline et al., 1990; Prasad and Edwards-Prasad, 1982; Slack and Proulx, 1989) and play a role in prostaglandin synthesis (Egan et al., 1979).

It is not clear, however, whether α-tocopherol inhibits cell growth by acting as an antioxidant or whether specific structural features of the molecule are required instead. We have now examined the effects of a series of tocopherols on proliferation of synchronized smooth muscle cells. Our findings indicate that no strong correlation can be observed between the antioxidant potency of the different tocopherols tested and their effect on cell proliferation.

To further elucidate the mechanism of action of α-tocopherol, we pursued a detailed analysis of the inhibitory effects of α-tocopherol on the cell cycle of smooth muscle cells. In vivo, smooth muscle cells are apparently arrested in a quiescent phase and start to proliferate after exposure to growth promoting factors like serum or PDGF. Entry in G_1 phase and further commitment to DNA synthesis is the result of a series of crucial events regulated by growth factors (Egan et al., 1979). Modulation of events in the late G_1 phase, where several restriction points like R or W are present (Campisi et al., 1982), could explain the mechanism of action of inhibitors of cell proliferation. Thus inhibitory responses of α-tocopherol in smooth muscle cells related to the stage of the cell cycle were studied by adding the compound at different times during the cycle, and it was established that the inhibition of cell proliferation by α-tocopherol occurred in the late G_1 phase.

PKC-dependent and independent pathways have been identified in smooth muscle cell proliferation (Griendling et al., 1989; Ran et al., 1986). To study the role of protein kinase C on the observed cellular effects of α-tocopherol and its analogs, a direct measurement of PKC activity in intact cells was employed (Alexander et al., 1989). The effect of α-tocopherol on the phosphorylation of a peptide substrate specific for PKC was studied in streptolysin-O permeabilized cells. Our findings indicate that the specific arrest of growth at the G_1/S boundary by α-tocopherol is associated with both an inhibition of protein kinase C activity and binding of phorbol esters to kinase C in the late G_1 phase.

Synchronized smooth muscle cells are a useful model for proliferation studies

The cell line used, A7r5 rat aortic smooth muscle (VSMC), was obtained from the American Type Culture Collection. Cells were grown in Dulbecco's modified Eagle medium (DMEM) containing 25 mM bicarbonate, 60 U/ml penicillin, 60 mg/ml streptomycin, and 10% fetal calf serum (FCS). Cells were usually seeded into 100 mm plastic culture dishes and grown to confluence at 37°C in a humidified atmosphere of 5% CO_2 (medium pH 7.4). Culture media were changed every 3 days. In all experiments, media and sera used were from the same batch number and source. Synchronous culture at the G_1/S boundary was obtained by a combination of serum deprivation and hydroxyurea treatment (Ashihara and Baserga, 1979). Briefly, cells were subcultured in 35 mm culture dishes containing 1.5 ml of DMEM supplemented with 2% FCS. When confluent, cells were made quiescent (G_0) by exposure to serum deficient medium (0.2% FCS). After 48 h, cells were re-stimulated by replacing the serum-deficient medium by DMEM containing 2% FCS during 8 h; then, hydroxyurea stock solution was added to each dish (final concentration 1.5 mM). After 14 h and when cells were blocked at the G_1/S boundary, medium was removed, cells were washed with PBS and transferred to fresh complete medium (DMEM, 2% FCS). This procedure allowed the cells to enter synchronously into the S-phase and progress through the cycle.

Onset and duration of the S-phase was determined by pulse-labeling with [^{3}H]thymidine for 1 h at different times after stimulation by serum, throughout the course of one complete cell cycle. Duplication on cell number occurs 24 h after serum stimulation indicating the entry of cells into mitosis (data not shown). Vascular smooth muscle cells growing exponentially in media containing FCS were made quiescent by incubating them for 48 h in low-serum media. Under these conditions, little amount of [^{3}H]-thymidine was incorporated into DNA but cells could be restimulated to growth when medium was supplemented with FCS or growth-factors were added.

To determine the onset and duration of the different phases of the cell cycle, cells were pulsed for 1 h with [^{3}H]thymidine at several times after releasing from the quiescent state (G_0). [^{3}H]thymidine incorporation started to increase approximately 11 h after application of the stimulus and reached the maximal value 5–6 h later. We have previously shown that addition of α-tocopherol when stimulating quiescent VSMC by serum, PDGF or endothelin, prevented specifically cell proliferation in a concentration-dependent manner ($IC_{50} = 35 \ \mu M$) (Table 1). To analyze cell cycle modifications produced by α-tocopherol, cells were synchronized at the G_1/S boundary by using a combination of serum deprivation and hydroxyurea (HU) treatment. Upon removal of HU,

Table 1. Effect of several compounds on the proliferation of three cell lines (A7r5 smooth muscle, Balb/3T3 fibroblasts and Saos-2 osteosarcoma) induced by five different stimuli. Cells were incubated in triplicate with the indicated concentrations of the compounds in the presence of the indicated stimuli for 24 h. Measurements of [^{3}H]thymidine incorporation were done as described in Methods and expressed as a percentage relative to control, no-treated cells

Addition/Cell line	Smooth muscle	Fibroblast	Osteosarcoma
1. Serum (2%)	100	100	100
+α-tocopherol (50 μM)	46	102	106
+α-tocopherol acetate (50 μM)	114	87	120
+trolox (50 μM)	92	76	85
+Butylated Hydroxy-Toluene (50 μM)	99	n.d.	n.d.
2. Lysophosphatidic acid (50 μM)	100	100	100
+α-tocopherol (50 μM)	90	87	118
3. Bombesin (20 nM)	100	100	100
+α-tocopherol (50 μM)	83	2	114
4. Endothelin (80 nM)	100	n.d.	n.d.
+α-tocopherol (50 μM)	30	n.d.	n.d.
+α-tocopherol (100 μM)	4	n.d.	n.d.
5. PDGF-BB (20 ng/ml)	100	n.d.	n.d.
+α-tocopherol (50 μM)	18	n.d.	n.d.
+α-tocopherol (100 μM)	5	n.d.	n.d.

n.d. = not determined

cells entered synchronously into the S-phase (Ashihara and Baserga, 1979). α-tocopherol (100 μM) did not alter the kinetics of the G_1–S transition, but when it was added at the time of growth stimulation, an inhibition of 50% in [^{3}H]thymidine incorporation was observed. Comparable results were obtained when cells were stimulated by 25 ng/ml PDGF.

Tocopherol analogs are differently effective on VSMC proliferation

Tocopherols were dissolved in absolute ethanol and added in required amounts to culture media such that the final ethanol concentration during growth of cells did not exceed 0.5%. Control media contained ethanol at similar concentrations. Following 48 h serum deprivation, cells were re-stimulated to growth by addition of FCS and the different compounds were added at the indicated times after re-stimulation. Cells were trypsinized and counted in a hemocytometer in triplicate after the completion of the cell cycle. Viability of the cells was assessed by the trypan blue dye exclusion method. Cell proliferation was studied by

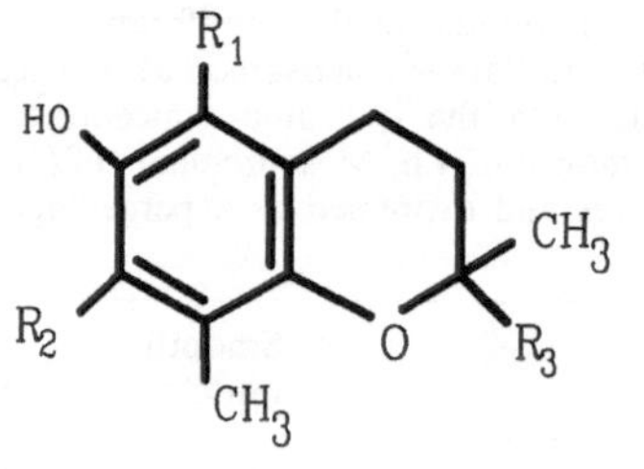

	R₁	R₂	R₃
α-tocopherol	-CH$_3$	-CH$_3$	-(CH$_2$-CH$_2$-CH$_2$-CH)$_3$ -CH$_3$ CH$_3$
β-tocopherol	-CH$_3$	-H	-(CH$_2$-CH$_2$-CH$_2$-CH)$_3$ -CH$_3$ CH$_3$
γ-tocopherol	-H	-CH$_3$	-(CH$_2$-CH$_2$-CH$_2$-CH)$_3$ -CH$_3$ CH$_3$
δ-tocopherol	-H	-H	-(CH$_2$-CH$_2$-CH$_2$-CH)$_3$ -CH$_3$ CH$_3$

Figure 1. Structural formulae of α-tocopherol and its homologues.

measuring the incorporation of [³H]thymidine into DNA (Boscoboinik et al., 1991). Cells were pulsed with [³H]-thymidine (0.5–1 mCi/well) for the first 6 h during S-phase following removal of hydroxyurea. After this period, cells were washed twice with PBS supplemented with 10 mg/ml glucose and 1 mg/ml bovine serum albumin, fixed for 30 min with ice-cold 5% trichloroacetic acid, and solubilized in 0.1 M NaOH/2% Na$_2$CO$_3$. The radioactivity incorporated into acid insoluble material was determined in a liquid scintillation analyzer. Cell proliferation was also studied by counting the cells after completion of their cycle.

The effect of the four major forms of tocopherol (α-, β-, γ- and δ-tocopherol, Fig. 1) on the proliferation induced by fetal calf serum is shown in Figure 2. At a concentration of 100 μM, α-tocopherol, γ-tocopherol and δ-tocopherol, all inhibited FCS-induced proliferation of VSMC; surprisingly, no effect of β-tocopherol was observed on both cell numbers (hatched column) or [³H]thymidine incorporation into DNA (blank column), compared to control. α-tocopherol was found to be slightly more potent than γ- or δ-tocopherols, the percentage of inhibition being 51%, 42% and 41% (compared to control), respectively. Moreover, α-, β-, γ and δ-tocopherols have never been found to be toxic for smooth muscle cells even at high concentrations, and cell viability was always greater than 95% as assessed by the trypan blue dye exclusion method.

To establish if the differential effects of the tocopherols on smooth muscle cell proliferation were due to their unequal uptake, the amount of these compounds present in 24 h-treated VSMC was measured. Cells

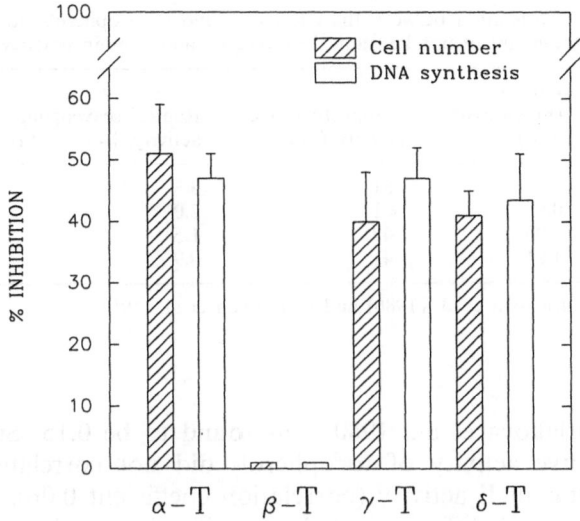

Figure 2. Effect of tocopherol homologues on cell growth. Quiescent cells were restimulated to growth and synchronized as described in Methods in the presence of 100 μM of the following compounds: α-tocopherol (α-T), β-tocopherol (β-T), γ-tocopherol (γ-T) and δ-tocopherol (δ-T). Cell number (hatched columns) and DNA synthesis (blank columns) were measured as described in Methods.

were incubated for 24 h with 100 μM of the different tocopherols (α-, β-, γ- and δ-tocopherols). After washing, measurements of tocopherol content were performed by reverse phase HPLC using a C-18 column (Waters, Inc.) with an in-line electrochemical detector essentially as described earlier (Lang et al., 1986). Uptake was expressed as nmol of incorporated tocopherol per 1×10^6 cells. Control cells were incubated in the absence of any compound.

As shown in Table 2, the content of added tocopherols (α-, β-, γ- and δ-tocopherols) was similar in smooth muscle cells. The amount of endogenous tocopherols present in untreated cells was found to be less than 10 pmol/10^6 cells. Thus, the lack of effect of β-tocopherol compared with the growth inhibition obtained with γ- and δ-tocopherol cannot be attributed to a smaller antioxidant potency of this tocopherol. Studies from several laboratories have shown that the antioxidant activity of several tocopherols strongly correlates with their biological activity (Burlakova et al., 1980; Burton and Ingold, 1981; Mukai et al., 1988; Niki et al., 1987; Bunyan et al., 1961). In fact, the sequence of antioxidant potency is $\alpha > \beta > \gamma > \delta$-tocopherol. However, no correlation was found between the antiproliferative activity of tocopherols and their antioxidant (peroxyl radical scavenging) activity (Table 2). The correlation coefficient between the antiproliferative effect of the analogs and their rate constants in the reaction with peroxyl

Table 2. Lack of correlation between the effect of different tocopherols on smooth muscle cells growth with amount of uptake, radical scavenging and vitamin potency

Compound	Uptake (nmoles/10^6 cells)	Antiproliferative activity (%)	Radical scavenging activity, $k_7 \times 10^6$ M/s	Vitamin E activity (IU/mg)
α-tocopherol	1.12	51	3.3	1.49
β-tocopherol	0.95	<1	2.0	0.75
γ-tocopherol	1.00	42	1.5	0.15
δ-tocopherol	1.81	41	0.9	0.05

*According to Burlakova et al. (1980) and to Bunyan et al. (1961)

radicals (Burlakova et al., 1980) was found to be 0.15. Similarly, the antiproliferative activity of tocopherols did not correlate with their biological vitamin E activity (correlation coefficient 0.06). We did not find a pronounced difference in the uptake of several tocopherols by smooth muscle cells. Thus, it appears that a hydroxyl group in position 6 of the chromane ring is not the only functional group involved in the inhibitory effect of these compounds on cell proliferation. Besides, the presence or absence of ring methyl groups in the 5-, 7-, and 8-positions appears to be important for the inhibitory effect.

Tocopherol analogs have different efficiencies as PKC inhibitors in permeabilized VSMC

Activity of PKC in permeabilized smooth muscle cells was performed according to a published procedure (Alexander et al., 1989) with minor modifications. Briefly, A7r5 cell populations at different stages of the cycle were preincubated with α-tocopherol or the indicated tocopherol, washed twice with PBS, resuspended in intracellular buffer (5.16 mM $MgCl_2$, 94 mM KCl, 12.5 mM Hepes, 12.5 mM EGTA, 8.17 mM $CaCl_2$, pH 7.4) and aliquoted in 220-ml portions (1.5×10^6 cells/ml). Assays were started by adding [γ^{32}P]ATP (40 cpm/pmol, final concentration 240 μM), peptide substrate (final concentration 250 μM), and streptolysin-O (0.6 i.u.). The reaction mixtures were incubated at 37°C for 5 min and stopped by adding 100 ml of 25% (w/v) trichloroacetic acid in 2 M acetic acid. After being kept on ice for 10 min, samples were centrifuged for 5 min and spotted on P81 ion-exchange chromatography paper (Whatman International); they were then washed several times with 30% (v/v) acetic acid containing 1% H_3PO_4 and once with ethanol. The P81 papers were dried, and the bound radioactivity was counted in a liquid scintillation analyzer. To estimate the background phosphorylation of the peptide due to kinase activity other than PKC, measurement of PKC was performed in cells treated for 24 h with 1 μM PMA. Down-regulation of PKC (Kraft et al., 1983) occurs under these conditions as established by PDBu binding. The amount of radioactive phosphate incorporated in the latter condition is subtracted from the experimental data to account for the specific activity.

We have previously shown that α-tocopherol inhibited PKC activity (Boscoboinik et al., 1991). To study the effect of different tocopherols on this reaction, streptolysin-O permeabilized VSMC and a PKC peptide substrate (Alexander et al., 1989) were employed. As can be seen in Figure 3, α-tocopherol had a very strong effect on PKC activity (more than 90% inhibition) whereas its analogs β-, γ-, and δ-tocopherols were completely ineffective.

Inhibition of protein kinase C in permeabilized VSMC was specific to α-tocopherol, while the other tocopherols had no effect. The inhibition was observed regardless of whether the cells were preincubated with α-tocopherol or the compound was added just before running the kinase C assay. This result is in favor of a direct ligand-protein interaction of PKC with α-tocopherol. Other tocopherols (γ- and δ-tocopherols) which were inhibitors of VSMC proliferation did not show any significant PKC inhibition indicating that molecular targets different to PKC are involved in the antiproliferative action of these tocopherols.

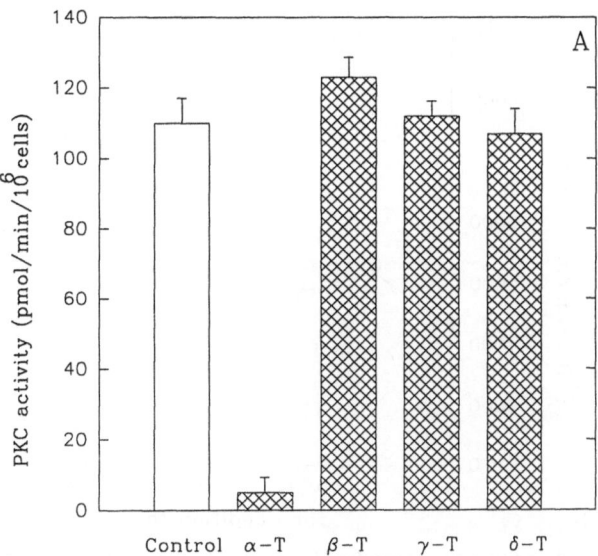

Figure 3. Effect of different tocopherols on PKC activity in permeabilized VSMC. Cells were incubated with 100 μM of the tocopherols as indicated and PKC activity was measured as described in Methods.

α-Tocopherol inhibits growth by acting in the late G_1 phase of the cycle

The inhibitory effect of α-tocopherol on smooth muscle cells-DNA synthesis was studied in greater detail by adding the compound at different times after stimulation of proliferation. The inhibitory response was observed when α-tocopherol was added both at the time of stimulation or at any time during the first 8 h in the G_1 phase. This result indicates that targets for α-tocopherol-mediated inhibition were not present in the early G_0/G_1 transition. However, when α-tocopherol was added 9 h after initial stimulation, no inhibitory effect on DNA synthesis was seen, indicating that once the cultures progressed beyond commitment to DNA synthesis (around 8 h after stimulation), cells became insensitive to the α-tocopherol effects (Fig. 4). The inhibitory effect of α-tocopherol on cellular proliferation appears to be reversible in that treated smooth muscle cells regain their capacity to grow within 30 h following removal of α-tocopherol. This indicates that cytostatic effects of α-tocopherol were not affecting the degree of cell proliferation and viability (Table 3).

PKC inhibition by α-tocopherol in permeabilized A7r5 cells is maximal in late G_1 phase of the cycle

Our previous results suggested that the mechanisms(s) by which α-tocopherol inhibited cellular proliferation was by modulating events

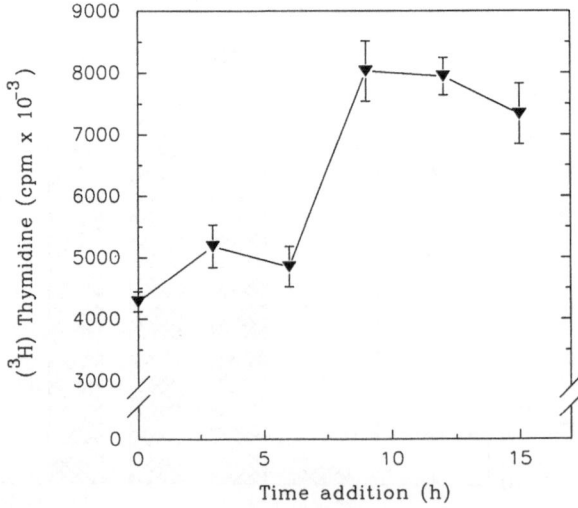

Figure 4. Delayed addition of α-tocopherol indicates an α-tocopherol-sensitive point in G_1. Quiescent cells were restimulated to grow in the presence of FCS. At the indicated times, 100 μM α-tocopherol was added to the cultures and then cells were synchronized by hydroxyurea treatment. After removal of HU, DNA synthesis was measured as described in Methods.

Table 3. Reversibility of the α-tocopherol effect on VSMC.: Low-serum arrested VSMC were restimulated to growth by the addition of 10% FCS, in the presence or absence of 100 μM α-tocopherol. After 6 h, cells were treated for 14 h with hydroxyurea, and subsequently cells pretreated with α-tocopherol were split in two sets of dishes: one received an additional dose of 100 μM α-tocopherol (α-tocopherol-treated cells), and the other received fresh medium-containing serum (washed cells). [³H]thymidine incorporation (3h-pulse) was determined at the indicated times after the hydroxyurea treatment. Recovery was calculated as follows: the difference between the counts in washed and treated cells divided by the difference between the counts in control and treated cells. The values represent [³H]thymidine uptake (c.p.m. × 10³/dish)

Time (h)	Control cells	Treated cells	Washed cells	Recovery (%)
3	121.5 ± 20.4	61.2 ± 0.8	80.6 ± 7.9	35
6	191.4 ± 6.3	110.9 ± 1.7	136.2 ± 10.7	35
9	33.0 ± 1.0	1.2 ± 0.3	24.7 ± 6.6	75
30	60.7 ± 5.5	24.5 ± 11.2	57.2 ± 2.6	92

in the late G_1 phase of the cell cycle. As shown previously (Boscoboinik et al., 1991), one possible target for α-tocopherol action might be protein kinase C. It is not known how kinase C activity is regulated during the smooth muscle cell cycle although it is speculated that both phosphorylation and dephosphorylation reactions may occur (Parker et al., 1989). Thus, we assayed protein kinase C activity at various stages of the cell cycle in permeabilized smooth muscle cells using a 12-amino acid peptide as a substrate, specific for kinase C. Maximal activity was observed during the late G_1 phase of the cycle (Fig. 5), indicating that

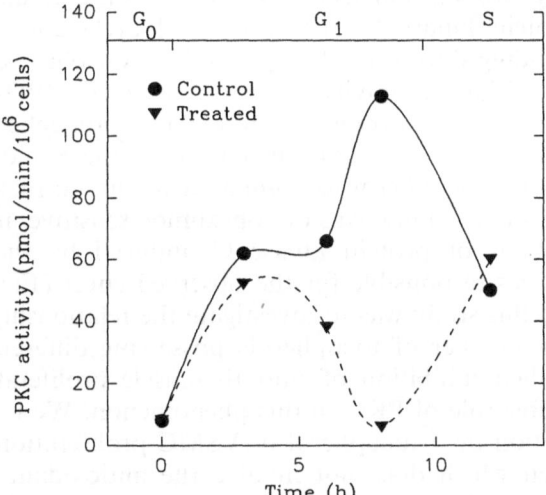

Figure 5. Effect of α-tocopherol on PKC activity during the cell cycle. Control (●) and α-tocopherol-treated (▼) cultures were taken at different stages of the cell cycle and activity of protein kinase C in permeabilized VSMC was determined as described in Methods. Results are the mean ±SD of three experiments.

the kinase activity is regulated in a cell cycle-dependent manner. Next, we measured the effects of α-tocopherol on PKC activity. Quiescent cells (G$_0$ phase) were restimulated to growth with serum in the presence of 50 μM α-tocopherol and kinase C activity was measured as described before. A significant inhibitory effect of α-tocopherol was observed in the late G$_1$ phase where maximal kinase activity was found in control cultures. Similar results were obtained when α-tocopherol was added directly to the assay mixture indicating that a preincubation of the cells with tocopherol was not a prerequisite to observe inhibitory effects.

The expression of [^{3}H]PDBu binding sites in A7r5 is affected by α-tocopherol

Cells were grown in 24-multi-well plates until confluent. All binding assays were done at 37°C in a total volume of 500 ml. Cells were incubated for 30 min with 10 nM [^{3}H]PDBu in the presence or absence of 10 μM unlabeled PDBu for determination of specific binding (Trilivas and Brown, 1989). Cells were washed, solubilized in NaOH/SDS/Na$_2$CO$_3$ and counted in a liquid scintillation analyzer. Protein kinase C is the major if not the only cellular receptor for tumor-promoting phorbol esters, and radioactive phorbol 12,13-dibutyrate has been used to demonstrate the existence of high-affinity specific receptors in a variety of tissues (Adams and Gullick, 1989). In smooth muscle cells, [^{3}H]PDBu was bound to a homogeneous class of binding sites as determined by Scatchard analysis. We analyzed the binding of tritiated PDBu to protein kinase C during the cell cycle and found a high increase in binding during the last part of the G$_1$ phase concomitantly with the peak of kinase activity. This increase of [^{3}H]PDBu binding in the late G$_1$ phase was strongly inhibited by α-tocopherol. However, increase of [^{3}H]PDBu binding occurred during the S and G$_2$ phases in the presence of α-tocopherol as compared to the untreated cells. This increase in phorbol binding was cycloheximide sensitive, indicating that *de novo* synthesis of protein kinase C, induced by the presence of α-tocopherol was responsible for the observed effect (Fig. 6).

The goal of this study was to investigate the relationship between the structure of a number of tocopherols possessing different antioxidant activity and their inhibition of smooth muscle proliferation, together with the possible role of PKC in this phenomenon. We have shown the specific inhibition by α-tocopherol of VSMC proliferation by a mechanism of action which does not involve the antioxidant properties of vitamin E. The inhibitory effect was cell cycle- and PKC dependent. Our finding of the modulation by α-tocopherol of cell growth opens the possibility of its use as an agent against smooth muscle cell proliferation, a hallmark in the arteriosclerotic process, and as an important tool

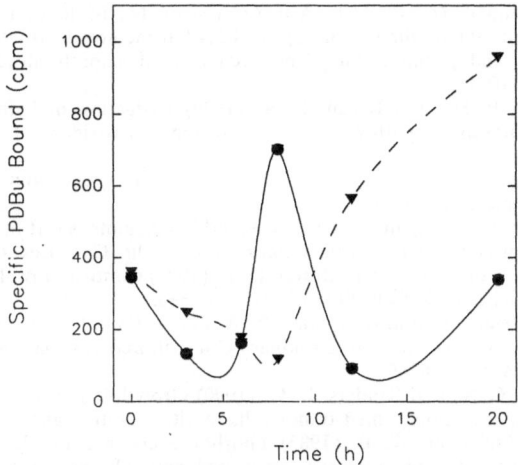

Figure 6. Effect of α-tocopherol on [³H]PDBu binding during the cell cycle. Phorbol binding was measured as described in Methods in cells at different stages of the cycle. Non-specific binding was determined by using an excess of cold PDBu. (●) Control, (▼) 100 μM α-tocopherol-treated cells.

for developing compounds capable of controlling pathways involved in the regulation of cell proliferation.

Acknowledgments. This work was supported by F. Hoffmann-La Roche AG. We greatly appreciated the helpful assistance of Ms. Barbara Conti and Mr. Peter Schmutz in part of the experiments described.

Adams, J. C., and Gullick, W. J. (1989) Differences in phorbol ester-induced down-regulation of protein kinase C between cell lines. Biochem. J. 257: 905–911.

Alexander, D. R., Hexham, J. M., Lucas, S. C., Graves, J. D., Cantrell, D. A., and Crumpton, M. J. (1989) A protein kinase C pseudosubstrate peptide inhibits phosphorylation of the CD3 antigen in streptolysin-O permeabilized human T lymphocytes. Biochem. J. 260: 893–901.

Ashihara, T., and Baserga, R. (1979) Cell Synchronization. Methods Enzym. 58: 248–262. (eds Jakoby and Pastan).

Bobik, A., Grooms, A., Miller, J. A., Mitchell, A., and Grinpukel, S. (1990) Growth factor activity of endothelin on vascular smooth muscle. Am. J. Physiol. 258: C408–C415.

Boscoboinik, D., Szewczyk, A., Hensey, C., and Azzi, A. (1991) Inhibition of cell proliferation by α-tocopherol. Role of protein kinase C. J. Biol. Chem. 266: 6188–6194.

Bunyan, J., McHale, D., Green, J., and Marcinkewicz, S. (1961) Brit. J. Nutr. 15: 253–257.

Burlakova, E. B., Kukhtina, E. N., Olkhovskaya, I. P., Sarycheva, I. K., Sinkina, E. B., and Khrapova, N. G. (1980) Biophysics 24: 989–993.

Burton, G. W., and Ingold, K. U. (1981) J. Amer. Chem. Soc. 103: 6472–6477.

Campisi, J. E., Medrano, E., Morreo, G., and Pardee, A. B. (1982) Restriction point control of cell growth by a labile protein: evidence for increased stability in transformed cells. Proc. Natl. Acad. Sci USA 79: 436–440.

Castellot, J. J., Pukac, L. A., Caleb, B. L., Wright, T. C., and Karnovsky, M. J. (1989) Heparin selectively inhibits a protein kinase C-dependent mechanism of cell cycle progression in calf aortic smooth muscle cells. J. Cell. Biol. 109: 3147–3155.

176

Desgranges, C., Campan, M., Gadeau, A-P, Geurineau, N., Mollard, P., and Razaka, G. (1991) Influence of 8-(N,N-diethylamino) octyl-3,4,5-trimethoxybenzoate (TMB-8) on cell cycle progression and proliferation of cultured arterial smooth muscle cells. Biochem. Pharm. 41: 1045–1054.

Egan, R. W., Gale, P. H., and Kuehl, F. A. (1979) Reduction of hydroperoxides in the prostaglandin biosynthesis pathway by a microsomal peroxidase. J. Biol. Chem. 254: 3295–3302.

Egan, R. W., Gale, P. H., and Kuehl, F. A. (1979) G_1 events and regulation of cell proliferation. Science 246: 603–608.

Geisterfer, A. A., Peach, M. J., and Owens, G. K. (1988) Angiotensin II induces hypertrophy, not hyperplasia, of cultured rat aortic smooth muscle cells. Circ. Res. 62: 749–756.

Gospodarowicz, D., Moran, J. S., and Braun, D. L. (1977) Control of proliferation of bovine vascular endothelial cells. J. Cell. Physiol. 91: 377–385.

Griendling, K. K., Tsuda, T., and Alexander, R. W. (1989) Endothelian stimulates diacylglycerol accumulation and activates protein kinase C in cultured vascular smooth muscle cells. J. Biol. Chem. 264: 8237–8240.

Kline, K., Cochran, G. S., and Sanders, B. G. (1990) Growth-inhibitory effects of vitamin E succinate on retrovirus-transformed tumor cells in vitro. Nutr. Cancer 14: 27–41.

Kraft, A. S., and Anderson, W. B. (1983) Phorbol esters increase the amount of Ca^{2+}, phospholipid-dependent protein kinase, associated with plasma membrane. Nature 301: 621–623.

Ku, G., Doherty, N. S., Wolos, J. A., and Jackson, R. L. (1988) Inhibition by probucol of interleukin-1 secretion and its implication in atherosclerosis. Am. J. Cardiol. 62: 77B.

Lang, J. K., Gohil, K., and Packer, L. (1986) Simultaneous determination of tocopherols, ubiquinols, and ubiquinones in blood, plasma, tissue homogenates, and subcellular fractions. Anal. Biochem. 157: 106–116.

Loesberg, C., van Wijk, R., Zandbergen, J., van Aken, W. G., van Mourik, J. A., and de Groot, Ph. G. (1985) Cell cycle-dependent inhibition of human vascular smooth muscle cell proliferation by prostaglandin E1. Exp. Cell Res. 160: 117–125.

Matsumoto, H., and Sasaki, Y. (1989) Staurosporine, a protein kinase C inhibitor, interferes with proliferation of arterial smooth muscle cells. Biochem. Biophys. Res. Commun. 158: 105–109.

Mukai, K., Fukuda, K., and Ishizu, K. (1988) Stopped-flow investigation of antioxidant activity of tocopherols. Finding of new tocopherol derivatives having higher antioxidant activity than α-tocopherol. Chem. Phys. Lipids 46: 31–36.

Munro, J. M., and Cochran, R. S. (1988) The pathogenesis of atherosclerosis: Atherogenesis and Inflammation. Lab Invest. 58: 249–261.

Niki, E., Tsuchiya, J., Yoshikawa, Y., Yamamoto, Y., and Kamiya, Y. (1987) Bull. Chem. Soc. Jpn. 60: 2161–2165.

Owens, G. K. and Reidy, M. A. (1985) Hyperplastic growth response of vascular smooth muscle cells following induction of acute hypertension in rats by aortic coarctation. Circ. Res. 57: 695–705.

Packer, L., and Landvik, S. (1989) Vitamin E: Introduction to biochemistry and health benefits. Ann. NY Acad. Sci 570: 1–6.

Parker, P. J., Kour, G., Marais, R. M., Mitchell, F., Pears, C., Schaap, D., Stabel, S., and Webster, C. (1989) Protein kinase C – a family affair. Mol. Cell. Endocrinol. 65: 1–11.

Prasad, K. N., and Edwards-Prasad, J. (1982) Effects of tocopherol (vitamin E) acid succinate on morphological alterations and growth inhibition in melanoma cells in culture. Cancer Res. 42: 550–555.

Raines, E. W., Dower, S. K., and Ross, R. (1989) Interleukin-1 mitogenic activity for fibroblasts and smooth muscle cells was due to PDGF-AA. Science 243: 393.

Ran, W., Dean, M., Levine, R. A., Henckle, C., and Campisi, J. (1986) Induction of c-fos and c-myc mRNA by epidermal growth on calcium ionophore is cAMP dependent. Proc. Natl. Acad. Sci. USA 83: 8216–8220.

Ross, R. (1986) The pathogenesis of atherosclerosis – An update. N. Engl. J. Med. 314: 488–500.

Ross, R., Glomset, J. A., Kariya, B., and Harker, L. (1974) A platelet-dependent serum factor that stimulates the proliferation of arterial smooth muscle cells in vitro. Proc. Natl. Acad. Sci. USA 71: 1207–1210.

Rozengurt, E. (1986) Early signals in the mitogenic response. Science 234: 161–166.

Schwartz, S. M., Campbell, G. R., and Campbell, J. H. (1986) Replication of smooth muscle cells in vascular disease. Circ. Res. 58: 427–444.

Slack, R., and Proulx, P. (1989) Studies on the effects of vitamin E on neuroblastoma N1E 115. Nutr. Cancer 12: 75–82.

Tappel, A. L. (1972) Vitamin E and free radical peroxidation of lipids. Ann. NY Acad. Sci. 203: 12–27.

Thyberg, J., and Palmberg, L. (1987) The calcium antagonist nisoldipine and the calmodulin antagonist W-7 synergistically inhibit initiation of DNA synthesis in cultured arterial smooth muscle cells. Biol Cell 60: 125–132.

Trilivas, I., and Brown, J. H. (1989) Increases in intracellular Ca^{2+} regulate the binding of [^{3}H]phorbol 12,13-dibutyrate to intact 132N1 astrocytoma cells. J. Biol. Chem. 264: 3102–3107.

Urano, S., Yano, K., and Matsuo, M. (1988) Membrane stabilizing effect of vitamin E: effect of α-tocopherol and its model compounds on fluidity of lecithin liposomes. Biochem. Biophys. Res. Commun. 150: 469–475.

Free Radicals and Aging
ed. by I. Emerit & B. Chance
© 1992 Birkhäuser Verlag Basel/Switzerland

Oxidative damage in Alzheimer's dementia, and the potential etiopathogenic role of aluminosilicates, microglia and micronutrient interactions

Peter H. Evans[a], Eiji Yano[b], Jacek Klinowski[c] and Ernst Peterhans[d]

[a]MRC Dunn Nutrition Unit, Milton Road, Cambridge, CB4 1XJ, England; [b]Department of Chemistry, University of Cambridge, Lensfield Road, Cambridge, CB2 1EW, England; [c]Department of Public Health, Tokyo University School of Medicine, Itabashi Ku, Kaga 2 Chome, Tokyo 173, Japan; and [d]Institute of Veterinary Virology, University of Berne, Länggass Strasse 122, CH-3012 Berne, Switzerland

Summary. While evidence implicating free radical oxidative processes in the etiopathogenesis of Alzheimer's dementia is accumulating, the specific cellular and biochemical mechanisms involved remain to be identified. The potential pathogenic role of microglial cells in neurodegenerative processes is indicated by the finding that purified murine microglial cells exposed in vitro to various model aluminosilicate particles stimulate the generation of tissue-injurious free radical reactive oxygen metabolites. Analogous inorganic aluminosilicate deposits have been reported to occur in the core of the characteristic senile plaques found in the brains of Alzheimer disease subjects. The possible modulation of free radial oxidative activity by antioxidant micronutrients and pharmacological agents, provides a rational basis for further preventative and therapeutic clinical investigations.

Introduction

"Unrespited, unpitied, unreprieved, Ages of hopeless end."

Paradise Lost   (Milton)

An early description of dementia by Aretaeus of Cappadocia in the 2nd century AD, referred to "An age related torpor of the senses, and stupefaction of the intellectual functions". Now, nearly two millenia later, and 84 years after the eponymous Alois Alzheimer described the pathognomic histopathological features in the brain, the disease continues to wreak its dreadful toll. Demographic trends in the coming decades predict an accelerating growth in the number and percentage of the elderly population. Since the prevalence of neurodegenerative diseases such as Alzheimer's dementia increases markedly after the age of about 75, the medical problems associated with severe cognitive dysfunction and the eventual destruction of the sufferer's personality and health, are anticipated to present major social, financial and ethical dilemmas in the not too distant future. Hence concerted research efforts to ameliorate the devastating effects of the disease are imperative.

While the etiology of Alzheimer's disease remains to be established, present evidence indicates an age-dependent and complex heterogeneous interaction of genetic and environmental influences (Murrell et al., 1991; Gautrin et al., 1989). Of the suspected environmental factors, aluminium, which comprises the most abundant metallic element in the earth's lithosphere, has been the most commonly implicated toxin. Although free radical oxidative species have long been mooted as agents in the aging process and in Alzheimer's disease (Harman, 1988), the precise etiopathogenic biochemical mechanisms involved have yet to be identified. With reference to this specific lacuna in our scientific understanding of the disease process, a proposed integrative hypothesis based on the finding of the capacity of particulate model aluminosilicates to stimulate the generation of brain microglial cell-derived free radical reactive oxidative metabolites (ROM), is presented. In addition, the potential implications of this hypothetical rationale for formulating possible preventative and therapeutic stratagems in the treatment of dementia, are explored.

Free radicals in the brain

The brain is especially vulnerable to oxidative damage in several respects. It has a high rate of oxidative metabolism, contains elevated concentrations of readily-oxidized polyunsaturated fatty acids, relatively low levels of antioxidants, and consists of post-mitotic cells. In addition, because of their elongated morphology and resultant high ratio of surface area to cell volume, neurons may be particularly prone to membrane-mediated damage and impaired repair processes. The generation of ROM within the brain arises from several sources, some of which are specific to the metabolic functions and cellular composition of brain tissue. ROM are generated from monoamine and xanthine oxidase activity, from arachidonate and neurotransmitter metabolism e.g. 6-tryptamine 4,5 dione (Volicer et al., 1990), and from reactive microglial cells, the stimulated brain macrophages (Sonderer et al., 1987). The effects of ROM-mediated damage to brain function include alteration in synaptic transmission (Colton et al., 1986), polymerization of tubulin (Davison et al., 1986), increased blood-brain barrier permeability (Wei et al., 1986), and enhanced release of excitotoxic amino acids, namely glutamate (Pelligrini-Giampietro et al., 1988). In Alzheimer's disease, evidence of free radical involvement is provided by enhanced cellular SOD activity (Zemlan et al., 1989), and in the overexpression of SOD in the 21 trisomy of Down's syndrome dementia (Kedziora et al., 1988). Further evidence is the increased brain G-6-PDH (Martins et al., 1986) and erthyrocyte glutathione peroxidase (Anneren et al., 1986) activity. An accumulation of lipofuscin (Dowson, 1982), of OH-proline in the neurofibrillary tangles (Vogelsang et al., 1989), and an increased level of serum lipid peroxides and related decrease in blood plasma levels of the antioxidant vitamins A, C and E (Burns et al., 1986; Jeandel et al., 1989), has also been found. These results have led to the suggestion that neurodegenerative disorders are due to a deficiency in the brain antioxidant system (Ceballos et al., 1990), although the gross brain vitamin E levels appear not to be decreased (Metcalfe et al., 1989). In addition, endogenous antioxidants, namely urate (Ames et al., 1981) and carnosine (Kohen et al., 1988) are decreased in the sera (Kasa et al., 1989) and brains (Perry et al., 1987) respectively, of Alzheimer subjects.

Aluminosilicates and reactive microglia

Of the potential environmental toxins implicated in the etiology of Alzheimer's disease aluminium (Ganrot, 1986) has been the prime suspect agent, although increased brain levels of mercury have also been detected (Wenstrup et al., 1990). Epidemiological studies have linked the aluminium content of drinking water to the prevalence of Alzheimer's disease (Martyn et al., 1989). Aluminium is commonly added to drinking water as a coagulant. The major intake of aluminium is from food, mainly plants, and as such is exacerbated by acid rain which leaches aluminium from the soil. A large intake of aluminium is also derived from food additives, and pharmaceutical products and incipients (Pennington, 1987). The identification, by ^{27}Al magic angle-spinning NMR, of aluminosilicate deposits within the cores of the characteristic neuritic plaques in the brain (Candy et al., 1986a), is of special significance to the etiopathogenesis of Alzheimer's disease. The finding of increased levels of silicon in the CSF of the senile but not the early onset types of Alzheimer's dementia (Hershey et al., 1984), indicates that the accumulation of such deposits within the brain is highly time-dependent. Since the olfactory areas of the brain are markedly affected in Alzheimer's disease, it has been suggested that the mode of entry of the aluminosilicates is directly into the brain from the nose via retrograde axonal transport along the olfactory nerve (Roberts, 1986). Experimental evidence for the transport of aluminium ions by this mechanism has been provided (Perl et al., 1987), although the finding of neurofibrillary tangles in a subject with olfactory dysgenesis has cast doubt on this suggestion (Arriagada et al., 1991). Ingested particulates may pass through the gastrointestinal tract (Pontefract et al., 1973) and have been detected in various body organs, including the brain (Auer-bach et al., 1980). Inhalation of aluminium dust, as an intended prophy-lactic to miner's silicosis, has been shown to be related to an increased incidence of cognitive dysfunction (Rifat et al., 1990), and shows the potential for pulmonary ingress. Particulates, of similar aluminium and silicon composition and morphology to kaolinite, have been identified in senile plaques (Henderson, W., personal communication), possibly transferred across the blood-brain barrier by phagocytosing monocytes (Williams et al., 1990). However, direct injection of silica particles into the brains of experimental animals did not duplicate the characteristic pathology of Alzheimer's disease (Rees et al., 1983). Apart from the translocation of preformed particulates, deposition of aluminosilicates within the brain may also be envisaged to occur by concretion of aluminium and silicon species in situ (Esiri et al., 1986), aluminosilicate colloids being highly insoluble (Birchall et al., 1988). Amorphous alumi-nosilicate deposits have been detected in the synovial fluid of dialysis patients (Netter et al., 1991), a finding of relevance to the reported increase in joint symptoms in subjects acutely exposed to high levels of water-borne aluminium (Clayton, 1989). Uptake of aluminium is en-hanced by citrate (Fulton et al., 1990) and by glutamate (Deloncle et al., 1990), whilst transfer of aluminium across the blood-brain barrier is mediated via brain receptors for the iron-cum-aluminium-binding protein transferrin (Roskams et al., 1990). Aluminosilicates are ubiqui-tous in the environment, comprising clays, feldspars and zeolite miner-als. Clays and zeolites exhibit a variety of absorptive, catalytic, ion-exchange and potential self-replicative properties (Cairns-Smith, 1986; Newsam, 1986) and are widely utilised and indeed synthesised for many industrial chemical purposes. The deposition of amyloid fibrils in association with aluminosilicate deposits in neuritic plaques (Masters et al., 1985), may be related to the capacity of minerals to act as nucleation centres for protein deposition (McPherson et al., 1988).

To investigate the etiopathogenic significance of the aluminosilicate deposits in Alzheimer plaques, whether they represent a mere inert epiphenomenon or contribute a potential nidus of toxic insult in the pathogenesis of the disease, in vitro investigations utilising purified human blood polymorphonuclear leukocytes (PMN) and murine brain microglia have been undertaken. Microglia have been shown, by a chemiluminescent technique, to generate ROM in response to various chemical and immunological agents (Sonderer et al., 1987). Using model aluminosilicate particulate zeolites of differing morphological, dimensional and chemical composition, it has also been demonstrated that both PMN (Evans et al., 1989a) and murine microglial cells (Evans et al., 1990; 1992), are also capable of being stimulated to generate ROM. Based on these findings, an hypothesis outlining the potential role of aluminosilicate-induced generation of phagocyte-derived ROM in the etiopathogenesis of Alzheimer's disease, so-called cephaloconiosis, has been proposed (Evans, 1988; Evans et al., 1991). Microglia, brain macrophage-type cells (Perry et al., 1988), have been identified within the senile plaque cores, suggesting that they may play an initiating role in plaque formation (Probst et al., 1987). A marked reactive microglial response in association with a neurotoxic reaction, is produced by the injection of Alzheimer amyloid plaque cores into rat brain (Frautschy et al., 1991). Clays (layered aluminosilicates) can also exert a direct toxic effect on cultured neurons (Banin et al., 1990). Subsequent chemiluminescent studies have indicated that soluble aluminium species may also stimulate the PMN respiratory burst (Stankovic et al., 1991). Evidence of the effect of free radical reactions is supported by the finding of increased brain lipid peroxidation in experimental rats fed aluminium in their diet (Ohtawa et al., 1983), a reaction possibly mediated by an enhancement of iron-mediated hydroxyl damage to membrane lipids (Gutteridge et al., 1985).

Micronutrient element and vitamin interactions

Calcium. The aging brain is characterised by altered calcium homeostasis (Gibson et al., 1987), the level of intracellular calcium being critical in the control of a multitude of cellular functions (Campbell, 1983). Of particular pertinence in Alzheimer's disease, is the role of Na/Ca cation shifts in mediating excitotoxic activity (Korf et al., 1986) and cytoskeletal function (Trump et al., 1981). Calcium is deposited within senile plaques (Candy et al., 1986a), and such increased local concentrations may have an adverse effect on phagocyte reactivity (Evans et al., 1990), and produce neurofibrillary degenerative changes, possibly mediated by activation of protein kinase C (Mattson et al., 1991). Furthermore, oxidative damage to brain synaptosomes may be exacerbated by calcium (Braughler et al., 1985).

Iron. Iron, a potential free radial catalyst, is reportedly increased in Alzheimer brains (Andorn et al., 1990), with local accretion in senile plaque cores (Candy et al., 1986b). Similarly, the focal accumulation of lactoferrin within plaques, together with SOD and glutathione peroxidase, has been interpreted as additional evidence of enhanced oxidative stress in Alzheimer's disease (Osmand et al., 1991). Iron is a common constituent of silicaceous mineral dust particles (Evans et al., 1987).

Zinc. Zinc dietary intake and nutritional status tends to be reduced in the elderly (Greger, 1977), and furthermore, adverse changes in zinc homeostasis have been proposed to be intimately linked to the aging process (Garfinkel, 1986). Zinc plays an important role in the central nervous system (Wallwork, 1987), aluminium uptake into the brain being enhanced in rats fed a zinc-deficient diet (Wenk et al., 1983). Zinc also exhibits an antioxidant function in inhibiting lipid peroxidation (Cao et al., 1991), possibly by countering iron-mediated free radical reactions (Wilson, 1989).

Vitamin E and selenium. Experimental studies have shown an increase in brain lipid peroxidation with age which is exacerbated by dietary vitamin E deficiency, the olfactory and hippocampus being especially affected (Noda et al., 1982). Similarly, intensive pretreatment with vitamin E retards peroxidative neural degeneration (Hall, 1987). Dietary intake of vitamin E in the elderly is generally somewhat low (Murphy et al., 1990), with the institutionalized elderly evidently at a greater risk of decreased vitamin E and selenium nutritional status (Campbell et al., 1989). Vitamin E protects against thyroxine-induced acceleration of lipid peroxidation (Asayama et al., 1989), which may be relevant to the report of thyroid-associated risk factors in Alzheimer's disease (Heyman et al., 1984). It is of interest to note that selenium deficiency alters thyroid function (Arthur et al., 1991), and supplementation with selenium has been used in the treatment of Down's subjects (Antila et al., 1990). Vitamin E inhibits phagocyte ROM generating activity (Sakamoto et al., 1990), an effect which is linked to protein kinase C-dependent activation of NADPH oxidase (Watson et al., 1991). Brain protein kinase C activity is also increased in rats fed aluminium (Johnson et al., 1990).

Vitamin C. Ascorbic acid may be useful in enhancing aluminium excretion, as shown in experimental rabbits (Fulton et al., 1990), and its use has been proposed to counteract the extra-cellular release of phagocyte-derived ROM (Anderson et al., 1987).

Antioxidant supplementation

Pharmacological agents. In a recent clinical study in Alzheimer subjects, an amelioration in the disease has been reported using the drug desferrioxamine to chelate and remove aluminium. However, since in addition desferrioxamine is a potent chelator of iron, it was also suggested that the beneficial effect may be mediated by a reduction in iron-dependent ROM generation (McLachlan et al., 1991). Nitric oxide is generated by brain tissue (Knowles et al., 1989), and the capacity of desferrioxamine to react with peroxynitrite (Beckman et al., 1990) offers an alternative mechanism for its therapeutic action. Head trauma is a risk factor in Alzheimer's disease (Mortimer, 1985), and the finding that iron-chelating 21-aminosteroid drugs (lazeroids) are useful in both treating experimental head injury (Hall et al., 1988) and in reducing iron-induced lipid peroxidation in Alzheimer autopsy brains (Subbarao et al., 1990), suggests that they may have a worthwhile role in Alzheimer's disease. Of special scientific significance to the postulated

184

pathogenic role of ROM, is the experimental finding in aging gerbils of a reversal of age-related protein oxidation and memory dysfunction produced by administration of a free radical spin trap (Carney et al., 1991).

Micronutrients. The influence of malnutrition in the etiopathogenesis of Alzheimer's disease is as yet hypothetical and remains to be substantiated (Abalan, 1984). A range of micronutrients exhibit promising antioxidant potential in vitro (Evans et al., 1989b), the bioflavonoid rutin for example shows potential value in countering particle-induced macrophage-derived ROM generation (Afanas'ev et al., 1989), the in vivo application of such experimental findings to humans is still at a rudimentary level. Whilst the advisability of the use of antioxidant vitamin supplementation is a matter of some debate (Machlin, 1989; Pryor, 1987), the practice of taking dietary supplements is widespread within the elderly population in the USA (Garry et al., 1982). However, whereas the efficacy in preventative terms of wide-scale micronutrient supplementation has yet to be demonstrated, the potential financial and ethical problems associated with the equivalent widespread prescription of pharmacological agents would pose considerable dilemmas. The controlled clinical trials which have been undertaken in elderly and dementia subjects using supplements of micronutrient antioxidants, namely: selenium and vitamin E (Tolonen et al., 1985); selenium, zinc, vitamins A, B6, C and E (Clausen et al., 1989); selenium in Down's syndrome (Antila et al., 1990); and zinc, selenium and gamma-linolenic acid (Van Rhijn et al., 1990), have been encouraging. Additional clinical investigations, carefully designed and larger in scale, are needed in this area of nutritional medicine.

Conclusion

Although there is accumulating evidence for an exacerbated oxidative stress in Alzheimer's disease, direct information on the cellular and biochemical mechanisms involved in the generation of the excess production of free radical oxidants is lacking. However, in the light of the model in vitro experiments described, it is proposed that reactive microglial cells may similarly act as potent in vivo sources of ROM, including nitric oxide and possibly other cytokines, as a result of chronic stimulation by analogous insoluble and persistent fibrillar aluminosilicate-amyloid particulates located in the Alzheimer plaque core nidus. Further research based on both limiting the intake of aluminium and aluminosilicates, and the use of antioxidant micronutrient supplements and drugs, provides a rational and feasible therapeutic strategy. If validated, one can hope that Milton's poetic despairing cry may be at least partly assuaged.

Abalan, F. (1984) Alzheimer's disease and malnutrition: a new etiological hypothesis. Med. Hypoth. 15: 385–393.

Afanas'ev, I. B., Korkina, L. G., Briviba, K. K., Gunar, V. I., and Velichkovskii, B. T. (1989) Protection of cells by rutin and iron-rutin complex against free radical damage, in: Medical, Biochemical and Chemical Aspects of Free Radicals. Hayaishi, O., Niki, E., Kondo, M. and Yoshikawa, T. eds. Elsevier, Amsterdam, pp. 515–518.

Ames, B. N., Cathcart, R., Schwiers, E., and Hochstein, P. (1981) Uric acid provides an antioxidant defence in humans against oxidant- and radical-caused aging and cancer: a hypothesis. Proc. Natl. Acad. Sci. 78: 6858–6862.

Anderson, R., and Lukey, P. T. (1987) A biological role for ascorbate in the selective neutralization of extracellular phagocyte-derived oxidants. Ann. NY Acad. Sci. 498: 229–247.

Andorn, A. C., Britton, R. S., and Bacon, B. R. (1990) Evidence that lipid peroxidation and total iron are increased in Alzheimer's brain. Neurobiol. Aging 11: 316.

Anneren, G., Gardner, A., and Lundin, T. (1986) Increased glutathione peroxidase activity in erthyrocytes in patients with Alzheimer's disease/senile dementia of Alzheimer's type. Acta Nerol. Scand. 73: 586–589.

Antila, E., Nordberg, U.-R., Syvåoja, E.-L., and Westermarck, T. (1990) Selenium therapy in Down syndrome (DS): a theory and clinical trial, in: Antioxidants in Therapy and Preventive Medicine. Emerit, I., Packer, L. and Auclair, C. eds. Plenum, NY, pp. 183–186.

Arriagada, P. V., Louis, D. N., Hedly-Whyte, E. T., and Hyman, B. T. (1991) Neurofibrillary tangles and olfactory dysgenesis. Lancet 337: 559.

Arthur, J. R., Nicol, F., and Beckett, G. J. (1991) The roles of selenium in thyroid hormone metabolism, in: Trace Elements in Man and Animals (TEMA-7). Momçilovic, B. ed. Inst. Med. Res., Zagreb, pp. 7; 3-7; 6.

Asayama, K., Dobashi, K., Hayashibe, H., and Kato, K. (1989) Vitamin E protects against thyroxine-induced acceleration of lipid peroxidation in cardiac and skeletal muscles in rats. J. Nutr. Sci. Vitaminol. 35: 407–418.

Auerbach, O., Conston, A. S., Garfinkel, L., Parks, V. R., Kaslow, H. D., and Hammond, E. C. (1980) Presence of asbestos bodies in organs other than the lung. Chest 77: 133–137.

Banin, E., and Meiri, H. (1990) Toxic effects of alumino-silicates on nerve cells. Neuroscience 39: 171–178.

Beckman, J. S., Beckman, T. W., Chen, J., Marshall, P. A., and Freeman, B. A. (1990) Apparent hydroxyl radical production by peroxynitrite: implications for endothelial injury from nitric oxide and superoxide. Proc. Natl. Acad. Sci. 87: 1620–1624.

Birchall, J. D., and Chappell, J. S. (1988) The chemistry of aluminium and silicon in relation to Alzheimer's disease. Clin. Chem. 34: 265–267.

Braughler, J. M., Duncan, L. A., and Goodman, T. (1985) Calcium enhances in vitro radical-induced damage to brain synaptosomes, mitochondria, and cultured spinal cord neurons. J. Neurochem. 45: 1288–1293.

Burns, A., and Holland, T. (1986) Vitamin E deficiency. Lancet 1: 805–806.

Cairns-Smith, A. G. and Hartman, H. eds. (1986) Clay Minerals and the Origin of Life. CUP, Cambridge.

Campbell, A. K. (1983) Intracellular Calcium: Its Universal Role as a Regulator. John Wiley, Chichester.

Campbell, D., Bunker, V. W., Thomas, A. J., and Clayton, B. E. (1989) Selenium and vitamin E status of healthy and institutionalized eldery subjects: analysis of plasma, erythrocytes and platelets. Br. J. Nutr. 62: 221–227.

Candy, J. M., Klinowski, J., Perry, R. H., Perry, E. K., Fairbairn, A., Oakley, A. E., Carpenter, T. A., Atack, J. R., Blessed, G., and Edwardson, J. A. (1986a) Aluminosilicates and senile plaque formation in Alzheimer's disease. Lancet 1: 354–357.

Candy, J. M., Oakley, A. E., Watt, F., Grime, G. W., Klinowski, J., Perry, R. H., and Edwardson, J. A. (1986b) A role for aluminium, silicon and iron in the genesis of senile plaques, in: Modern Trends in Aging Research. EURAGE, John Libbey, 147: 443–450.

Cao, G., and Chen, J. (1991) Effects of dietary zinc on free radical generation, lipid peroxidation, and superoxide dismutase in trained mice. Archs Biochem. Biophys. 291: 147–153.

Carney, J. M., Starke-Reed, P. E., Oliver, C. N., Landum, R. W., Cheng, M. S., Wu, J. F., and Floyd, R. A. (1991) Reversal of age-related increase in brain protein oxidation, decrease in enzyme activity, and loss in temporal and spatial memory by chronic administration of the spin-trapping compound N-tert-butyl-α-phenylnitrone. Proc. Natl. Acad. Sci. 88: 3633–3636.

186

Ceballos, I., Javoy-Agid, F., Delacourte, A., Defossez, A., Nicole, A., and Sinet, P. M. (1990) Parkinson's disease and Alzheimer's disease: neurodegenerative disorders due to brain antioxidant system deficiency? in: Antioxidants in Therapy and Preventive Medicine. Emerit, I., Packer, L. and Auclair, C. eds. Plenum, NY, pp. 493–498.

Clausen, J., Nielson, S. A., and Kristensen, M. (1989) Biochemical and clinical effects of an antioxidative supplementation of geriatric patients. A double blind study. Biol. Trace. Element Res. 20: 135–151.

Clayton, B. (1989) Water pollution at Lowermoor North Cornwall. Report of the Lowermoor Incident Health Advisory Group. Cornwall and Isles of Scilly District Health Authority.

Colton, C. A., Colton, J. S., and Gilbert, D. L. (1986) Changes in synaptic transmission produced by hydrogen peroxide. J. Free Rad. Biol. Med. 2: 141–148.

Davison, A. J., Legault, N. A., and Steele, D. W. (1986) Effect of 6-hydroxydopamine on polymerization of tubulin. Biochem. Pharmacol. 35: 1411–1417.

Deloncle, R., Guillard, O., Clanet, F., Courtois, P., and Piriou, A. (1990) Aluminium transfer as glutamate complex through the blood-brain barrier. Biol. Trace Element Res. 25: 39–45.

Dowson, J. H. (1982) Neuronal lipofuscin accumulation in ageing and Alzheimer dermentia: a pathogenic mechanism? Br. J. Psychiat. 140: 142–148.

Esiri, M. M., and Williams, R. J. P. (1986) Comments on an olfactory source for an environmental influence and possible involvement of aluminium in the development of Alzheimer's disease. Neurobiol. Aging 7: 582–583.

Evans, P. H., Campbell, A. K., Yano, E., and Goodman, B. (1987) Phagocytic oxidant stress and antioxidant interactions in the pneumoconioses and dust-induced tumourigenic lung disease, in: Free Radicals, Oxidant Stress and Drug Action. Rice-Evans, C. ed. Richelieu Press, London, pp. 213–235.

Evans, P. H. (1988) Alzheimer's senile dementia: a radical case of cephaloconiosis. Neurobiol. Aging 9: 225–226.

Evans, P. H., Klinowski, J., Yano, E., and Urano, N. (1989a) Alzheimer's disease: a pathogenic role for aluminosilicate-induced phagocytic free radicals. Free Rad. Res. Comms. 6: 317–321.

Evans, P. H., Campbell, A. K., Yano, E., and Morgan, L. G. (1989b) Environmental cancer, phagocytic oxidant stress and nutritional interactions, in: Nutritional Impact of Food Processing. Somogyi, J. C. and Muller, H. R. eds: Karger, Basel. Bibl. Nutr. Dieta 43: 313–326.

Evans, P. H., Peterhans, E., Bürge, T., and Klinowski, J. (1990) Aluminosilicate-induced free radical generation by murine brain glial cells in vitro: potential pathogenic and nutritional interactions in Alzheimer's dementia. Neurobiol. Aging 11: 288.

Evans, P. H., Klinowski, J., and Yano, E. (1991) Cephaloconiosis: a free radical perspective on the proposed particulate-induced aetiopathogenesis of Alzheimer's dementia and related disorders. Med. Hypoth. 34: 209–219.

Evans, P. H., Peterhans, E., Bürge, T., and Klinowski, J. (1992) Aluminosilicate-induced free radical generation by murine brain glial cells in vitro: potential significance in the aetio-pathogenesis of Alzheimer's dementia. Dementia 3: 1–6.

Frautschy, S. A., Baird, A., and Cole, G. M. (1991) Effects of injected Alzheimer β-amyloid cores in rat brain. Proc. Natl. Acad. Sci. 88: 8362–8366.

Fulton, B., and Jeffery, E. H. (1990) Absorption and retention of aluminium from drinking water. 1. Effect of citric and ascorbic acid on aluminium tissue levels in rabbits. Fundam. Appl. Toxicol. 14: 788–796.

Ganrot, P. O. (1986) Metabolism and possible health effects of aluminium. Environ. Health Perspect. 65: 363–441.

Garfinkel, D. (1986) Is aging inevitable? The intracellular zinc deficiency hypothesis of aging. Med. Hypoth. 19: 117–137.

Garry, P. J., Goodwin, J. S., Hunt, W. C., Hooper, E. M., and Leonard, A. G. (1982) Nutritional status in a healthy elderly population: dietary and supplemental intakes. Am. J. Clin. Nutr. 36: 319–331.

Gautrin, D., and Gauthier, S. (1989) Alzheimer's disease: environmental factors and etiologic hypotheses. Can. J. Neurol. Sci. 16: 375–387.

Gibson, G. E., and Peterson, C. (1987) Calcium and the aging nervous system. Neurobiol. Aging 8: 329–343.

Greger, J. L. (1977) Dietary intake and nutritional status in regard to zinc of institutionalized aged. J. Gerontol. 32: 549–553.

Gutteridge, J. M. C., Quinlan, G. J., Clark, I., and Halliwell, B. (1985) Aluminium salts accelerate peroxidation of membrane lipids stimulated by iron salts. Biochim. Biophys. Acta 835: 441–447.

Hall, E. D., Yonkers, P. A., McCall, J. M., and Braughler, J. M. (1988) Effects of the 21-aminosteroid U74006F on experimental head injury in mice. J. Neurosurg. 68: 456–461.

Hall, E. D. (1987) Intensive anti-oxidant pretreatment retards motor nerve degeneration. Brain Res. 413: 175–178.

Harman, D. (1988) Free radicals in aging. Mol. Cell Biochem. 84: 155–161.

Hershey, L. A., Hershey, C. O., and Varnes, A. W. (1984) CSF silicon in dementia: a prospective study. Neurology 34: 1197–1201.

Heyman, A., Wilkinson, W. E., Stafford, J. A., Helms, M. J., Sigmon, A. H., and Weinberg, T. (1984) Alzheimer's disease: a study of epidemiological aspects. Ann. Neurol. 15: 335–341.

Jeandel, C., Nicolas, M. B., Dubios, F., Nabet-Belleville, F., Penin, F., and Cuny, G. (1989) Lipid peroxidation and free radical scavengers in Alzheimer's disease. Gerontology 35: 275–282.

Johnson, G. V. W., Cogdill, K. W., and Jope, R. S. (1990) Oral aluminium alters in vitro protein phosphorylation and kinase activities in rat brain. Neurobiol. Aging. 11: 209–216.

Kasa, M., Bierma, T. J., Waterstraat, F., Corsaut, M., and Singh, S. P. (1989) Routine blood chemistry screen: a diagnostic aid for Alzheimer's disease. Neuroepidemiol. 8: 254–261.

Kedziora, J., and Bartosz, G. (1988) Down's syndrome: a pathology involving the lack of balance of reactive oxygen species. Free Rad. Biol. Med. 4: 317–330.

Knowles, R. G., Palacios, M., Palmer, R. M. J., and Moncada, S. (1989) Formation of nitric oxide from L-arginine in the central nervous system: a transduction mechanism for stimulation of the soluble guanylate cyclase. Proc. Natl. Acad. Sci. 86: 5159–5162.

Kohen, R., Yamamoto, Y., Cundy, K. C., and Ames, B. N. (1988) Antioxidant activity of carnosine, homocarnosine, and anserine present in muscle and brain. Proc. Natl. Acad. Sci. 85: 3175–3179.

Korf, J. Gramsbergen, J. B. P., Prenen, G. H. M., and Go, K. G. (1986) Cation shifts and excitotoxins in Alzheimer and Huntington disease and experimental brain damage. in: Progress in Brain Research. Swaab, D. F., Fliers, E., Mirmiran, M., Van Gool, W. A. and Van Haaren, F. eds. Elsevier, Amsterdam, pp. 213–226.

Machlin, L. J. (1989) Use and safety of elevated dosages of vitamin E in adults, in: Elevated Dosages of Vitamins. Benefits and Hazards. Walter, P., Brubacher, G. and Stahelin, H. eds. Hans Huber, Toronto, pp. 56–68.

Martins, R. N., Harper, C. G., Stokes, G. B., and Masters, C. L. (1986) Increased cerebral glucose-6-phosphate dehydrogenase activity in Alzheimer's disease may reflect oxidative stress. J. Neurochem. 46: 1042–1045.

Martyn, C. N., Osmond, C., Edwardson, J. A., Barker, D. J. P., Harris, E. C., and Lacey, R. F. (1989) Geographical relation between Alzheimer's disease and aluminium in drinking water. Lancet 1: 59–62.

Masters, C. L., Multhaup, G., Simms, G., Pottgiesser, J., Martins, R. N., and Bayreuther, K. (1985) Neuronal origin of a cerebral amyloid: neurofibrillary tangles of Alzheimer's disease contain the same protein as the amyloid of plaque cores and blood vessels. EMBO J. 4: 2757–2763.

Mattson, M. P., Rychlik, B., and Engle, M. G. (1991) Possible involvement of calcium and inositol phospholipid signaling pathways in neurofibrillary degeneration, in: Alzheimer's Disease: Basic Mechanisms, Diagnosis and Therapeutic Strategies. Iqbal, K., McLachlan, D. R. C., Winblad, B. and Wisniewski, H. M. eds. Wiley, Chichester, pp. 191–198.

McPherson, A., and Shlichta, P. (1988) Heterogeneous and epitaxial nucleation of protein crystals on mineral surfaces. Science 239: 385–387.

McLachlan, D. R. C., Dalton, A. J., Kruck, T. P. A., Bell, M. Y., Smith, W. L., Kalow, W., and Andrews, D. F. (1991) Intramuscular desferrioxamine in patients with Alzheimer's disease. Lancet 337: 1304–1308.

Metcalfe, T., Bowen, D. M., and Muller, D. P. R. (1989) Vitamin E concentrations in human brain of patients with Alzheimer's disease, fetuses with Down's syndrome, centenarians and controls. Neurochem. Res. 14: 1209–1212.

Mortimer, J. A., French, L. R., Hutton, J. T., and Schuman, L. M. (1985) Head injury as a risk factor for Alzheimer's disease. Neurology 35: 264–267.

Murphy, S. P., Subar, A. F., and Block, G. (1990) Vitamin E intakes and sources in the United States. Am. J. Clin. Nutr. 52: 361–367.

Murrell, J., Farlow, M., Ghetti, B., and Benson, M. D. (1991) A mutation in the amyloid precursor protein associated with hereditary Alzheimer's disease. Science 254: 97–99.

Netter, P., Steinmetz, J., Gillet, P., Kessler, M., Bardin, T., Fener, P., Burnel, D., Gaucher, A., Pourel, J., and Bannwarth, B. (1991) Amorphous aluminosilicates in synovial fluid in dialysis-associated arthropathy. Lancet 337: 554–555.

Newsam, J. M. (1986) The zeolite cage structure. Science 231: 1093–1099.

Noda, Y., McGeer, P. L., and McGeer, E. G. (1982) Lipid peroxides in brain during aging and vitamin E deficiency: possible relations to changes in neurotransmitter indices. Neurobiol. Aging 3: 173–178.

Ohtawa, M., Seko, M., and Takayama, F. (1983) Effect of aluminum ingestion on lipid peroxidation in rats. Chem. Pharm. Bull. 31: 1415–1418.

Osmand, A. P., and Switzer, R. C. (1991) Differential distribution of lactoferrin and Alz-50 immunoreactivites in neuritic plaques and neurofibrillary tangles, in: Alzheimer's Disease: Basic Mechanisms, Diagnosis and Therapeutic Strategies. Iqbal, K., McLachlan, D. R. C., Winblad, B. and Wisniewski, H. M. eds. Wiley, Chichester, pp. 219–228.

Pelligrini-Giampietro, D. E., Cherici, G., Alesiani, M., Carla, V., and Moroni, F. (1988) Excitatory amino acid release from rat hippocampal slices as a consequence of free-radical formation. J. Neurochem. 51: 1960–1963.

Pennington, J. A. T. (1987) Aluminium content of foods and diets. Food Additiv. Contam. 5: 161–232.

Perl, D. P., and Good, P. F. (1987) Uptake of aluminium into central nervous system along nasal-olfactory pathways. Lancet 1: 1028.

Perry, T. L., Yong, V. W., Bergeron, C., Hansen, S., and Jones, K. (1987) Amino acids, glutathione, and glutathione transferase activity in the brains of patients with Alzheimer's disease. Ann. Neurol. 21: 331–336.

Perry, V. H., and Gordon, S. (1988) Macrophages and microglia in the nervous system. TINS 11: 273–277.

Pontefract, R. D., and Cunningham, H. M. (1973) Penetration of asbestos through the digestive tract of rats. Nature 243: 352–353.

Probst, A., Brunnschweiler, H., Lautenshlager, C., and Ulrich, J. (1987) A special type of senile plaque, possibly an initial stage. Acta Neuropathol. 74: 133–141.

Pryor, W. A. (1987) Views on the wisdom of using antioxidant vitamin supplements. Free Rad. Biol. Med. 3: 189–191.

Rees, S., and Cragg, B. (1983) Is silica involved in neuritic (senile) plaque formation? Acta Neuropathol. 59: 31–40.

Rifat, S. L., Eastwood, M. R., McLachlan D. R. C., and Corey, P. N. (1990) Effect of exposure of miners to aluminium powder. Lancet 336: 1162–1165.

Roberts, E. (1986) Alzheimer's disease may begin in the nose and may be caused by aluminosilicates. Neurobiol. Aging 7: 561–567.

Roskams, A. J., and Connor, J. R. (1990) Aluminium access to the brain: a role for transferrin and its receptor. Proc. Natl. Acad. Sci. 87: 9024–9027.

Sakamoto, W., Fujie, K., Handa, H., Ogihara, T., and Mino, M. (1990) In vivo inhibition of superoxide production and protein kinase C activity in macrophages from vitamin E-treated rats. Internat. J. Vit. Nutr. Res. 60: 338–342.

Sonderer, B., Wild, P., Wyler, R., Fontana, A., Peterhans, E., and Schwyzer, M. (1987) Murine glia cells in culture can be stimulated to generate reactive oxygen. J. Leukocyte Biol. 42: 463–473.

Stankovic, A., and Mitrovic, D. R. (1991) Aluminum salts stimulate luminol-enhanced chemiluminescence production by human neutrophils. Free Rad. Res. Comms. 14: 47–55.

Subbarao, K. V., Richardson, J. S., and Ang, L. C. (1990) Autopsy samples of Alzheimer's cortex show increased peroxidation in vitro. J. Neurochem. 55: 342–345.

Tolonen, M., Halme, M., and Sarna, S. (1985) Vitamin E and selenium supplementation in geriatric patients. A double-blind preliminary clinical trial. Biol. Trace Element Res. 7: 161–168.

Trump, B. F., Berezesky, I. K., and Phelps, P. C. (1981) Sodium and calcium regulation and the role of the cystoskeleton in the pathogenesis of disease: a review and hypothesis. Scan. Electron Microscopy 11: 435–454.

Van Rhijn, A. G., Prior, C. A., and Corrigan, F. M. (1990) Dietary supplementation with zinc sulphate, sodium selenite and fatty acids in early dementia of Alzheimer's type. J. Nutr. Med. 1: 259–266.

Vogelsang, G. D., Zemlan, F. P., and Dean, G. E. (1989) Hyperpurification of paired helical filaments reveals elevations in hydroxyproline content and a core structure related peptide fragment, in: Alzheimer's Disease and Related Disorders. Iqbal, K., Wisniewski, H. M. and Winblad, B., eds. Alan Liss Inc, NY, pp. 791–800.

Volicer, L., and Crino, P. B. (1990) Involvement of free radicals in dementia of the Alzheimer type: a hypothesis. Neurobiol. Aging 11: 567–571.

Wallwork, J. C. (1987) Zinc and the central nervous system. Prog. Food Nutr. Soc. 11: 203–247.

Watson, F., Robinson, J., and Edwards, S. W. (1991) Protein kinase C-dependent and -independent activation of the NADPH oxidase of human neutrophils. J. Biol. Chem. 266: 7432–7439.

Wei, E. P., Ellison, M. D., Kontos, H. A., and Povlishock, J. T. (1986) $O_2 \cdot$ radicals in arachidonate-induced increased blood-brain barrier permeability to proteins. Am. J. Physiol. 251: H693–699.

Wenk, G. L., and Stemmer, K. L. (1983) Suboptimal dietary zinc intake increases aluminum accumulation into the rat brain. Brain Res. 288: 393–395.

Wenstrup, D., Ehmann, W. D., and Markesbery, W. R. (1990) Trace element imbalances in isolated subcellular fractions of Alzheimer's disease brains. Brain Res. 533: 125–131.

Williams, A. E., and Blakemore, W. F. (1990) Monocyte-mediated entry of pathogens into the central nervous system. Neuropathol. Appl. Neurobiol. 16: 377–392.

Willson, R. L. (1989) Zinc and iron in free radical pathology and cellular control, in: Zinc in Human Biology. Mills, C. F. ed. Springer-Verlag, London, pp. 147–172.

Zemlan, F. P., Thienhaus, O. J., and Bosmann, H. B. (1989) Superoxide dismutase activity in Alzheimer's disease: possible mechanism for paired helical filament formation. Brain Res. 476: 160–162.

Free Radicals and Aging
ed. by I. Emerit & B. Chance

The role of alterations in free radical metabolism in mediating cognitive impairments in Down's syndrome

Karen L. Brugge, Sharon Nichols, Dean Delis, Tsunao Saitoh and Doris Truaner

Department of Neuroscience's and Department of Psychiatry, University of California, San Diego, CA 92039, USA

Summary. Down's syndrome (DS) is a genetic disorder involving an excess of chromosome 21 (trisomy 21) in approximately 96% of the cases and comprises approximately 15% of the population with mental retardation (Heller, 1969). In addition to the constitutional mental deficiencies associated with the syndrome many DS patients develop dementia associated with Alzheimer's disease (AD) in their later years of life (Thase et al., 1984). The genetic locus for Cu,Zn-superoxide dismutase (SOD1), a key enzyme in free radical metabolism, is located on chromosome 21, and the activity level of this enzyme is elevated by approximately 50% in a variety of cells of DS patients (see Kedziora and Bartosz, 1988; Sinet, 1982). Because alterations in free radical metabolism may be involved in neuronal death and may be associated with a number of pathological manifestations of DS, it is important to understand the role of free radical metabolism in cognitive impairments of DS, the topic discussed in this chapter.

Alterations of free radical metabolism in Down's syndrome

In addition to an elevation of SOD1 activity in various cells of DS patients a number of studies show an elevation of glutathione peroxidase (GSHPx) activity as well (Kedziora and Bartosz, 1988; Sinet, 1982). SOD catalyzes the reaction which converts superoxide radicals to peroxide, and GSHPx metabolizes peroxide to water and oxygen. In vitro and in vivo studies demonstrate that elevations of SOD1 results in an elevation of GSHPx. Conditions of oxidative stress such as paraquat toxicity (Groner, et al., 1986) or ionization radiation (Petkau, 1978; Misra et al., 1976), which generate free radicals, is associated with elevations of GSHPx as well as SOD1 activity. Cells transfected with human SOD1 gene (Groner et al., 1986), and various cells of the transgenic mouse (Ceballos et al., 1991a), which contain an extra copy of the SOD1 gene, show an elevation of GSHPx activity as well as an increase in SOD1 activity. Furthermore, the SOD1 activity in transfected neuroblastoma cells or mouse L cells was found to be significantly correlated with GSHPx activity at $r = 0.99$ and $r = 0.88$, respectively (Ceballos et al., 1988). However, the balance in the synthesis/activity of these two enzymes, particularly in conditions of oxidative stress, may be altered such that elevations in SOD activity would exceed that of GSHPx. Although SOD1 activity is elevated by approximately 50% in DS patients, GSHPx activity is reported to be elevated within a range of 30 to 50%, and sometimes, to show no elevation compared to controls (Kedziora and Bartosz, 1988). The activity of catalase, another enzyme which catalyzes the reaction for converting peroxide to water and oxygen is not elevated in DS. Furthermore, SOD1 activity is not significantly correlated with GSHPx among DS patients (Brugge et al., 1991, manuscript in preparation), unlike studies involving more controlled laboratory conditions such as in SOD1-transfected cells (Ceballos et al., 1988). Thus, submaximal elevations of GSHPx activity in DS may, in contrast to SOD activity, be attributed to factors such as environmental conditions.

The results of Groner and colleagues (1991) are consistent with the hypothesis that the balance between SOD and GSHPx activity is critical in the prevention of toxic effects of free radicals. These investigators studied the effects of paraquat toxicity in human HeLa and mouse L cells transfected with the human SOD1 gene. Although cells with elevated SOD1 activity showed some resistance to the acute toxic effects of paraquat, cells with even greater elevations of SOD1 showed a reduction in resistance to paraquat toxicity as measured by the percentage of cells surviving exposure to the toxin. The authors of this study hypothesized that the reduced resistance of cells with further elevations of SOD1 was due to an increased production of peroxide. They noted that cells showing resistance to paraquat toxicity no longer showed resistance with more prolonged exposure, even at the lowest concentrations employed. Although these investigators did not measure GSHPx activity, one might suppose that in the cells with reduced resistance, GSHPx activity was not sufficiently elevated to compensate for the elevation of SOD1 in response to an increased genetic dosage as well as to paraquat-induced elevation.

Similar effects of elevations in SOD, as reported to occur in transfected cells under conditions of oxidative stress, such as paraquat toxicity, may exist in DS in which the genetic dosage for GSHPx is not in excess. Thus an imbalance in the free radical metabolic system may exist, and suboptimal elevations in GSHPx may play an important role in the pathological manifestations of DS. Based on results obtained by Sinet and colleagues (1979), these investigators proposed that GSHPx may play an important role in the cerebral function of DS patients. Due to the elevation of peroxides that may exist in DS, at least during conditions of oxidative stress, increased lipid peroxidation may occur (Kedziora and Bartosz, 1988; Brooksbank and Balazs, 1984). Several studies show that in response to conditions of oxidative stress such as paraquat toxicity (Groner et al., 1986) or ionization radiation (Southern and Powis, 1988), an elevation of thiobarbituric acid reactive products occurs, which is believed to reflect the presence of lipid peroxidation.

Free radical toxicity in the brain

It appears that the aging brain may be particularly susceptible to alterations of free radical metabolism, as suggested by Harman (1983) and Floyd et al. (1984). While GSHPx activity increases in a variety of cells exhibiting an elevation of SOD as discussed above, GSHPx activity does not appear to increase in brain homogenates from human SOD1 transgenic mice (Ceballos et al., 1991a), trisomy 16 mice (Anneren and Epstein, 1987), which have an excess of the mouse SOD gene, or in DS fetal brains (Brooksbank and Balazs, 1984). Yet these brain homogenates exhibit the predicted elevation of SOD activity. Furthermore, an increase in thiobarbituric acid reactive products occurs, suggesting an elevation of lipid peroxidation, and may be due to peroxides generated by an imbalance between the enzymatic activity of GSHPx and SOD. Polyunsaturated fatty acids, a major constituent of brain membranes, particularly synaptosomal membranes, can undergo lipid peroxidation as described elsewhere (Clausen, 1984). In DS patients polyunsaturated fatty acid content is reduced, particularly in the synaptosomes, in contrast to normal brain tissue (Shah, 1979), possibly due to increased lipid peroxidation. Thus, one might speculate that mental deficiency in DS may be due to an imbalance in the oxidative system resulting from an elevation of SOD activity with the lack of sufficient elevations of GSHPx activity in response to oxidative stress. Further investigation in this area of research is required to elucidate possible mechanisms in mediating the neuropathological manifestations of DS.

Free radical metabolism and cognitive impairment in Down's syndrome

Sinet and colleagues (1979) found that erythrocyte GSHPx activity was significantly positively correlated with the intelligence quotient (IQ) in DS patients, while IQ or GSHPx activity was not correlated with age. Thus, an elevation of GSHPx may reduce oxidative damage in a variety of cells in DS, including the brain. In an older population of 14 DS subjects, erythrocyte GSHPx activity was inversely correlated with severity of a specific type of memory deficit, the number of intrusion errors, as shown in Figure 1A (Brugge et al., 1991, manuscript in preparation). Thus fewer intrusion errors were associated with greater GSHPx activity among the DS patients, which is consistent with results

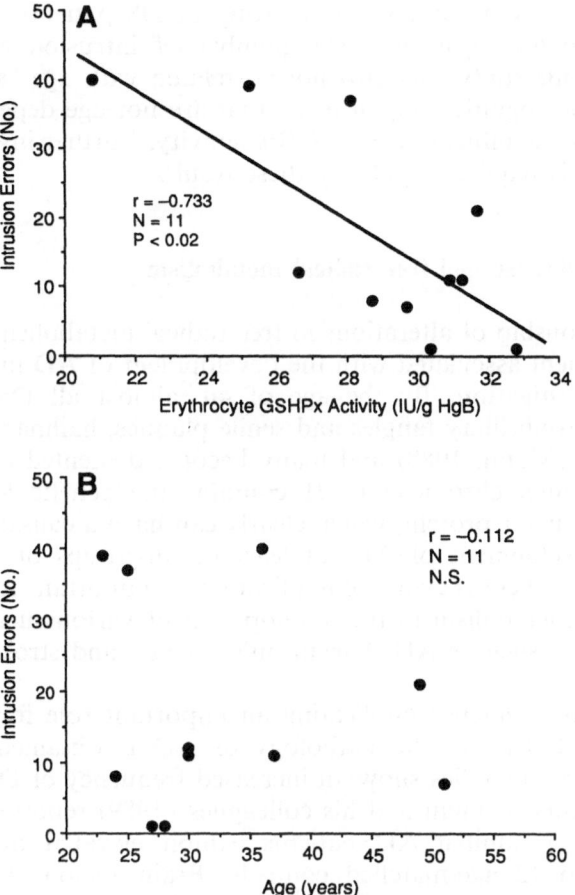

Figure 1. Correlations between number of intrusion errors and A) erythrocyte glutathione peroxidase (GSHPx) activity and B) age. Each panel shows the results of a Pearson product correlation analysis.

of Sinet's group and suggests a reduction of cognitive impairment associated with an increase in GSHPx activity in DS. However, IQ was not correlated with GSHPx activity in the DS patients of the later study (Brugge et al., 1991). The discrepancy in results on IQ in these two studies may be due to differences in psychometric tests employed for determination of IQ or to differences in the ages of the subjects (76% of the patients in the former study were under 15 years and the subjects of the later study were between 23 and 51 years old). In older DS subjects, IQ may be influenced by variables associated with aging, which may be accelerated in DS (Martin, 1976), or with the development of AD. In our study, verbal IQ was significantly correlated with age while IQ in the SINET study using a younger group of DS patients and different psychometric tests was not. The number of intrusion errors in the patients of our study was also not correlated with age (see Fig. 1B). Thus, specific cognitive impairments that are not age dependent in DS patients may be influenced by GSHPx activity. Further investigation in DS patients is required to clarify these results.

Alzheimer's disease and free radical metabolism

The relationship of alterations in free radical metabolism with cognitive impairment associated with the development of AD in DS patients requires investigation. By the age of 40, almost all DS individuals develop neurofibrillary tangles and senile plaques, hallmarks of AD, in their brains (Mann, 1988) and many become demented (Thase et al., 1984). Although chromosome 21 contains the genetic locus for the amyloid precursor protein, which clearly can have a causal relationship with the development of AD, at least in subgroups of AD patients (Marx, 1991), there is evidence implicating an important role of altered free radical metabolism in the development of various neurodegenerative disorders, such as AD, Parkinson's disease, and stroke (Halliwell, 1989).

Most of the evidence implicating an important role for altered free radical metabolism in the pathology of AD is obtained in non-DS patients. Several studies show an increased frequency of DS in families of AD patients. Zemlan and his colleagues (1989) reported that fibroblasts from 8 familial AD patients exhibit elevated SOD1 activity compared to 12 age-matched controls. Brain regions known to be affected early in the progression of AD, such as the frontal cortex and hippocampus (Brun, 1983), show preferential alterations in free radical metabolism relative to other less affected brain regions such as the cerebellum (see Mizuno and Ohta, 1986), which is relatively free of neuropathology in AD patients. Immunohistochemical techniques using antisera against SOD1 proteins show intense staining in neurons affected in AD; the large pyramidal cells of the hippocampus, neurons of the association cortex, and moderate staining in granule cells of the dentate gyrus, in contrast to other cells in these brain regions of postmortem human brains of normal patients (Delacourte et al., 1988). In situ hybridization techniques in the same brains (Ceballos et al., 1991b) revealed high levels of SOD transcripts in an almost identical pattern to that observed with the immunohistochemical technique.

Thus, an elevation of SOD may be associated with neurons that are vulnerable to neurodegenerative diseases such as AD. Similar results to those on brains of non-AD patients, as described above, were obtained in brains from AD patients (Delacourte et al., 1988 and Ceballos et al.,

1991b). Furthermore, some neurons staining positive with antisera against SOD1 also showed positive staining with anti-paired helical filament antisera, suggesting an elevation in SOD1 content in degenerating as well as non-degenerating neurons of AD brains. The enzyme activity of SOD is also reported to be generally elevated in regions of AD brains compared to controls (Marklund et al., 1985). An elevation of SOD1 may be associated with lipid peroxidation as discussed earlier, particularly since the brain may show an imbalance between elevations in SOD1, or free radicals and changes in compensatory defense mechanisms, such as GSHPx, against the toxic effects of free radicals or peroxides. Consistent with this hypothesis, lipid peroxidation is reported to be increased only in those brain regions associated with neurodegenerative changes of AD patients (Subbarao et al., 1990) as well as in fetal DS brains (Brooksbank and Balazs, 1989) that are known to develop the neuropathological changes of AD.

A few studies have investigated activity levels of SOD1 and GSHPx in erythrocytes of AD patients and controls, and some investigators have proposed that the differences between these two groups may be confounded by other variables such as age or diet. Jaendel et al. (1989) attributed the observed decrease in GSHPx activity in AD patients, compared to controls, to malnutrition among the AD patients. However, in this study malnutrition was defined by physical parameters without reference to actual diet. Since AD is reported to be a systemic disorder (Katzman et al., 1991) in which various enzyme systems are altered, the physical changes measured in the AD patients could be attributed, at least in part, to physiological changes of the disease rather than simply to malnutrition. Thus, the reduction in GSHPx may reflect the systemic manifestations of AD. Erythrocyte SOD1 activity showed no significant difference between the 24 AD patients and controls in this study. Perrin and colleagues (1990) reported a significant increase in erythrocyte SOD1 in AD patients with a slight increase in GSHPx that was nonsignificant. These results are consistent with the hypothesis that an alteration in free radical metabolism may exist in AD patients. Sulkava and colleagues (1986) failed to report differences between AD patients and controls in SOD1 or GSPHx activities, although a trend for an increase in SOD1 activity in the AD groups was observed. Only four AD patients were included in the Sulkava study, which may not be a sufficient number to reveal significant differences.

Free radical metabolism and Alzheimer's disease in Down's syndrome

Little data exists concerning the potential role of alterations of free radical metabolism in the pathogenesis of AD in DS patients. Percy and her colleagues (1990) studied a population of "demented" and "nondemented" DS patients where measures of erythrocyte SOD1, GSHPx, and catalase activity were contrasted to nondemented age-matched controls. In this study, demented DS patients failed to show a significant increase in SOD1 activity compared to age-matched healthy controls while nondemented DS patients revealed an increase in SOD activity. The elevation in erythrocyte GSHPx activity in the demented DS patients, compared to age-matched controls, only reached a $p < 0.042$ level of significance, while the nondemented group showed an elevation at the $p < 0.0042$ level. It is difficult to interpret these results since dementia was not clearly defined, and dementia is difficult to detect in a population in which baseline cognitive function is already dramatically impaired. Furthermore, a longitudinal design, which would likely provide a more accurate distinction between demented and nondemented patients, was not employed. The nondemented patients were possibly exhibiting cognitive changes associated with AD, but more sensitive measures were required to detect the change, particularly since by age 40, all DS patients have the neuropathology of AD and the mean age of the nondemented group was 46.0 years. Furthermore, the pattern of cognitive impairment associated with development of AD may be different from that of the non-DS population. Intrusion errors are characteristic of dementia associated with AD in the non-DS population and are virtually absent in healthy controls (Butters et al., 1987). However, intrusion errors occur among DS patients ranging from 23 to 51 years of age and do not correlate with age (Brugge et al., 1991). Because the prevalence of neuropathologic changes of AD is age-dependent (Mann, 1988) in DS patients, it seems that the appearance of intrusion errors is not associated with the development of dementia associated with AD in this population. Thus, further investigation in this area is required to elucidate the role of free radical metabolism in the development of AD in DS.

Ammeren, K., and Epstein, C. (1987) Lipid peroxidation and superoxide dismutase-1 and glutathione peroxidase activities in trisomy 21 fetal mice and human trisomy 21 fibroblasts. Pediatry. Res. 21: 88–92.

Brooksbank, B., and Balazs, R. (1984) Superoxide dismutase, glutathione peroxides and lipoperoxidation in Down's syndrome fetal brain. Dev. Brain Res. 16: 37–44.

Brun, A. (1983) An overview of light and electron microscopic changes, in: Alzheimer's Disease. The Stanford Reference. B. Reisberg, ed. The Free Press, A Division of MacMillan Inc., pp. 37–47.

Butters, N., Granholm, E., Salmon, D., and Grant, I. (1987) Episodic and semantic memory: a comparison of amnesic and demented patients. J. Clin. Exp. Neuropsychol. 5: 479–497.

Ceballos, I., Delabar, J., Nicole, A., Lynch, R., Hallewell, R., Kamoun, P., and Sinet, P. (1988) Expression of transfected human CuZn superoxide dismutase gene in mouse L cells and NS20Y neuroblastoma cells induces enhancement of glutathione peroxidase activity. Biochim biophys. Acta 949: 58–64.

Ceballos, I., Nicole, A., Briand, P., Grimber, G., Delacourte, A., et al. (1991a) Expression of human Cu-Zn superoxide dismutase gene in transgenic mice: model for gene dosage effect in Down syndrome. Free Rad. Res. Comm. 12–13: 581–589.

Ceballos, I., Javoy-Agid, F., Delacourte, A., Defossez, A., et al. (1991b) Neuronal localization of copper-zinc superoxide dismutase protein and mRNA within the human hippocampus from control and Alzheimer's disease brains. Free Rad. Res. Comm. 12–13: 571–580.

Clausen, J. (1984) Demential syndromes and the lipid metabolism. Acta Neurol. Scand. 70: 345–355.

Delacourte, A., Defossez, A., Ceballos, I., Nicole, A., and Sinet, P. (1988) Preferential localization of copper zinc superoxide dismutase in the vulnerable cortical neurons in Alzheimer's disease. Neuroscience Lett. 92: 247–253.

Floyd, R., Zaleska, M., and Harmon, J. (1984) Possible involvement of iron and oxygen free radicals in aspects of aging in brain, in: Free Radicals in Human Biology, Aging and Disease. Armstrong, D., Sohal, R. and Slater, T., eds. Raven Press, pp. 143–161.

Groner, Y., Elroy-Stein, O., Bernstein, Y., Dafni, N., Levanon, D., Danciger, E., and Neer, A. (1986) Molecular genetics of Down's syndrome: overexpression of transfected human Cu/Zn-superoxide dismutase gene and the consequent physiological changes. Cold Spring Harbor Symposia on Quant. Bio. LI: 381–393.

Harman, D. (1983) Free radical theory of aging: consequences of mitochondrial aging. Age 6: 86–94.

Halliwell, B. (1989) Oxidants and the central nervous system: some fundamental questions. Acta Neurol. Scand. 126: 23–33.

Heller, J. (1969) Human chromosome aberrations as related to physical and mental dysfunction. J. Hered. 60: 239–248.

Jaendel, C., Nicolas, M., Dubois, F., Nabet-Belleville, F., Penin, F., and Cuny, G. (1989) Lipid peroxidation and free radical scavengers in Alzheimer's disease. Gerontology 35: 275–282.

Katzman, R., and Saitoh, T. (1991) Advances in Alzheimer's disease. FASEB 5: 278–286.

Kedziora, J., and Bartosz, G. (1988) Down's syndrome: a pathology involving the lack of balance of reactive oxygen species. Free. Rad. Bio. Med. 4: 317–330.

Mann, D. (1988) The pathological association between Down's syndrome and Alzheimer's disease. Mech. Ageing Dev. 43: 99–136.

Marklund, S., Adolfsson, R., Gottries, C. G., and Winblad, B. (1985) Superoxide dismutase isoenzymes in normal brains and in brains from patients with dementia of Alzheimer type. J. Neurol. Sci. 67: 319–325.

Martin, G. (1976) Genetics syndrome in man with potential relevance to the pathobiology of aging, in: Genetic Effects on Aging. Bergsma, P. and Harrison, D., Eds. Alan Liss Inc., NY, pp. 5–39.

Marx, J. (1991) Mutation identified as a possible cause of Alzheimer's disease. Science 251 (4996): 876–877.

Misra, H., and Fridovich, I. (1976) Superoxide dismutase and the oxygen enhancement of radiation lethality. Archs Biochem. Biophys. 176: 577–581.

Mizuno, Y., and Ohta, K. (1986) Regional distributions of thiobarbi turic acid-reactive products, activities of enzymes regulating the metabolism of oxygen free radicals, and some of the related enzymes in adult and aged rat brains. J. Neurochem. 46(5): 1344–1352.

Percy, M., Dalton, A., Markovic, V., Crapper McLachlan, D., Hummel, J. Rusk, K., and Andrews, D. (1990) Red cell superoxide dismutase, glutathione peroxidase and catalase in Down syndrome patients with and without manifestations of Alzheimer disease. Amer. J. Med. Genetics 35: 459–467.

Perrin, R., Briancon, S., Jeandel, C., Artur, Y., Minn, A., Penin, F., and Siest, G. (1990) Blood activity of Cu/Zn superoxide dismutase, glutathione peroxidase and catalase in Alzheimer's disease: a case-control study. Gerontology 36: 306–313.

Petkau, A. (1978) Radiation protection by superoxide. Photobiology 28: 765–774.

Shah, S. (1979) Fatty acid composition of lipids of human brain myelin and synaptosomes: changes in phenylketonuria and Down's syndrome. Int. J. Biochem. 10: 477–482.

Sinet, P., Lejeune, J., and Jerome, H. (1979) Trisomy 21 (Down's syndrome) glutathione peroxidase, hexose monophosphate shunt and I.Q. Life Sci. 24: 29–34.

Sinet, P. (1982) Metabolism of oxygen derivatives in Down's syndrome. N.Y. Acad. Sci. 396: 83–94.

Southern, P., and Powis, G. (1988) Free radicals in medicine. I. Chemical nature and biological reactions. Mayo Clin. Proc. 63: 381–389.

198

Sulkava, R., Nordberg, U., Erkinjuntti, T., and Westermarck, T. (1986) Erythrocyte glutathione peroxidase and superoxide dismutase in Alzheimer's disease and other dementias. Acta Neurol. Scand. 73: 487–489.

Subbarao, K., Richardson, S., and Ang, L. (1990) Autopsy samples of Alzheimer's cortex show increased peroxidation in vitro. J. Neurochem. 55 (1): 342–345.

Thase, M., Tigner, R., Smeltzer, D., and Liss, L. (1984) Age-related neuropsychological deficits in Down's syndrome. Biol. Psychiat. 19(4): 571–585.

Zemlan, F., Thienhaus, O., and Bosmann, H. (1989) Superoxide dismutase activity in Alzheimer's disease: possible mechanism for paired helical filament formation. Brain Res. 476: 160–162.

Free Radicals and Aging
ed. by I. Emerit & B. Chance

Free radicals and neurotransmitters in gerbil brain. Influence of age and ischemia reperfusion insult

Bernard Delbarre, Gisèle Delbarre and François Calinon

Faculty of Medicine, 2 bis Bv. Tonnellé, F-37032 Tours, France

Summary. In gerbil brain, levels of hydroxyl radicals (OH') and neurotransmitters such as glutamate, aspartate, GABA (gamma aminobutyric acid) are low at birth, reach a plateau and decrease with age. On the other hand, when gerbils are exposed to an ischemia reperfusion insult (IRI) the older animals have a higher stroke index and hydroxyl radical as well as glutamate and other neuromediators are concomitantly increased. This discrepancy is probably due to differences in the ability of old individuals to respond to oxidative stress.

The still incompletely understood relationship between oxidative damage to proteins and accumulation of amino acids, which have an important role as neurotransmitters at physiologic concentrations, but may become neurotoxic at high concentrations, is discussed.

Introduction

Considerable attention has been given to the possible role of oxidative damage in the aging process as a consequence of the free radical theory of aging (Harman, 1982). The brain is particularly sensitive to oxidative damage because of its high concentration of polyunsaturated fatty acids and its high rate of oxygen consumption. The brain is also low in antioxidant protective agents such as the antioxidant enzymes, alpha-tocopherol and glutathione (Floyd et al., 1984). Furthermore, certain brain regions are high in total iron content (Hill and Switzer, 1984). Several studies on age-related events such as ischemia-reperfusion injury and stroke have reported indirect evidence for free radical involvement given the protective effect of superoxide dismutase (SOD), catalase and drugs acting as lipid peroxidation inhibitors (for review see Jacobsen et al., 1990).

Only few studies have directly measured oxygen radical formation in senescent brain or after ischemia/reperfusion injury. Utilizing several methods to assess oxidative events, including salicylate trapping of hydroxyl free radicals and quantitation of protein oxidation, it could be demonstrated that oxidative damage occurs in the ischemia/reperfusion lesioned mongolian gerbil brain (Oliver et al., 1980). These authors could show that hydroxyl radical formation damages proteins and in particular the enzyme glutamine synthetase. Loss of enzyme activity may then lead to build-up of glutamate, the substrate for this enzyme. Glutamate belongs to the group of excitatory amino acids, which are endowed with neurotoxic effects and strongly implicated in the neuronal death observed in hippocampus following an ischemia/reperfusion insult (IRI) (Olney, 1989). A cooperation between free radicals and excitatory amino acids has been suggested also by Pellegrini-Giampietro et al. (1988; 1990) for the genesis of ischemia-induced neuronal damage. Moreover, it could be shown that chronic administration of the spin trapping compound N-tert-butyl-alpha-phenylnitrone (PBN) can reverse the age-related protein oxidation and the associated loss in temporal and spatial memory of the gerbil (Carney et al., 1991). These observations suggested a relationship between oxidative damage and brain function with the hope that age-related disorders such as impaired learning and memory might be influenced by antioxidants and free radical scavengers.

On the other hand, when oxygen radical formation was measured in rat brain, using the probe 2′, 7′-dichlorofluorescein diacetate, an age-dependent decrease in the formation rate of oxygen radicals was observed (LeBel and Bondy, 1991). Also no difference in oxygen radical formation was apparent between age groups following an in vitro challenge with ascorbate/$FeSO_4$. The age-dependent decreases in cerebral oxygen radical generation coincided with age-dependent increases in SOD. In contrast to free radical generation, there was an increase in protein degradation and overall proteolytic activity with age. The authors concluded that their findings do not support the free radical theory of aging and that modifications in proteins and activated protein catabolic pathways are the major contributing factors in the normal process of senescence.

In our laboratory, we studied hydroxyl radical formation and amino acid accumulation in gerbils of advanced age. In addition, we determined the levels of monoamines and indolamines, since these neurotransmitters are studied by many authors in models of cerebral ischemia (for review see Meldrum, 1980; Delbarre et al., 1990b). The biochemical results were compared with clinical signs, expressed as the stroke index. We also studied the same parameters after exposure to IRI in gerbils of different age groups since it has been reported that IRI causes death in older gerbils more readily than in younger gerbils (Floyd and Carney, 1991).

The gerbil, animal model for human cerebral ischemia

Cerebral ischemia has to be studied on monkeys, since all other laboratory animals have collaterals between the vertebral and carotid artery before these blood vessels enter the brain. In the gerbil (*Meriones*

SENSITIVE GERBIL. LEFT FUNDUS

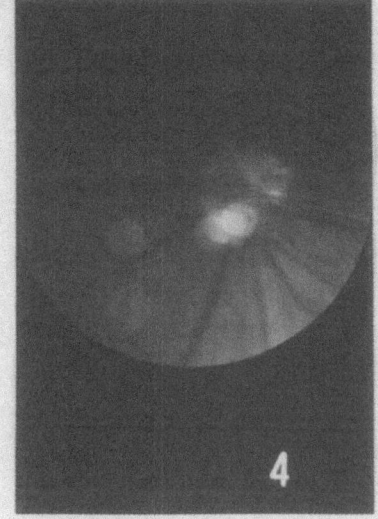

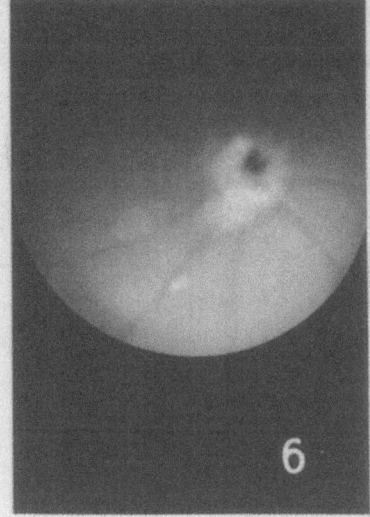

BEFORE LIGATION

1 min AFTER LIGATION

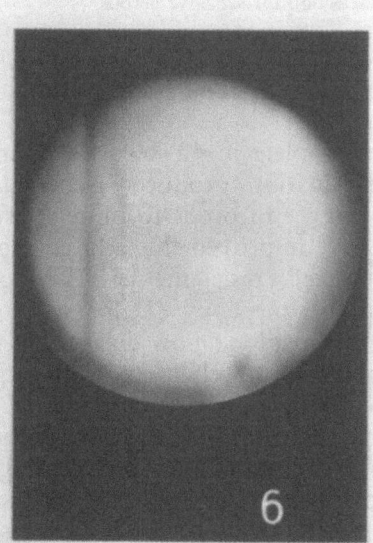

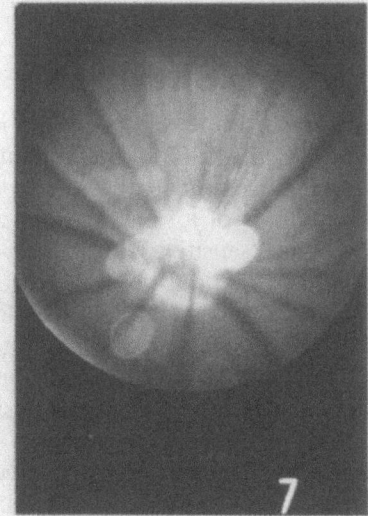

LEFT FUNDUS

RIGHT FUNDUS

5 min AFTER LEFT CAROTID LIGATION

Figure 1.

202

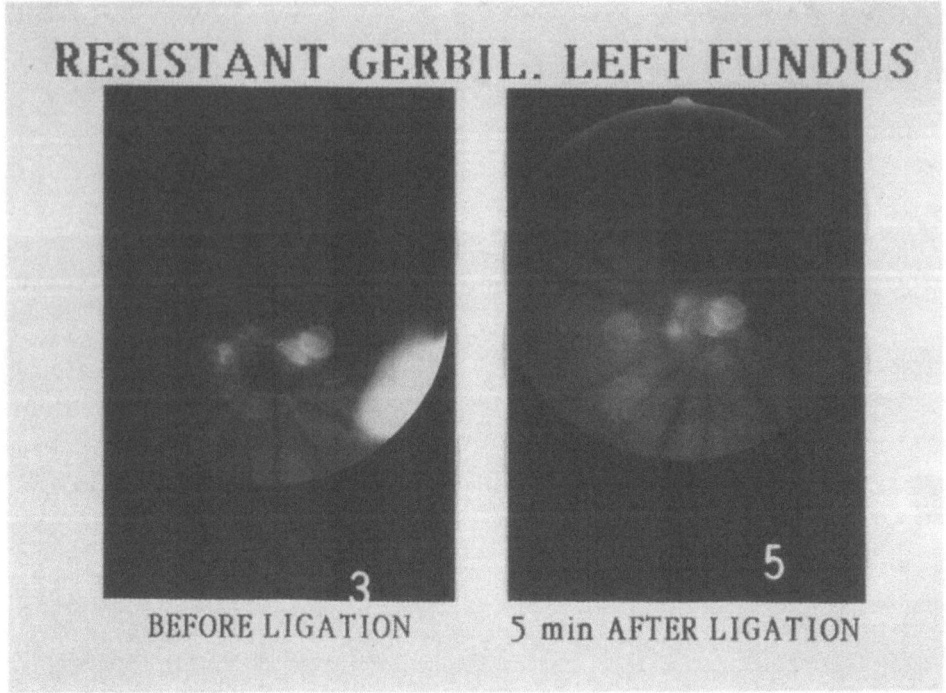

RESISTANT GERBIL. LEFT FUNDUS

BEFORE LIGATION   5 min AFTER LIGATION

Figures 1 and 2. Ocular fundus examination for selection for sensitive gerbils.

unguiculatus), the circle of Willis is incomplete in 40–60% of animals (Levine et al., 1966), and unilateral ligation produces ischemia in ipsilateral forebrain in the same percentage. In order to obtain results for all animals of a study group, we have improved this animal model by selecting "sensitive" and "non-sensitive" (resistent) gerbils by ocular fundus examination with direct ophtalmoscopy (Delbarre et al., 1988). In "sensitive" gerbils, the left retinal blood flow disappears five minutes after occlusion, while in "resistent" gerbils, the left retinal blood flow persists after occlusion of the left carotid (Fig. 1 and 2). Previous work using electrophysiological and biochemical methods has confirmed the reliability of this selection. Indeed 5 min after left carotid ligation, the b wave of the retinogram disappears in sensitive gerbils (Delbarre et al., 1990a), while this wave persists in resistent gerbils. Abnormal biochemical data were noted in all gerbils selected as "sensitive" on the ligated side of the brain only (Delbarre et al., 1990b). According to the clinical signs observed, ischemia reperfusion injury may be localized in various regions of the brain (Table 1).

Table 1. Stroke index scale

Stroke Clinical signs	Localization	Index
Ptosis	Medulla oblongata (SNS*)	2
Piloerection	Medulla oblongata (SNS*)	1
Loss of equilibrium	Pons, Mesencephalon	1
Head cocked	Labyrinths	3
Analgesia	Brain stem (opioid)	3
Rearing	Pons, Mesencephalon	3
Rotation	Caudate nucleus	2
or Rotation (circling behavior)	Caudate nucleus	4
Epilepsy or Rolling seizure	Rhinencephalon, Cortex	3
Splayed out hind limb	Cortex	3
Extreme rotation of hind limb	Cortex	3
Distance between hind paws		2
(>4 and >6 cm)	Motor Cortex	
or Distance between hind paws (<4 cm)	Motor Cortex	4
Paucity of movements	Cortex	1
Coma: $\Sigma + 6$		37
Death: Coma $+ 3$		40

*SNS: Sympathetic nervous system.

Experimental data according to age

Gerbils of advanced age were studied in subgroups of 8 animals. Free radical formation was evaluated by salicylate trapping of hydroxyl radicals according to Floyd et al. (1986). As may be seen in Figure 3, the values for OH radical generation are low in young animals, increase up to the age of 10–12 months and then decrease with aging. The amino acids glutamate, aspartate, GABA and the neuromediators dopamine, dopac (dihydroxyphenyl acetic acid), HVA (homovanillic acid) and 5-HT (serotonin) behave in the same way. The findings for the latter are in accordance with previous work on rats (Pradham et al., 1980; Wenk et al., 1989; Erdo et al., 1988).

Hydroxyl radicals and glutamate values were also determined in the retina and again showed the same pattern; increase up to the age of 12 months, decrease thereafter for animals in age groups 15 to 30 months (Fig. 4).

Based on these data, we decided to study IRI in gerbils aged 3, 9 and 15 months (Fig. 5).

Ischemia reperfusion insult

Cerebral ischemia was induced by left carotid occlusion (60 min) in sensitive gerbils of the three age groups. Stroke index, levels of neurotransmitters and hydroxyl radicals were studied 30 min after release of the clip in ipsilateral brain or retina. Caudate nucleus, hippocampus and retina were chosen for the study, since modifications in these regions are well-documented for IRI. The results were compared to those of sham-treated controls of the same age groups.

As may be seen in Figure 6, the stroke index was significantly higher in the old animals (age group 15 months). Dopamine, 5-HT, glutamate and hydroxyl radicals were increased after IRI compared to sham-operated animals and this increase was again higher in the old animals (Fig. 7). The same was true, when glutamate and hydroxyl radicals were studied in the retina (Delbarre et al., 1991).

Conclusions

There has been increasing evidence implicating the involvement of oxygen free radicals in the injury which develops in tissue that has

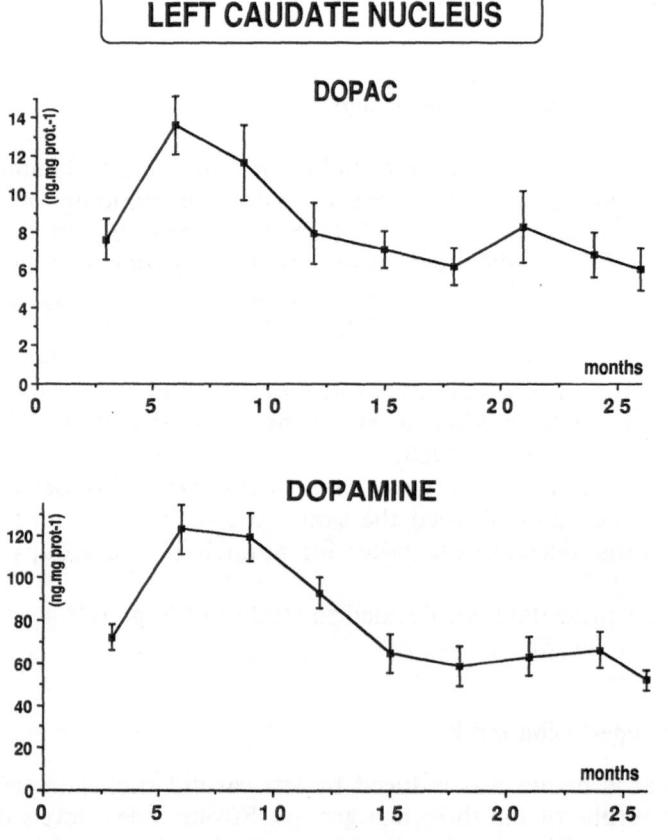

Figure 3. (continued)

LEFT CAUDATE NUCLEUS

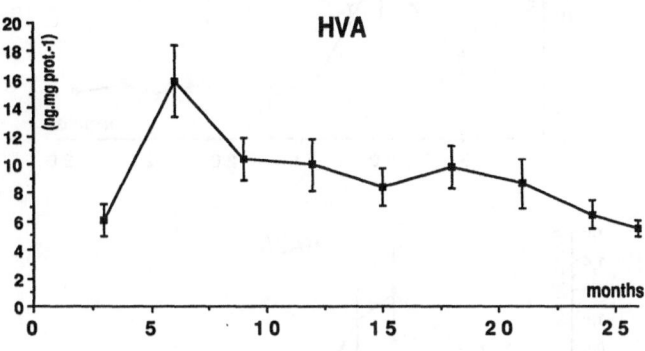

LEFT HIPPOCAMPUS of GERBIL

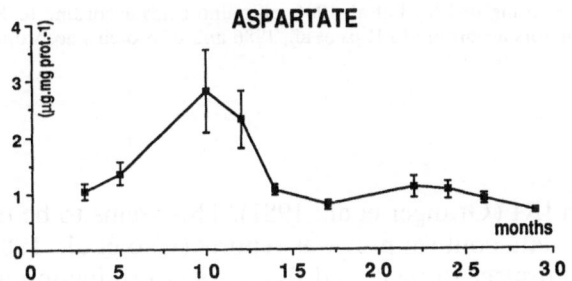

Figure 3. (continued)

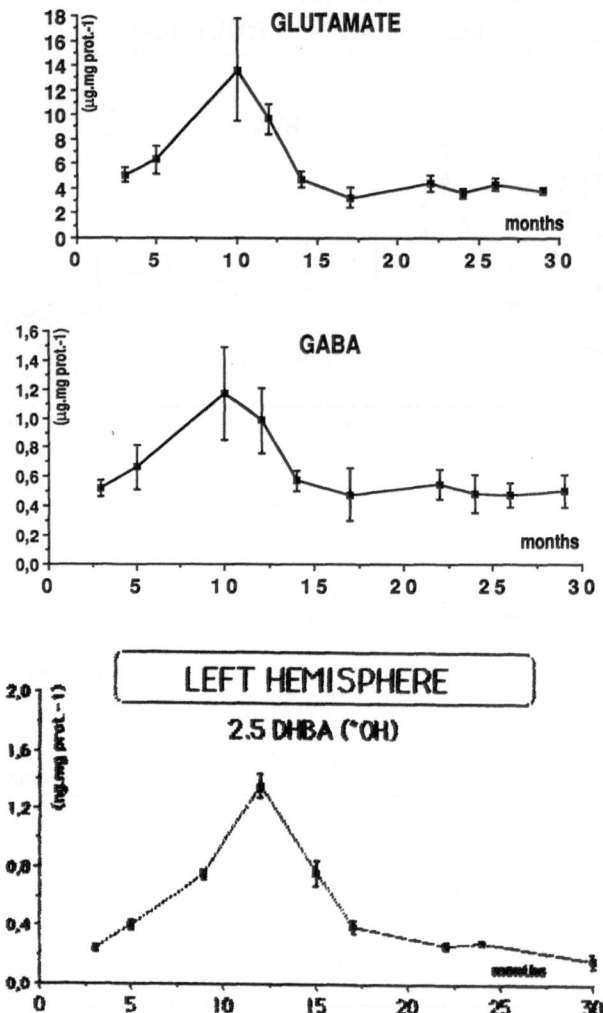

Figure 3. Measurements of hydroxyl radical formation and of neurotransmitters in gerbil brain according to age. Each age group consists of 6–8 animals. Levels of 2.5 DHBA(OH') were determined according to Floyd et al. 1986, of amino acids according to Xu et al., 1986, of other neuromediators according to Rips et al., 1986 and of proteins according to Markwell et al., 1978.

experienced an IRI (Granger et al., 1981). This seems to be true also for the brain. Our data confirm previous reports (Cao et al., 1988; Oliver et al., 1990) with respect to increased free radical production after IRI to the brain of gerbils. Not only glutamate, but also other neuromediators showed simultaneous increase in the IRI lesioned side of the gerbil

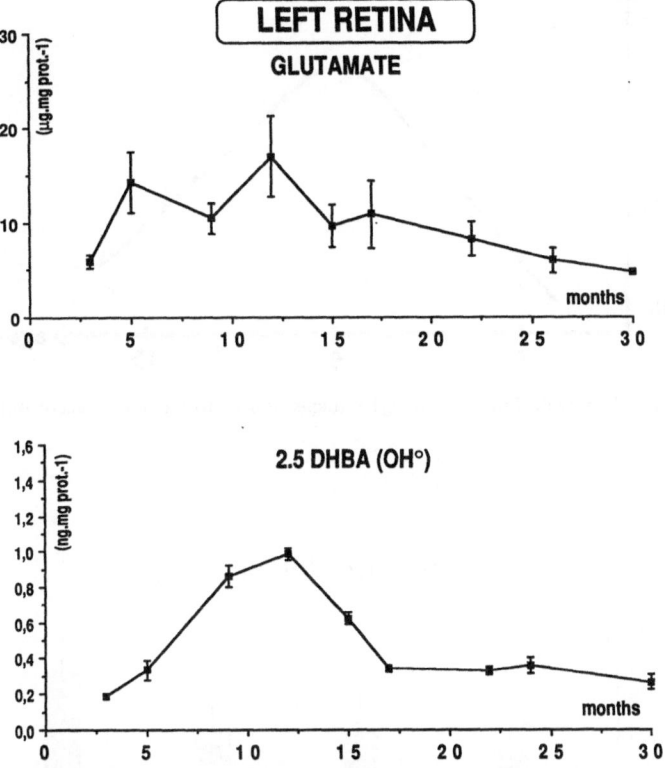

Figure 4. Hydroxyl radical formation and glutamate levels in the retina according to age. Methods used as indicated for the brain.

brain. The IRI had more serious consequences as the age of the animals progressed, as could be deduced from the stroke index.

In contrast to IRI, the innate aging process was not accompanied by increased free radical production, since hydroxyl radicals were even lower in old than in young animals. This was a confirmation of the findings reported by LeBel and Bondy (1991) for aging rats. It is possible that a concomitant increase in SOD was responsible, as suggested by these authors. This may represent a protective antioxidant response of the organism, which is sufficient in the slow intrinsic aging process to prevent oxidative damage to proteins and subsequent glutamate accumulation. It is overwhelmed in the acute and important oxidative stress imposed by IRI, in particular in the already delicate pro-antioxidant balance of older organisms.

The relation between oxygen free radicals, protein damage and excitatory neurotransmitters needs further study. Too high glutamate levels,

208

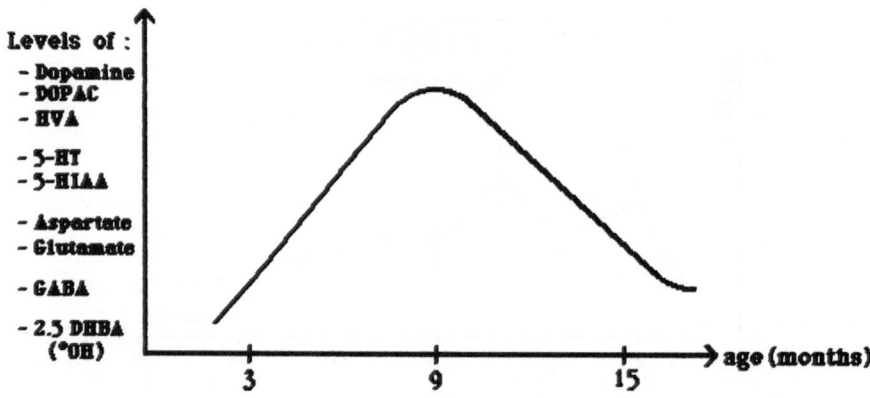

Figure 5. Global presentation of hydroxyl radicals and neuromediators according to age.

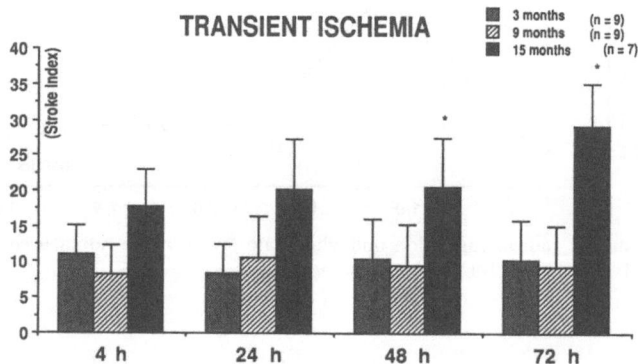

Figure 6. Stroke index for 'sensitive' gerbils exposed to unilateral carotid occlusion. Mann-Whitney U-test ($p < 0.05*$). Comparison 9 and 15/3 months.

as they may arise from glutamine synthetase insufficiency, are toxic and may contribute to the neuronal losses associated with aging. On the other hand, too low glutamate levels may be deleterious, since glutamate is involved in memory. This shows the need for an equilibrium not only for the pro-antioxidant state, but also for excitotoxic molecules. The elucidation of these interrelated mechanisms is an exciting challenge, in particular for the development of protective agents. In this regard our findings are also of practical relevance, since new drugs should be tested on gerbils of different age groups.

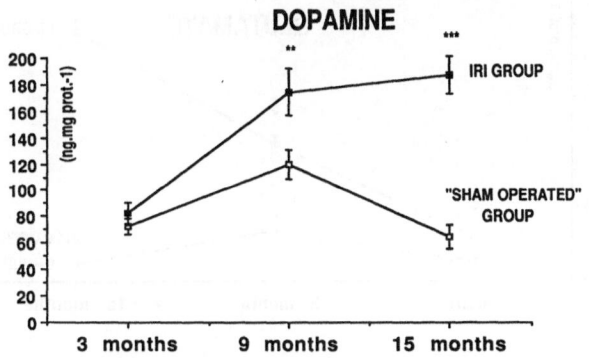

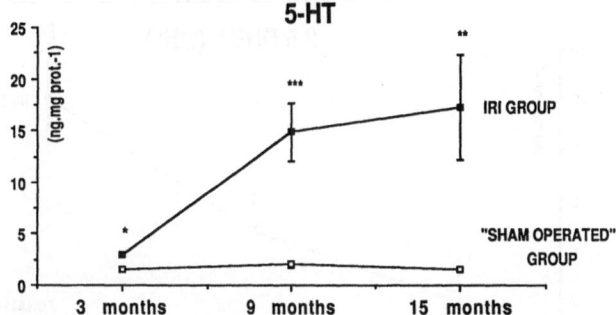

Figure 7. (continued)

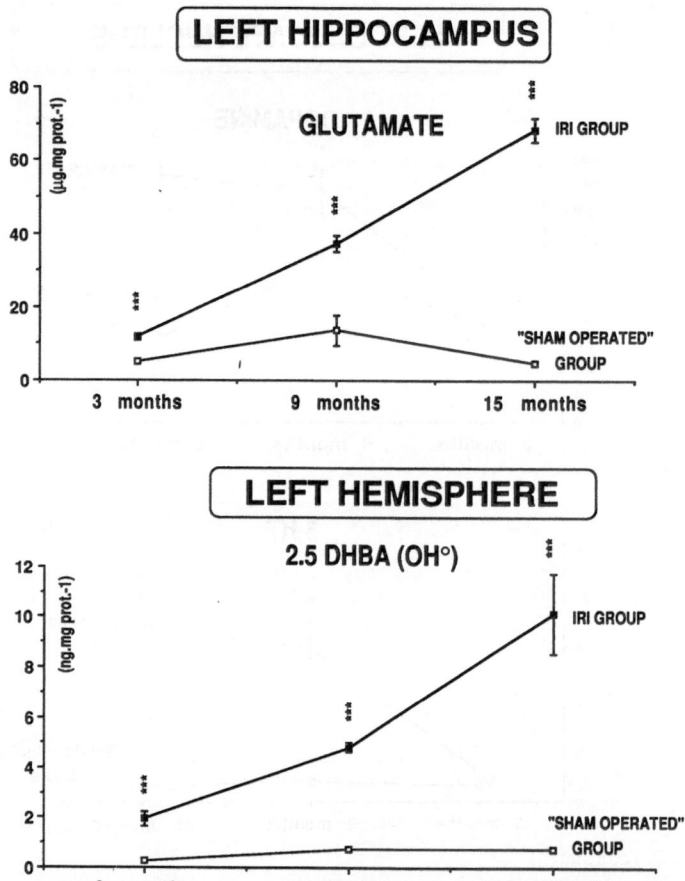

Figure 7. (continued)

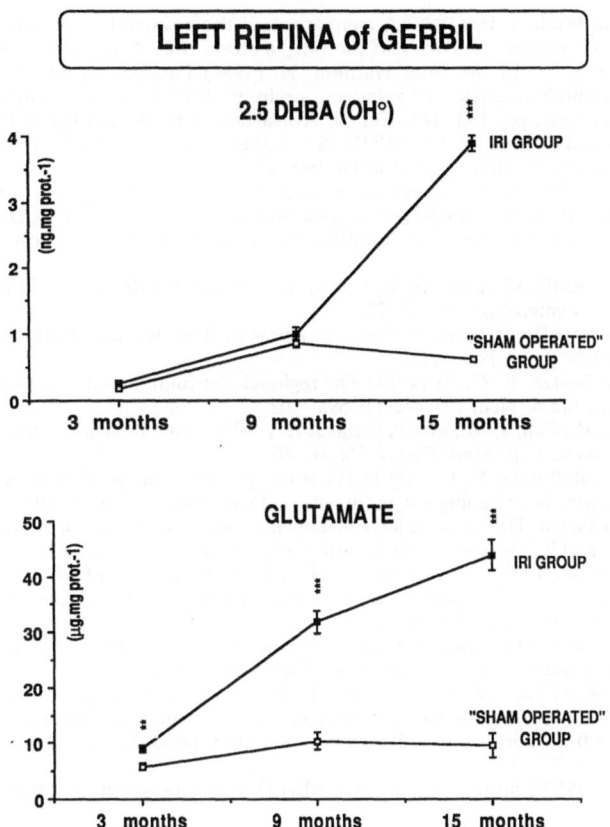

Figure 7. Comparison between sham-operated and IRI-exposed animals of the three age groups (Mean of 6 animals each). Unpaired Student t-test. $p < 0.05^*$; $p < 0.01^{**}$; $p < 0.001^{***}$.

Cao, W., Carney, J. M., Duchon, A., Floyd, R. A., and Chevion, M. (1988) Oxygen free radical involvement in ischemia and reperfusion injury to brain. Neurosci. Lett. 88: 233–238.

Carney, J. M., Starke-Reed, P. E. Oliver, C. N., Landum, R. W., Cheng, M. S., Wu, J. F., and Floyd, R. A. (1991) Reversal of age-related increase in brain protein oxidation, decrease in enzyme activity, and loss in temporal and spatial memory by chronic administration of the spin-trapping compound N-tert-butyl-α-phenylnitrone. Proc. Natl. Acad. Sci. 88: 3633–3636.

Delbarre, G., Delbarre, B., and Barrau, Y. (1988) A suitable method to select gerbils with incomplete circle of Willis. Stroke 19, 1: 126.

Delbarre, G., Delbarre, B., Calinon, F., and Ferger, A. (1991) Accumulation of amino-acids and hydroxyl free radicals in brain and retina of gerbil after transient cerebral ischemia. J. Ocul. Pharmacol. 7: 2, 147–155.

Delbarre, G., Delbarre, B., and Ferger, A. (1990a) Electroretinogram of gerbil after transient ischemia. Soc. Neurosc. Abstr. 16: 408.

Delbarre, G., Delbarre, B., Ferger, A., Calinon, F., and Loiret, C. (1990b) Metabolism of monoamines and indolamines in gerbil cerebral ischemia. Biogenic Amines, vol. 7, 6: 557–568.

212

Erdö, S. L., and Wolff, J. R. (1988) A comparison of the postnatal changes in aspartate and glutamate levels in cerebral cortex of the rat. Neurosci. Res. Comm. 4, 1: 51–56.

Floyd R. A., Zaleska, M. M., and Harmon, H. (1984) Possible involvement of iron and oxygen free radicals in aspects of aging and brain, in: Free Radicals in Molecular Biology, Aging, and Disease, pp. 143–161. Eds D. Armstrong et al., Raven Press, New York.

Floyd, R. A., and Carney, J. M. (1991) Age influence on oxidative events during brain ischemia/reperfusion. Arch. Gerontol. Geriatr. 12: 155–177.

Floyd, R. A., Henderson, R., Watson, J. J., and Wong, P. K. (1986) Use of salicylate with high pressure liquid chromatography and electrochemical detection (LCED) as a sensitive measure of hydroxyl free radicals in adriamycin treated rats. Free Radicals Biol. Med. 2: 13–18.

Granger, D. N., Rutili, G., and McCord, J. M. (1981) Superoxide radicals in feline intestinal ischemia. Gastroenterology 81: 22–29.

Harman, D. (1982). The free-radical theory of aging; in: Free Radicals in Biology. Ed. W. A. Pryor, Academic Press, New York.

Hill, J. M., and Switzer R. C. III (1984) The regional distribution and cellular localization of iron in the rat brain. Neuroscience 11: 595–603.

Jacobsen, E. J., Mc Call, J. M., and Panetta, J. A. (1990) Acute ischemic and traumatic injury to the CNS. Ann. Rep. Med. Chem. 25: 31–40.

LeBel, C. P., and Bondy, S. C. (1991) Persistent protein damage despite reduced oxygen radical formation in the aging rat brain. Int. J. Devl. Neurosci. 9, 2: 139–146.

Levine, S., and Payan, H. (1966) Effects of ischemia and other procedures on the brain and retina of the gerbil (Meriones unguiculatus). Exp. Neurol. 16: 255–262.

Markwell, M. A. K., Haas, S. M., Beiber, L. L., and Tolbert, N. E. (1978) A modification of the Lowry procedure to simplify protein determination in membrane and lipoprotein samples. Analyt. Biochem. 87: 206–210.

Meldrum, B. (1990) Protection against ischaemic neuronal damage by drugs acting on excitatory neurotransmission. Cereb. Brain Metab. Rev. 2: 27–57.

Oliver, C. N., Starke-Reed, P. E., Stadtman, E. R., Liu, G. J., Carney, J. M., and Floyd, R. A. (1990) Oxidative damage to brain proteins, loss of glutamine synthetase activity, and production of free radicals during ischemia/reperfusion-induced injury to gerbil brain. Proc. Natl. Acad. Sci. 87: 5144–5147.

Olney, J. W. (1989) Excitotoxicity and N-methyl-D-aspartate receptors. Drug Dev. Res. 17: 299–319.

Pelligrini-Giampietro, D. E., Cherici, G., Alesiani, M., Carla, V., and Moroni, F. (1988) Excitatory amino acid release from rat hippocampal slices as a consequence of free-radical formation. J. Neurochem. 51: 1960–1963.

Pelligrini-Giampietro, D. E., Cherici, G., Alesiani, M., Carla, V., and Moroni, F. (1990) Excitatory amino acid release and free radical formation may cooperate in the genesis of ischemia-induced neuronal damage. J. Neurosci. 10: 1035–1041.

Pradham, S. N., (1980) Central neurotransmitters and aging. Life Sci. 26: 1643–1656.

Rips, R., and Morier, E. (1986) Automatic assay of biogenic amines and their metabolites in mice using high pressure liquid chromatography and electrochemical detection LCED. Prog. in HPLC 2: 375–394.

Wenk, G. L., Pierce, D. J., Struble, R. G., Price, D. L., and Cork, L. C. (1989) Age-related changes in multiple neurotransmitter systems in the monkey brain. Neurobiol. Aging 10: 11–19.

Xu, X., L'Helgoualc'h, A., Morier-Teissier, E., and Rips, R. (1986) Determination of γ-aminobutyric acid in the mouse hypothalamus and hippocampus using liquid chromatography/electrochemistry. J. Liq. Chrom. 9, 10: 2253–2267.

Free Radicals and Aging
ed. by I. Emerit & B. Chance
© 1992 Birkhäuser Verlag Basel/Switzerland

Free radicals, lipid peroxidation, SOD activity, neurotransmitters and choline acetyltransferase activity in the aged rat brain

M. Hiramatsu, R. Edamatsu[a] and A. Mori[a]

Yamagata Technopolis Foundation, Yamagata 990, Japan, and [a]Department of Neuroscience, Institute of Molecular and Cellular Medicine, Okayama University Medical School, Okayama 700, Japan

Summary. The mechanism of aging is suggested to be related to oxygen free radicals. Free radicals, lipid peroxidation and SOD activity have been reported to be increased in the aged brain. A Japanese herbal medicine, Sho-saiko-to-go-keishi-ka-shakuyaku-to (TJ-960), which has scavenging activities against hydroxyl radicals, superoxide, 1,1-diphenyl-2-picrylhydrazyl radicals, carbon-centered radicals and alpha-tocopheroxyl radicals, decreased carbon-centered radicals and thiobarbituric acid reactive substances (TBARS) levels in the aged rat brain after a 3-week oral administration of 5% TJ-960 solution. TJ-960 elevated superoxide dismutase (SOD) activity in the cytosol fraction of the hippocampus and hypothalamus of aged rats. It decreased norepinephrine and 5-hydroxytryptamine (5-HT) levels in the hypothalamus and increased the 5-HT level in the cerebellum. TJ-960 treatment increased choline acetyltransferase activity in aged rats. As herbal medicines do not generally have harmful side effects, antioxidant TJ-960 appears to be a suitable prophylactic agent against some neuronal symptoms of aging.

Introduction

Many reports have been accumulated on subjects related to free radicals and aging. The levels of antioxidants such as carotenoids, α-tocopherol and uric acid in plasma, serum or liver have been reported to increase with aging (Cutler, 1991). It is considered that these antioxidants are synthesized to scavenge generated free radicals with aging. Thus it is important to find other excellent antioxidants against aging. A Japanese herbal medicine Sho-saiko-to-go-keishi-ka-shakuyaku-to (TJ-960, Tsumura & Co., Tokyo) is a vacuum-concentrated extract of nine herbs in the following ratio: 7.0 Bupleuri radix (*Bupleurum falcatum* L.); 5.0 Pinelliae tuber (*Pinellia ternata* Breitenbach); 3.0 Scutellariae radix (*Scutellaria baicalensis* Georgi); 4.0 Zizyphifructus (*Zizyphus Vulgaris* Lamarck var. *inermis* Bunge); 3.0 Ginseng radix (*Panax ginseng* C. A. Meyer); 2.0 Glychyrrhizae radix (*Glycyrrhiza glabra* L. var. *glandulifera* Regel et Herder, *Glycyrrhiza uralensis* Fisher); 1.0 Ziniberis rhizoma (*Zingiber officinale* Roscoe); 6.0 Paeoniae radix (*Paeonia albiflora* Pallas var. *trichocarpa* Bunge); and 4.0 Cinnamomi cortex (*Cinnamomum cassia* Blume). The scavenging activity of TJ-960 on free radicals and its effect on free radicals and lipid peroxides, SOD activity, neurotransmitter levels and choline acetyltransferase activity in the aged rat brain are described here.

Free radicals and lipid peroxidation

1,1-Diphenyl-2-picrylhydrazyl (DPPH) radicals are stable lipid soluble radicals. TJ-960 scavenged 30 μM DPPH radicals in the range of 125 μg/ml to 5 mg/ml dose-dependently (Hiramatsu et al., 1988b). Superoxide radicals are generated by the hypoxanthine and xanthine oxidase system. These radicals of twelve signals were detected as dimethyl-1-pyrroline-1-oxide (DMPO) spin adducts using ESR spectrometry. TJ-960 scavenged superoxide dose-dependently in the range of 12.5 μg/ml to 12.5 mg/ml (Hiramatsu et al., 1988b).

Hydroxyl radicals were generated in a α-guanidinoglutaric acid solution and quartet signals of hydroxyl radicals were detected as DMPO spin adducts by ESR spectrometry. Carbon-centered radicals were also generated in this solution. Five mg/ml of TJ-960 significantly quenched hydroxyl and carbon-centered radicals about 45% of each in 80 mM α-guanidinoglutaric acid solution (Hiramatsu et al., 1988b).

Carbon-centered radicals are generated by ascorbic acid and $FeCl_2$ in the rat brain homogenate after 15 minutes incubation at 37°C, and carbon-centered radicals are detected as DMPO spin adducts using ESR spectrometry. TJ-960 solution of 22.7 mg/ml completely quenched the carbon-centered radicals and other free radicals (Hiramatsu et al., 1988b). α-Tocopheroxyl radicals could not be detected in microsomes and subparticle membranes (SPM) of rat liver, whereas lipoxygenase and arachidonic acid generated these radicals in both membranes of rat liver after the provision of a vitamin E supplemented diet. TJ-960 scavenged α-tocopheroxyl radicals in microsomes and SPM of rat liver and further decreased the loss of α-tocopherol by lipoxygenase and arachidonic acid (Hiramatsu et al., 1990).

Because of its excellent scavenging activity on free radicals TJ-960 was further subjected to neurochemical study using aged male Wistar rats. The 3.5- and 24-month-old rats were divided into two groups; one group was administered a 5% TJ-960 aqueous solution orally for 3 months and the other group was given water as control. After 3 months, i.e., at their 6.5th and 27th month respectively, the animals were sacrificed by decapitation and the cerebrum was rapidly taken out.

While brain lipid peroxide and superoxide radical formation have been reported to increase with age (Sawada and Carlson, 1987), the chronic treatment of TJ-960 decreased the level of carbon-centered radicals and thiobarbituric acid reactive substance (TBARS) formation, which is used as an index of lipid peroxidation, in the cerebrum of aged rat brain compared with the control group (Hiramatsu et al., 1988a).

In summary, the above in vitro experiments showed that TJ-960 has potent scavenging activities on hydroxyl radicals, superoxide and DPPH radicals. It was further shown that TJ-960 has a quenching action on α-tocopheroxyl radicals and carbon-centered radicals in rat liver membranes and mouse brain homogenates, respectively. The decrease in carbon-centered radicals and inhibition of TBARS formation in aged rat brain may be due to its excellent antioxidant activity against oxygen free radicals and other free radicals.

SOD activity

Some papers have reported and compared total SOD activity, Cu, Zn-SOD activity and Mn-SOD activity in the aged rodent brain with those of a young adult brain and have found no coincidence in the change of brain SOD activity with aging (Kellogg et al., 1976; Mavelli et al., 1978; Geremia et al., 1990; Danh et al., 1983; De Quiroga, 1990). SOD activity was measured using the ESR spectrometry method (Hiramatsu et al., 1987) and the results showed that SOD activity in the mitochondrial fraction from the cortex, hippocampus, striatum, hypothalamus, midbrain, pons-medulla oblongata and cerebellum were markedly increased in aged rats compared with adult rats. Similarly, SOD activity in the cytosol fraction was elevated in all parts except the hypothalamus in aged rats (Hiramatsu et al., 1992). However, the administration of TJ-960 to aged rats resulted in a decrease in SOD activity in the mitochondrial fraction of the striatum, and an increase in the hippocampus and hypothalamus. In adult rats, TJ-960 increased SOD activity in the cytosol fraction of the cortex and had no effect on the other fractions (Hiramatsu et al., 1988a).

Neurotransmitters

In aged rats and mice, dopamine and norepinephrine levels are decreased (Hirschorn et al., 1982; Estes and Simpkins, 1980, 1984; Osterburg et al., 1981; Simpkins et al., 1977). The hypothalamic norepinephrine level is low in aging animals (Estes and Simpkins, 1980; Osterburg et al., 1981; Simpkins et al., 1977) and in Alzheimer patients (Yates et al., 1981; Gottfries et al., 1983; Yates et al., 1983). The dopamine-β-hydroxylase (DBH) activities in the cerebrospinal fluid of aged people and patients with Alzheimer's disease are decreased (Fujita et al., 1982; Cross et al., 1981; Perry et al., 1981). In addition, cell loss in the locus coeruleus, reduced nucleolar volume in locus coeruleus neurons, and loss of norepinephrine are reported in the norepinephrinergic system of Alzheimer patients (Hardy et al., 1985).

The administration of TJ-960 for 3 weeks increased the norepinephrine level in the hypothalamus but not in any other brain regions. The dopamine level in seven parts of the aged rat brain was not affected by the treatment of TJ-960 (Hiramatsu et al., 1986). However, the 5-hydroxytryptamine level in the hypothalamus was decreased and the level in the cerebellum was increased after chronic administration of TJ-960 (Hiramatsu et al., 1988a).

There have been few reports on the amino acid level in the aged rat brain. Our studies have shown that overall brain amino acid levels are lower in aged than in adult rats. Cerebral cortical γ-aminobutyric acid, glutamine, and lysine levels are significantly higher in aged than in adult rats, while in the cerebellum, glutamate levels are higher and alanine levels lower. Administration of TJ-960 to rats resulted in a significant increase in cortical levels of taurine, serine and alanine in aged rats and a significant increase in cortical taurine, glutamate, glutamine, glycine and alanine in adult rats. In the cerebellum, TJ-960 produced increases in glycine and γ-aminobutyric acid levels in adult rats but no significant changes were seen in aged rats. (Hiramatsu et al., 1988c). Thus, TJ-960 increased only cerebral cortical taurine, serine and alanine in aged rats. It also increased taurine levels in adult rats. Taurine levels are not altered with age.

CAT activity

Previous reports have shown decreased CAT activity in the striatum and hippocampus of aged rats (Allen et al., 1983; Haba et al., 1988; Sims et al., 1982). CAT activity was measured according to a modification of Fonnum's method (1974) as described by Haba et al. (1988). In comparison with adult rat brain, CAT activity in the hippocampus, pons-medulla oblongata and striatum was lower and the activity in the cerebellum was higher. The enzyme activity in the cortex of aged rats also appeared to be lower than that in adult rats but it was not significant statistically. The administration of TJ-960 increased the CAT activity in the hippocampus and striatum but it did not affect the activity in the seven brain parts of adult rats (Hiramatsu et al., 1989). While the mechanism underlying this activity still remains to be clarified, the observed increase in CAT activity may suggest a protective effect of TJ-90 against aging in the brain.

Conclusion

TJ-960 was first developed as an anticonvulsant using an experimental animal model for epilepsy (Hiramatsu et al., 1981; Hiramatsu et al., 1988c; Sugaya et al., 1988). The authors described here that TJ-960 has a free radical scavenging action in vitro and in vivo. Paeoniflorin, which is one element of Paeoniae radix, a component of TJ-960, does not have any scavenging action of free radicals. Traditionally, it has been believed that Chinese herbal medicine exhibits the effect only as a cocktail of complex plant extracts. It therefore remains an open question as to which specific component of TJ-960 may exert the effect seen as an antiaging effect, a free radical scavenging action and an increase in TBARS levels, SOD activity and CAT activity. Anyhow, it can be said at this point that antioxidants are excellent prophylactive agents against aging as shown in TJ-960.

Allen, S., Benton, J. S., Goodhardt, M. J., Haan, E. A., Sims, N. R., Smith, C. C. T., Spillane, J. A., Bowen, D. M., and Davison, A. N. (1983) Biochemical evidence of selective nerve cell changes in the normal ageing human and rat brain. J Neurochem. 41: 256–265.

Cutler, R. G. (1991) Antioxidants and aging. Am. J. Clin. Nutr. 53: 373s–379s.

Cross, A. J., Crow, T. J., Perry, E. K., Perry, R. H., Blessed, G., and Tomlinson, B. E. (1981) Reduced dopamine-beta-hydroxylase activity in Alzheimer's disease. Br. Med. J. 282: 93–94.

Danh, H. C., Benedetti, M. S., and Dostert, P. (1983) Differential changes in superoxide dismutase activity in brain and liver of old rats and mice. J. Neurochem. 40: 1003–1007.

De Quiroga, G. B., Perez-Campo, R., and Lopezz Torres, M. (1990) Anti-oxidant defences and peroxidation in liver and brain of aged rats. Biochem. J. 272: 247–250.

Estes, K. S., and Simpkins, J. W. (1980) Age-related alterations in catecholamine concentrations in discrete preoptic area and hypothalamic regions in the male rat. Brain Res. 194: 556–560.

Estes, K. S., and Simpkins, J. W. (1984) Age-related alterations in dopamine and norepinephrine activity within microdissected brain regions of ovariecotomized long evans rats. Brain Res. 298: 209–218.

Fonnum, F. (1974) A rapid radiochemical method for the determination of choline acetyltransferase. J. Neurochem. 24: 407–409.

Fujita, K., Maruta, K., Teradaira, R., Beppu, H., Ikegame, M., and Kawai, K. (1982) Dopamine beta-hydroxylase activity in human cerebrospinal fluid from various age groups. Clin. Chem. 28: 1403–1404.

Geremia, E., Baratta, S. D., Zafarana, S., Giodarno, R., Pinizzotto, M. R., La Rossa, M. D. G., and Garozzo, A. (1990) Antioxidant enzymatic systems in neuronal and glial cell-enriched fractions of rat brain during aging. Neurochem. Res. 15: 719–723.

Gottfries, C. G., Adolfsson, R., Aquilonius, S. M., Carlsson, A., Eckernas, S. A., Nordberg, A., Oreland, L., Svennerholm, L., Wiberg, A., and Winblad, B. (1983) Biochemical changes in dementia disorders of Alzheimer type (AD/SDAT). Neurobiolog. Aging 4: 261–271.

Hardy, J., Adolfsson, R., Alafusoff, I., Bucht, G., Marcusson, J., Nyberg, P., Pedrahl, E., Wester, P., and Winblad, B. (1985) Transmitter deficits in Alzheimer's disease. Neurochem. Int., 7: 545–563.

Haba, K., Ogawa, N., Kawata, M., and Mori, A. (1988) A method for parallel determination of choline acetyltransferase and muscarinic cholinergic receptors: Application in aged-rat brain. Neurochem. Res. 13: 951–955.

Hiramatsu, M., Edamatsu, R., Kabuto, H., and Mori, A. (1988a) Effect of Sho-saiko-to-go-keishi-ka-shakuyaku-to (TJ-960) on monoamines, amino acids, lipid peroxides, and superoxide dismutase in brains of aged rats in: Recent Advances in the Pharmacology of KAMPO (Japanese herbal medicines) (E. Hosoya and Y. Yamamura, Eds). Excerpta Medica, pp. 128–135.

218

Hiramatsu, M., Edamatsu, R., Kohno, M., and Mori, A. (1988b) Scavenging of free radicals by Sho-saiko-to-go-keishi-ka-shakuyaku-to (TJ-960) in: Recent Advances in the Pharmacology of KAMPO (Japanese herbal medicines) (E. Hosoya and Y. Yamamura, Eds). Excerpta Medica, pp. 120–127.

Hiramatsu, M., Haba, K., Edamatsu, R., Hamada, H., and Mori, A. (1989) Increased choline acetyltransferase activity by Chinese herbal medicine Sho-saiko-to-go-keishi-ka-shakuyaku-to in aged rat brain. Neurochem. Res. 14: 249–251.

Hiramatsu, M., Kabuto, H., and Mori, A. (1986) Effects of shosaiko-to-go-keishi-ka-shakuyaku-to (TJ-960) on brain catecholamine level of aged rats. IRCS Med. Sci. 14: 189–190.

Hiramatsu, M., Kabuto, H., and Mori, A. (1988c) Effects of Sho-saiko-to-go-keishi-ka-shakuyaku-to (TJ-960) on convulsions and brain 5-hydroxytryptamine in El mice, in: Recent Advances in the Pharmacology of KAMPO (Japanese herbal medicines) E. Hosoya and Y. Yamamura, Eds. Excerpta Medica, pp. 69–73.

Hiramatsu, M., and Kohno, M. (1987) Determination of superoxide activity by electron spin resonance spectrometry using the spin trap method. JEOL News 23A: 7.

Hiramatsu, M., Kohno, M., Edamatsu, R., and Mori, A. (1992) Increased superoxide dismutase activity in aged human cerebrospinal fluid and rat brain by electron spin resonance spectrometry using the spin trap method. J. Neurochem. 58: 1160–1164.

Hiramatsu, M., Velasco, R. D., and Packer, L. (1990) Vitamin E radical reaction with antioxidants in rat liver membranes. Free Rad. Biol. Med. 9: 459–464.

Hischorrn, I. D., Marman, M. H., and Sharpless, N. S. (1982) Dopamine receptor sensitivity following nigrostriatal lesion in the aged rat. Brain Res. 234: 357–368.

Kellogg, E. W., and Fridovich, I. (1976) Superoxide dismutase in the rat and mouse as a function of age and longevity. J. Gerontol. 31: 405–408.

Mavelli, I., Mondovi, B., Federico, R., and Rotilio, G. (1978) Superoxide dismutase activity in developing rat brain. J. Neurochem. 31: 363–364.

Osterburg, H. H., Donahue, H. G., Severson, J. A., and Finch, C. E. (1981) Catecholamine levels and turnover during aging in brain regions of male C57BL/6J mice. Brain Res. 224: 337–352.

Perry, E. K., Blessed, G., Tomlinson, B. E., Perry, R. H., Crow, T. J., Cross, A. J., Dockray, G. J., Dimaline, R., and Arregui, A. (1981) Neurochemical activities in human temporal lobe related to aging and Alzheimer-type changes. Neurobiol. Aging 2: 251–256.

Sawada, M., and Carlson, J. C. (1987) Changes in superoxide radical and lipid peroxide formation in the brain, heart and liver during the lifetime of the rat. Mech. Aging Dev. 41: 125–137.

Simpkins, J. W., Mueller, G. P., Huang, H. H., and Meites, J. (1977) Evidence for depressed catecholamine and enhanced sereotonin metabolism in aging male rats: possible relation to gonadotropin secretion. Endocrinology 100: 1672–1678.

Sims, N. R., Marek, K. L., Bowen, D. M., and Davison, A. N. (1982) Production of [^{14}C]acetylcholine and [^{14}C]carbon dioxide from [U-^{14}C]glucose in tissue prisms from aging rat brain. J. Neurochem. 38: 488–492.

Sugaya, E., Ishige, A., Sekiguchi, K., Iizuka, S., Sugimoto, A., Yuzurihara, M., and Hosoya, E. (1988) Inhibitory effect of mixture of herbal drugs (TJ-960, SK) on pentylenetetrazol-induced convulsions in El mice. Epilepsy Res. 2: 337–339.

Yates, C. M., Ritchie, I. M., Simpson, J., Maloney, A. F. J., and Gordon, A. (1981) Noradrenaline in Alzheimer-type dementia and Down syndrome. Lancet ii: 39–40.

Yates, C. M., Simpson, J., Gordon, A., Maloney, A. F. J., Allison, Y., Ritchie, I. M., and Urquhart, A. (1983) Catecholamines and cholinergic enzymes in pre-senile and senile Alzheimer-type dementia and Down's syndrome. Brain Res. 280: 119–126.

Free Radicals and Aging
ed. by I. Emerit & B. Chance
© 1992 Birkhäuser Verlag Basel/Switzerland

Evidence for drug metabolism as a source of reactive species in the brain

Jean-François Ghersi-Egea and Marie-Hélène Livertoux

Centre du Médicament, Université de Nancy I, CNRS URA 597, 30 Rue Lionnois, F-54000 Nancy, France

Summary. Several pathways for reactive species formation involving xenobiotic metabolism exist in the brain. They include oxidative activation by different enzymatic systems like cytochrome P-450 and monoamine oxidases, and superoxide radical production issued from reductive xenobiotic metabolism. They may contribute to cellular impairment observed in various physiopathological situations.

Introduction

Drug metabolism is achieved by several enzymatic systems, mainly localized in the liver. By catalyzing functionalization or conjugation reactions, these enzymes lead to the inactivation and elimination of numerous drugs and other xenobiotics (Testa and Jenner, 1976). However, in some cases, highly reactive and therefore toxic molecules like oxygen and xenobiotic radicals or electrophilic products, can be generated from these biological reactions, by different mechanisms.

The main enzymes implicated in drug oxidation are the cytochrome P-450-dependent monooxygenases, largely active in hepatic microsomes. The insertion of an oxygen molecule into a substrate results from a multistep process, which includes the transfer of 2 electrons by electron transfer chains (mainly NADPH-cytochrome P-450 reductase), from a nucleotidic cofactor to cytochrome P-450. During this catalytic cycle, the electrons are introduced in two sequential one-electron steps, resulting in the formation of several active complexes. As a consequence, substrate-independent reactions occur, which lead to the formation of superoxide and hydrogen peroxide, even in the presence of a substrate (Ortiz de Montellano, 1986). An isoenzyme of cytochrome P-450 which is induced by ethanol, cytochrome P-450 2E1, has a high rate of oxygen consumption and is especially efficient in producing such species (Ekström and Ingelman-Sundberg, 1989).

In several cases, drug-metabolizing enzymes catalyze a metabolic activation rather than producing inactive products. The electrophilic metabolites formed can be toxic per se, or generate free radicals. A well-known example is the biotransformation of polycyclic aromatic hydrocarbons (PAH) by a combination of an isoenzyme of cytochrome P-450 called 1A1 and epoxide hydrolase, into highly reactive diol-epoxide metabolites which are responsible for cancer promotion (Sims et al., 1974).

Another pathway for the formation of reactive species during drug metabolism is through reduction reactions.

During the reductive pathway of nitro-derivatives to their corresponding amines, reactive nitroso- and hydroxylamine intermediates, which possess greater cytotoxic properties than the parent nitroheterocycle, can be produced (Holtzman et al., 1981; Kedderis and Miwa, 1988).

One-electron reduction of xenobiotics leading to radical species can also be achieved by several mechanisms during drug metabolism. This occurs via the P-450 cycle under low oxygen tension. In this case the first electron is transferred to the substrate bound on the catalytic site rather than to the oxygen-protein complex, as demonstrated for solvents like halothane (De Groot and Noll, 1983). Cellular damages can also occur, in the absence or presence of oxygen, through a cytochrome P-450-independent enzymatic single electron transfer from nucleotidic cofactors to several exogenous molecules, such as quinone, iminium or nitroaromatic compounds. These reductive pathways are catalyzed by several flavoenzymes such as NADPH cytochrome P-450 reductase. In addition to a direct reaction with macromolecules, and depending upon the extent of oxygenation, the formed exogenous free radicals can either undergo a dismutation, a further enzymatic reduction by accepting a second electron, or a redox cycling leading to a reoxidation into the parent molecule with a concomitant formation of superoxide anion radical from molecular oxygen (Mason, 1990). The ability of exogenous compounds for such redox cycling in biological systems will depend on the thermodynamics of electron transfer reactions, and can be reasonably predicted from the difference between one-electron reduction potentials of the electron donor (the reductase), the xenobiotic and/or its reduced metabolites, and oxygen.

To some extent, several drug metabolizing enzymes are also present in the brain (Minn et al., 1991), and recently some evidence of side effects resulting from this metabolism was pointed out.

Implication of cerebral drug metabolism in reactive species formation

In addition to inactive metabolites, a production of free radicals or reactive metabolites was demonstrated in the brain, mostly by in vitro experiments, both during xenobiotic oxidative or reductive pathways. Examples of exogenous substrates and enzymes implicated in such processes are described in Table 1, and will be examined later on.

Table 1. Example of exogenous substrates able to undergo a metabolic activation by brain preparations

Substrate	Enzymes implicated	Selected references
Parathion	Cytochrome P-450	Norman and Neal, 1976 Chambers et al., 1989
Polycyclic aromatic hydrocarbon	Cytochrome P-450	Cohn et al., 1977 Das et al., 1981 Other references in text
Alcohols	Catalase Alcohol dehydrogenase	Tampier, 1978 Raskin and Sokoloff, 1974
Para chloro-amphetamine	Cytochrome P-450?	Miller et al., 1986
MPTP	Monoamine oxidase	Chiba et al., 1985 Arora et al., 1988
Morphine	Cytosolic dehydrogenase	Nagamatsu et al., 1983
Nitro-compounds Quinones Pyridinium	Flavine reductases (NADPH cytochrome P-450 reductase)	Köchli et al., 1980 Ghersi-Egea et al., 1991

Oxidation reactions

Cytochrome P-450 is present within different brain regions of various species. The amounts detected in microsomal fractions from cerebral tissue are low, but its main localization is mitochondrial (Walther et al., 1986). The relative contribution of this hemoprotein to the basal production of active oxygen species is yet to be established. Recently, however, a cytochrome P-450 isoenzyme especially effective in reducing dioxygen to superoxide anion, cytochrome P-450 2E1, which is also responsible for the metabolic activation of exogenous compounds to cytotoxic or carcinogenic metabolites, was found in different regions of the rat brain (Hansson et al., 1990). The role of this enzyme in drug metabolic activation remains to be investigated in the brain.

Among the reactive metabolites produced by cerebral cytochrome P-450-dependent monooxygenases, paraoxon and sulfur were identified in vitro by Nordman and Neal, as early as 1976, as a result of the oxidative desulfuration of parathion. Later on, this desulfuration was shown to occur also in vivo in the rat brain (Chambers et al., 1989). Paraoxon and the activated sulfur atom produced, both able to undergo covalent binding to various proteins, are extremely toxic.

A cytochrome P-450-dependent aryl-hydrocarbon hydroxylase activity, responsible for the activation or precarcinogens, was evidenced both in brain microsomes and mitochondria (Cohn et al., 1977; Das et al., 1981), with qualitative and quantitative differences in the metabolites formed during perinatal development (Rouet et al., 1984). The presence of the cytochrome P-450 isoenzyme 1A1/2, responsible for this metabolic activation of PAH, has been identified in different rat brain regions by several groups, both by immunological techniques and by the use of specific substrates (Kapitulnik et al., 1987; Perrin et al., 1990). The biochemical pathway ultimately leading to the formation of carcinogenic metabolites by a combination of cytochrome P-450 1A1 and epoxide hydrolase reactions can also occur in the brain, as epoxide hydrolase is present in brain fractions (Rouet et al., 1984; Ghersi-Egea et al., 1988).

A metabolic activation of xenobiotics to chemically reactive intermediates seems to be involved also in several selective neurotoxic pathways. The toxicity of para-chloroamphetamine towards serotoninergic neurons seems related to its cerebral biotransformation, possibly mediated by cytochrome P-450, into reactive metabolites which covalently bind to proteins (Miller et al., 1986). 1-methyl-4-phenyl-1,2,3,6-tetrahydropyridine (MPTP) induces a Parkinson-like illness in human and primates (Snyders and D'Amato, 1986). The enzymatic oxidation of MPTP by brain mitochondrial MAO-B into its dihydropyridinium form, which is further oxidized to the toxic metabolite 1-methyl-4-phenylpyridinium cation (MPP^+), is required for the development of its nigrostriatal toxicity. This was demonstrated both by in vitro and in vivo experiments (Chiba et al., 1985; Harik et al., 1987; Arora et al., 1988). The formation of MPP^+ can also be mediated by a cerebral N-methyltransferase from the corresponding pyridine precursors (Ansher et al., 1986). Several mechanisms of toxicity for the formed pyridinium ions were proposed. Among them, the enzymatic and non-enzymatic oxygen-dependent production of free radicals, and particularly superoxide anion, was shown to be mediated by these metabolites (Rossetti et al., 1988).

The oxidation of several alcohols to aldehydes can, as least in vitro, be achieved by different cerebral enzymatic systems, including alcohol dehydrogenase (Raskin and Sokoloff, 1974), catalase (Tampier, 1978), and possibly cytochrome P-450 2E1. The toxicological relevance of this metabolic pathway is linked to the ability of aldehydes to chemically react with monoamines, thus leading to alkaloid-like condensation products such as tetrahydroisoquinoline (TIQ) or tetrahydro-beta-carboline derivatives. These substances can act as false transmitters or as neurotoxins, and thus exert considerable effects on brain functions (Melchior and Collins, 1982). Furthermore, N-methyl-isoquinolinium ion ($NMIQ^+$), a potent inhibitor of catecholamine metabolism, structurally similar to MPP^+ and possibly involved in the etiology of Parkinson disease, can be synthetized in human brain from TIQ. The biotransformation of TIQ by N-methyltransferase leads to the formation of N-methyl-TIQ, which is oxidized by monoamine oxidase into $NMIQ^+$ (Naoi et al., 1989).

The last example of cerebral oxidative activation will be the biotransformation of morphine into morphinone by a cytosolic dehydrogenase in the mouse brain. The formed metabolite covalently binds to proteins and glutathione (Nagamatsu et al., 1983).

In addition to inactive or reactive species, active metabolites, i.e. compounds with pharmacological properties, can be formed in the brain during drug metabolism. They will not be reviewed here.

Reduction reactions

The ability of brain tissue to catalyze the pathways for the one-electron reduction of xenobiotics was suggested by a relatively high cerebral activity of NADPH-cytochrome P-450 reductase (Ghersi-Egea et al., 1989), an enzyme implicated in the reductive activation of several xenobiotics into free radicals (Trush et al., 1982). Indeed, a covalent binding of catecholestrogens to microsomal preparations from rat brain was described, which was mediated by NADPH-cytochrome P-450 reductase (Sasame et al., 1977). We recently provided evidence of a superoxide anion radical production by rat brain preparations during the reductive metabolism of exogenous molecules able to undergo a single electron reduction, like quinones, nitroheterocycles and pyridinium compounds (Ghersi-Egea et al., 1991). This production was largely associated with the microsomal fraction in a NADPH-dependent pathway, and was probably mediated by NADPH-cytochrome P-450 reductase. A NADH-dependent xenobiotic-mediated production of superoxide radicals, mostly associated with the mitochondrial fraction, was also observed. The enzymatic one-electron reduction of the molecules used led to their corresponding free radical, which, in the presence of oxygen, is reoxidized to reform the parent compound and superoxide radical. Not only superoxide production, but also the consumption of nucleotidic cofactors can be deleterious during this redox cycling.

Whether or not reactive intermediates other than free radicals are produced in the brain during the biotransformation of xenobiotics remains to be clarified, but this is suggested by several findings. First, various reductases with low substrate specificity were found in the brain (for a review, see Mesnil et al., 1984). Secondly, we have shown that during the one-electron reduction of xenobiotics in the rat brain under aerobic conditions, the ratio of oxidized nucleotidic cofactor to formed superoxide was lower than the expected theoretical ratio resulting from a complete reoxidation reaction of the corresponding radical metabolite (Ghersi-Egea et al., 1991). This indicates than the primary radical formed can be further reduced, rather than reoxidized. Finally, a mitochondrial reductase activity toward aromatic nitrocompounds, which catalyze the reduction to the hydroxylamine but not to the amine, has been evidenced in the rat brain (Köchli et al., 1980).

Drug-metabolizing enzymes in the blood-brain interfaces

The blood-brain barrier, which restricts and regulates the bidirectional exchange of molecules between the plasma and the central nervous system, is composed of capillary endothelial cells. In addition to several biochemical characteristics (transport systems and enzymatic content), the endothelial cells possess tight intercellular junctions and few pinocytic vesicles (Joò, 1985). Thus, before reaching the brain parenchyma, a drug must cross the plasma membranes and the cytoplasm of these cells. Other minor ways of entry include the passage through choroid plexuses and circumventricular organs, which lack a typical blood-brain barrier.

Recently, evidence for the localization of drug-metabolizing enzymes within the rate cerebral microvessels (Ghersi-Egea et al., 1988), and pituitary gland (Ghersi-Egea et al., 1992) was provided. This supports the hypothesis of a protective role for these enzymes against chemical aggression. Biotransformation of chemicals before their entry in the brain tissue, i.e. at the blood-brain interfaces, can lead to inactive, and/or polar metabolites unable to cross the abluminal membrane of cerebral endothelial cells. Indeed the metabolism of MPTP by endothelial MAO results in a protective effect, and explains the differences in susceptibility among mammalian species to MPTP intoxication which are related to capillary MAO activities (Riachi et al., 1988).

However, we have to keep in mind that an alteration in the blood-brain barrier properties by a drug metabolic activation would have dramatic consequences on normal brain function. We have demonstrated the ability of cerebral microvessels to produce superoxide radicals by interaction with various xenobiotics. Although the protective capacity of brain capillaries against oxidative alteration seems to be high (Tayarani et al., 1987), a drug-mediated increase of superoxide radical production can lead to cellular damage. Further works are needed to explore drug-induced blood-brain barrier dysfunctions.

Conclusion

The occurrence of cellular damage in the brain due to the metabolism of some drugs will depend on the ability of the molecule to cross the blood-brain barrier, on its reactivity (related to its redox potential), as well as on the specific activity and the cellular localization of the enzyme catalyzing the metabolic activation. The exact significance of such mechanisms in the brain needs to be established by further investigation, especially by in vivo experiments, but deserves much attention, since in the brain the lipid content is high and the neuronal cells cannot regenerate (Bourre, 1988). So, even if only few data concerning cerebral xenobiotic metabolism are available in humans, care has to be taken in case of chronic intake of drugs or exposition to pollutants, as these molecules, through initiation or increase in the cerebral production of reactive species like free radicals, might participate in pathological diseases as well as in 'normal aging'.

Ansher, S. S., Cadet, J. L., Jakoby, W. B., and Baker, J. K. (1986) Role of N-methyl-transferases in the neurotoxicity associated with the metabolites of 1-methyl-4-phenyl-1,2,3,6-tetrahydropyridine (MPTP) and other 4-substituted pyridines present in the environment. Biochem. Pharmacol. 35: 3359–3364.

Arora, P. K., Riachi, N. J., Harik, S. I., and Sayre L. M. (1988) Chemical oxidation of 1-methyl-4-phenyl-1,2,4,6-tetrahydropyridine (MPTP) and its in vivo metabolism in rat brain and liver. Biochem. Biophys. Res. Comm. 152: 1339–1347.

Bourre, J. M. (1988) The effect of dietary lipids on the central nervous system in aging and disease: Importance of protection against free radicals and peroxidation, in: Crossroads in Aging, pp. 141–167. Academic-Press, London.

Chambers, J. E., Munson, J. R., and Chambers, H. W. (1989) Activation of the phosphothionate insecticide parathion by rat brain in situ. Biochem. Biophys. Res. Comm. 165: 327–333.

Chiba, K., Peterson, L. A., Castagnoli., K. P., Trevor., A. J., and Castagnoli, N. (1985) Studies on the molecular mechanism of bioactivation of the selective nigrostriatal toxin 1-methyl-4-phenyl-1,2,3,6-tetrahydropyridine. Drug Metab. Dispos. 13: 344–347.

Cohn, J. A., Alvares, A. P., and Kappas A. (1977) On the occurence of cytochrome P-450 and aryl hydrocarbon hydroxylase activity in rat brain. J. Exp. Med. 145: 1607–1611.

Das, M., Seth, P. K., and Mukhtar, H. (1981) NADH-dependent inducible aryl hydrocarbon hydroxylase activity in rat brain mitochondria. Drug Metab. Dispos. 9: 69–71.

De Groot, H., and Noll, T. (1983) Halothane hepatotoxicity: relation between metabolic activation, hypoxia, covalent binding, lipid peroxidation and liver cell damages. Hepatology 3: 601–606.

Ekström, G., and Ingelman-Sundberg, M. (1989) Rat liver microsomal NADPH-supported oxidase activity and lipid peroxidation dependent on ethanol-inducible cytochrome P-450 (P450IIE1). Biochem. Pharmacol. 38: 1313–1319.

Ghersi-Egea, J. F., Minn, A., and Siest, G. (1988) A new aspect of the protective functions of the blood-brain barrier: activities of four drug metabolizing enzymes in isolated brain microvessels. Life Sci. 42: 2515–2523.

Ghersi-Egea, J. F., Minn, A., Daval, J. L., Jayyosi, Z., Arnould, V., Souhaili-El Amri, H., and Siest, G. (1989) NADPH cytochrome P-450 (c) reductase: biochemical characterization in rat brain and cultured neurones, and evolution of activity during development. Neurochem. Res. 14: 883–888.

Ghersi-Egea, J. F., Livertoux, M. H., Minn, A., Perrin, R., and Siest, G. (1991) Enzyme mediated superoxide radical formation initiated by exogenous molecules in rat brain preparations. Toxicol. Appl. Pharmacol. 110: 107–117.

Ghersi-Egea, J. F., Leininger, B., Minn, A., and Siest, G. (1992) Drug metabolizing enzymes in the rat pituitary gland, in: Progress in Brain Research: Circumventricular organs and brain fluid environment: Molecular and Functional Aspects. Vol. 91, pp. 373–378.

Hannsson, T., Tindberg, N., Ingelman-Sundberg, M., and Kohler, C. (1990) Regional distribution of ethanol-inducible cytochrome P-450 IIE1 in the rat central nervous system. Neuroscience 34: 451–463.

Harik, S. I., Schmidley, J. W., Iacofano, L. A., Blue, P., Arora, P. K., and Sayre L. M. (1987) On the mechanisms underlying 1-methyl-4-phenyl-1,2,3,6-tetrahydropyridine toxicity: The effect of perinigral infusion of 1-methyl-4-phenyl-1,2,3,6-tetrahydropyridine, its metabolite and their analogs in the rat. J. Pharm. Exp. Ther. 241: 669–676.

Holtzman, J. L., Crankshaw, D. L., Peterson, F. J., and Polnaszek, C. F. (1981) The kinetics of aerobic reduction of nitrofurantoin by NADPH-cytochrome P-450 (c) reductase. Mol. Pharmacol. 20: 669–673.

Joò, F. (1985) The blood-brain barrier in vitro: ten years of research on microvessels isolated from the brain. Neurochem. Int. 7: 1–25.

Kapitulnik, J., Gelboin, H. V., Guengerich, F. P., and Jacobowitz, D. M. (1987) Immunohistochemical localization of cytochrome P-450 in rat brain. Neuroscience 20: 829–833.

226

Kedderis, G. L., and Miwa, G. T. (1988) The metabolic activation of nitroheterocyclic therapeutic agents. Drug Metab. Rev. 19: 33–62.

Köchli, H. W., Wermuth, B., and Von Wartburg, J. P. (1980) Characterization of a mitochondrial NADH-dependent nitro reductase from rat brain. Biochim. Biophys. Acta 616: 133–142.

Mason, R. P. (1990) Redox cycling of radical anion metabolites of toxic chemicals and drugs and the Marcus theory of electron transfer. Envir. Hlth Perspect. 87: 237–243.

Melchior, C., and Collins M. A. (1982) The route and significance of endogenous synthesis of alkaloids in animals. Crit. Rev. Toxicol. 9: 313–356.

Mesnil, M., Testa, B., and Jenner, P. (1984) Xenobiotic metabolism by brain monooxygenases and other cerebral enzymes. Adv. Drug Res. 13: 95–207.

Miller, K. J., Anderholm, D. C., and Ames, M. M. (1986) Metabolic activation of the serotoninergic neurotoxin para-chloroamphetamine to chemically reactive intermediates by hepatic and brain microsomal preparations. Biochem. Pharmacol. 35: 1737–1742.

Minn, A., Ghersi-Egea, J. F., Perrin, R., Leininger, G., and Siest, G. (1991) Drug metabolizing enzymes in rat brain and cerebral microvessels. Brain Res. Rev. 16: 65–82.

Nagamatsu, K., Kido, Y., Terao, T., Ishida, T., and Toki, S. (1983) Studies on the mechanism of covalent binding of morphine metabilites to proteins in mouse. Drug. Metab. Disp. 11: 190–194.

Naoi, M., Matsuura, S., Takahashi, T., and Nagatsu, T. (1989) A N-methyltransferase in human brain catalyses N-methylation of 1,2,3,4-tetrahydroisoquinoline into N-methyl-1,2,3,4-te-trahydroisoquinoline, a precursor of a dopaminergic neurotoxin, N-methylisoquinolinium ion. Biochem. Biophys. Res. Commun. 161: 1213–1219.

Norman, B. J., and Neal, R. A. (1976) Examination of the metabolism in vitro of parathion (diethyl-p-nitrophenylphosphorothionate) by rat lung and brain. Biochem. Pharmacol. 25: 37–45.

Ortiz de Montellano, P. R. (1986) Oxygen activation and transfer, in: Cytochrome P-450, Structure, Mechanism, and Biochemistry, pp. 217–272. Ed. P. R. Ortiz de Montellano. Plenum Press, New York and London.

Perrin, R., Minn, A., Ghersi-Egea, J. F., Grassiot, M. C., and Siest, G. (1990) Distribution of cytochrome P-450 activities towards alkoxyresorufin derivatives in rat brain regions, subcellular fractions and isolated cerebral microvessels. Biochem. Pharmacol. 40: 2145–2151.

Raskin, N. H., and Sokoloff, N. (1974) Changes in brain alcohol dehydrogenase activity during chronic ethanol ingestion and withdrawal. J. Neurochem. 22: 427–434.

Raichi, N. J., Harik, S. I., Kalaria, R. N., and Sayre, L. M. (1988) On the mechanisms underlaying 1-methyl-4-phenyl-1,2,3,6-tetrahydropyridine neurotoxicity. II. Susceptibility among mammalian species correlates with the toxin's metabolic patterns in brain microvessels and liver. J. Pharmacol. Exp. Ther. 244: 443–448.

Rossetti, Z. L., Sotgiu, A., Sharp, D. E., Hadjiconstantinou, M., and Neff, N. H. (1988) 1-methyl-4-phenyl-1,2,3,6-tetrahydropyridine (MPTP) and free radicals in vitro. Biochem. Pharmacol. 37: 4573–4574.

Rouet, P., Dansette, P., and Frayssinet, C. (1984) Ontogeny of Benzo(a)pytene hydroxylase, epoxide hydrolase and glutathione-S-transferase in the brain, lung and liver of C57B1/6 mice. Dev. Pharmacol. Ther. 7: 245–258.

Sasame, H. A., Ames, M. M., and Nelson, S. D. (1977) Cytochrome P-450 and NADPH cytochrome c reductase in rat brain: formation of catechols and reactive metabolites. Biochem. Biophys. Res. Commun. 78: 919–926.

Sims, P., Grover, P. L., Swaisland, A., Pal, K. and Hewer, A. (1974) Metabolic activation of benzo(a)pyrene proceeds by a diol-epoxide. Nature 252: 326–328.

Snyder, S. H., and D'Amato, R. J. (1986) MPTP: A neurotoxin relevant to the pathophysiology of Parkinson's disease. Neurology 36: 250–258.

Tampier, L. (1978) Methanol oxidation by rat brain homogenates. Pharmacol. Res. Commun. 10: 823–829.

Tayarani, I., Chaudière, J., Lefauconnier, J. M., and Bourre, J. M. (1987) Enzymatic protection against peroxidative damage in isolated brain capillaries. J. Neurochem. 48: 1399–1402.

Testa, B., and Jenner, P. (1976) Drug metabolism: chemical and biochemical aspects, in: Drugs and the Pharmaceutical Sciences, vol. 4. Marcel Dekker Inc., New York and Basel.

Trush, M. A., Mimnaugh, E. G., and Gram, T. (1982) Activation of pharmacologic agents to radical intermediates. Biochem Pharmacol. 31: 3335–3346.

Walther, B., Ghersi-Egea, J. F., Minn, A., and Siest, G. (1986) Subcellular distribution of cytochrome P-450 in the brain. Brain Res. 375: 338–344.

Free Radicals and Aging
ed. by I. Emerit & B. Chance

Anticarcinogenic activities of carotenoids in animals and cellular systems

Norman I. Krinsky

Department of Biochemistry, Tufts University School of Medicine, 136 Harrison Avenue, Boston, MA 02111-1837, USA

Summary. A large number of studies have indicated that carotenoid pigments act as anticarcinogenic agents in animals treated with either ultraviolet light, ultraviolet light with chemicals, or with chemical carcinogens alone. Although pharmacological doses of cartenoids were used in the early experiments, more recent evidence indicates that relatively small doses can be effective. These studies have been complemented by investigations in bacteria and mammalian tissue, either in cell culture or in organ culture, where it has been demonstrated that various carotenoid pigments can prevent mutagenesis, genotoxic effects, or malignant transformation. It would appear that these effects are intrinsic to the carotenoid molecule, and not necessarily due to the metabolic conversion to retinoids. Partially based on these observations, it has been suggested that carotenoid pigments may function as chemopreventive agents for reducing the risk of cancer in humans. Numerous studies are underway to test this hypothesis.

Introduction

The actions of carotenoids in bacteria, in cell cultures, and in animals that support the hypothesis that carotenoids have anticarcinogenic properties have been reviewed earlier (Krinsky, 1989; Krinsky, 1991). Many of these studies served as the experimental basis for the proposal that β-carotene might function to reduce human cancer rates (Peto et al., 1981). Since those reviews were prepared, evidence continues to accumulate that in many systems, carotenoids are effective in delaying the onset of or in preventing tumorigenesis in animals, and genotoxicity or malignant transformation in cell and organ cultures. In addition, more evidence has appeared that carotenoids exhibit immunoenhancing action. Since these reports include observations that carotenoids without any provitamin A activity, such as canthaxanthin, cryptoxanthin, crocetin, fucoxanthin, and lycopene (Fig. 1), have similar activities, it would appear that the results are attributable to properties of the intact carotenoid molecule, and not necessarily to metabolites, such as the retinoids, retinol, retinal, or retinoic acid. Such observations suggest that the activities reported here are related to the known chemical and biological functions of carotenoids, which include, among others, photoprotection, radical quenching, and antioxidant behavior. The connections between the known functions and the activities reported here, however, remain to be elucidated.

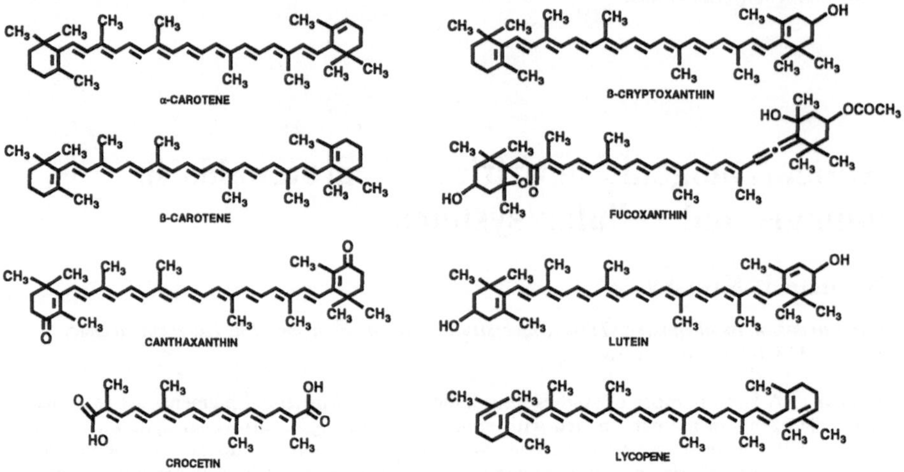

228

Figure 1. Structures of α-carotene, β-carotene, canthaxanthin, crocetin, cryptoxanthin, lutein, lycopene, and fucoxanthin.

Effects in animal systems

Early studies of the effects of carotenoids in preventing tumor formation in rats and mice have been reviewed recently (Krinsky, 1989; Krinsky, 1991). Basically, several groups demonstrated that tumors induced in mice or rats by either ultraviolet (UV) light along (Mathews-Roth, 1983; Mathews-Roth and Krinsky, 1987), a combination of UV light and carcinogens such as dimethylbenzanthracene (DMBA) (Mathews-Roth, 1982), benzo[a]pyrene (BP) (Santamaria et al., 1981; Santamaria et al., 1983) or 8-methoxypsoralen (8-MOP) (Santamaria et al., 1984), or administration of N'-N-methylnitro-nitrosoguanidine (MNNG) (Santamaria et al., 1985) could be delayed by the administration of either β-carotene or canthaxanthin (Fig. 1). UV-B light has continued to be used as a carcinogenic insult, and canthaxanthin, at 10 gm/kg, can reduce the tumor burden, without influencing the tumor incidence (Gensler et al., 1990). This effect is even more striking when a combination of canthaxanthin and retinyl palmitate (120 IU/g) is added to the diet.

Other groups have used dietary or environmental carcinogens in experimental animals supplemented with carotenoids, and have observed protection against either DMBA (Alam and Alam, 1987; Alam et al., 1988; Alam et al., 1984) or dimethylhydrazine (DMH) (Basu et al., 1988; Temple and Basu, 1987) with β-carotene or canthaxanthin. Topical administration of β-carotene not only inhibited (Suda et al., 1986), but also reversed (Schwartz et al., 1986) squamous cell carcinoma produced in the hamster buccal pouch by treatment with topical

DMBA. Similar results were reported with oral administration of either β-carotene, canthaxanthin, or an algal extract containing carotenoids (Schwartz et al., 1988). Recently, this work has been confirmed and extended by 2 independent laboratories. One report indicates that topically applied β-carotene not only prevents DMBA-induced cheek tumors in hamsters, but also prevents the accompanying stomach tumors (Gijare et al., 1990). Also, the β-carotene-treated animals retained a normal SDS-polyacrylamide electrophoretic pattern of cheek pouch keratin, whereas keratin from DMBA-treated hamsters had an abnormal pattern. In another report, a similar decrease in tumor incidence in DMBA-treated hamsters, as well as a decrease in polyamine levels in erythrocytes and urine in the β-carotene-treated animals, was observed (Hibino et al., 1990).

There have been other examples of β-carotene acting as an anticarcinogen. It is possible to induce tumors fairly rapidly by using both the carcinogen, DMBA, as well as the tumor promoter, phorbol myristyl acetate (PMA). When this system is used with either Skh or Sencar mice and treated with 3% β-carotene in their diets, either in the form of beadlets (containing 10% β-carotene), or by adding the crystalline pigment directly to the chow, both preparations protect the Skh mice, but the protection in the Sencar strain is not significant (Lambert et al., 1990). This experiment points out the very great importance of strain differences in animals, and this difference may be exaggerated when one compares experiments in different species. A similar protocol with DMBA/PMA-induction of skin tumors in Skh mice, using β-carotene at 175 μg/day, reported a significant decrease in the number of skin papillomas, but no effect on the ultimate development of malignant tumors (Steinel and Baker, 1990). Based on these observations, these authors concluded that β-carotene was working during the PMA-induced promotional phase of tumor formation.

In addition to the β-carotene experiments reported above, canthaxanthin at 1.1–3.4 mg (2–6 mmoles)/kg was fed to rats for 3 weeks prior to treatment with DMBA, and a 65% decrease in the incidence of mammary tumors was observed (Grubbs et al., 1991). They have also compared these doses of canthaxanthin to 1 mmole/kg of retinyl acetate after methylnitrosourea (MNU) treatment, and found no difference, therefore concluding that the carotenoid pigment does not inhibit promotion. Crocetin has also been tested, where it was originally reported that intraperitoneal injection of this carotenoid inhibited the growth of C-6 glial cells in rats (Wang et al., 1989). Crocetin has also been tested against the carcinogen, aflatoxin B_1(AFB), where a 3-day oral treatment with 2–6 mg/kg prevented the formation of AFB-DNA adducts, and increased the liver levels of GSH, GSH-transferase, and GSH peroxidase (Wang et al., 1991b).

Drug metabolizing effects. Several groups have postulated that carotenoids might be acting by affecting the metabolism of carcinogens, either by increasing the normal metabolic pathways, or by inducing new pathways that render the carcinogen inactive (Basu et al., 1987; Edes et al., 1989; Menon et al., 1987). A variation of that hypothesis has appeared recently (Edes et al., 1991), indicating that BP treatment in rats leads to a decrease in the retinol levels in liver and small intenstine, and that β-carotene administration, at 2 g/kg, prevents this depletion. This type of observation would suggest that it is the retinoid depletion that is associated with carcinogenesis, and that would certainly indicate that a follow-up experiment with the non-provitamin A, canthaxanthin, is warranted in this system.

Negative effects. As already mentioned, Sencar mice are not significantly protected by 3% β-carotene from tumors induced with DMBA/PMA (Lambert et al., 1990). Using DMH followed by MNU in F344 rats, β-carotene (0.2%) had only weak, organ-specific effect in preventing tumor formation (Imaida et al., 1990). In addition, 2 reports have appeared on the effects of carotenoids on BBN-induced bladder cancer. In one case, β-carotene fed to rats at 1.6 gm/kg for 42 weeks colored the organs, but did not protect against tumor formation (Pedrick et al., 1990). In another report, β-carotene or canthaxanthin was fed at 1 gm/kg for 5 weeks before, and 26 weeks after treatment with hydroxy-BBN, and only the mice receiving the β-carotene supplement showed significant protection against the development of bladder cancer (Mathews-Roth et al., 1991). The different results reported in these 2 papers may be a result of species differences in response to either the carotenoid or the carcinogen.

Immunological effects. In a preliminary report, liposomes containing β-carotene induced an active inflammatory infiltrate in oral tumor-bearing hamsters that was not observed in hamsters lacking tumors (Schwartz et al., 1990a). This infiltrate, which consisted of macrophages, lymphocytes, mast cells, and PMNs, stained positively for TNF-α, and that observation could partially explain the cytotoxicity associated with the carotenoids.

Antimutagenic actions in bacteria

Mutagenesis in *S. typhimurium* can be assayed readily, and this system has been used to demonstrate the protective effects of β-carotene against the mutagenic potential of 8-MOP and UV-A (Santamaria et al., 1984), or of cyclophosphamide (Belisario et al., 1985). Several other carotenoids have now been tested in this system, using AFB-induced mutagenesis, where it was observed that cryptoxanthin was more effective than β-carotene or canthaxanthin, and lycopene had no effect (He and Campbell, 1990).

Effects on cells and cell cultures

The very interesting observations that both β-carotene and canthaxanthin could inhibit malignant transformation brought about by either methylcholanthrene (MCA) or x-radiation in C3H10T1/2 cells has been reviewed recently (Bertram et al., 1990), and the work has been extended to other carotenoids. In addition to β-carotene and canthaxanthin, α-carotene and lycopene were also effective in inhibiting MCA-induced malignant transformation (Bertram et al., 1991). Interestingly, lutein was inhibitory at 10^{-5} M, but at lower concentrations, this dihydroxy-xanthophyll actually increased the number of transformants. When α-tocopherol was tested, it also inhibited malignant transformations, but was only about 10% as active as lycopene.

Following earlier reports of the effectiveness of crocetin in animals, this compound has been tested in C3H10T1/2 cells exposed to AFB. At 100 μM, crocetin treatment results in an elevation in the concentration of cytosolic GSH, and in increase in the activity of GSH S-transferase and GSH peroxidase (Wang et al., 1991a). These effects might explain the action of crocetin on altering the activity of microsome-activated AFB, resulting in decreased cytotoxicity, and a decrease in AFB-DNA adducts.

Fucoxanthin (Fig. 1), derived from brown algae, has also been reported to inhibit tumor cell growth (Okuzumi et al., 1990). The growth of the human neuroblastoma cell line, GOTO, was inhibited when exposed to fucoxanthin over a 3-day period, and the expression of N-*myc* was inhibited within 4 h of exposure.

There is some additional evidence that carotenoids can specifically inhibit the growth of tumor cells in culture. For example, proliferation of cultured human squamous cells (SK-MES lung carcinoma or SCC-25 oral carcinoma) was inhibited by the addition of as little as 70 μM β-carotene or canthaxanthin (Schwartz et al., 1990b). This concentration of carotenoid had no effect on the growth of normal human keratinocytes. Additionally, these authors reported that the β-carotene effect in tumor cells was accompanied by a rapid appearance of a unique 70 kD protein, analogous to heat shock proteins (Schwartz et al., 1990b).

Some confusion still exists with respect to the possible mechanisms of action of the carotenoids in these systems. Since prostaglandins (PG) have been demonstrated to be strong tumor promoters (Cerutti, 1988), an investigation has been initiated into the effects of carotenoids and retinoids on PG formation. In this study (ElAttar and Lin, 1991), the conversion of ^{14}C-arachidonic acid to PG was monitored in squamous carcinoma cells of the tongue (SCC-25), in the presence of retinoic acid, N-(4-hydroxyphenyl) retinamide (N-4-HPR), canthaxanthin, or β-carotene. Both the retinoids and canthaxanthin inhibited the formation

232

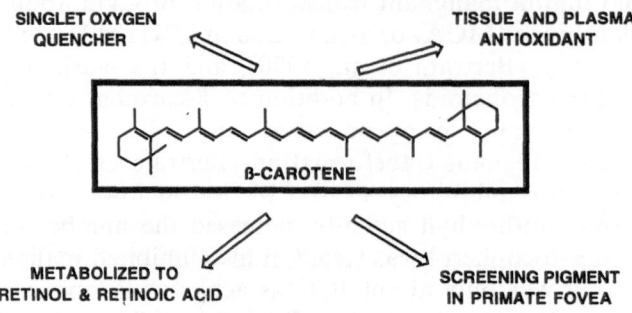

Figure 2. Biological properties of carotenoids.

of PG, thus appearing to have anticarcinogenic activity. However, β-carotene actually increased the conversion of arachidonic acid to PG, acting therefore to stimulate tumor promotion.

Conclusions

Some of the available evidence dealing with carotenoid pigments acting as antimutagenic, chemopreventive, or immunoenhancing agents have been described here. The animal studies supporting carotenoid involvement as anticarcinogenic agents continue to appear, although there are negative findings as well. In most cases, investigators are working with animals that normally do not absorb carotenoids from their diets, and therefore, high concentrations of the pigments must be added to the diet to elevate plasma and tissue levels. Even in the bacterial and cell systems reported here, the delivery of the extremely lipophilic carotenoids to the cells is problematic and it is difficult to know what the concentrations of the pigments are in the medium bathing the cells.

Nevertheless, many of the reports are striking with respect to the effects observed. What will be more striking will be the connection between the biological properties of carotenoids (Fig. 2) and the effects reported here.

Acknowledgment. Much of the work in the author's laboratory has been supported by the National Cancer Institute, grant number CA 51506.

Alam, B. S., and Alam, S. Q. (1987) The effect of different levels of dietary β-carotene on DMBA-induced salivary gland tumors. Nutr. Cancer 9: 93–101.
Alam, B. S., Alam, S. Q., and Weir, J. C. Jr. (1988) Effects of excess vitamin A and canthaxanthin on salivary gland tumors. Nutr. Cancer 11: 233–241.

Alam, B. S., Alam, S. Q., Weir, J. C. Jr., and Gibson, W. A. (1984) Chemopreventive effects of β-carotene and 13-*cis*-retinoic acid on salivary gland tumors. Nutr. Cancer 6: 4–12.

Basu, T. K., Temple, N. J., and Hodgson, A. M. (1988) Vitamin A, beta-carotene and cancer, in: Nutrition, Growth, and Cancer. G. P. Tryfiades and K. N. Prasad, eds. Alan R. Liss, New York, pp. 217–228.

Basu, T. K., Temple, N. J., and Ng, J. (1987) Effect of dietary β-carotene on hepatic drug-metabolizing enzymes in mice. J. Clin. Biochem. Nutr. 3: 95–102.

Belisario, M. A., Pecce, R., Battista, C., Panza, N., and Pacilio, G. (1985) Inhibition of cyclophosphamide mutagenicity by β-carotene. Biomed. Pharmacother. 39: 445–448.

Bertram, J. S., Pung, A., Churley, M., Kappock, T. J. IV, Wilkins, L. R., and Cooney, R. V. (1991) Diverse carotenoids protect against chemically induced neoplastic transformation. Carcinogenesis 12: 671–678.

Bertram, J. S., Rundhaug, J. E., and Pung, A. (1990) Carotenoids inhibit chemically- and physically-induced neoplastic transformation during the post-initiation phase of carcinogenesis, in: Nutrients and Cancer Prevention. K. N. Prasad and F. L. Jr. Meyskens, eds. Humana Press, Clifton, NJ, pp. 99–111.

Cerutti, P. A. (1988) Oxidant tumor promoters, in: Growth Factors, Tumor Promoters, and Cancer Genes. N. H. Coburn, H. L. Moses and E. J. Stanbridge, eds. Alan R. Liss, New York, pp. 239–247.

Edes, T. E., Gysbers, D. G., Buckley, C. S., and Thornton, W. H. Jr. (1991) Exposure to the carcinogen benzopyrene depletes tissue vitamin A: β-carotene prevents depletion. Nutr. Cancer 15: 159–166.

Edes, T. E., Thornton, W. Jr., and Shah, J. (1989) β-Carotene and aryl hydrocarbon hydroxylase in the rat: an effect of β-carotene independent of vitamin A activity. J. Nutr. 119: 796–799.

ElAttar, T. M. A., and Lin, H. S. (1991) Effects of retinoids and carotenoids on prostaglandin formation by oral squamous carcinoma cells. Prostaglandins, Leukotrienes, Essent. Fatty Acids 43: 175–178.

Gensler, H. L., Aickin, M., and Peng, Y. M. (1990) Cumulative reduction of primary skin tumor growth in UV-irradiated mice by the combination of retinyl palmitate and canthaxanthin. Cancer Lett. 53: 27–31.

Gijare, P. S., Rao, K. V. K., and Bhide, S. V. (1990) Modulatory effects of snuff, retinoic acid, and β-carotene on DMBA-induced hamster cheek pouch carcinogenesis in relation to keratin expression. Nutr. Cancer 14: 253–259.

Grubbs, C. J., Eto, I., Juliana, M. M., and Whitaker, L. M. (1991) Effect of canthaxanthin on chemically induced mammary carcinogenesis. Oncology 48: 239–245.

He, Y., and Campbell, T. C. (1990) Effects of carotenoids on aflatoxin B_1-induced mutagenesis in *S. typhimurium* TA 100 and TA 98. Nutr. Cancer 13: 243–253.

Hibino, T., Shimpo, K., Kawai, K., Chihara, T., Maruta, K., Arai, M., Nagatsu, T., and Fujita, K. (1990) Polyamine levels of urine and erythrocytes on inhibition of DMBA-induced oral carcinogenesis by topical beta-carotene. Biogenic Amines 7: 209–216.

Imaida, K., Hirose, M., Yamaguchi, S., Takahashi, S., and Ito, N. (1990) Effects of naturally occurring antioxidants on combined 1,2-dimethylhydrazine- and 1-methyl-1-nitrosourea-initiated carcinogenesis in F344 male rats. Cancer Lett. 55: 53–59.

Krinsky, N. I. (1989) Carotenoids in medicine, in: Carotenoids: Chemistry and Biology. N. I. Krinsky, M. M. Mathews-Roth and R. F. Taylor, eds. Plenum Press, New York and London, pp. 279–292.

Krinsky, N. I. (1991) Effects of carotenoids in cellular and animal systems. Am. J. Clin. Nutr. 53: 238S–246S.

Lambert, L. A., Koch, W. H., Wamer, W. G., and Kornhauser, A. (1990) Antitumor activity in skin of Skh and sencar mice by two dietary β-carotene formulations. Nutr. Cancer 13: 213–221.

Mathews-Roth, M. M. (1982) Antitumor activity of β-carotene, canthaxanthin and phytoene. Oncology 39: 33–37.

Mathews-Roth, M. M. (1983) Carotenoid pigment administration and delay in development of UV-B-induced tumors. Photochem. Photobiol. 37: 509–511.

Mathews-Roth, M. M., and Krinsky, N. I. (1987) Carotenoids affect development of UV-B-induced skin cancer. Photochem. Photobiol. 46: 507–509.

Mathews-Roth, M. M., Lausen, N., Drouin, G., Richter, A., and Krinsky, N. I. (1991) Effects of carotenoid administration on bladder cancer prevention. Oncology 48: 177–179.

234

Menon, R., Bartley, J., Som, S., and Banerjee, M. R. (1987). Metabolism of 7,12-dimethyl-benz[a]anthracene by mouse mammary cells in serum-free organ culture medium. Eur. J. Cancer Clin. Oncol. 23: 395–400.

Okuzumi, J., Nishino, H., Murakoshi, M., Iwashima, A., Tanaka, Y., Yamane, T., Fujita, Y., and Takahashi, T. (1990) Inhibitory effects of fucoxanthin, a natural carotenoid, on N-*myc* expression and cell cycle progression in human malignant tumor cells. Cancer Lett. 55: 75–81.

Pedrick, M. S., Turton, J. A., and Hicks, R. M. (1990) The incidence of bladder cancer in carcinogen-treated rats is not substantially reduced by dietary β-carotene (BC). Int. J. Vit. Nutr. Res. 60: 189–190.

Peto, R., Doll, R. J., Buckley, J. D., and Sporn, M. B. (1981) Can dietary β-carotene materially reduce human cancer rates? Nature 290: 201–208.

Santamaria, L., Bianchi, A., Adreoni, L., Santagati, G., Arnaboldi, A., and Bermond, P. (1984) 8-Methoxypsoralen photocarcinogenesis and its prevention by dietary carotenoids. Preliminary results. Med. Biol. Environ. 12: 533–537.

Santamaria, L., Bianchi, A., Arnaboldi, A., and Andreoni, L. (1981) Prevention of the benzo[a]pyrene photocarcinogenic effect by β-carotene and canthaxanthin. Med. Biol. Environ. 9: 113–120.

Santamaria, L., Bianchi, A., Arnaboldi, A., Andreoni, L., and Bermond, P. (1983) Benzo[a]pyrene carcinogenicity and its prevention by carotenoids. Relevance in social medicine, in: Modulation and Mediation of Cancer by Vitamins. F. L. Meyskens and K. N. Prasad, eds. Karger, Basel, pp. 81–88.

Santamaria, L., Bianchi, A., Ravetto, C., Arnaboldi, A., Santagati, G., and Andreoni, L. (1985) Supplemental carotenoids prevent MNNG induced cancer in rats. Med. Biol. Environ. 13: 745–750.

Santamaria, L., Bianchi, L., Bianchi, A., Pizzala, R., Santagati, G., and Bermond, P. (1984) Photomutagenicity by 8-methoxypsoralen with and without singlet oxygen involvement and its prevention by beta-carotene. Relevance to the mechanism of 8-MOP photocarcinogenicity and to PUVA application. Med. Biol. Environ. 12: 541–546.

Schwartz, J., Shklar, G., Reid, S., and Trickler, D. (1988) Prevention of experimental oral cancer by extracts of Spirulina-Dunaliella Algae. Nutr. Cancer 11: 127–134.

Schwartz, J., Suda, D., and Light, G. (1986) Beta carotene is associated with the regression of hamster buccal pouch carcinoma and the induction of tumor necrosis factor in macrophages. Biochem. Biophys. Res. Commun. 136: 1130–1135.

Schwartz, J. L., Flynn, E., and Shklar, G. (1990a) The effect of carotenoids on the antitumor immune response in vivo and in vitro with hamster and mouse immune effectors. Ann. N.Y. Acad. Sci. 587: 92–109.

Schwartz, J. L., Singh, R. P., Teicher, B., Wright, J. E., Trites, D. H., and Shklar, G. (1990b) Induction of a 70KD protein associated with the selective cytoxicity of beta-carotene in human epidermal carcinoma. Biochem. Biophys. Res. Comm. 169: 941–946.

Steinel, H. H., and Baker, R. S. U. (1990) Effects of β-carotene on chemically-induced skin tumors in HRA/Skh hairless mice. Cancer Lett. 51: 163–168.

Suda, D., Schwartz, J., and Shklar, G. (1986) Inhibition of experimental oral carcinogenesis by topical beta-carotene. Carcinogenesis 7: 711–715.

Temple, N. J., and Basu, T. K. (1987) Protective effect of β-carotene against colon tumors in mice. JNCI 78: 1211–1214.

Wang, C.-J., Chou, M.-Y., and Lin, J.-K. (1989) Inhibition of growth and development of the transplantable C-6 glioma cells inoculated in rats by retinoids and carotenoids. Cancer Lett. 48: 135–142.

Wang, C.-J, Shiah, H.-S., and Lin, J.-K. (1991a) Modulatory effect of crocetin on aflatoxin B_1 cytotoxicity and DNA adduct formation in C3H10T1/2 fibroblast cell. Cancer Lett. 56: 1–10.

Wang, C.-J., Shiow, S.-J., and Lin, J.-K. (1991b) Effects of crocetin on the hepatotoxicity and hepatic DNA binding of aflatoxin B_1 in rats. Carcinogenesis 12, 459–462.

Free Radicals and Aging
ed. by I. Emerit & B. Chance
© 1992 Birkhäuser Verlag Basel/Switzerland

Aging and cancer: Plasma antioxidants and lipid peroxidation in young and aged breast cancer patients

Mariette Gerber and Claire Ségala

Groupe d'Epidémiologie métabolique, INSERM-CRLC, F-34094 Montpellier Cedex 5, France

Summary. The relationship between aging and cancer is complex because the intrication takes place at the cell, the organism and the environment levels. On the other hand, carcinogenesis is a multi-step process, and different mechanisms may be involved in each step. For example, oxidants and antioxidants may play a different role depending upon the phase considered. Tumors in older patients are generally described as slow growing. The difference in tumor aggressiveness between young and older patients is especially obvious in breast cancer patients. The age specificity of some breast risk factors suggests that breast cancer which has been diagnosed in an aged woman was induced late in her life. We address the question whether the characteristics of a senescent organism with regards to oxidant–antioxidant status could be causally related to the slow evolution of tumors in old patients.

Introduction

The relationship between age and cancer is highly intricated and has to be considered from several points of view. From the cellular point of view, transformation equals immortalization, contrary to the entire organism level which is mortally affected by cancer. Tumor development induces inflammatory reactions, like tumor necrosis factor (TNF) which are ultimately deleterious for the subject, when the cachectic effects predominate. Moreover, tumors affect the functions of an organ and consequently the survival of the body. On the other hand, cancer incidence increases with age, but tumors in aged patients are less aggressive. A hypothesis is proposed for these apparently paradoxical observations and will be discussed and exemplified by our work on breast cancer in young and aged women.

Aging and cancer at the cellular level

Aging, for a normal animal cell, appears to follow a pathway characterized by physiological changes ultimately resulting in death. This process of senescence is unavoidable as shown by the pioneer work of Hayflick on diploid human cell lines (Hayflick and Moorhead, 1965). For fibroblasts, it resembles a state of terminal differentiation (Goldstein, 1990). Cancerous cell lines, on the contrary can be propagated indefinitely and present characteristics opposed to the aged cell.

Slow growing and aged cells are characterized by increased fibronectin content, increased adhesiveness and decreased plating efficiency. There exists also a low glutathione peroxidase (GSHpx) activity. Fast growing clones display the reverse properties (Kelner, 1990). In normal liver cell regeneration, Slater and his group found that the lipid peroxidation level correlates negatively and the vitamin E level positively with the rate of DNA synthesis (personnal communication). GSHpx was also found to be correlated with DNA synthesis in human bone narrow (Smaaland et al., 1991). These findings suggest that lipid peroxidation could provide some type of regulation for proliferation, which, in turn, could be counteracted by antioxidants.

Cells originating from tumor mass are characterized by several macromolecular changes: the fatty acid composition of the membrane is modified (Masotti et al., 1988) and the lipid peroxidation is lower than in normal cells (Burton et al., 1983). Other authors have shown that various trace elements and enzymes involved in the antioxidant defense are increased in cancer cells. A higher level of Zn (Muller and Iffand, 1981; Rogulji'c et al., 1980) and of Se, Zn, Cu and Se-dependent GSHpx has been described in tumor tissues (Collechi et al., 1987; Di Ilio et al., 1985). Moreover, the lowest proportion of polyunsaturated fatty acid is found in the most aggressive breast cancers (Lanson et al., 1990), and ovarian tumor cells develop resistance to drug-induced oxidative stress by increasing the synthesis of the glutathione system enzymes (Lee et al., 1989).

Slater and his group have shown in two different animal tumors (Cheeseman et al., 1986; Cheeseman et al., 1988) that the level of lipid peroxidation products is low and the level of vitamin E high, when compared to normal non-proliferating cells. The protective effect of Zn has been demonstrated in animal experiments where a variety of transplantable tumors were inhibited when implanted into animals maintained on a Zn-deficient diet (Dewis et al., 1970; Dewis and Pories, 1972; Fenton et al., 1980; Fenton et al., 1985; Mills et al., 1981; Mills et al., 1984; Minkel et al., 1979). We could demonstrate that a very high level of vitamin E intake increases tumor growth and decreases the survival of tumor-transplanted mice (Gerber et al., 1990).

It seems that a fast growing cell and, moreover, "transformed" cell tends to eliminate peroxidation risk and accumulate antioxidant defense. It is very likely that macromolecular adaptative responses are mounted in cancer cells in order to ensure an unlimited capacity of replication. Thus, at the cellular level, aging and cancer are antagonistic.

Aging and cancer at the organism level

With regard to individuals and populations, it is well known that cancer incidence increases with age (Crawford et al., 1987; Mor, 1987; Newel et al., 1989; Stewart, 1989). There is no clear indication that age *per se* is responsible for this observation, or the increased length of time of exposure at exogenous risk factors, or both.

On the other hand, whereas mortality seems to be higher in old cancerous patients (Adami et al., 1986; Crawford et al., 1987; Mueller et al., 1978; Yancik et al., 1989) for the reasons exposed above, relapse-free survival has been reported to be better (Baranovski et al., 1986; Herbsman et al., 1981; Simpson et al., 1988). Indeed, tumor progression and invasion have been reported to be slower in aged subjects than in young (Charlson et al., 1982; Erschler, 1986; Gentili et al., 1981).

Several mechanisms have been proposed: tumors in aged subjects are intrinsically different either in histology type (Allen et al., 1986; Dixon et al., 1985; Teeter et al., 1987; Yancik et al., 1989) or in other cellular characteristics, like the expression of hormonal receptors in breast cancer (Cox, 1987). Host-related factors can also play a role: a direct or indirect regulation of the cell proliferation may be accomplished by the physiological characteristics of a senescent organism. It has been suggested (Erschler, 1986) that tumors in aged patients lose the growth advantage confered by blocking antibodies found in young individuals, because the depressed immune system in old subjects is unable to respond to tumor antigen. Other host factors like the status oxidant–antioxidant can be involved. Generally, aged subjects, have a high level of lipid peroxidation (Poubelle et al., 1982) and a lower level of antioxidants (Tolonen et al., 1985). How can this situation, often proposed as a mechanism for tumor induction (Ames, 1988), play an opposite role with regard to the proliferation of tumor cells? The understanding of carcinogenesis, the molecular characteristics at the different phases and the natural history of cancer, help in answering this question.

The natural history of cancer

It is well known that cancer is a multi-step, polygenic process. Various events involved in the process of carcinogenesis are described for each successive phase.

The induction phase is made up of two steps, initiation and promotion. At the initiation step, a change occurs in the cell genes resulting in a deregulation of the normal pathways. The cell is now endowed with the capacity to proliferate without control, it is "transformed" or "initiated". This occurs after a mutation, an amplification or a translocation of a cellular oncogene. It might also be the loss of a repressor gene. Chemical or physical mutagens can be involved in this step, and free radicals are a common link in this process. Therefore protection against transformation can be provided by antioxidants. Generally, transformation is firmly settled by successive mitoses during the promotion step. Any mechanism inducing the cell cycle will act as a promotor; growth factors, for instance launch cells from the G1 to the S phase. Since reactive oxgyen species are known to play a part in cell activation (Dornand and Gerber, 1989; Schreck et al., 1991), free radicals are expected to act as promotors. H202 and PMA are classically used both in vitro and in vivo, respectively (Cerutti, 1985; Nakamura et al., 1985; Kinzel et al., 1986). However, to complicate the picture, in some experiments, in order to avoid the toxic effect of the free radicals onto the cell, antioxidants are necessary to ensure promotion (Cerutti, 1989).

Tumor progression can be defined as the phase during which tumor cells become organized as a growing solid mass. At this phase, replication is the main goal of the tumor cells, and as shown in the previous paragraph, adaptative mechanisms are implicated in this process. It might be hypothesized that a senescent environment characterized by a high lipid peroxidation and a low antioxidant level may negatively influence tumor growth, compared to a young organism which is characterized by a more favorable oxidant–antioxidant status.

Invasion also appears to be the result of a gene mutation, which could be responsible for the synthesis of enzymes, like proteases favoring the dissemination of the tumor cells throughout the organism via blood and lymphatic circulation. At this stage as well, a high proliferating capacity of tumor cells is present.

Thus, depending on the phase considered, free radicals can differently affect the process of carcinogenesis; they are implicated in the causal mechanism at induction, and in the regulatory mechanism at progression and maybe invasion. Symmetrically, antioxidants eventually display opposite action. At the organism level, the higher level of free radicals produced in aged subjects might increase initiation and promotion, but on the other hand it might hinder progression in a second step.

Breast cancer in young and aged patients

We became interested in this question after our findings from a case-control study on breast cancer conducted jointly in France and Italy which showed that vitamin E was higher in cases than in controls (Gerber et al., 1988) and plasma lipid peroxidation products were lower in cases than in controls (Gerber et al., 1989). The difference in vitamin E between cases and controls was even higher when studied in young subjects (Gerber et al., 1989). Other antioxidants and some trace elements involved in antioxidant defense were also found elevated in breast cancer (Cavallo et al., 1991; Gerber et al., 1991).

Vitamin E and trace elements were shown to be increased in the serum of breast cancer patients (Garofalo et al., 1980; Rougereau et al., 1987). Increased levels of Se was demonstrated in erythrocytes of patients with an early diagnosis of breast cancer (Meyer and Verreault, 1987). Thus, it seems that the changes in the oxidant–antioxidant status of the cancerous cell are also found in the blood of cancerous patients, at least at the beginning of the clinical disease.

These considerations prompted us to undertake a study [This study, which is still in progress, is being conducted with the collaboration of J. Simony-Lafontaine (pathologist), C. Astre (biochemist), and A.-V. Guizart and H. Mathieu-Daude (statisticians), all at the CRLC-Val d'Aurelle, 34094 Montpellier Cedex 5, France] comparing the oxidant–antioxidant status of breast cancer patients according to the characteristics of aggressivity of the tumor and to the age of the patient. The histologic type, the stage of differentation, DNA index and synthesis are related to tumor aggressivity (Gentili et al., 1981; Meyer et al., 1979; Strauss et al., 1980; Tubiana, 1989). Tumor size is related to tumor cell proliferation and node involvement to invasion and proliferation. We measured MDA, cholesterol and vitamin E in plasma. Vitamin E is expressed by the ratio vitamin E on total cholesterol. GSHpx was evaluated in erythrocytes and Se in serum. Because of its association with Se-dependent GSHpx, Se is included under the term antioxidant. In order to fit our hypothesis, MDA should be lower and antioxidants higher in the more aggressive tumors, and in younger patients.

As a preliminary remark it should noted that breast cancer in old women is generally diagnosed later than in young women who are more informed of screening methods and more prone to take advantage of them. Consequently, characteristics related to the time of diagnosis will be unevenly distributed between the two groups. Stratification on two variables which depend upon the time of diagnosis, namely tumor size and node involvement will be performed when applicable with respect to the size of the samples.

Levels of MDA and measured antioxidants according to histologic type of the tumor

In the young age group ($N = 42$), three histologic types were analyzed: invasive adenocarcinomas, invasive lobular carcinomas and carcinoma in situ. Colloid-mucinous carcinoma, which is a low-evoluting tumor, is very rare in young patients; only one case was recorded and not considered in this analysis. Women with the two invasive types have higher antioxidants and lower MDA than the carcinoma in situ, with

Table 1. Blood analytes in young breast cancer patients according to histologic type

	Invasive adenocarcinoma ($N = 25$)	Invasive lobular carcinoma ($N = 6$)	Carcinoma in situ ($N = 9$)	p^1	p^2
Vit. E/TC	3.39 ± 0.74	3.63 ± 0.99	2.69 ± 1.15	0.09	0.18
Se	138.2 ± 50.9	142.2 ± 45.5	91.0 ± 23.5	0.14	0.11
GSHpx	66.1 ± 27.6	73.0 ± 21.4	56.7 ± 27.8	NS	NS
MDA	0.37 ± 0.17	0.40 ± 0.17	0.53 ± 0.25	0.14	NS

p^1 = Invasive adenocarcinoma versus carcinoma in situ; p^2 = Invasive lobular carcinoma versus carcinoma in situ.
Vit. E/TC: ratio vitamin E on total cholesterol. Se: selenium, mg/l; GSHpx: glutathione peroxidase.
Units: g/Hb in erythrocytes; MDA: malonyl-dialdehyde after HPLC, μ mole/l.
Vitamin E (plasma) measured by HPLC with UV detection. Selenium (serum) measured by atomic absorption spectrometry. Glutathione peroxidase (erythrocytes) measured by enzymatic assay. Malonyl-dialdehyde (plasma): TBA complex detected by fluorimetry after extraction by HPLC.

some values indicating a trend toward significance (Table 1). When stratified on tumor size, the GSHpx levels of young patients with the smallest tumors (< 2 cm) are higher for both invasive adenocarcinomas and invasive lobular carcinomas compared to the carcinomas in situ (76.8 ± 40.0 and 88.0 ± 15.5, respectively versus 52.0 ± 26.9, $p = 0.17$ and 0.13, respectively).

In the aged group ($N = 47$), only two cases of carcinoma in situ were recorded and not analyzed. On the contrary, colloid-mucinous carcinoma is more frequent. Except for vitamin E in invasive adenocarcinomas, the results fit the picture when comparing the invasive carcinomas with the colloid-mucinous carcinomas but do not reach significance (Table 2). When stratified on tumor size, the GSHpx levels of aged patients with the larger tumors (> 2 cm) are higher for both invasive adenocarcinomas and invasive lobular carcinomas compared to colloid-mucous carcinoma (63.6 ± 20.1 and 66.6 ± 26.2 respectively versus 48.0 ± 9.0, $p = 0.08$ and 0.12 respectively).

When comparing the level of each blood analyte in the invasive lobular carcinoma, GSHpx and Se are not significantly higher and MDA is lower in young than in older women.

Levels of MDA and measured antioxidants according to the differentiation stage

None of the young patients showed a well-differentiated tumor. Aged patients with well-differentiated tumors have a lower vitamin E, Se level and a higher level of MDA than the patients with less differentiated tumors. However, the difference does not reach statistical significance.

Table 2. Blood analytes in aged breast cancer patients according to histologic type

	Invasive adenocarcinoma (N = 32)	Invasive lobular carcinoma (N = 8)	Coll. mucinous carcinoma (N = 6)	p^1	p^2
Vit. E/TC	3.60 ± 1.36	4.15 ± 1.46	3.79 ± 0.80	NS	NS
Se	123.2 ± 49.1	105.0 ± 49.5	100.0 ± 4.2	NS	NS
GSHpx	65.8 ± 19.9	66.6 ± 26.2	52.0 ± 12.7	0.12	NS
MDA	0.37 ± 0.16	0.50 ± 0.45	0.64 ± 0.31	0.05	0.16

p^1 = Invasive adenocarcinoma versus colloid mucinous carcinoma; p^2 = Invasive lobular carcinoma versus colloid mucinous carcinoma.
Vit. E/TC: ratio vitamin E on total cholesterol. Se: selenium, mg/l; GSHpx: glutatione peroxidase.
Units: g/Hb in erythrocytes; MDA: malonyl-dialdehyde after HPLC, μmole/l.
Vitamin E (plasma) measured by HPLC with UV detection. Selenium (serum) measured by atomic absorption spectrometry. Glutathione-peroxidase (erythrocytes) measured by enzymatic assay. Malonly-dialdehyde (plasma): TBA complex detected by fluorimetry after extraction by HPLC.

Levels of MDA and measured antioxidants according to the DNA index

Beforehand, it should be noted that only 35.7% of the young patients showed diploid tumors versus 69.2% in the aged group (chi square 7.5, $p = 0.02$).

Except for GSHpx which is higher in young patients with aneuploid (an indication of aggressiveness) tumors than in those with diploid tumors (70.4 ± 28.0 versus 49.9 ± 26.1, $p = 0.06$), there is no noticeable difference with respect to DNA index.

When stratified on basis of node involvement, for the class with node involvement, MDA is lower in patients with aneuploid tumors both in young and aged patients (0.23 ± 0.07 versus 0.31 ± 0.02, NS and 0.18 – only one patient – versus 0.25 ± 0.07, NS, respectively) and Se is higher in older patients with aneuploid tumors (196.7 – only one patient – versus 141.3 ± 18.9, NS). For the class without node involvement, GSHpx is higher in the aged patients with multiploid tumors (70.4 ± 10.8 versus 60.6 ± 13.5, $p = 0.2$).

In young patients with aneuploids tumors GSHpx level is higher than in older ones (70.4 ± 28.0 versus 64.3 ± 13.7, NS).

Levels of MDA and measured antioxidants according to proliferation index

Vitamin E is higher in the proliferating (ki > 15%) tumors of the 5 young patients recorded than in the non-proliferating tumors of 23 other young patients (4.03 ± 0.04 versus 3.26 ± 0.12, $p = 0.04$). GSHpx is non-significantly higher (69.2 ± 15.4 versus 61.3 ± 22.2, NS and

Table 3. Blood analytes in young breast cancer patients according to tumor size

	Tumor < 2 cm (N = 14)	Tumor > 2 cm (N = 27)	p
Vit. E/TC	2.86 ± 1.14	3.59 ± 1.05	0.04
Se	100.3 ± 28.0	143.3 ± 50.6	0.11
GSHpx	68.2 ± 31.9	63.8 ± 25.0	NS
MDA	0.53 ± 0.20	0.34 ± 0.17	0.01

Vit. E/TC: ratio vitamin E on total cholesterol. Se: selenium, mg/l; GSHpx: glutathione peroxidase.
Units: g/Hb in erythrocytes; MDA: malonyl-dialdehyde after HPLC, μmole/l.
Vitamin E (plasma) measured by HPLC with UV detection. Selenium (serum) measured by atomic absorption spectrometry. Glutathione peroxidase (erythrocytes) measured by enzymatic assay. Malonyl-dialdehyde (plasma): TBA complex detected by fluorimetry after extraction by HPLC.

69.5 ± 17.8 versus 55.27 ± 21.8, $p = 0.17$, when restricted to large tumors), but MDA is also higher (0.63 ± 0.04 versus 0.40 ± 0.18, $p = 0.05$). We had no data for Se. There are no noticeable differences among the levels of the various blood analytes in the aged group.

The vitamin E level of the 5 proliferating tumors of the young patients is higher than that of the 4 proliferating tumors of the old patients (4.03 ± 0.04 versus 3.35 ± 0.06, $p = 0.11$).

Levels of MDA and measured antioxidants according to tumor size

When comparing patients with tumors smaller than 2 cm to those with tumors larger 2 cm at palpation, the findings in the young group fit the hypothesis (Table 3). In the aged group, the picture is more confusing with a higher Se in the large-tumor group (123.8 ± 42.7) versus 77.7 ± 18.5, $p = 0.13$), but a lower GSHpx (61.7 ± 20.8 versus 73.2 ± 5.9, $p = 0.06$).

Although the differences are in agreement with the expected trend between young and aged groups for GSHpx and Se, they never reach statistical significance.

Levels of MDA and measured antioxidants in node involvement

Se and MDA are the blood analytes which appear more interesting with respect to node involvement. When metastatic nodes are present, Se is higher in the 9 young and the 16 aged patients than in patients without metastatic nodes (144.0 ± 55.0 versus 114.2 ± 39.1, $p = 0.2$ and 151.9 ± 26.8 versus 79.4 ± 17.5, $p = 0.006$, respectively). In the same categories, MDA is lower in young and aged patients (0.25 ± 0.15 versus 0.47 ± 0.19, $p = 0.01$ and 0.33 ± 0.17 versus 0.48 ± 0.3, $p = 0.08$, respectively).

Se is higher in young than in aged patients without node involvement (114.2 ± 39.1 versus 79.4 ± 17.5, $p = 0.04$). MDA is lower in young than in aged patients with node involvement (0.25 ± 0.15 versus 0.33 ± 0.17, $p = 0.13$).

Preliminary remarks and comments

As a whole, there is a general tendency for markers of aggressiveness and proliferation to be correlated with high levels of antioxidants and low levels of MDA. This is somehow stronger for young patients than for aged patients. The only significant differences are associated to proliferation. In young patients, they bear on vitamin E, for tumor size and DNA synthesis, and on MDA, for tumor size and node involvement. In aged patients, Se shows a significantly higher value in patients with node involvement than without.

There are other values which are borderline significant, or with a trend, but do not reach significance. This is especially true for GSHpx. It might be because of the large variability in the level of this erythrocyte enzyme within the population of the study. Very likely, the small sample size is also responsible for the lack of significance of the results. Moreover, this small size did not allow for thorough statistical analysis, including multivariate analysis and interactions. However, these preliminary findings justify a further investigation, although it is impossible to say at this stage whether the oxidant–antioxidant status can be used as a marker of aggressiveness and/or as a prognostic factor.

It is clear that breast cancer in young patients displays the features of more aggressive tumors: this is shown in our samples according to histologic type, level of differentiation and DNA index. Moreover, although we have few cases in this class, it is interesting to note that none of the tumors from the aged group showing a $Ki > 15\%$ are multiploid whereas 66% of the tumors from young patients presenting a $Ki > 15\%$ are multiploid. We have had some evidence (Segala et al., 1991) that breast cancer in an aged woman was initiated late in her life. Therefore, it is reasonable to think that the difference of characteristics between breast cancer in young and aged women is likely due to the distinct physiological environment. Whether the disparity in the oxidant–antioxidant status between young and aged women plays a role in the proliferation rate of breast cancer at the progression and invasion phases is difficult to tell from these preliminary findings, since there is only one statistically significant difference between young and aged patients, regarding selenium, in the class of patients without node involvement. Again, it might be that the sample size is too small. It might be that the age classes are not well selected; in our previous study, (Gerber et al., 1989) on a population limited to 65 years of age, there

244

was a slightly higher level of vitamin E in pre-menopausal than in post-menopausal women which is not found here. It might also be that another important factor like the hormonal status plays a prominent role in this age-related difference. Further investigation on a complete sample (100 in each group plus study of the intermediate age group) and taking into consideration the hormonal receptors, should help us in resolving this question.

Acknowledgment. This work has been financially supported by ARC.

Adami, H., Malker, B., Holmberg, L., et al. (1986) The relation between survival and age at diagnosis in breast cancer. N. Engl. J. Med. 315: 559–563.

Allen, C., Cox., E. B., Manton, K. G., and Cohen, H. J. (1986) Breast cancer in the elderly: current patterns of care. J. Am. Geriatr. Soc. 34: 37–42.

Ames, B. N. (1988) Measuring oxidative damage in humans: relation to cancer and ageing. I.A.R.C. Sci. Publ. 89: 407–416.

Baranovsky, A., and Myers, M. (1986) Incidence and survival in patients 65 years of age and older. CA 36: 27–41.

Burton, G. W., Cheeseman, K. H., Doba, T., et al. (1983) Vitamin E as an anti-oxydants in vitro and in vivo, in: Ciba Foundation symposium 101, Biology of Vitamin E, London, Pitman, pp. 4–14.

Cavallo, F., Gerber, M., Marubini, E., Richardson, S., et al. (1989) Zinc and Copper in breast cancer. A joint study in Northern Italy and Southern France. Cancer 64: 2347–2353.

Cerutti, P. A. (1985) Pro-oxidant state and tumor promotion. Science 227: 375–381.

Cerutti, P. A. (1989) Response modification in carcinogenesis. Env. Health Persp. 81: 39–43.

Charilon, M. E., and Feinstein, A. R. (1982) Rapid growth rate in breast cancer: a confounding variable in adjuvant and chemotherapy trials. Lancet 1343–1345.

Cheeseman, K. H., Collins, M., Proudfoot, K., et al. (1986) Studies on lipid peroxidation in normal and tumor tissues. The Novikoff rat liver tumour. Biochem. J. 235: 507–514.

Cheeseman, K. H., Emery, S., Maddix, S. P., et al. (1988) Studies on lipid peroxidation in normal and tumour tissues. The Yoshida rat liver tumour. Biochem. J. 250: 247–252.

Collechi, P., Cafiero, F., Vercchio, C., et al. (1987) Erythrocyte, plasma and serum level of zinc and copper in normal subjects and in patients with rectal cancer. Medical Science Research 15: 1073–1074

Cox, E. B. (1987) Breast cancer in the elderly. Clin. Geriatr. Med. 3: 695–713.

Crawford, J., and Cohen, H. J. (1987) Relationship of cancer and ageing. Clin. Geriatr. Med. 3: 419–432.

Dewis, W., and Pories, W. J. (1972) Inhibition of spectrum of animal tumors by dietary zinc deficiency. J. Natl. Cancer Inst. 48: 375–381.

Dewis, W., Pories, W. J., Richter, M. C., et al. (1970) Inhibition of Walker 256 carcinosarcoma growth by dietary zinc deficiency. Proc. Soc. Exp. Biol. Med. 135: 17–22.

Di Ilio, C., Saccheta, P., Del Boccio, G., et al. (1985) Glutathione peroxidase, glutathione S-transferase and glutathione reductase activities in normal and neoplastic human breast tissue. Cancer Lett. 29: 37.

Dixon, J. M., Page, D. L., Anderson, T. J., et al. (1985) Long-term survivors after breast cancer. Br. J. Surg. 72: 445–448.

Dornand, J., and Gerber, M. (1989) Inhibition of murine T cell responses by antioxidants: the targets of lipoxigenase pathway inhibitors. Immunology 68: 384–391.

Ershler, W. B. (1986) Why tumors grow more slowly in old people. J.N.C.I. 77: 837–839.

Fenton, M. R., and Burke, J. P. (1985) Subcellular zinc distribution in livers and tumors of plasmocytoma-bearing mice. Nutr. Res., 5: 1383–1391.

Fenton, M. R., Burke, J. P., Tursi, F. D., and Arena, F. P. (1980) Effect of a zinc deficient diet on the growth of a IgM secreting plasmocytoma (TEPC-183). JNCL 65: 1271–1282.

Garofalo, J., Ashikari, H., Lesser, M. L., Budrick, S. J., and Bodey, G. P. (1980) Serum zinc, copper, and the Cu/Zn ratio in patients with benign and malignant breast lesions. Cancer 46: 2682–2685.

Gentili, C., Sanfilippo, O., and Silvestrini, R. (1981) Cell proliferation and its relationship to clinical features and relapse in breast cancer. Cancer 48: 974–979.

Gerber, M., Cavallo, F., Marubini, E., et al. (1988) Liposoluble vitamins and lipid parameters in breast cancer. A joint study in Northern Italy and Southern France. Int. J. Cancer 42: 489–494.

Gerber, M., Richardson, S., Crastes de Paulet, P., et al. (1989) Relationship between vitamin E and poly-unsaturated fatty acids in breast cancer. Nutritional and metabolic aspects. Cancer 64: 2347–2353.

Gerber, M., Richardson, S., Favier, F., and Crastes de Paulet, A. (1990) Vitamin E and tumor growth, in: Antioxidants in Therapy and Medicine. Ed. I. Emerit et al. p. 129–132.

Gerber, M., Richardson, S., Salkeld, R., and Chappuis, P. (1991) Anti-oxydants in female breast cancer patients. Cancer Inv. 9: 421–428.

Goldstein, S. (1990) Replicative senescence: The human fibroblast comes of age. Science 249: 1129–1132.

Hayflick, L., and Moorhead, P. S. (1961) Human diploid cell lines. Exp. cell. Res. 25: 585–589.

Herbsman, H., Feldman, J., Seldera, J., et al. (1981) Survival following breast cancer surgery in the elderly. Cancer 47: 2358–2363.

Kelner, M. J. (1990) Abnormal cell growth induced by antioxidant enzyme imbalances. Free Rad. Biol. Med. 9: Suppl 1, 132.

Kinzel, V., Fürstenberger, G., Loehrke, H., et al. (1986). Three-stage tumorigenesis in mouse skin: DNA synthesis as a prerequisite for the conversion stage induced by TPA prior to initiation. Carcinogenesis 7: 779–782.

Lanson, M., Bougnoux, P., Besson, P., et al. (1990) n-6 Polyunsaturated fatty acids in human breast carcinoma phophatylethanolamine in early relapse. Brit. J. Cancer 61: 776–778.

Lee, F. Y. F., Veyssey, A., Rofstad, E., et al. (1989) Heterogeneity of glutathione content in human ovarian cancer. Cancer Res. 49: 5244–5248.

Masotti, L., Casali, E., and Galeotti, T. (1988) Lipid peroxidation in tumour cells. Free Radical Biol. Med. 4: 377–386.

Meyer, F., and Verreault, R. (1987) Erythrocyte selenium and breast cancer risk. Amer. J. Epidemiol. 125: 917–919.

Meyer, J. S., and Hixon, B. (1979) Advanced stage and early relapse of breast carcinomas associated with high thymidine labeling indices. Cancer Res. 39: 4042–4047.

Mills, B. L., Broghamer, W. L., Higgins, P. J., and Lindeman, R. D. (1981) A specific dietray zinc requirement for the growth of the Walkner 256/M1 tumor in the rat. Am. J. Clin. Nutr. 34: 1661–1669.

Mills, B. L., Broghamer, W. L., Higgins, P. J., and Lindeman, R. D. (1984) Inhibition of tumor growth by zinc depletion of rats. J. Nutr. 114: 746–756.

Minkel, D. T., Dolhum, P. J., Calhoun, B. L., Saryan, L. A., and Petering, D. H. (1979) Zinc deficiency and growth of Ehrlich ascites tumors. Cancer Res. 39: 2451–2456.

Mor, V. (1988) Malignant disease and the elderly. Research and the ageing population. Wiley, Chichester (Ciba Foundation Symposium 134) pp. 160–176.

Mueller, C. B., Ames, F., and Anderson, G. D. (1978) Breast cancer in 3558 women. Age as a significant determination in the rate of dying and causes of death. Surgery (St Louis) 83: 123–132.

Muller, W., and Iffand, R. (1981) Studies metals in meningiomas measured by atomic absorption spectrometry. Acta neuropathol. 55: 53–58.

Nakamura, Y., Colbur, N. H., and Gindhart, T. D. (1985) Role of reactive oxygen in tumor promotion: Implication of superoxide anion in promotion of neoplastic transformation in JB6 cells by TPA. Carcinogenesis 6: 229–235.

Newel, G. R., Spitz, M. R., and Silder, J. G. (1989) Cancer and age. Seminars in Oncology 16: 3–9.

Poubelle, P., Chaintreuil, J., and Bensadou, J. (1982) Plasma lipoperoxides and ageing. Biomedicine 36: 164–166.

Rogulji'c, A., Roth, A., Kolari, C. K., and Marici, C. Z. (1980) Iron, copper and zinc liver tissue levels in patients with malignant lymphoma. Cancer 46: 565–569.

Rougereau, A., Person, O., and Rougereau, G. (1987) Fat soluble vitamins and cancer localization associated to an abnormal ketone derivative of D3 vitamin: Carcinomedin. Internat. J. Vit. Nutr. Res. 57: 367–373.

246

Schreck, R., Rieber, P., and Baeuerle, P. A. (1991) Reactive oxygen intermediates as apparently widely used messengers in the activation of the NF-kappaB transcription factor and HIV-1. EMBO 10: 2247–2248.

Segala, C., Gerber, M., and Richardson, S. (1991) The pattern of risk factors for breast cancer in a Southern France population, Interest for a stratified analysis by age at diagnosis. Brit. J. Cancer 64: 919–925.

Simpson, H. W., Pauson, A. W., and Griffiths, K. (1988) Genesis of breast cancer is in the premenopause. The Lancet 9: 74–76.

Smaaland, R., Svardal, A. M., Lote, K., et al. (1991) Glutathione content in bone marrow and circadian stage relation to DNA synthesis. J. Natl. Cancer Inst. 83: 1092–1098.

Stewart, J. A. (1989) Breast cancer and ageing. Seminars in Oncology 16: 41–50.

Strauss, M. J., and Moran, R. E. (1980) The cell cycle kinetics of human breast cancer. Cancer 46: 2634–2639.

Teeter, S. M., Holmes, F. F., and McFarlane, M. J. (1987) Lung carcinoma in the elderly population. Cancer 60: 1331–1336.

Tolonen, M., Halme, M., and Sarna, S. (1985) Vitamin E and selenium supplementation in geriatric patients. A double blind perliminary clinical trail. Biol. Trace Element. Res. 7: 161–168.

Tubiana, M. (1989) Tumor cell proliferation kinetics and tumor growth rate. Acta Oncol. 28: 113–121.

Yancik, R., Ries, L. G., and Yates, J. W. (1989) Breast cancer in ageing women. 63: 979–81.

Free Radicals and Aging
ed. by I. Emerit & B. Chance
© 1992 Birkhäuser Verlag Basel/Switzerland

DNA damage in mammalian cell lines with different antioxidant levels and DNA repair capacities

Bernadette M. Hannigan, Shirley-Ann M. Richardson and
P. Gerald McKenna

Biomedical Sciences Research Centre, Department of Biological and Biomedical Sciences, University of Ulster, Coleraine BT52 1SA, Northern Ireland

Summary. A wide range of DNA damage is known to be caused by reactive oxygen species (ROS). Defence against the effects of such damage include damage prevention (e.g. antioxidant activity) and the removal of damaged moieties from DNA (DNA repair). Radiation (X-ray) sensitive murine lymphoma (LY) cells were seen to be more susceptible to ROS-induced damage than were radiation resistant cells. This difference was unlikely to be due to the marginally decreased DNA excision repair capacity of the sensitive cells. Radiation sensitive cells did, however, have lower endogenous antioxidant enzyme levels. Thus, the importance of assessing all levels of a cell's response to ROS, in determining the major factors leading to increased mutagen sensitivity, is emphasised.

Introduction

Many investigations on reactive oxygen species (ROS)-induced damage to cellular DNA have documented the appearance of oxidatively-modified bases following their excision from genomic or mitochondrial DNA (Fraga et al., 1990). Other studies have shown the persistence of ROS-induced base damage, DNA strand breaks and chromosome aberrations in intact cells (Emerit et al., 1983). Thus the potential toxicity of ROS towards a cell may be modified both by the cell's capacity to prevent DNA damage occurring, i.e. its antioxidant potential, its ability to tolerate such damage and by the effectiveness of its DNA repair mechanisms (Strain et al., 1991).

The volume of literature on the relationship between cellular antioxidant activity and ROS-induced damage is extensive (e.g. Halliwell and Gutteridge, 1989). On the relationship between DNA repair and damage many studies have utilised cellular irradiation, and hence irradiation-induced ROS, as the damaging modality. Few studies have made use of other ROS sources such as may pertain in vivo, e.g. the phagocyte respiratory burst or the xanthine oxidase (xo) catalysed conversion of xanthine (x) to uric acid and the superoxide anion (O_2^-) to assess damage and repair.

Several types of DNA repair may, potentially, be operable in the ROS-exposed cell. These include post-replication repair, nucleotide excision repair and base excision repair. It is highly likely that the excision repair capacity of cells exposed to potentially damaging ROS doses may be highly relevant to the ability of such cells to resist lethal or sub-lethal damage. The availability of mammalian cell lines with compromised DNA repair capacities has facilitated a number of studies describing the enhanced clastogenic activities of several known mutagens on such cells. In this laboratory two cell lines have been used extensively in DNA repair and mutation studies. These are the radiation (X-ray) resistant and sensitive variants (LY-R and LY-S, respectively) of the L5178Y mouse lymphoma cell line. McKenna et al. (1985) showed that L5178Y cells exhibited considerably less unscheduled DNA synthesis (UDS, an indication of excision repair) following UV treatment than did DNA repair proficient cells of the murine Friend Erythroleukaemia (FEL) line. Beer et al. (1983) had shown an inverse cross-sensitivity to X-ray treatment on the one hand and UV treatment on the other between LY-R and LY-S cells. Thus, analysis of the response of L5178Y cell types to ROS could be a useful model in which to assess the contribution of tertiary defence mechanisms to minimising the effect of ROS-mediated DNA damage.

The effects of ROS on radiation resistant and sensitive cells

The x + xod reaction was used as a source of ROS. In this laboratory, mitotic indices (Ml) cell growth and ^{3}H-thymidine (^{3}H-dT) incorporation into de novo synthesised DNA have been measured in both cell lines at various times up to 48 h following a period (1–1.5 h) in the presence of a range of concentrations of x and xod (up to 2.0 mM x + 2.0 mU xod). LY-S cells showed significant decreases (p < 0.02) in Ml values, in cell growth and in ^{3}H-dT incorporation relative to LY-R cells on exposure to increasing concentrations of x + xod. These differences were still evident at 48 h following ROS treatment. With mean chromosome numbers of 39 and 40 for LY-S and LY-R cells, respectively, the two cell lines were considered to present similar genetic targets such that any assessment of DNA damage would be meaningful. LY-S cells had a marginally increased frequency of chromosome aberrations relative to LY-R cells and, although this could be accounted for, in part, by their higher frequency of spontaneous aberrations, such a result agrees with the X-ray experiments of Bocian et al. (1982). LY-S cells showed a significantly (p < 0.01) increased susceptibility to the induction of mutations leading to 6-thioguanine resistance, i.e. point-mutations at the HGPRT locus.

The influence of DNA repair

The possible contribution of DNA repair deficiency to this increased DNA damage was assessed by measurement of DNA excision repair as unscheduled DNA synthesis (UDS) observable following autoradiography in cells exposed to x + xod as before (Rasmussen and Painter, 1964). LY-R cells exhibited marginally more excision repair than did LY-S (Richardson et al., 1990a,b). Such a small difference is unlikely to represent the major factor determining the differential sensitivities of the two cell lines. Overall, the level of UDS detected in L5178Y cells appears to be significantly less than that of repair-proficient FEL cells also exposed to x + xod-induced ROS (Richardson et al., 1990c; 1991).

From the results outlined above, the increased sensitivity of LY-S cells to ROS-mediated damage, relative to LY-R, has been established using a number of independently measured genetic end-points and ROS produced in a reaction system of in vivo relevance. This correlates well with the known X-ray sensitivity of LY-S. Conversely, LY-R cells are sensitive to UV irradiation, with LY-S cells being relatively resistant. Hagan et al. (1988) showed that LY-S have the higher excision repair capacity following UV-damage. Hence, UV-specific (nucleotide) excision repair is unlikely to be important for the removal of ROS-induced damage.

The role of endogenous antioxidants

To determine whether the endogenous antioxidant capacity of LY-R and LY-S cells could have contributed to their different levels of ROS-induced point mutations, endogenous activities of the antioxidant enzymes superoxide dismutase (SOD), catalase (Cat) and glutathione peroxidase (GPx) were measured in these cell types by the methods of Jones and Suttle (1981), Aebi (1983) and Paglia and Valentine (1967) respectively, and are shown in Table 1. LY-R cells can be seen to possess significantly more of each antioxidant than do LY-S cells. An increase in a single antioxidant is unlikely to lead to cell protection against ROS but to cause an imbalance in one ROS type with potentially adverse consequences. With higher levels of all three antioxidants measured it is likely that LY-R cells have a greater overall antioxidant

Table 1. Endogenous superoxide dismutase (SOD), catalase (Cat) and glutathione peroxidase (GPx) levels per 10^6 cells. Results are the mean of three independent measurements $\pm$ SD

Cell line	SOD (nmoles)	Cat (Units)	GPx (Units)
L5178Y(R)	0.40 ± 0.002	1.6 ± 0.0005	10.4 ± 0.0015
L5178Y(S)	0.20 ± 0.002	0.35 ± 0.001	5.2 ± 0.0012

250

capacity than have LY-S which may be a major factor in their increased relative resistance to ROS. Measurement of a greater range of intracellular antioxidants could, perhaps, further strengthen this argument.

Conclusion

It is evident that, in order to establish the cause of a cell's relative resistance or susceptibility to any mutagen capable of giving rise to intracellular or extracellular ROS, it is necessary that both DNA repair capacity and antioxidant status be evaluated.

Acknowledgment. This work was supported by a grant from the Ulster Cancer Foundation. S-A. M. Richardson was a D.E.N.I. Distinction Award student.

Aebi, H. (1983) Catalase in vitro, in: Methods of Enzymatic Analysis. H. U. Bergmeyer, ed. 3rd Ed. Verlag Chemie, Weinheim, Germany.

Beer, J., Budzicka, E., Niepokojoczycka, E., Rosiek, O., Szumiel, I., and Walicka, M. (1983) Loss of tumourigenicity with simultaneous changes in radiosensitivity and photosensitivity during in vitro growth of L5178Y murine lymphoma cells. Cancer Res. 43: 4736–4742.

Bocian, E., Bouzyz, E., Rosiek, O., and Ziemba-Zoltowska, B. (1982) Chromosomal aberrations induced by X-rays in two mouse lymphoma (L5178Y) sublines of different radiosensitivity. Int. J. Radiat. Biol. 42: 347–352.

Emerit, I., Levy, A., and Cerutti, P. (1983) Suppression of tumour promoter phorbol myristate acetate induced chromosome breakage by antioxidants and inhibitors of arachidonic acid metabolism. Mutat. Res. 110: 327–335.

Fraga, C. G., Shigenaga, M. K., Park, J-W., Degan, P., and Ames, B. N. (1990) Oxidative damage to DNA during ageing: 8-Hydroxy-2-deoxyguanosine in rat organ DNA and urine. Proc. Natl. Acad. Sci. 87: 4533–4537.

Hagan, M. P., Dodgen, D. P., and Beer, Z. J. (1988) Impaired repair of UVC induced DNA damage in L5178Y-R cells: sedimentation studies with the use of bromodeoxyuridine photolysis. Photochem. Photobiol. 47: 815–821.

Halliwell, B., and Gutteridge, J. M. C. (1989) Free Radicals in Biology and Medicine. (2nd ed.) Clarendon Press, Oxford.

Jones, D. G., and Suttle, N. F. (1981) The effect of copper deficiency on leukocyte function in sheep and cattle. Res. in Vet. Sci. 31: 151–156.

McKenna, P. G., Yasseen, A. A., and McKelvey, V. J. (1985) Evidence for direct involvement of thymidine kinase in excision repair processes in mouse cell lines. Somat. Cell Mol. Genet. 11: 239.

Paglia, D. E., and Valentine, W. N. (1967) Studies on the quantitative and qualitative characterisation of erythrocyte glutathione peroxidase. J. Lab. Clin. Med. 70: 158–161.

Rasmussen, R. E., and Painter, R. B. (1964) Evidence for repair of ultraviolet damaged deoxyribonucleic acid in cultured mammalian cells. Nature, 203: 1360–1362.

Richardson, S.-A. M., Hannigan, B. M., and McKenna, P. G. (1990a) DNA repair of oxidative damage in the cell line L517Y(S). Int. J. Rad. Biol. 57: 1257.

Richardson, S.-A. M., Hannigan, B. M., and McKenna, P. G. (1990b) Assessment of unscheduled DNA synthesis (UDS) in murine lymphoma cells exposed to enzymatically generated oxidants. Br. J. Cancer 62: 530.

Richardson, S.-A. M., Hannigan, B. M., and McKenna, P. G. (1990c) Assessment of DNA repair (by UDS) of oxidative damage in two murine leukaemia cell lines. Free Rad. Biol. Med. 9: 52.

Richardson, S.-A. M., Hannigan, B. M., and McKenna, P. G. (1991) DNA repair induced by reactive oxygen species (ROS) in the Friend erythroleukaemia cell line 707 Ir. J. Med. Sci. 160: 149.

Strain, J. J., Hannigan, B. M., and McKenna, P. G. (1991) The pathophysiology of oxidant damage. J. Biomed. Sci. 2: 19–24.

Free Radicals and Aging
ed. by I. Emerit & B. Chance
© 1992 Birkhäuser Verlag Basel/Switzerland

Variable α-tocopherol stimulation and protection of glutathione peroxidase activity in non-transformed and transformed fibroblasts

C. E. P. Goldring, W. L. Hu, N. R. Rao, C. Rice-Evans, R. H. Burdon[a] and A. T. Diplock

Free Radical Research Group, Division of Biochemistry, UMDS, Guy's Hospital, London SE1 9RT, U.K., and [a]Department of Bioscience and Biotechnology, University of Strathclyde, Glasgow G4 0NR, U.K.

Summary. Studies on glutathione metabolism in an established baby hamster kidney cell line (BHK-21/C13) and in its polyoma virus-transformed counterpart (BHK-21/PyY), have revealed a significant stimulation of intracellular glutathione peroxidase activity (Se-independent plus Se-dependent) by α-tocopherol supplementation (14 μM). This stimulation was found to be much greater in the transformed cells. Other GSH-requiring enzyme activities (namely glutathione reductase and glutathione transferase) were unaltered by α-tocopherol treatment, suggesting a degree of specificity in its action on GSHpx. In unsupplemented growth media, the GSHpx activity in both cell lines was significantly decreased by an oxidative stress. However, the same stress applied to the α-tocopherol-supplemented cells had no effect on the stimulated GSHpx activity, suggesting a protection afforded by the α-tocopherol.

Introduction

The results of human studies indicate an inverse relationship between the level of intake of certain antioxidant nutrients and risk of certain diseases, particularly cancer and heart disease (Gey et al., 1987; Diplock, 1991; Riemersma et al., 1991). Thus, it is felt that consideration should be given to increasing the recommended daily allowance of some of them (Diplock, 1987).

Using a specific model of one stage in the progression of a cell from a normal to a transformed phenotype, namely the culture of an established parent cell line and its polyoma virus-transformed counterpart, we have been seeking to explore the relationships between oxidative stress, antioxidants, and lipid peroxidation in cancer.

The intracellular antioxidant system is highly complex, and free radical scavengers might not work independently, but to a certain extent in synchrony, in protecting cells from oxidative damage. The results of our work seem to bear this out, and may point to some specificity in the relationships between the various antioxidants. Specifically, we have investigated the relationship between α-tocopherol and various enzymes involved in glutathione metabolism in our cell lines, i.e. glutathione peroxidase (GSHpx), glutathione reductase (GSHR), and glutathione transferase (GST); only the GSHpx was influenced by α-tocopherol supplementation.

Methods

Non-transformed baby hamster kidney fibroblasts (BHK-21/C13) and baby hamster kidney fibroblasts *transformed using polyoma virus* (BHK-21/PyY) (Burdon et al., 1990) were routinely grown in Dulbecco's modified Eagle's medium (DMEM), containing fetal calf serum (FCS) at a concentration of 10%, under an atmosphere of air containing 5% carbon dioxide at 37°C. Supplementation with 14 μM α-tocopherol of the standard growth medium was by the addition of α-tocopherol dissolved in ethanol. The α-tocopherol content of the growth medium (DMEM plus 10% FCS) was estimated to be 0.00072 μM; the same batch of FCS was used throughout this work.

In the enzyme determinations, the fibroblasts were grown to near confluency in the standard growth medium (either unsupplemented or supplemented with α-tocopherol). The cell monolayers were then washed twice with DMEM. Where the cells were stressed, DMEM containing 374 μM $FeCl_3$/10 mM ADP was added to the monolayers and the flasks were incubated under normal growth conditions for 2 h. The unstressed cells were treated identically, except for the omission of the $FeCl_3$/ADP. The cell monolayers from all treatments were then washed twice using 307 buffer solution, trypsinised, washed once in growth medium, twice in phosphate buffered saline (PBS), suspended in PBS, and counted using a Coulter counter.

Cells were homogenised using sonication, and the homogenates were then analysed as follows:

1) Total (using cumene hydroperoxide as substrate) and Se-dependent (using H_2O_2 as substrate) GSHpx activities were determined using the coupled assay procedure (Paglia and Valentine, 1967).

2) GSHR activity was determined using the NADPH oxidation method (Racker, 1955).

3) GST activity was determined using chlorodinitrobenzene as substrate (Habig and Jakoby, 1981).

Results were tested for significance using the Mann-Whitney U test, and expressed as the mean $\pm$ SEM.

Results and discussion

The effects of α-tocopherol on non-transformed (BHK-21/C13) and transformed (BHK-21/PyY) fibroblasts with respect to various antioxi-

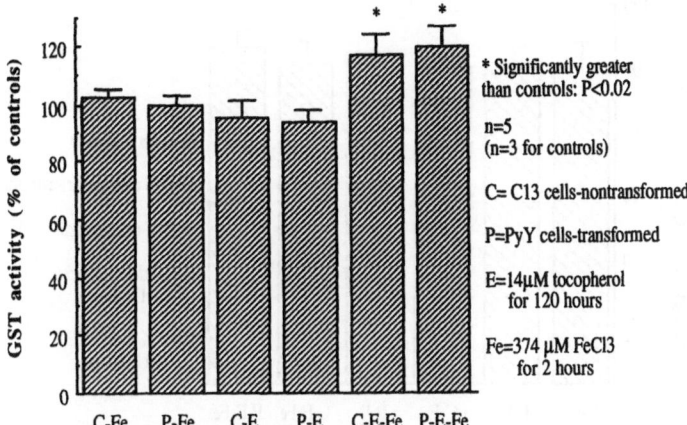

Figure 1. Influence of α-tocopherol supplementation and iron-catalysed oxidative stress on GST activity in non-transformed (C13) and transformed (PyY) fibroblasts.

dant enzymes (glutathione tranferase, glutathione reductase, and glutathione peroxidase) were followed with or without oxidative stress using the Fe^{3+}-ADP system. Figure 1 depicts the glutathione transferase activity in both cell lines. The levels were similar, with no significant differences seen either with α-tocopherol supplementation alone or with oxidative stress. Interestingly the glutathione transferase activity was marginally increased in both cell types after α-tocopherol supplementation and oxidative stress compared to that of the controls. This increase could be an α-tocopherol-mediated mechanism to ensure conjugation of some of the aldehydic products (e.g. 4-hydroxy-2,3-*trans*-nonenal) formed during oxidative stress.

The results of α-tocopherol supplementation and iron-catalysed oxidative stress on glutathione reductase activity in the non-transformed and transformed fibroblasts are shown in Figure 2. No appreciable difference was seen in glutathione reductase activity in transformed fibroblasts compared to that of non-transformed fibroblasts. In addition, glutathione reductase activity was unaffected by the various treatments in both cell lines.

The changes in glutathione peroxidase activity in non-transformed and transformed fibroblasts on oxidative stress with and without α-tocopherol supplementation in the growth medium are seen in Figure 3. The non-transformed fibroblasts had only marginal increases in total glutathione peroxidase activity on supplementation with α-tocopherol for 96 h, while their transformed counterparts showed a very significant rise in the enzyme activity. The enzyme activity dropped in both cell lines to a similar extent after oxidative stress without α-tocopherol supplementation. However, the enzyme levels remained unaltered in both cell types

254

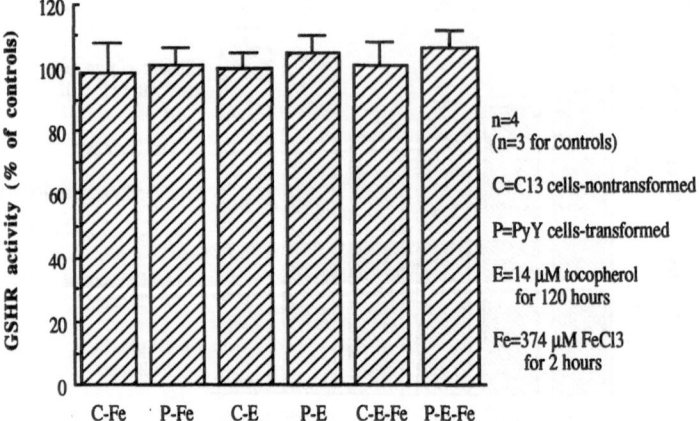

Figure 2. Influence of α-tocopherol supplementation and iron-catalysed oxidative stress on GSHR activity in non-transformed (C13) and transformed (PyY) fibroblasts.

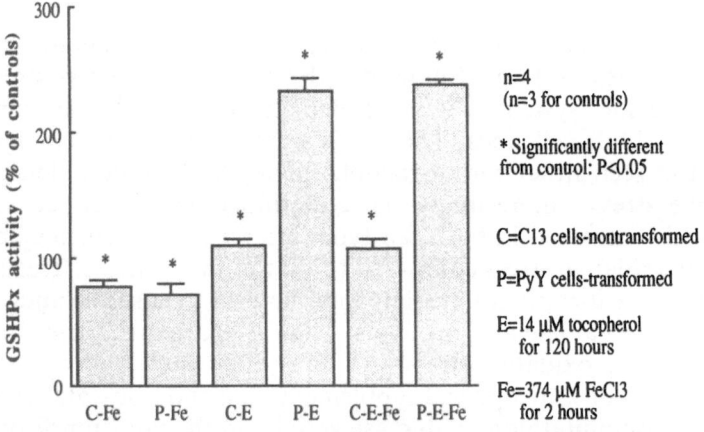

Figure 3. Influence of α-tocopherol supplementation and iron-catalysed oxidative stress on GSHpx activity in non-transformed (C13) and transformed (PyY) fibroblasts.

after oxidative stress with α-tocopherol supplementation compared to the activity seen in cells grown only with α-tocopherol. The Se-dependent glutathione peroxidase activity also followed this pattern after such treatments.

It is known that there is not necessarily a parallel variation in the activities of the selenium-containing glutathione peroxidase and glutathione reductase between various non-transformed and transformed cells; the activity of the former is generally higher and the activity of the latter lower, even though glucose-6-phosphate dehydrogenase, the supplier of reducing equivalents for glutathione reductase is increased. In fact, the activities of many of the intracellular antioxidant enzymes are highly variable in cancer cells. Such abnormal antioxidant enzyme activities have been detected in a variety of studies on tumour cells including SV40-transformed cells, various hepatoma cells, melanoma tumors and a range of ascites tumor cells (Goldring et al., 1992).

It was previously found that peroxides and hydroxyl radicals can inactivate glutathione peroxidase and SOD (Pigeolet and Remacle, 1991). We have observed here that exposure of our non-transformed and transformed cells to Fe^{3+}-ADP stress depleted the glutathione peroxidase activity, with no appreciable differences in other glutathione-dependent enzyme activities between the cell lines. Pigeolet and Remacle (1991) have shown that glutathione peroxidase appears to be similar to SOD in that it does not seem to be well protected against free radical damage. These workers postulated that if the free radical attack is not too strong, or nullified by supplemented antioxidants, a swift removal of the altered enzyme might occur with the concomitant resynthesis of new enzyme. The changes in glutathione peroxidase activity in our cell lines after α-tocopherol supplementation alone, or after α-tocopherol supplementation plus oxidative stress, could be due to a protection against the oxidation of selenocysteine residues of the glutathione peroxidase enzyme, or a sparing effect on the glutathione peroxidase by the preferential oxidation of α-tocopherol. This selenocysteine oxidation might not necessarily be via the direct action of the Fe^{3+}-ADP prooxidant system, but could instead be due to the action of Fe^{3+}-ADP in stimulating reactive aldehydes via the lipid peroxidation pathway. These aldehydes are fairly hydrophilic and could easily oxidise the Se-H group in the GSHpx molecule. The differences in glutathione peroxidase activity seen between the two cell lines could not be attributed to a difference in uptake; at 14 μM α-tocopherol, the transformed cells took up less of the available α-tocopherol (1.3 nmoles/million cells) than the transformed cells (2.9 nmoles/million cells). It is possible that this protein exhibits some heterogeneity between these cell lines. However, an unequal utilisation of the α-tocopherol between these cells, once it had been incorporated intracellularly, might also explain these differences.

Acknowledgements. Financial support by the Association for International Cancer Research is gratefully acknowledged. The authors thank Sue Peach and Vera Gill for expert technical assistance.

Burdon, R. H., Gill, V., and Rice-Evans, C. (1990) Oxidative stress and tumour cell proliferation. Free Rad. Res. Comm. 11: 65–76.

Diplock, A. T. (1987) Dietary supplementation with antioxidants. Is there a case for exceeding the recommended dietary allowance? Free Rad. Biol. Med. 3: 199–201.

Diplock, A. T. (1991) Antioxidant nutrients and disease prevention; an overview. Am. J. Clin. Nutr. 53: 189S–193S.

Gey, K. F., Brubacher, G. B., and Stähelin, H. B. (1987) Plasma levels of antioxidant vitamins in relation to ischemic heart disease and cancer. Am. J. Clin. Nutr. 45: 1368–1377.

256

Goldring, C. E. P., Rice-Evans, C., Burdon, R. H., Rao, R., Haq, I., and Diplock, A. T. (1992) Uptake of α-tocopherol and its influence on cell proliferation and lipid peroxidation in transformed and non-transformed fibroblasts. Archs. Biochem. Biophys. (in press).

Habig, W. H., and Jakoby, W. B. (1981) Assays for differentiation of glutathione S-transferases. Methods Enzymol. 77: 398–405.

Paglia, D. E., and Valentine, W. N. (1967) Studies on the quantitative and qualitative characteristics of erythrocyte glutathione peroxidase. J. Lab. Clin. Med. 70: 158–169.

Pigeolet, E., and Remacle, J. (1991) Susceptibility of glutathione peroxidase to preteolysis after oxidative alteration by peroxide and hydroxyl radicals. Free Rad. Biol. Med. 11: 191–195.

Racker, E. (1955) Glutathione reductase from bakers' yeast and beef liver. J. Biol. Chem. 217: 855–865.

Riemersma, R. A., Wood, D. A., Macintyre, C. C. A., Elton, R. A., Gey, K. F., and Oliver, M. F. (1991) Risk of angina pectoris and plasma concentrations of vitamins A, C and E and carotene. Lancet. 337: 1–5.

Free Radicals and Aging
ed. by I. Emerit & B. Chance
© 1992 Birkhäuser Verlag Basel/Switzerland

Assessing and counteracting the prooxidant effects of anticancer drugs

Pierre Bienvenu, Laurent Caron, Didier Gasparutto and
Jean-François Kergonou

Centre de Recherches du Service de Santé des Armées, Unite de Radiobiochimie, 24 Avenue des Maquis du Grésivaudan, B.P. 87, F-38700 La Tronche, Grenoble, France

Summary. The relationship between peroxide generation and respective cellular damage, triggering various biochemical consequences is first discussed. Then we review the prooxidant effects of various anticancer drugs including anthracyclines and bleomycin, platinum derivatives and the N- and S-mustards. We present and discuss some experimental results on peroxidase inhibition by drugs such as zinc salts, almitrine, deferoxamine, which had previously been tested as efficient in vivo treatment on chlormethine intoxication. In an overview we propose that not only ionizing radiations and anticancer drugs, but also promoters and initiators of cancer might all generate free radicals, in turn triggering oxidative processes generating endogenous peroxides, then probably amplifying the deleterious biological response. The possible limitations of drug therapies decreasing peroxide generation are presented.

Introduction

In keeping with the free radical theory of aging, some deleterious effects involving oxygen-derived free radicals produced by irradiation or anticancer drug treatments, may potentially decrease life expectancy.

Even if some organ or tissue specificity is recognized, we hypothesize that both cytotoxicity and genotoxicity exerted by anticancer drugs might to some extent depend on oxygen radical generation and perhaps also involve some relatively stable organic peroxides, mainly produced in a relatively long-lasting process by over stimulated endogenous oxidases.

Therefore after some considerations on the potential role of peroxides in cytotoxicity and genotoxicity, we review some prooxidant effects of several anticancer drugs with particular stress on the nitrogen mustards and some insights into the various biochemical and pathological effects probably related to oxygen activation. Then we take into account the relationship between previous experimental results on therapeutic efficiency of some drugs against chlormethine intoxication and the respective antioxidant potency of such drugs. We also deal with modelization of the prooxidant effects of chlormethine.

Peroxide generation and cellular damage

There is an apparent paradox in the therapeutic applications of anticancer drugs: their usefulness is based on their cytotoxic properties, but they are also "positively involved in the causation of secondary tumors, including AML, myelodystrophic syndromes, solid tumors" (Bailey and Blackledge, 1990) which could probably be ascribed to gene mutations, and/or chromosonal aberrations (Heddle et al., 1991).

Such delayed side effects are considered as relatively infrequent and have been attributed to genotoxicity, i.e. DNA damage. Both cytoxicity and DNA damage may be initiated not only by ionizing radiation but also either by exogenous peroxides such as cumene hydroperoxide or by endogenous peroxides such as H_2O_2 and lipid peroxides which might be generated at increased rates during oxidative stress e.g. after intoxication by redox active drugs such as menadione, diquat, BCNU, (= 1,3-bis (2-chloroethyl)-1 nitrosourea) or adriamycin and perhaps also by several if not all anticancer drugs. Besides lipid peroxidation, oxidative stress is accompanied by oxidation of thiols and pyridine nucleotides, depression of the glycolytic pathway and intracellular ATP levels, DNA damage with base hydroxylation, strand scission and cross-linking, loss of intracellular NAD^+ and activation of poly(ADP-ribose) polymerase, rise in cytoplasmic calcium, perturbation of the cytoskeleton and mitochondrial swelling (Atzori et al., 1990).

t-Butyl hydroperoxide is able to kill cultured hepatocyes by peroxidizing their membrane lipids (Masaki et al., 1989) whereas the more lipophilic cumene hydroperoxide was more cytotoxic but exerted less DNA damage than t-Bu OOH. Sandström (1991) also proved in vitro on cell culture that the most hydrophilic H_2O_2 was both more cytotoxic and more deleterious towards DNA than these two other peroxides. This fact is particularly interesting as far as H_2O_2 is produced during water irradiation (Kimball et al., 1955), and its mechanism of cytotoxicity seems similar to the cell damage exerted by ionizing radiation (Link, 1992). Hydrogen peroxide and adriamycin semiquinone are both able to cause DNA strand breaks by damaging deoxyribose (Bates and Winterbourn, 1982).

Anthracyclines and bleomycin

Much work has been devoted to the prooxidant effect of the anthracyclines (review by Powis, 1987), and of bleomycin (review: Lazo et al., 1987). The cardiotoxicity of adriamycin has been related to lipid peroxidation which could be shown in vitro on mouse heart microsomes containing low levels of alpha-tocopherol (D'Alessandro et al., 1980).

This oxidative process was NADPH-dependent in rat heart and liver microsomes (Mimnaugh et al., 1981), and produced by other anthracycline drugs (Mimnaugh et al., 1982). Furthermore, the involvement of oxygen-derived free radicals has been proposed because of inhibition by specific scavengers (Mimnaugh et al., 1983) and of stimulation exerted on this process by iron ions (Daugherty et al., 1982), which might be given in vivo by ferritin (Demant, 1984).

Direct evidence of superoxide and hydroxyl radicals generation has been presented for the bleomycin-Fe^{2+} complex, on the basis of spin-trapping experiments (Lown and Sim, 1977), and such radicals might explain the bleomycin-induced cleavage of DNA.

Platinum derivatives

The previously examined drugs are able to complex iron efficiently. Cis-platin comprises a metal and therefore probably does not need chelation to induce lipid peroxidation (Gemba and Sugihara, 1986).

Some salient facts presented here are probably related to the radical generation by cis-platin and carbo-platin (see Borch, 1987, for review). Cis-platin reacts with DNA and exerts hematological, renal, gastrointestinal toxicity, and also neurotoxicity clinically indistinguishable from that caused by vitamin E deficiency (Mollman, 1990), and Yuhas et al. (1980) have shown that a well-known radioprotector and chemoprotector against mustard toxicitiy (Yuhas et al., 1979), the aminothiol, WR 2721, protects cancer patients against such side effects. This protection, as well as that given by sodium thiosulfate might remain efficient even if these drugs are given 30 min after cis-platin (in Borch, 1987). An ACTH analog protects against its neurotoxicity (Gerritsen Van Der Hoop et al., 1990) and selenite is efficient against the nephrotoxicity.

Cis-platin inhibits critical enzymes such as ATPase, gamma-glutamyltranspeptidase and most important metalloenzymes such as cytochrome P-450 (Bompart, 1990) and alpha-2-macroglobulin, the zinc carrier (Gonias et al., 1984) which is also inhibited by gamma irradiation (Weiqiao et al., 1990), as is cytochrome P-450 (Zajac and Bernard, 1982). The cross-linking of alpha-2-macroglobulin might explain hyperzincuria and hypozincemia observed in patients treated with cis-platin (Sweeney et al., 1989). A platinum complex similar to cis-platin is able to inhibit lipoxygenase (Michaud-Soret and Chottard, 1992).

Such similar biochemical effects are likely, but have not been established, for the increased cytotoxicity of the alkylating agents melphalan, cyclophosphamide, and CCNU, when cis-platin was also administered in tumor bearing mice (Horsman et al., 1984).

Nitrogen and sulfur mustards

Papirmeister et al. (1991) haver presented and discussed three main mechanisms of mustard cytotoxicity.

The first one, called 'polymerase hypothesis', involves sulfur mustard (=SM)-induced DNA damage, leading to poly(ADP-ribose) polymerase activation, in turn depleting cellular NAD^+, inhibiting glycolysis, activating hexose monophosphate shunt, and finally releasing proteases, causing subsequent cell necrosis, and skin injury.

The second, or thiol-Ca^{2+} hypothesis, is based on sulfur mustard oxidative or electrophilic stress, causing glutathione depletion and thiol homeostasis disturbation, which may then either alter the cytoskeleton, or inhibit calcium transport, thereby activating calcium-dependent enzymatic catabolic processes resulting in protein degradation, DNA breaks, and phospholipid hydrolysis, which might all potentially lead to cell death.

The last one, or lipid peroxidation hypothesis, is also based on glutathione depletion, but the main consequence of it is lipid peroxidation, due to increased H_2O_2 generation. In this case, as after phospholipid hydrolysis, a loss of membrane integrity may cause cell death.

The well-known radiomimetic efficiency of mustards, which is manifested by the inhibition of cell proliferation, and associated with genotoxicity, has led numerous researchers to studies dealing with DNA.

As Papirmeister et al. wrote: "inordinate emphasis may have been placed on the highly potent genotoxicity of SM because of the presumption that knowledge of the most SM-sensitive cellular function (i.e., DNA replication and normal cell division) would a priori lead to the elucidation of the acute injury mechanism." However, for these authors, "the relevance of the polymerase hypothesis to acute SM toxicity has not been established", and "no critical tests of the thiol-Ca^{2+} or lipid peroxidation hypotheses have been performed."

Not only the mechanisms evoked, but also the efficiency of some antioxidant therapies discussed by these authors clearly point to a central role played by the oxidative stress exerted by the S or N mustards.

The probably spontaneous activation of S and N mustards yields the respective episulfonium and imonium ions which are highly reactive electrophiles and may in turn activate various biological activation and detoxication systems. Most probably, some microsomal systems including NADPH or NADH as electron donors are simulated by such groups, much like what happens with compounds bearing quinone, dipyridyle, nitro and azo groups (Ziegler and Kehrer, 1990).

A previous experiment done with a system including microsomes, a NADPH enzymatic generator and a linoleic acid emulsion has revealed the appearance of the characteristic peak of conjugated dienes, typical of lipid peroxidation, stimulated by chlormethine addition (Bienvenu and Gasparutto, unpublished observation). The lipid peroxides arising in that way might in turn exert several deleterious effects such as inhibition of ADP phosphorylation and glycolysis (Hyslop et al., 1988), or even DNA damage (Schaich and Borg, 1984; Tait et al., 1984; Kobayashi et al., 1988).

Peroxidase inhibition

Since oxygen-derived radicals and lipid peroxidation are involved in mustard effects, we tested the antioxidant potency of drugs which might inhibit these prooxidant effects. Some of these drugs had already been tested in vivo as effective drugs against bone marrow damage triggered either by irradiation (Bienvenu and Kergonou, 1988) or by intoxication with the chlormethine mustard (Bienvenu et al., 1990). The results were evaluated on the basis of the decreasing efficiency of the drugs against conjugated diene generation, as measured at 234 nm, on a linoleic acid emulsion and either soybean lipoxygenase or xanthine/xanthine-oxidase.

The results presented in Table 1 clearly show that almitrine, deferoxamine and to a lesser extent zinc salts are powerful inhibitors of lipoperoxidation, either promoted by lipoxygenase, or by xanthine-

Table 1. Inhibition of two enzymatic systems of various antioxidants. The relative efficiencies are expressed as percentages of the control values; IC 50 = concentration for 50% inhibition

		Enzymatic systems	
		Lipoxidase	Xanthine/xanthine oxidase
Inhibitors and concentrations	Allopurinol	22% for $2.5 \cdot 10^{-5}$ M	21% for $1.78 \cdot 10^{-5}$ M
	Zinc sulphate	IC 50 = $5.6 \cdot 10^{-5}$ M	—
	Zinc chloride	IC 50 = $5.9 \cdot 10^{-5}$ M	IC 50 = $1.74 \cdot 10^{-4}$ M
	Deferoxamine	26.8% for $8.52 \cdot 10^{-6}$ M	33.7% for $1.21 \cdot 10^{-6}$ M
	EDTA	23.6% for $3.57 \cdot 10^{-5}$ M	28% for $2 \cdot 10^{-5}$ M
	Magnesium sulphate	—	13.2% for $7.11 \cdot 10^{-5}$ M
	Almitrine	IC 50 = $5.2 \cdot 10^{-6}$ M	2.37% for $3.26 \cdot 10^{-6}$ M
	Vitamin E	IC 50 = $8.7 \cdot 10^{-6}$ M	—

oxidase. Furthermore, vitamin E inhibited the effect of lipoxygenase, whereas magnesium sulphate inhibited that of xanthine oxidase. These results were well correlated with the biological effectiveness of almitrine, deferoxamine, and zinc salts, which were able to protect mice against the deleterious effects of both gamma radiation injury and chlormethine mustard intoxication. These drugs were administered about fifteen minutes after the agressions, so that, despite their probably smaller relative efficiency, they could be more realistic therapies than the classical radioprotective drugs, which must be given before irradiation.

One must pay attention to some important pharmacological parameters, such as the relative hydrophobicity of the drug, when the in vitro potency is tentatively correlated with biological potency in such therapeutic attempts. EDTA, e.g. is more lipophilic than the other drugs, and this fact might explain its reduced therapeutic effect, possibly because some oxidative processes take place in the bulk lipidic phase of membranes.

As far as hydroperoxide-dependent enzymatic peroxidations may take place during drug biotransformations, the inhibitory effects of some anticancer drugs have also been tested on arachidonic acid metabolism, either by soybean lipoxygenase or by prostaglandin synthase (Kolodziejczyk and Lown, 1990). Surprisingly, some anthracycline anticancer drugs, such as daunorubicin and mitoxantrone exerted inhibitory effects in these systems, possibly because, despite the absence of the initial activation step, they nevertheless possessed metal-chelating properties, therefore sharing paradoxical antioxidant effects.

Overview and discussion

Roe (1989) claimed that "it is no longer tenable to believe in the existence of "pure initiators" and "pure promoters"." Hence, a unified view about cytotoxicity and carcinogenicity may be proposed on the basis of prooxidant states involved not only in the so-called tumor promotion (Cerutti, 1985), e.g. by phorbol myristate acetate (Emerit and Cerutti, 1982), but also in tumor initiation and in various physical (radiations either ionizing or not; hyperoxia, hypoxia) or chemical factors (peroxides, anticancer drugs, redox-cycling drugs (Cohen and d'Arcy, 1987), radiation sensitizers, etc) triggering oxidative stress (Sies, 1991), and involving free radicals (Halliwell and Gutteridge, 1989).

The various molecular effects recalled by Atzori et al. (1990) are triggered e.g. by ionizing radiation, and by anticancer drugs, either "alkylating" or not, which may by themselves increase radiosensitivity (Kovacs et al., 1988) and as we already mentioned may share some antioxidant antagonist treatments such as WR-2721.

Among these molecular effects, besides the already mentioned oxidative alterations bearing on lipids, nucleic acids, thiols and other cofactors such as NAD^+, selenium, and tocopherol, there are some enzymatic systems involved in xenobiotic oxidation, such as cytochrome P-450, which are often depleted.

All these effects can easily be traced to the initial involvement of free radicals, and to the amplification probably triggered by endogenous hydroperoxides generated during lipid peroxidation. As we tried to counteract the enzymatic generation of such peroxides, it should be stressed that such potential therapies might be limited because hydroperoxides share a physiological role (Ursini et al., 1991), e.g. in arachidonic acid metabolism. The important role played by metal ions in oxidative stress and in potential therapies has been discussed elsewhere (Bienvenu and Kergonou, 1990, Bienvenu et al., 1990).

Atzori, L., Gotgreave, I., and Moldeus, P. (1990) Critical events in the toxicity of redox active drugs, in: Selective Activation of Drugs by Redox Processes, pp. 229–235. Eds G. E. Adams, A. Breccia, E. M. Fielden and P. Wardmam. Plenum Press, New York.

Bailey, G., and Blackledge, G. (1990), Cytostatics and immunosuppressive drugs, in: Side Effects of Drugs Annual, Vol. 14, pp. 397–420. Eds M. N. G. Dukes and L. Belley. Elsevier, Amsterdam.

Bates, D. A., and Winterbourn, C. C. (1982) Deoxyribose breakdown by the adriamycin semiquinone and H_2O_2: evidence for hydroxyl radical participation. FEBS Lett. 145: 137–142.

Bienvenu, P., and Kergonou, J. F. (1990) Antagonism or synergy among iron, copper and zinc, in: Metal Ions in Biology and Medicine, pp. 545–548. Eds Ph. Collery et al. J. Libbey-eurotext, Paris.

Bienvenu, P., Herodin, F., Therin, J.-Y., and Fatome, M. (1990) Antioxidant effects in radioprotection, in: Antioxidants in Therapy and Preventive Medicine, pp. 291–300. Eds I. Emerit, L. Packer and C. Auclair. Plenum Press, New York.

Bompart, G. (1990) Cisplatin-induced changes on cytochrome P-450, lipid peroxidation and some P-450 related specific catalytic activities in rat liver. J. Toxicol. Clin. Expérim. 10: 375–383.

Borch, R. F. (1987) The platinum anti-tumor drugs, in: Metabolism and Action of Anticancer Drugs, pp. 162–193. Eds G. Powis and R. A. Prough. Taylor and Francis, London.

Cerutti, P. A. (1985) Prooxidant states and tumor promotion. Science 227: 375–381.

Cohen, G. M., and D'Arcy Doherty, M. (1987) Free radical mediated cell toxicity by redox-cycling chemicals. Cancer 55 Suppl. VIII: 46–52.

D'Alessandro, N. D., Dusonehet, L., Crosta, L., Crescimanno, M., and Rausa, L. (1980) Does catalase play a role in Adriamycin-induced cardiotoxicity. Pharmacol. Res. Commun. 12: 441–446.

Daugherty, J. P., Wheat, M., Conley, S., Cooley, E., Vanzant, C., Loggins, L., and Durant, J. R. (1982) Involvement of reactive oxygen species in Adriamycin (ADR) cardiotoxicity Proc. Am. Ass. Cancer Res. 23: 171, cited in: Archakov, A. I. and Bachmanova, G. I. (1990), Cytochrome P-450 and Active Oxygen. Taylor and Francis, London.

Demant, E. J. F. (1984) Transfer of ferritin-bound ion to adriamycin. FEBS Lett. 176: 97–100.

Emerit, I., and Cerutti, P. A. (1982) Tumor promoter phorbol 12-myristate 12-acetate induces a clastogenic factor in human lymphocytes. Proc. Natl. Acad. Sci. USA 79: 7509–7513.

Gemba, M., and Sugihara, K. (1986) Evaluation of cisplatin-induced lipid peroxidation in rat renal tissues by means of slice technique, in: Nephrotoxicity of Antibiotics and Immunosuppressants, pp. 167–177. Eds. T. Tanabe, J. B. Hook and H. Endou. Elsevier, Amsterdam.

264

Gerritsen Van der Hoop, R., Vecht, C. J., Van der Burg, M. E. L., Elderson, A., Boogerd, W., Heimans, J. J., Vries, E. P., Van Houweling, J. C., Jennekens, F. G. I., Gipsen, W. H., and Neijt, J. P. (1990) Prevention of cisplatin neurotoxicity with an ACTH (4–9) analogue in patients with ovarian cancer. N. Engl. J. Med. 322: 89–94.

Gonias, S. L., Oakley, A. C., Walther, P. J., and Pizzo, S. V. (1984) Effects of diethyldithiocarbamate and nine other nucleophiles on the intersubunit protein cross-linking and inactivation of purified human alpha-2-macroglobulin by cis-diamminedichloroplatinum (II). Cancer Res. 44: 5764–1673.

Halliwell, B., and Gutteridge, J. M. C. (1989) Free Radicals in Biology and Medicine, Clarendon Press, Oxford.

Heddle, J. A., Khan, M. A., Urlando, C., and Pagura, M. E. (1991) Measuring gene mutation in vivo, in: New Horizons in Biological Dosimetry, pp. 281–289. Eds B. L. Gledhill and F. Mauro. Wiley-Liss, Inc, New York.

Horsman, M. R., Hirst, D. G., Brown, D. M., and Brown, J. M. (1984) Modification of alkylating agent cytotoxicity by cisplatin. Int. J. Radiat. Oncol. Biol. Phys, 10: 1669–1673.

Hyslop, P. A., Hinshaw, I. D. B., Halsey, W. A., Jr, Schraufstatter, I. U., Sauerheber R. D., Spragg, R. G., Jackson, R. G., and Cochrane, C. G. (1988) Mechanisms of oxidant mediated cell injury. J. Biol. Chem. 263: 1665–1675.

Kalyanaraman, B., Perez-Reves, E., and Mason, R. P. (1989) Spin-trapping and direct electron spin resonance investigations of the redox metabolism of quinone anticancer drugs. Biochim. Biophys. Acta 630: 119–130.

Kimball, R. F., Haron, J. Z., and Gaither, N. (1955) Tests for a role of H_2O_2 in X-ray mutagenesis II. Attempts to induce mutations by peroxide. Radiation Res. 3: 435–443.

Kobayashi, S., Ueda, K., Morita, J., and Komano, T. (1988) DNA damage induced by free radicals generated from hydroperoxides, in: Medical, Biochemical and Chemical Aspects of Free Radicals, pp. 1513–1516. Eds O. Hayaishi, E. Niki, M. Kondo and T. Yoshiawa. Elsevier, Amsterdam.

Kolodziejczyk, P., and Lown, I. W. (1990) Peroxidase-induced metabolism and lipid peroxide scavenging by antitumor agents, in: Antioxidants in Therapy and Preventive Medicine, pp. 323–338. Eds I. Emerit, L. Packer and C. Auclair. Plenum Press, New York.

Kovacs, C. J., Evans, M. J., Hooker, J. L., and Jonhke, R. M. (1988) Long-term consequences of chemotherapeutic agents on hematopoiesis: development of altered radiation tolerance. NCl Monogr. 6: 45–49.

Lazo, J. S., Sebti, S. M., and Filderman, A. E. (1987) Metabolism of bleomycin and bleomycin-like compounds, in: Metabolism and Action of Anticancer Drugs, pp. 194–210. Eds G. Powis and R. A. Prough. Taylor and Francis, London.

Link, E. M. (1992) Mechanism of H_2O_2 cytotoxicity and its resemblance to cell damage by ionzing radiation. Abstracts of the SFRR/ARR Meeting, Manchester, Jan. 6–8: 1992.

Lown, J. W., and Sim, S. K. (1977) The mechanism of the bleomycin-induced cleavage of DNA. Biochim. Biophys. Acta 77: 1150–1157.

Masaki, N., Kyle, M. E., and Farber, J. L. (1989) Tert-butylhydroperoxide kills cultured hepatocytes by peroxidizing membrane lipids, Archs Biochem. Biophys. 269: 390–399.

Michaud-Soret, I., and Chottard, J. C. (1992) Investigation of sulfur containing amino acids at the lipoxygenase active site using a platinum complex. Biochem. Biophys. Res. Commun. 182: 779–785.

Mimnaugh, E. G., Gram, T. E., and Trush, M. A. (1983) Stimulation of mouse heart and liver microsomal lipid peroxidation by athracycline drugs: characterization and effects of reactive oxygen scavengers. J. Pharmacol. Exp. Ther. 226: 806–816.

Mimnaugh, E. G., Trush, M. A., Ginsburgl, E., and Gram, T. E. (1982) Differential effects of anthracycline drugs on rat heart and liver microsomal reduced nicotinamide dinucleotide phosphate-dependent lipid peroxidation. Cancer Res. 42: 3574–3582.

Mimnaugh, E. G., Trush, M. A., and Gram. T. E. (1981) Stimulation of rat heart and liver microsomal NADPH-dependent lipid peroxidation. Biochem. Pharmacol. 30: 2797–2804.

Mollman, J. E. (1990) Cisplatin neurotoxicity. N. Engl. J. Med. 322: 126–127.

Papirmeister, B., Feister, A. J., Robinson, S. I., and Ford, R. D. (1991), Medical Defence Against Mustard Gas: Toxic Mechanisms and Pharmacological Implications. CRC Press, Boca Raton.

Pihl, A., Eldjarn, L., and Bremer, J. (1957) On the mode of action of X-ray protective agents III. The enzymatic reduction of disulfides. J. Biol. Chem. 227: 339.

Powis, G. (1987) Anthracycline metabolism and free radical formation, in: Metabolism and Action of Anticancer Drugs, pp. 211–260. Eds G. Powis and R. A. Prough. Taylor and Francis, London.

Roe, F. J. C. (1989) What is wrong with the way we test chemicals for carcinogenic activity, in: Advances in applied toxicology, pp. 1–17. Eds A. D. Dayan and A. J. Paine. Taylor and Francis, London.

Sandström, B. E. (1991) Induction and rejoining of DNA single-strand breaks in relation to cellular growth in human cells exposed to three hydroperoxides to 0°C and 37°C. Free Rad. Res. Comms. 15: 79–89.

Schaich, K. M., and Borg, D. C. (1984) Radiomimetic effects of peroxidizing lipids on nucleic acids and their bases, in: Oxygen Radicals in Chemistry and Biology, pp. 603–606. Eds W. Bors, M. Saran and D. Tait. de Gruyter, Berlin.

Sies, H. (1991) Oxidative Stress: Oxidants and Antioxidants. Academic Press, London.

Sweeney, J. D., Ziegler, P., Pruet, C., and Spaulding, M. B. (1989) Hyperzincuria and hypozincemia in patients treated with cisplatin. Cancer 63: 2093–2095.

Tait, D., Martin-Bertram, H., Hagen, U., and Bors, W. (1984) Influence of arachidonic acid on irradiation effects in a DNA model substance, in: Oxygen Radicals in Chemistry and Biology, pp. 607–610. Eds W. Bòrs, M. Saran and D. Tait. de Gruyter, Berlin.

Ursini, F., Maiorino, M., and Sevanian, A. (1991) Membrane Hydroperoxides, pp. 319–336, in: Sies, H., op. cit.

Wieqiao, J., Shijie, C., Suzhen, M., Peifang, F., Min., L. Hao, Z., Yonghe, L., and Jiewei, H. (1990) Changes of cathepsin D and alpha 2 – macroglobulin in rats after acute total body irradiation. Nucl. Sci. Tech. 1: 187–192.

Yuhas, J. M., Spellman, J. M., Jordan, S. W., Pardini, M. C., Afzal, S. M. J., and Culo, F. (1980) Treatment of tumours with the combination of WR 2721 and cis-diammine dichloroplatinum (II) or cyclophosphamide. Br. J. Cancer 42: 574–585.

Yuhas, J. M. (1979) Differential protection of normal and malignant tissues against the cytotoxic effects of mechlorethamine. Cancer treat. Rep. 63: 971–976.

Zajac, J. M., and Bernard, P. (1982) Effects of whole-body irradiation on the microsomal enzyme system and on cytochrome P-450 of rat liver. Enzymes 27: 19–24.

Ziegler, D. M., and Kehrer, J. P. (1990) Oxygen radicals and drugs: in vitro measurements. Meth. Enzymol. 1986: 621–626.

Free Radicals and Aging
ed. by I. Emerit & B. Chance
© 1992 Birkhäuser Verlag Basel/Switzerland

Effect of photooxidation on the eye lens and role of nutrients in delaying cataract

Allen Taylor

Laboratory for Nutrition and Vision Research, USDA Human Nutrition Research Center on Aging at Tufts University, 711 Washington St., Boston MA 02111, USA

Summary. The function of the eye lens is to collect and focus light on the retina. To do so, it must remain clear during the decades of life. Upon aging, lens constituents are damaged and precipitate in opacities called senile cataracts. Laboratory and epidemiologic data indicate that the damage is due in part to light and active forms of oxygen. Antioxidant nutrients – ascorbate, carotenoids, and tocopherol – appear to offer protection against cataract.

Fifty million persons worldwide are blind due to cataract, and, in the U.S., there are 1.2 million cataract surgeries performed at an annual cost (including physician visits) of over $3.2 billion. It has been estimated that a 10-year delay in the development of cataract would eliminate the need for half the surgeries. Since it will not be possible to replace most of the damaged lenses, it is essential to determine the efficacy of supplying adequate levels of antioxidant nutrients early in life to preserve lens function.

Introduction

This review will address relationships between pathobiology of cataract and nutritional approaches to the delay of this disorder. Only studies which found associations between nutritional status and cataract are reviewed here. Readers are referred to other recent reviews or summaries for data, which are not treated here (Taylor et al., 1991; Jacques and Taylor, 1991; West, 1991; Taylor, 1989; Bunce et al., 1989; Varma, 1984; Sperduto, 1990, Third National Eye Institute Symposium on Eye Disease Epidemiology, 1991).

The function of the eye lens is to focus light on the retina (Figs. 1a, d). In order to do so, it must remain clear throughout the decades. Dysfunction of the lens due to partial or complete opacification is called cataract (Fig. 1b, c, e). The term "senile cataract" is used to distinguish lens opacification associated with old age from other forms of cataract, such as cataract associated with trauma, congenital disorders, etc. (reviewed in Jacques and Taylor, 1991). Despite differences in nomenclature and recent indications that various types of senile cataracts have different etiology, many forms of cataract are similar with regard to impact on vision, pathology, and some aspects of biochemistry.

Approximately 50 million persons in the world are blind from cataracts. More than 700,000 cataract extractions are performed annu-

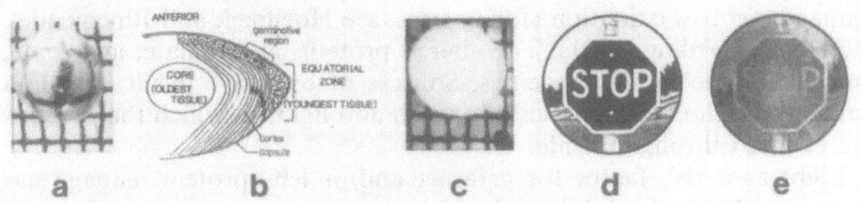

Figure 1. The eye lens. a) clear lens. Note that the wire grid behind the lens is clearly visible. b) morphology of the lens. The anterior surface of the lens has a unicellular layer of epithelial cells (youngest tissue). Cells at the anterior equatorial region divide and the maturing cells move to the equator. As cells begin to express crystallins, the major lens gene products, they are overlayed by less mature cells and are thus found in the cortex. As development and maturation procede the cells denucleate and elongate. Tissue originally present in the embryonic lens is found at the core (oldest tissue) of the mature lens. c) cataractous lens. Note that the grid behind the lens is no longer visible. d) view through a clear lens. The image is crisp and clear and the field uncolored. e) view through a cataractous lens. The image is partially obscured and the field is colored.

ally. Cataract extraction is the most frequently performed surgery among the elderly and accounts for the largest single item in Medicare expenditures. Annual costs for cataract extraction and associated visits to physicians are over $3.5 billion in the United States alone (McBride, 1985). Although an awesome problem in the industrialized nations, the situation is exacerbated in lesser-developed countries where 1) there is a dearth of ophthalmologists to perform the lens extractions and replacements, 2) cataracts are 3–4 times as common, and 3) cataracts develop earlier in life than in the U.S. (West, 1991).

In the U.S. the prevalence of senile cataract which impacts on vision is approximately 4.5%, 18%, and 45% among persons aged 52–64, 65–74, and 75–85 years, respectively (Leske and Sperduto, 1983). If early lens changes are also included, the prevalence is twice as high. It has been estimated that if cataracts could be delayed by only 10 years, the need for cataract extraction would be diminished by half (U.S. Dept. of Health and Human Services, 1983). In addition to considerable financial benefit, a delay in cataractogenesis would lead to gains in productivity and enhancement in life quality of our elderly. A variety of laboratory tests and epidemiological studies suggest that this objective might be approached with proper nutrition. This material is reviewed below.

Etiology of senile cataract

About 98% of the solid mass of the lens is protein. Lens proteins are extremely long lived, and it is not surprising that in the aged lens a vast majority of the protein matter is extensively damaged. Much of this damage involves oxidation (for reviews, see Hoenders and Bloemendal, 1981; and Harding, 1981). The altered proteins accumulate, aggregate, and precipitate in lens opacities. Sources of oxidative insult include a variety of high-energy species of oxygen and light, to which the proteins are exposed throughout life.

Light as a risk factor for cataract and/or lens protein damage has been established by epidemiological and laboratory studies. Various studies show that persons living or working in environments with higher intensities of incident and/or reflected ultraviolet light (Taylor et al., 1988; Bochow, 1989; Brilliant et al., 1983), living closer to the equator, (Zigman, 1983), and possibly living at higher elevations (Guo-min et al., 1990; but also see Brilliant et al., 1983) have elevated risk of various forms of cataract. These epidemiological data have been corroborated or anticipated in vivo by exposure of squirrels to ultraviolet (Zigman, 1991) and in many experiments in vitro. The latter indicate that exposure of lens constituents to various wavelengths of light results in alterations which are quite similar to those found in cataract (Blondin et al., 1986; Blondin and Taylor, 1987; Zigler, 1984; Varma et al., 1984; Zigler and Goosey, 1984; Taylor et al., 1991b). They also indicate that both structural proteins, proteases, and many other enzymes are inactivated upon exposure to ultraviolet light.

There is also evidence to support a role for oxidation in cataractogenesis. Nuclear cataract was observed in patients treated with hyperbaric oxygen therapy (Palmquist et al., 1984). Markedly elevated levels of mature cataract were observed in mice that survived exposure to 100% oxygen twice weekly for 3 h (Schocket et al., 1972). A decline in glutathione (GSH) and increase in glutathione disulfide-oxidative changes normally related to aging or cataract were also noted. A higher incidence of cataract was noted in lenses exposed to hyperbaric oxygen in vitro (Giblin et al., 1988). Smoking appears to induce oxidative stress and has been associated with diminished levels of antioxidants, ascorbate and carotenoids (Schectman et al., 1989) and enhanced cataract at a young age (Flaye et al., 1989; West et al., 1989). Conversely, incorporating the industrial antioxidant and food preservative, 0.4% butylated hydroxytoluene in diets of galactose fed (50% of diet) rats, diminished prevalence of cataract (Srivastava and Ansari, 1988).

The accumulation of damaged proteins in lens opacities indicates that protective systems are not keeping pace with the insults that damage lens proteins. Protection against photooxidative insult can be conceived as a two-stage process. First, there is protection of proteins and other constituents by lens antioxidants and antioxidant enzymes. Secondary defenses include proteolytic systems, which in other tissues serve to selectively remove damaged or obsolete proteins in a timely fashion (Taylor and Davies, 1987). Most of these capabilities exist in young lens

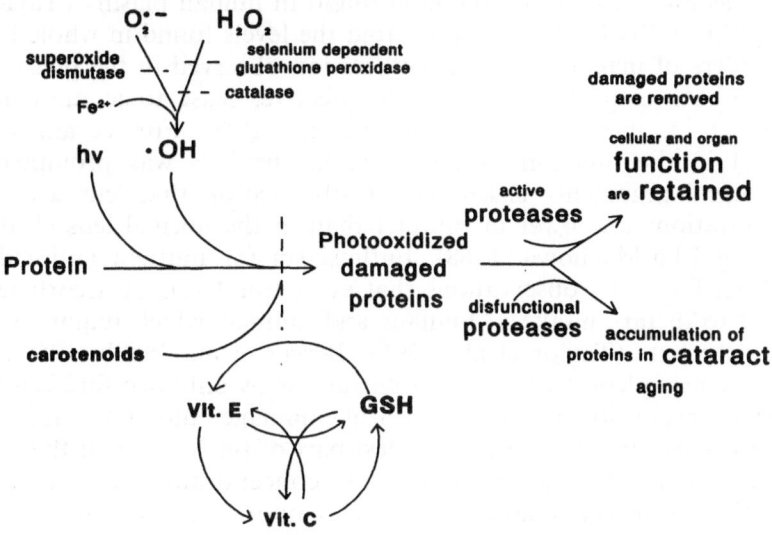

LIGHT & OXYGEN DAMAGE PROTEIN & PROTEASES; ANTIOXIDANTS DELAY DAMAGE

Figure 2. Proposed interaction between lens proteins, oxidants, light, antioxidants, and antioxidant enzymes. Lens proteins are extremely long lived. They are subject to alteration by light and various forms of oxygen. They are protected indirectly by antioxidant enzymes superoxide dismutase, catalase, and glutathione reductase/peroxidase. These enzymes convert active oxygen to less damaging species. Direct protection is offered by antioxidants: glutathione (GSH), ascorbate (vitamin C), tocopherol (vitamin E), and carotenoids. Levels of reduced and oxidized forms of these molecules are determined by interaction between the three and with the environment (Wefers and Seis, 1988). Proteins which are damaged may accumulate and precipitate in cataract if there is insufficient proteolytic capability. When the proteolytic capability is sufficient obsolete and damaged proteins may be reduced to their constituent amino acids. Upon aging some of the eye antioxidant supplies are diminished, antioxidant enzymes inactivated, and proteases less active. This appears to be related to the accumulation, aggregation, and eventual precipitation in cataractous opacities, of damaged proteins.

tissue (Eisenhauer et al., 1988; Berger et al., 1988b; Jahngen-Hodge et al., 1991; Jahngen-Hodge et al., 1992; Huang et al., 1992). A pictorial summary of insults and protective species and a proposal of their interactions are indicated in Figure 2.

Antioxidant capabilities change during aging

Shortly after the discovery of the antioxidant capabilities of nutrients, the search began for antioxidants in the lens (Muller and Buschke, 1934; Bellows, 1936). This and subsequent work revealed that the major aqueous antioxidants in the lens are ascorbate (Taylor et al., 1991a) and GSH (Kuck, 1974; Bunce et al., 1990; Reddy, 1990). Both are present in the lens at approximately mM concentrations. The lens level of ascorbate is as much as 60-fold the level found in human plasma (Taylor et al., 1991a). GSH levels are several-fold the levels found in whole blood and orders of magnitude the concentration observed in the plasma.

Ascorbate is probably the most effective, least toxic antioxidant identified in mammalian systems (Levine, 1986; Frei et al., 1988). Interest in the function of ascorbate in the lens was prompted by teleological arguments based on the observation that lens ascorbate concentrations are lower in cataract than in the normal lens (Wilczek and Zygulska-Machowa, 1968). Enthusiasm for nutrient antioxidants has been fueled by observations that eye tissue levels of ascorbate are related to dietary intake in humans and animals which require exogenous ascorbate (Taylor et al., 1991a; Berger et al., 1988b, 1989), and that increasing lens ascorbate concentrations by only two-fold is associated with protection against cataract-like damage (Blondin et al., 1990).

In the lens core (Fig. 1b), the oldest part of the lens – and the region involved in much senile cataract – the concentration of ascorbate is only 25% of the surrounding cortex (Nakamura and Nakamura, 1935). Furthermore, ascorbate levels in the lens are significantly lower in old guinea pigs than in young animals with the same dietary intake of ascorbate (Berger et al., 1989). It is likely that the same pertains in humans as well (Taylor, Nutritional Status Survey of the Elderly: The Boston study). These data suggest that either there is age-related depletion of ascorbate in the lens or that the bioavailability of this compound changes with age. GSH levels also diminish in the older and cataractous lens (Reddy, 1990).

Tocopherol and carotenoids are lipid-soluble antioxidants with probable roles in maintaining membrane integrity (Machlin and Bendich, 1987) and GSH recycling (Costagliola et al., 1986). Concentrations of tocopherol in the whole lens are in the μM range. Since most of the compound is found in the membranes, in the membrane fraction these concentrations can be orders of magnitude higher. Age-related changes in levels of tocopherol and carotenoids have not been documented.

The lens also contains elevated levels of antioxidant enzymes: glutathione peroxidase/reductase, catalase, and superoxide dismutase. These interact via the forms of oxygen (Fig. 2). They also interact with the nonenzymic antioxidants, i.e. GSH is a substrate for glutathione peroxidase. The activities of many antioxidant enzymes are compromised upon development, aging, and cataract formation (Berman, 1991).

From these data it is clear that the young lens has significant antioxidant protection. However, age-related compromises in the function of antioxidant enzymes, concentrations of the antioxidants, and activities of secondary defenses may lead to diminished protection against oxidative insults. This diminished protection leaves the long-lived proteins and other constituents vulnerable.

Relationships between nutrition and cataract in laboratory studies

Water-soluble antioxidants. Feeding elevated ascorbate delayed progress of – or prevented: galactose cataract in guinea pigs (Kosegarten and Maher, 1978) and rats (Vinson et al., 1986), selenite-induced cataracts in rats (Devamanoharan et al., 1991), lens opacification in GSH-depleted chick embryos (Nishigori et al., 1986), and delayed UV-induced protein and protease damage in guinea pig lenses (Blondin et al., 1986, 1987).

There have been several attempts to exploit the reducing capabilities of GSH. Injection of GSH-OMe was associated with delayed buthionine sulfoximine-induced (Martenssen et al., 1989) and naphthalene cataract (Rathbun et al., 1990; also see Vina et al., 1989; Reddy, 1990). Preliminary evidence from studies with galactose-induced cataract also indicates some advantage of maintaining elevated GSH status in rats (Meydani, Blumberg and Taylor, unpublished). However, it is not clear that feeding GSH is associated with higher eye tissue levels of this antioxidant.

Lipid-soluble antioxidants. Tocopherol is reported to be effective in delaying a variety of induced cataracts in animals including galactose- (Creighton et al., 1985; Bhuyan et al., 1983; for review see Jacques and Taylor, 1991) and aminotriazole-induced cataracts in rabbits (Bhuyan and Bhuyan, 1984). Although elevated carotenoid intake is frequently associated with health benefits, including delay of cataract and age-related degeneration of the macula (Seddon, personal communication), little experimental work has been done regarding lens changes in response to variations in levels of this nutrient.

Considering the paradigm that extended lens function is associated with prolonged antioxidant protection, results in calorically restricted animals were surprising. The Emory mouse develops cataracts which closely resemble human senile cataract (Taylor et al., 1989). Restriction of caloric intake extends youth and delays cataract (as well as many other late-life diseases) in these animals. Since cataract is associated with oxidative stress, it might be anticipated that the delay in cataract would be accompanied by elevated ascorbate levels in the protected animals. Nevertheless, plasma ascorbate concentrations were lower than in the restricted mice (Taylor, 1991b, Taylor, Jacques and Dorey, (1992) in press).

Epidemiological evidence which indicates a benefit of maintaining elevated antioxidant levels

In this section, anticataractogenic effects of individual nutrients are considered first. Since there may be synergistic effects of nutrients, the

Table 1. Risk of developing cataract in relationship to plasma or dietary levels of nutrients

Nutrient	Percentile cut off		Adjusted odds ratio[1] for cataract in percentiles			
	20th	80th	20	21–79	80	Cataract type[5]
Plasma						
Vitamin C (μmol/l)	40	90	3.5	3.7[2]	1	Any
			3.7	3.3[2]	1	CX
			11.3[3]	8.2[3]	1	PSC
Carotenoids (μmol/l)	1.7	3.3	5.6[3]	1.4	1	Any
			7.2[2]	1.6	1	CX
			5.7	0.6	1	PSC
Vitamin E (μmol/l)	21.0	35	1.2	1.2	1	Any
			1.2	1.1	1	CX
			3.0	3.3	1	PSC
Index			5.8[2]	3.6[3]	1	Any
			6.2[2]	3.2	1	CX
Intake[4]						
Vitamin C (mg/day)	125	490	4.0[3]	1.6	1	Any
			3.7[3]	1.4	1	CX
			11.0[2]	2.8	1	PSC
Vitamin E (mg/day)	8.4	35	2.2	2.3	1	Any
			2.7	2.7[3]	1	CX
			2.5	1.4	1	PSC
Index			2.2	0.73	1	Any
			2.7	0.90	1	CX
			3.0	0.42	1	PSC

[1]Adjusted for age, gender, race, and diabetes; [2]$p < 0.05$; [3]$0.05 < p < 0.10$; [4]Carotenoid intake is not included in this intake summary because the relationships were not significant. [5]Any = cataract in any part of lens. CX = cortical cataract. PSC = posterior subcapsular cataract.

effect on cataract of groups of nutrients are also considered. Several epidemiological studies corroborate the inverse relationship between dietary, plasma, or tissue levels of ascorbate, and protection against cataracts. Jacques and Chylack (1991) classified subjects into three plasma-ascorbate categories: <40, 40–90, > 90 μmol/l (Table 1). Risk of cataracts for subjects with low- and moderate-ascorbate levels were estimated relative to subjects with high levels. Persons with lowest ascorbate status had 11-fold the risk of developing cataract in the posterior region of the lens as persons with the highest ascorbate status. When all types of cataracts are considered together, subjects in the mid-range plasma-ascorbate group had over three times the risk of developing cataract as those with high-lens ascorbate.

Evaluation of relative risk of developing cataract for persons with specified dietary nutrient intake (Jacques and Chylack, 1991) corroborated many of the relationships observed for plasma nutrient status. The protective effects of vitamin C intake are similar to the effects observed using plasma vitamin C measures (Table 1). The risk of developing any form of cataract is four-fold higher among subjects with low vitamin C intake. It should be noted that the terms low, moderate, and high are only relative. For example, a subject with vitamin C intake of 120 mg/ day (twice the RDA) would be classified as low. This is because, in this cohort population, most volunteers reported high ascorbate intake. In another case control study, Robertson et al. (1989) found approximately 1/3 the risk of developing advanced cataracts in persons who consumed daily supplements of ascorbate (> 300 mg/day). Leske and coworkers (Leske et al., 1991) corroborated these data. In a large case-control study, elevated ascorbate intake was associated with a statistically significant decrease in odds ratio (0.48) for developing nuclear cataract, and similar trends were noted for other types of cataract as well.

Protective effects of nutrients are not confined to aqueous soluble moieties. Jacques and Chylack (1991) observed that the risk of developing cataract among persons with low-plasma carotenoid levels (< 1.7 μmol/l) was over five times the risk than among persons with higher plasma-carotenoid concentrations. Leske examined diet in terms of vitamin A. Vitamin A is a derivative of β carotene. The risk of developing nuclear- and mixed-type cataracts decreases with higher vitamin A intake (odds ratio = 0.45 and 0.6 for nuclear and mixed type cataracts, respectively (Leske et al., 1991).

Robertson et al. (1989) found that persons who consumed tocopherol supplements (400 I.U./day) had about 1/3 the risk of developing cataracts as compared with control cases. Since tocopherol is lipid soluble and ascorbate is aqueous soluble, it might be anticipated that greater benefit would be observed for persons who consumed supplements of both vitamins. No synergistic effect was noted. Leske et al. (1991) also found elevated tocopherol intake protection against cataract (cortical odds ratio = 0.59, mixed type odds ratio = 0.58). The same association pertained, but significance was not achieved, for the association between elevated tocopherol intake and cataract in studies by Jacques and Chylack (1991).

In order to evaluate synergistic effects of the antioxidant nutrients, various antioxidant indices were constructed. Jacques proposed one antioxidant index based on plasma-nutrient levels and another based on nutrient-intake levels (Table 1). The subject was scored high if his/her nutrient levels fell into the highest quintile for at least two of the antioxidant nutrients and was not in the lowest quintile for the third. A low score was assigned if a subject was in the lowest quintile for at least one of the nutrients but not in the highest quintile for either of the others. All other subjects were assigned to the intermediary or moderate index scores. Risk of senile cataract and micronutrient index were related in a dose-dependent fashion. In contrast with the data of Robertson et al. (1989), a possible additive effect of antioxidant nutrients is indicated by the decreased risk of cataract for persons in the higher plasma nutrient group (Jacques et al., 1988b). The risk of developing any type of cataract was 5.8 and 3.6 times higher for subjects with low- and moderate-plasma nutrient index scores, respectively, as compared with persons with high scores.

To evaluate the role of nutrient intake, serving values (servings per day) that approximated the 20th percentile were used as the cut-off points, and the relative risks of cataract for those with low fruit or vegetable intake were again estimated by odds ratios. Significant increases are indicated in the risk of cataract for persons who consume an average of fewer than one and a half servings of fruit (odds ratio 3.4), fewer than two servings of vegetables (odds ratio 3.6), or fewer than three and a half servings of fruit and vegetables per day (odds ratio 5.7). Posterior subcapsular cataract appears to be most strongly affected by fruit and vegetable intake, but cortical cataract risk is also significantly elevated in persons with low fruit and vegetable intake. Using a similar index, Leske et al. (1991) found that for cataract development the odds ratios comparing persons with high-index scores to those with low-index scores were 0.4 for both cortical and mixed cataract.

These data are further supported when Leske et al. (1991) found that the use of multiple vitamins is associated with decreased risk for each type of cataract; odds ratios for posterior subcapsular, cortical, nuclear, and mixed cataracts were 0.4, 0.52, 0.55, and 0.7, respectively. Mohan et al. (1989) constructed a somewhat more complex antioxidant scale, which included red blood cell levels of glutathione peroxidase, glucose-6-phosphatedehydrogenase, and plasma levels of ascorbate and tocopherol. They found positive, inverse associations between antioxidant status and cataracts involving the posterior subcapsular region or mixed cataracts (with posterior and nuclear component). Correlations were also sought between cataract and levels of individual nutrients, as well as with other combinations of nutrients and enzyme activities. These did not indicate significant relationships.

Secondary defenses

It would appear that, as a functional aggregate, proteases constitute a secondary defense in that they selectively remove damaged proteins (Taylor and Davies, 1987). We and others established the presence in young lens tissue of most proteolytic pathways which are found in other cell types (Jahngen et al., 1990; Jahngen-Hodge et al., 1991; Huang et al., manuscript submitted; Eisenhauer et al., 1988; Berger et al., 1988a; Taylor and Davies, 1987). These are summarized briefly in Table 2. When functioning properly, these proteases might rid the lens of damaged proteins. However, upon aging some of these enzymatic capabilities are found in a state of reduced activity (reviewed in Taylor and

Table 2. Effect of aging and development on proteolytic activities in lens tissues and cultured lens cells[1]

Activity	Source	Effect of aging development
Neutral proteinase/ Proteasome/ High molecular weight protease[2-4]	Tissue	↓
Endopeptidase[5]	BLEC[10]	↓
LAP[5,6]	Tissue BLEC	↓
Cathepsins[5]	BLEC	↓
Calpain[7-9]	Tissue BLEC	↑ N.D.[11]

[1]In cultured cells – aging was simulated by progressive passage of cells in culture.[2]Fleshman, K. R., and Wagner, B. J. 1984; [3]Ray, K., and Harris, H. 1985; [4]Murakami, K. et al., 1990; [5]LAP leucine amino peptidase, Eisenhauer, D. A. et al., 1988; [6]Taylor, A. et al., 1983; [7]Varnum, M. D. et al., 1989; [8]Yoshida, H. et al., 1989; [9]Lipman et al., 1991; [10]BLEC beef lens eptithelial cells; [11]Not determined.

Davies, 1987). Elevated antioxidant intake extends function of some of these proteolytic capabilities in the UV-exposed lens (Blondin et al. 1987).

Conclusion

Epidemiological and laboratory studies indicate that cataractogenesis involves a significant level of photooxidative stress. Lifestyle is associated with these stresses. Several studies noted a correlation between enhanced dietary antioxidant intake or elevated plasma antioxidant levels and diminished risk of developing cataract. This is heartening since it appears possible to increase eye levels of some antioxidants by dietary practice. Further work should be done on secondary defenses. The proteolytic systems should be further defined and their relationship with lens substrates elucidated.

It is difficult to compare the various studies. That the correlations between diet and risk of cataract were not always with the same form of cataract may indicate, in addition to the conclusions reached, that the cataracts were graded differently and/or that there are common etiological features of each of the forms of cataract described. It may also be that insufficient data have been gathered. Each of the studies noted above utilized case-control designs, and each assessed nutrient status only once. Since these nutrient intake or dietary nutrient status measures are highly variable and the effects of diet are likely to be cumulative, studies should be performed on populations for which long-term dietary records are available. Thus, hindsight suggests a need for more uniform methods of

276

cataract evaluation, diet recording, blood testing, etc., and a need for longitudinal and natural history studies.

Nevertheless, the consensus is impressive. It is possible to adjust normal dietary practice to obtain the level of ascorbate ($\gtrsim 500$ and $\gtrsim 200$ mg/day for men and women, respectively) needed to achieve plasma levels of 80 μM. Slightly higher intakes may be required in the elderly to optimize levels of ascorbate. It is also possible to achieve 1.7 μM plasma carotenoids by dietary means. It may not be possible to achieve optimal levels of tocopherol without supplements. The intervention trials now underway will be of interest in this regard. Prudence would indicate that prescription of megadoses of vitamins for extension of visual function is not necessary for populations not at risk.

In most studies some of the most significant correlates of cataract were poor educational status and poverty (Harding and van Heyningen, 1987; McLaren, 1980; Leske, 1991; Mohan et al., 1989). Since it will not be possible to surgically remove most of the cataracts and since it is preferable to avoid debilities, devising strategies to delay cataract should assume a high priority in both the developed and unindustrialized parts of the world.

Acknowledgments. We acknowledge Dr. R. D. Lipman for preparation of parts of Figure 1, Ms. E. Epstein for editorial assistance, and Dr. M. Lahav for helpful discussions. Support for this work was in part with Federal funds from the USDA under contract no. 53-3K06-0-1, The Daniel and Florence Guggenheim Foundation, Hoffman-La Roche Inc., and NIH Grant no. 522360.

Bellows, J. (1936) Biochemistry of the lens. V. Cevitamic acid content of the blood and urine of subjects with senile cataract. Arch. Ophthalmol. 15: 78–83.

Berger, J., Eisenhauer, D., and Taylor, A. (1988a) Intracellular protein degradation in cultured bovine lens epithelial cells. In Vitro Cell. Dev. Biol. 24: 990–994.

Berger, J., Shepard, D., Morrow, F., Sadowski, J., Haire, T., and Taylor, A. (1988b) Reduced and total ascorbate in guinea pig eye tissues in response to dietary intake. Curr. Eye Res. 7: 681–686.

Berger, J., Shepard, D., Morrow, F., and Taylor, A. (1989) Relationship between dietary intake and tissue levels of reduced and total vitamin C in the nonscorbutic guinea pig. J. Nutr. 119: 734–740.

Berman, E. R. (1991) in: Biochemistry of the Eye. Plenum Press, New York.

Bhuyan, D. K., Podos, S. M., Machlin, L. T., Bhagavan, H. N., Chondhury, D. N., Soja, W. S., and Bhuyan, K. C. (1983) Antioxidant in therapy of cataract II: Effect of all-roc-alpha-tocopherol (vitamin E) in sugar-induced cataract in rabbits. Invest. Ophthalmol. Vis. Sci. 24: 74.

Bhuyan, K. C., and Bhuyan, D. K. (1984) Molecular mechanism of cataractogenesis: III. Toxic metabolites of oxygen as initiators of lipid peroxidation and cataract. Curr. Eye Res. 3: 67–81.

Blondin, J., Baragi, V. J., Schwartz, E., Sadowski, J., and Taylor, A. (1986) Delay of UV-induced eye lens protein damage in guinea pigs by dietary ascorbate. Free Radical Biol. Med. 2: 275–281.

Blondin, J., and Taylor, A. (1987a) Measures of leucine aminopeptidase can be used to anticipate UV-induced age-related damage to lens proteins: ascorbate can delay this damage. Mech. Ageing Dev. 41: 39–46.

Blondin, J., Baragi, V. J., Schwartz, E., Sadowski, J., and Taylor, A. (1987b) Dietary vitamin C delays UV-induced age-related eye lens protein damage, in: Vitamin C. Ann. N.Y. Acad. Sci., New York, vol. 498, pp. 460–463.

Bochow, T. W., West, S. K., Azar, A., Mouqoz, B., Sommer, A., and Taylor, H. R. (1989) Ultraviolet light exposure and risk of posterior subcapsular cataracts. Arch. Ophthalmol. 107:369–372.

Brilliant, L. B., Grasset, N. C., Pokhrel, R. P., Kolstad, A., Lepkowski, J. M., Brilliant, G. E., Hawks, W. N., and Pararajasegaram, R. (1983) Associations among cataract prevalence. Sunlight hours, and altitude in the Himalayas. Am. J. Epidemiol. 118: 250–264.

Bunce, G. E., Kinoshita, J., and Horwitz, J. (1990) Nutritional factors in cataract. Ann. Rev. Nutr. vol. 10 pp. 233–254.

Costagliola, C., Ivliano, G., Menzione, M., Rinaldi, E., Vito, P., and Auricchhio, G. (1986) Effect of vitamin E on glutathione content in red blood cells, aqueous humor and lens of humans and other species. Exp. Eye Res. 43: 905–914.

Creighton, M. O., Ross, W. M., Stewart-DeHaan, P. J., Sanwai, M., and Trevithick, J. R. (1985) Modeling cortical cataractogenesis. VII: Effects of vitamin E treatment on galactose-induced cataracts. Exp. Eye Res. 40: 213–222.

Devamanoharan, P. S., Henein, M., Morris, S., Ramachandran, S., Richards, R. D., and Varma, S. D. (1991) Prevention of selenite cataract by vitamin C. Exp. Eye Res. 52: 563–568.

Eisenhauer, D. A., Berger, J. J., Peltier, C. Z., and Taylor, A. (1988) Protease activities in cultured beef lens epithelial cells peak and then decline upon progressive passage. Exp. Eye Res. 46: 579–590.

Flaye, D. E., Sullivan, K. N., Cullinan, T. R., Silver, J. H., and Whitelocke, R. A. F. (1989) Cataracts and cigarette smoking: the City Eye Study. Eye 3: 379–384.

Fleshman, K. R., and Wagner, B. J. (1984) Changes during aging in rat lens endopeptidase activity. Exp. Eye Res. 39: 543–551.

Frei, B., Stocker, R., and Ames, B. N. (1988) Antioxidant defenses and lipid peroxidation in human blood plasma. Proc. Natl. Acad. Sci., USA 85: 9748–9752.

Giblin, F. J., Schrimscher, L., Chakrapani, B., and Reddy, V. N. (1988) Exposure of rabbit lens to hyperbaric oxygen in vitro: regional effects on GSH level. Invest. Ophthalmol. Vis. Sci. 29: 1312–1319.

Guo-min, W., Spector, A., Lus, C.-q., Tang, L.-q., Xu, L.-h., Guo, W.-Y., and Huang, Y.-q. (1990) Prevalence of age-related cataract in Ganzi and Shanghai. Chinese Med. J. 103: 945–951.

Harding, J. J., and van Heyningen, R. (1987) Epidemiology and risk factors for cataract. Eye 1: 537–541.

Harding, J. J. (1981) Changes in lens proteins in cataract, in: Molecular and Cellular Biology of the Eye Lens. pp. 327–366. Ed. H. Bloemendal, John Wiley and Sons, New York.

Hoenders, H. J., and Bloemendal, H. (1981) Aging of lens proteins, in: Molecular and Cellular Biology of the Eye Lens, pp. 279–326. Ed. H. Bloemendal, John Wiley and Sons, New York.

Huang, L. L., Jahngen-Hodge, J., and Taylor, A. (1992) Bovine lens epithelial cells have a ubiquitin-dependent proteolysis system. (Manuscript submitted).

Jacques, P. F., Chylack, L. T., Jr., McGandy, R. B., and Hartz, S. C. (1988) Antioxidant status in persons with and without senile cataract. Arch. Ophthalmol. 106: 337–340.

Jacques, P. F., and Chylack, L. T., Jr. (1991) Epidemiologic evidence of a role for the antioxidant vitamins and carotenoids in cataract prevention Am. J. Clin. Nutr. 53: 352S–355S.

Jacques, P. F., and Taylor, A. 1991. Micronutrients and age-related cataracts, in: Micronutrients in Health and in Disease Prevention. Eds A. Bendich and C. E. Butterworth. Marcel Dekker, Inc., New York.

Jahngen, J. H., Lipman, R. D., Eisenhauer, D. A., Jahngen, E. G. E., Jr., and Taylor, A. (1990) Aging and cellular maturation cause changes in ubiquitin-eye lens protein conjugates. Arch. Biochem. Biophys. 276: 32–37.

Jahngen-Hodge, J. H., Laxman, E., Zuliani, A., and Taylor, A. (1991) Evidence for ATP ubiquitin-dependent degradation of proteins in cultured bovine lens epithelial cells. Exp. Eye Res. 52: 341–347.

Jahngen-Hodge, J. H., Cyr, D., Laxman, E., and Taylor, A. Ubiquitin and ubiquitin conjugates in human lens. (1992) Exp. Eye. Res. in press.

Kosegarten, D. C., and Mayer, T. J. (1978) Use of guinea pigs as model to study galactose-induced cataract formation. J. Pharm. Sci. 67: 1478–1479.

278

Kuck, J. F. R., Jr. Composition of the lens. (1974) in: Cataract and Abnormalities of the Lens, vol. 26, pp. 69–96. Ed. J. G. Bellows. Grune & Stratton, Inc., New York.

Leske, M. C., and Sperduto, R. D. (1983) The epidemiology of senile cataracts: a review. Am. J. Epidem. 118: 152–165.

Leske, M. C., Chylack, L. T., and Wu, S.-Y. (1991) The lens opacities case-control study risk factors for cataract. Arch. Ophthalmol. 109: 244–251.

Levine, M. (1986) New concepts in the biology and biochemistry of ascorbic acid. New Eng. J. Med. 314: 892–902.

Lipman, R. D., Cyr, D. E., David, L. L., and Taylor, A. (1991) Calpain in cultured bovine lens eptithelial cells. Curr. Eye Res. 10: 11–17.

Machlin, L. J. and Bendich, A. (1987) Free radical tissue damage: Protective role of antioxidant nutrients. FASEB J. 1: 441–445.

Martenssen, J., Steinherz, R., Jain, A., and Meister, A. (1989) Glutathione ester prevents buthionine sulfoximine-induced cataracts and lens epithelial cell damage. Biochem. 86: 8727–8731.

McBride, J. (1985) Congressional Record, Congressional Subcommittee on Health and Long Term Care. October 1985.

McLaren, D. S. (1980) in: Nutritional Ophthamology, 2nd ed., Academic Press, London.

Mohan, M., Sperduto, R. D., Angra, S. K., Milton, R. C., Mathur, R. L., Underwood, B., Jaffery, N., and Pandya, C. B. (1989) India-US case-control study of age-related cataracts. Arch. Ophthalmol. 107: 670–676.

Muller, H. K., and Buschke, W. (1934) Vitamin C in linse, kammerwasser und blut normalem und pathologischem linsentstoffwech. Arch. F. Augenh. 108: 368–390.

Murakami, K., Jahngen, J. H., Lin, S. W., Davies, K. J. A., and Taylor, A. (1990) Lens proteasome shows enhanced rates of degradation of hydroxy radical modified alpha-crystallin. Free Radical Biol. Med. 8: 217–222.

Nakamura, B., and Nakamura, O. (1935) Ufer das vitamin C in der linse und dem kammerwasser der menschlichen katarakte. Graefes Arch. Clin. Exp. Ophthalmol. 134: 197–200.

Nishigori, H., Lee, J. W., Yamauchi, Y., and Iwatsuru, M. (1986) The alteration of lipid peroxide in glucocorticoid-induced cataract of developing chick embryos and the effect of ascorbic acid. Curr. Eye Res. 5: 37–40.

Palmquist, B., Philipson, B., and Barr, P. (1984) Nuclear cataract and myopia during hyperbaric oxygen therapy. Br. J. Ophthalmol. 68: 113–117.

Rathbun, W. B., Holleschau, A. M., Murray, D. L., Buchanan, A., Sawaguchi, S., and Tao, R. V. (1990) Glutathione synthesis and glutathione redox pathways in naphthalene cataract in the rat. Curr. Eye Res. 9: 45–53.

Ray. K., and Harris, H. (1985) Purification of neutral lens endopeptidase: close similarity to neutral proteinase in pituitary. Proc. Nat. Acad. Sci. USA 82: 7545–7549.

Reddy, V. N. (1990) Glutathione and its function in the lens – an overview. Exp. Eye Res. 150: 771–778.

Robertson, J. McD., Donner, A. P., and Trevithick, J. R. (1989) Vitamin E intake and risk for cataracts in humans. Ann. N.Y. Acad. Sci. 570: 372–382.

Schectman, G., Byrd, J. C., and Gruchow, H. W. (1989) The influence of smoking on vitamin C status in adults. Am. J. Health 79: 158–162.

Schocket, S. S., Esterson, J., Bradford, B., Michaelis, M. R., and Richards, R. D. (1972) Induction of cataracts in mice by exposure to oxygen. Israel J. Med. 8: 1596–1601.

Srivastava, S. K., and Ansari, N. H. (1988) Prevention of sugar-induced cataractogenesis in rats by butylated hydroxytoluene. Diabetes 37: 1505–1508.

Taylor, A., Daims, M. A., Brown, M. J., and Cohen, J. (1983) Localization of leucine aminopeptidase in hog lenses using immunofluorescence and activity assays. Invest. Ophthalmol. Vis. Sci. 24: 1172–1181.

Taylor, A., and Davies, K. J. A. (1987) Protein oxidation and loss of protease activity may lead to cataract formation in the aged lens. Free Radical Biol. Med. 3: 371–377.

Taylor, A. (1989) Associations between nutrition and cataract. Nutr. Rev. 47: 225–234.

Taylor, A., Zuliani, A. M., Hopkins, R. E., Dallal, G. E., Treglia, P., Kuck, J. F. R., and Kuck, K. A. (1989) Moderate caloric restriction delays cataract formation in the Emory mouse. FASEB J. 3: 1741–1746.

Taylor, A., Jacques, P. F., Nadler, D., Morrow, F. Sulsky, S. I., and Shepard, D. (1991a) Relationship in humans between ascorbic acid consumption and levels of total and reduced ascorbic acid in lens, aqueous humor, and plasma. Curr. Eye Res. 10: 751–759.

Taylor, A., Jahngen-Hodge, J., Huang, L. L., and Jacques, P. (1991b) Aging in the eye lens: Roles for proteolysis and nutrition in formation of cataract. AGE 14: 65–71.

Taylor, A. (1992) Vitamin C, in: Nutritional Status Survey of the Elderly: The Boston Study. Eds S. C. Hartz, R. M. Russell, and I. H. Rosenberg. Smith Gordon Limited, London (in press).

Taylor, A., Jacques, P. F., and Dorey, K. D. (1992) Oxidation and aging: Impact on vision, in: Proceedings of the International Conference on Antioxidants. Princeton Scientific Press, Princeton, NJ.

Taylor, H. R., West, S. K., Rosenthal, F. S., Munoz, B., Newland, H. S., Abbey, H., and Emmett, E. A. (1988) Effect of ultraviolet radiation on cataract formation. New Eng. J. Med. 319: 1429–1433.

Third National Eye Institute Symposium on Eye Disease Epidemiology. (1991) National Eye Institute, National Institutes of Health.

U.S. Dept. of Health and Human Services, Report of the Cataract Panel. (1983) Vol. 2, Pt. 3. Vision Research, A National Plan. NIH publication no. 83-2473. Washington, DC: U.S. Department of Health and Human Services.

Varma, S. D., Chand, O., Sharma, Y. R., Kuck, J. F., and Richards, K. D. (1984) Oxidative stress on lens cataract formation. Role of light and oxygen. Curr. Eye Res. 3: 35–57.

Varnum, M. D., David L. L., and Shearer, T. R. (1989) Age-related changes in calpain II and calpastatin in rat lens. Exp. Eye Res. 49: 1053–1065.

Vina, J., Perez, C., Furukawa, T., Palacin, M., and Vina, J. R. (1989) Effect of oral glutathione on hepatic glutathione levels on rats and mice. Br. J. Nutr. 62: 683–691.

Vinson, J. A., Possanza, C. J., and Drack, A. V. (1986) The effect of ascorbic acid on galactose-induced cataracts. Vol. 33, Nutr. Reports Int. pp. 665–668.

Wefers, H., and Seis, H. (1988) The protection of ascorbate and glutathione against microsomal lipid peroxidation is dependent on vitamin E. Eur. J. Biochem. 174: 353–357.

West, S. K., Munoz, B., Emmett, E. A., and Taylor, H. R. (1989) Cigarette smoking and risk of nuclear cataracts. Arch. Ophthalmol. 107: 1166–1169.

West, S. K. (1991) Who develops cataracts? Arch. Ophthalmol. 109: 196–197.

Wilczek, M., and Zygulska-Machowa, H. (1968) Zawartosc witaminy C W. roznych typack zaem. J. Klin. Oczna 38: 477–480.

Yoshida, H., Murachi, T., and Tsukahara, I. (1989) Distribution of calpain I, calpain II, and calpastatin in bovine lens. Invest. Ophthalmol. Vis. Sci. 26: 953–956.

Zigler, J. S., and Goosey, J. D., (1984) Singlet oxygen as a possible factor in human senile nuclear cataract development. Curr. Eye Res. 3: 59–65.

Zigman, S. (1983) The role of sunlight in human cataract formation. Surv. Ophthalmol. 27: 317–326.

Zigman, S., Paxhia, T., McDaniel, T., Lou, M. F., and Yu, N. -T. (1991) Effect of chronic near-ultraviolet radiation on the gray squirrel lens in vivo. Inv. Ophthalmol. Vis. Sci. 32: 1723–1732.

Free Radicals and Aging
ed. by I. Emerit & B. Chance

Carotenoids in the retina – A review of their possible role in preventing or limiting damage caused by light and oxygen

Wolfgang Schalch

Human Nutrition Research, Vitamins & Fine Chemical Division, F. Hoffmann – La Roche, Basel, Switzerland

Summary. Two of the circa 600 naturally occurring carotenoids, zeaxanthin and lutein, the major carotenoids of maize and melon respectively, are the constituents of the macula lutea, the yellow spot in the macula, the central part of the retina in primates and humans. Of the circa ten carotenoids found in the blood these two are specifically concentrated in this area, which is responsible for sharp and detailed vision. This paper reviews the ideas that this concentration of dietary carotenoids in the macula is not accidental, but that their presence may prevent or limit damage due to their physicochemical properties and their capability to quench oxygen free radicals and singlet oxygen, which are generated in the retina as a consequence of the simultaneous presence of light and oxygen. Additionally, in vitro and in vivo animal experiments are reviewed as well as observational and epidemiological data in humans. These show that there is enough circumstantial evidence for a protective role of carotenoids in the retina to justify further research. Some emphasis will be put on age-related macular degeneration (AMD), a multifactorial degenerative retinal disease for which the exposure to light and thus photochemical damage has been suggested as one of the etiological factors. Recent attempts at nutritional intervention in this condition will also be reviewed.

Introduction

One of the most important irreversible blinding diseases is a disease of the retina (Bressler et al., 1988): age-related macular degeneration (AMD). Unlike cataract where vision can be restored by replacing the diseased lens, AMD does not provide the possibility of replacing the diseased retina. Therefore, and because no treatment for this disease exists, prevention appears to be the only way to deal with this severe ophthalmological condition. This review discusses ideas on the role carotenoids may play to limit or prevent light- and oxygen-induced damage in the retina. Experimental in vitro work, in vivo work in animals, observational data as well as epidemiological investigations in humans and the first attempts at nutritional intervention will be presented.

Carotenoids are among the most important pigments occurring in living organisms (Isler, 1971) including fruits and vegetables (Gross,

1987, 1991). The idea that these substances of dietary origin could play an important role in the prevention of degenerative diseases of the retina was formulated when it was discovered that the two carotenoids zeaxanthin, the carotenoid of maize, and lutein, which occurs i.e. in melon, form the macular yellow pigment in the center of the retina (Nussbaum, 1981). Due to its yellow color, the macular pigment absorbs blue light and thus shields this sensible area of sharpest vision in the center of the retina from those visible wavelengths which are believed to be the most damaging. The idea of a preventive potential of carotenoids has been further substantiated by the discovery, that carotenoids are efficient quenchers of singlet oxygen and oxygen free radicals, which can be continuously generated within the retina by the simultaneous presence of light and oxygen.

Occurrence of carotenoids in systems exposed to light and oxygen

The principle to exploit the properties of carotenoids for protection against the deleterious effects of the simultaneous presence of light and oxygen is not new (Krinsky, 1968). It was developed by photosynthetic bacteria about $1.2 \cdot 10^9$ years ago. Recently the molecular structure of the photosynthetic reaction center of the purple bacterium *Rhodopseudomonas viridis* was established (Deisenhofer and Michel, 1989), demonstrating the presence of dihydroneurosporene, a carotenoid similar to lycopene, the major carotenoid of tomatoes and a potent singlet oxygen quencher (Conn et al., 1991). Carotenoids have also been used in visual systems. The compound eye of *Drosophila* contains carotenoids and appears to be particularly dependent on them. Recently it was demonstrated that their rhabdomeres, the functional visual entities of insect eyes, undergo drastic structural changes when this species is deprived of carotenoids (Sapp et al., 1991). These changes are reversible as soon as the carotenoids are replaced. Carotenoids were also identified spectroscopically in the photoreceptors of the lateral eyes of two barnacle species (Minke and Kirschfeld, 1978) and the crab limulus (Benolken, 1976). The corneas of puffer fish, which are clear in the dark but become yellow when exposed to light due to the migration of carotenoids into the corneal chromatophores (Appleby and Muntz, 1979), are another example where carotenoids fulfil a protective role in the eye. Recently the presence of a carotenoid in the lens of the deep-sea hatchet fish was discovered using resonance raman detection (Yu et al., 1991). Carotenoids also occur in the oil droplets of the retinae of birds (Jane and Bowmaker, 1988). These carotenoids might not only have a role in color vision, but also in photoprotection (Kirschfeld, 1982).

Reviews on the subject

The possible role of carotenoids in systems exposed to light and oxygen has attracted considerable attention as shown by the number of reviews on the subject. Two classical reviews deal with the protective functions of carotenoids against harmful photooxidation in a variety of different organisms (Krinsky, 1968) and with the theoretical basis of their action (Krinsky, 1979). Our knowledge of the macular pigment, 200 years after its discovery, was reviewed by Nussbaum et al. (1981). This was followed by reviews specifically directed to the implications of the presence of carotenoids in visual systems (Kirschfeld, 1982) and a very elaborate general discussion of the role of antioxidants in the retina (Handelman and Dratz, 1986). The most recent review (Gerster, 1991) has discussed the protective role of carotenoids and the antioxidant vitamins C and E in the macula.

The physicochemical basis for the role of carotenoids in preventing and limiting photochemical damage

Before dealing with the protective action of carotenoids in the eye, the mechanisms of photochemical reactions which can induce damage will briefly be reviewed. Detailed discussions of the reactions involved can be found elsewhere (Krinsky, 1979). If light of appropriate wavelength is absorbed by a sensitizer molecule, this molecule can be excited to its first excited singlet state (^1S). This singlet state has a very short lifetime ($< 10^{-7}$ sec) and dissipates its energy either by interacting with the solvent, emitting a photon in the process of fluorescence or by a radiationless transition to its lowest excited triplet state (^3S). This triplet state is of central importance for photochemical reactions. The efficiency of any photosensitizer is determined by its ability to form a long-lived triplet state in high quantum yield. The retina contains a large number of molecules which absorb visible or near-ultraviolet light and can therefore act as photosensitizers (Dayhaw-Barker et al., 1986). The triplet states generated by the sensitization process have basically two possibilities of initiating photochemical reactions (Krinsky, 1979). Type I reactions are redox reactions initiated by the direct reaction of the triplet with suitable substrates not involving oxygen to produce radicals which can cause cellular damage. Type II reactions, which are usually called photodynamic reactions, are characterized by its reaction with molecular oxygen to produce either singlet oxygen or the superoxide anion radical $O_2^{\cdot-}$. In either Type I or Type II reactions, the net result is the production of active free radicals, with the Type II reactions predominantly yielding singlet oxygen. All these reactive molecules have the potential of causing temporary or permanent damage to retinal tissue (Andley, 1987).

Theoretically, there are two ways by which the yellow carotenoids can interefere with the initiation of photochemical reactions: by physical absorption of blue light thus preventing light from reaching the sensitizing molecules or by chemical quenching of either the sensitizer triplet state or the singlet oxygen subsequently generated. It has long been known that carotenoids can quench singlet oxygen, and earlier data (Foote, 1979; Burton and Ingold, 1984; di Mascio et al., 1989) have recently been corroborated by Conn et al. (1991). When only considering some of the more common carotenoids, e.g. lycopene, the major carotenoid of tomatoes, appears to be the most efficient singlet oxygen quencher. The next best quencher is astaxanthin, the pink carotenoid of salmon, followed by all-trans beta-carotene, which occurs in carrots. The singlet oxygen quenching efficiencies of zeaxanthin and all-trans beta-carotene are similar, whereas lutein, which has one conjugated double bond less than zeaxanthin, has a lower efficiency. This is consistent with data that the number of conjugated double bonds determines the efficiency of singlet oxygen quenching (Krinsky, 1968). Truscott (1991) has also studied the quenching of singlet oxygen by carotenoids incorporated into Triton X-100 micelles. In organic solvents the cis isomers of beta-carotene had similar quenching efficiencies as the all-trans isomer, but in micelles the quenching efficiency of the all-trans isomer is higher. This phenomenon is explained by the linear shape of all-trans beta-carotene which, in contrast to the bent shape of the cis isomers, allows the carotenoid to extend to the border of the micelle and thus reduces the distance to singlet oxygen generated outside of the micelle. This difference in singlet oxygen quenching of cis/trans forms in micelles demonstrates that orientation may be an important factor to consider and it is interesting that the non-random orientation of carotenoids in the retina was hypothesized to make them more effective blue filters (Handelman et al., 1991a).

The retina – a special compartment (see also Fig. 1)

The retina is the only organ which is continuously exposed to high levels of focused radiant energy in a highly oxygenated environment. For example, vertebrate retinae maintained in vitro show a higher oxygen consumption per mg protein than any other tissues tested (Sickel, 1972). Accordingly, this simultaneous presence of light and oxygen together with the abundance of molecular sensitizers gives the potential for oxygen free radicals and singlet oxygen to be generated. Because of their high content of phospholipids the cells most vulnerable by the attack via activated oxygen species are the functional entities of photoreceptors, their outer segment membranes. In vertebrates, photoreceptors typically contain about 50% docosahexaenoic acid (Stone et

284

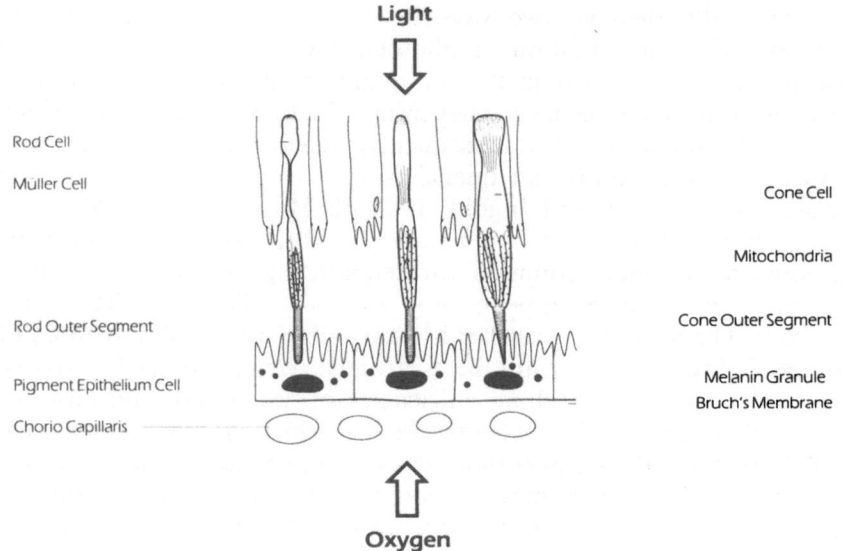

Light

Rod Cell

Müller Cell

Cone Cell

Mitochondria

Rod Outer Segment

Cone Outer Segment

Pigment Epithelium Cell

Melanin Granule
Bruch's Membrane

Chorio Capillaris

Oxygen

Figure 1. Schematic presentation of the outer layers of the retina. Light (arrow) crosses the inner layers of the retina (not shown) and reaches the photoreceptors where the nerve impulses necessary for vision are generated. Photoreceptors are continually renewed and old cells are phagocytosed by the retinal pigment epithelium. Oxygen (arrow) is supplied by blood circulating in the chorio capillaries. Larger molecules have access to the retina only via specific transport mechanisms through Bruch's membrane which constitutes one part of the blood retina barrier.

al., 1979), a n-3 fatty acid, which is the most highly unsaturated fatty acid occurring in nature. To compensate for the continuous oxidative damage, the photoreceptors have the capacity of continuous renewal. While new outer segment disks are being generated at the base of the photoreceptor outer segments, old disks at the tip are being phagocytosed by the cells of the retinal pigment epithelium (RPE), which also have a very active metabolism, but unlike the photoreceptors do not have any potential to regenerate so that damaged cells cannot be replaced (Marshall, 1985). In addition, there are other ultrastructural features, which make particularly the macula vulnerable to damage by light. This area of sharpest vision has a high density of cone photoreceptors with the center of the macula consisting exclusively of cones. Cones have been shown to be more prone to damage by blue light than rods, but do not have as active a renewal system as the rod photoreceptors (Marshall, 1985). The macula is also less efficiently shielded from incoming light because its special anatomy does not provide as many cell layers lying between the incoming light and the photoreceptors as in other retinal areas (Krebs and Krebs, 1991). Furthermore, the center of the macula is avascular; thus it does not contain any retinal blood vessels; the blood circulating in these could contribute to the absorption of light (Muntz, 1972).

The macula – prime target for carotenoids in retinae of monkeys and man

General description of the macular yellow pigment

If carotenoids can indeed have protective functions in the retina, it is not surprising that they are concentrated in the macula. The yellow coloration of the macula lutea was first noted in postmortem retinas of humans and nonhuman primates and later confirmed to be also an in vivo feature (Nussbaum, 1981). According to its spectral characteristics its chemical identity was tentatively described as the xanthophyll lutein (Wald, 1945). The definitive chemical identification of the individual components of macular yellow using HPLC demonstrated that lutein was only one component of macular yellow pigment. The other carotenoid identified was its structural isomer zeaxanthin (Bone et al., 1985). Their chemical structures are shown in Figure 2. To investigate where the pigment is localized in the macula, its vertical distribution within the macula of monkeys was determined using microspectropho-tometry (Snodderly et al., 1984a,b). This demonstrated that the macular carotenoids are most densely localized between the incoming light and the photoreceptors where they could efficiently act as a blue light filter, shielding the most sensitive part of the retina from this radiation. At the same time, however, carotenoids are close enough to the photoreceptors to allow for direct chemical quenching (Kirschfeld, 1982). The horizontal distribution of the macular pigment was determined in monkeys (Snodderly et al., 1991) using HPLC and an elaborate method of dissecting the macular region and showed that it is not uniformly distributed with more zeaxanthin than lutein being present in the center of the macula. Zeaxanthin declines very rapidly toward the

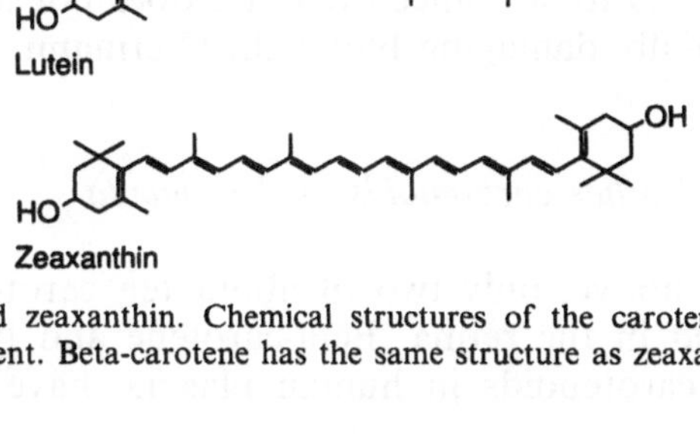

Figure 2. Lutein and zeaxanthin. Chemical structures of the carotenoids constituting the macular yellow pigment. Beta-carotene has the same structure as zeaxanthin but without the OH-groups.

edge of the macula, so that lutein becomes the dominant carotenoid at the periphery. The carotenoid content of the entire macula shows considerable interindividual variation with mean values around 70 ng for the combined masses of lutein and zeaxanthin (Handelman et al., 1988).

Dietary origin and preferential concentration in the macula

As the carotenoids of the macular yellow pigment are of dietary origin, their nutritional supply may be an important factor. The absolute configuration of the carotenoids in the human macula appears to be the same as that of the most common naturally occurring isomers of zeaxanthin and lutein in fruits and vegetables (Cains et al., 1991). This indicates that there is no metabolic or postabsorptive alteration of the ingested carotenoids; additionally it also may indicate that the uptake mechanism even discriminates according to chiral configuration. Furthermore, monkeys maintained on a diet free of carotenoids for up to five years do not show a yellow coloration of their macula (Malinow et al., 1980). In human plasma, lutein and zeaxanthin have mean concentrations of around 225 μg/l and 67 μg/l compared to 338 μg/l and 203 μg/l for beta-carotene and lycopene, respectively (Krinsky et al., 1990). Thus, while in plasma lutein clearly dominates over zeaxanthin, the opposite is true for the center of the macula. This suggests the existence of a very efficient transport mechanism which first selects two out of about ten carotenoids present in plasma and then specifically accumulates zeaxanthin, the more efficient singlet oxygen quencher of the two carotenoids (di Mascio et al., 1979), in the center of the macula. Interestingly, carotenoids in the macula start to build up during prenatal life, with a yellow spot becoming detectable postnatally when the combined masses of the two carotenoids have reached about 5 ng with no age dependency being apparent afterwards (Bone et al., 1988). In the RPE, on the other hand, the activity of catalase, an important element of the enzymatic antioxidant defence system, decreases with ages (Liles et al., 1991). Thus, the presence of carotenoids in the macula may be most important in older age and also from infancy to about 15 years of age when the lens is totally uncolored and does not have any capacity to absorb potentially damaging blue light (Lermann, 1983).

The occurrence of other carotenoids in the macula

As mentioned above, only two of about ten carotenoids present in plasma are found in the retina. Beta-carotene and lycopene, the two most abundant carotenoids in human plasma, have so far not been detected in the retina, and beta-carotene was even absent in the retina of a person who had ingested for a long time high doses of beta-carotene and canthaxanthin for medical purposes (Daicker et al., 1987). Whether the retina, in which the blood retina barrier is destroyed by disease, can be accessible to beta-carotene as suggested by the observation of crystalloid particles in the retinae of retinitis pigmentosa patients who had been treated with high doses of beta-carotene (Yoser and Heckenlively, 1989) remains to be proven. On the other hand, it has been known that one other carotenoid, canthaxanthin, can reach the retina and induce the formation of crystalline deposits in the inner layers of the retina after high-dose and long-term ingestion. The presence of these deposits does not appear to have any clinically significant consequences, nor does it induce degenerative changes in the retina. The deposits start to disappear slowly after the high-dose intake of canthaxanthin has been stopped (Arden and Barker, 1991). It remains to be established how far the polarity of carotenoid molecules determines their potential to reach the healthy retina.

The role of the yellow color of the macular pigment

Zeaxanthin and lutein, due to their yellow color absorb blue light with absorption, peaks around 460 nm. The classical explanation for the presence of macular yellow pigment is that it acts as an optical blue-filter to reduce chromatic aberration, thus improving visual acuity (Nussbaum et al., 1981). However, blue light is especially damaging to the retina and it was estimated that about 20 times less energy is needed to produce photic damage of the retina with 440-nm (blue) light than with 533-nm (red) light (Ham and Mueller, 1989). Although UV-light is even more damaging, cornea and lens are effective filters for those wavelengths. However, visible blue light can and has to reach the retina. Therefore the possible protective function of the macular yellow pigment has increasingly been considered, and it was pointed out that the absorption spectrum of the macular pigment almost perfectly matched the action spectrum of the "Blue Light Hazard" so that it could filter out the blue portion of the spectrum before extensive damage would be caused in the light-sensitive portions of the photoreceptor cells (Marshall, 1985).

Carotenoid deficiency in monkeys – effects on the retina

Macaque monkeys which were raised on a carotenoid-free diet for up to almost six years showed interesting effects in addition to the absence of macular yellow pigment (Malinow et al., 1980). When tested with fluorescein angiography the monkeys on the carotenoid-free diet showed significantly more hyperfluorescence than the control monkeys. Furthermore, they had more drusen which are also an important feature in AMD. This indicates that probably due to the lack of yellow macular pigments some degenerative processes may have already started. It could be speculated that because of the relatively short duration of carotenoid deficiency and the relatively young age (seven years) of the animals, these processes had not manifested themselves in drastic clinical pictures yet. It is known that AMD is a disease of old age which in humans normally does not manifest itself before the 5th decade of life (Ferris, 1983) which would be equivalent to an age of approximately 20 years in macaque monkeys.

Age-related macular degeneration (AMD)

AMD is the major cause of severe irreversible vision loss in the western world among persons older than 50 (Bressler et al., 1988). Its prevalence in the United States was estimated by two population-based studies. The Framingham study found AMD in at least one eye in 8.8% of 2477 persons older than 51 years. AMD is strongly dependent on age, with prevalence numbers of 1.6% for the age group 52–64 years old, 11% in the 65–74-year-olds and 27.9% for subjects older than 75 (Kini et al., 1978). The National Health and Nutrition Examination Survey (NHANES) revealed similar figures and additionally provided data on cataract in 65–74-year-olds, in whom this condition has a prevalence of 31% (Klein and Klein, 1982).

The majority of authors (Bressler et al., 1988) understand AMD as a complex clinical picture which in the presence of some degree of visual loss is manifested by drusen, yellowish-white, elevated, and often confluent nodules consisting of abnormal glycoproteins and glycolipids at the base of the cells of the RPE (Frank, 1989), regionally localized atrophy of the RPE and retinal changes associated with choroidal neovascularization. While the first two manifestations are classifying factors of the non-exudative form of AMD, the latter defines the exudative or neovascular form of AMD. The lack of an appropriate animal model for AMD has long been a significant impeding factor for research into etiology and treatment of AMD. Recently data were presented indicating that conditions very similar to human AMD can be observed in aged rhesus monkeys (Ordy et al., 1991). It remains to be shown whether this diurnal primate model can be used as an analogy to humans. The detailed pathogenetic mechanisms leading to AMD are not yet completely clear. Recent work has implicated solar radiation (Young, 1988) and an autoimmune mechanism (Gurne et al., 1991). One hypothesis discussed the role of the integrity of the complex of photoreceptor and RPE cells (Marshall, 1985). Throughout life the RPE cells continuously phagocytose photoreceptor disks in the process of their continuous renewal. This life-long phagocytosis leads to a progressive accumulation of improperly degraded debris which can appear in the form of lipofuscin granules, autofluorescent particles visible microscopically in the RPE. If the load of improperly degraded material within the RPE cells is too large, the cells try to extrude it through Bruch's membrane (see Fig. 1) towards the external blood circulation. With advancing age, however, this process gets more and more difficult due to the increasing impermeability of this membrane. As a consequence, non-extruded debris can build up towards the inner layers of the retina. This can result in the formation of drusen and the initiation of subretinal neovascularization, which ultimately leads to a disintegration of the complex of photoreceptors and RPE cells. These changes are confined to the macular area, and the resulting visual impairment is therefore characterized by the loss of central vision which is equivalent to the inability to read.

Chronic exposure to sunlight has long been suggested to be important in the etiology of AMD (Young, 1988). Thus the long-term effects of chronic exposure to near UV or blue light of the retina in rats are comparable to the histological changes seen in AMD (Tso, 1985). Additionally, there are a number of observational data in humans. Data collected in subjects, in whom the lenses were removed and were either not replaced or replaced by non-UV-absorbing intraocular lenses, show that these subjects have a greater loss of short-wavelength-sensitive cones than those with UV-absorbing implants (Werner et al., 1989). Another suggestion that light-induced damage is associated with an apparent increase of AMD in eyes without an artificial lens was put forward by a recent reanalysis (Liu et al., 1989) of the NHANES. Furthermore, a reanalysis of a population-based study of Chesapeake Bay watermen, which originally did not indicate any association between the exposure to UV light and risk for AMD (West et al., 1989), now suggests a significant association between exudative AMD and blue light exposure history in the preceding twenty years (Taylor et al., 1990). It is the extent of this exposure to blue light which could be attenuated by carotenoids. If this could be substantiated, these substances would qualify as at least one component of a preventive strategy to deal with this severe condition for which no established treatment yet exists. In the following, evidence which points into this direction will be presented.

Scientific findings suggesting that carotenoids may protect the retina

Findings in in vitro experiments

Delmelle (1977) has reported that beta-carotene can inhibit the photochemical destruction of dimethylfuran. In the presence of beta-carotene, he has illuminated with light of 360 nm a mixture of all-trans retinal, one of the possible photosensitizers in the eye, and dimethyl-furan, a chemical known to be disintegrated by singlet oxygen. He found less disintegration of dimethylfuran as monitored by the decrease of its absorption peak at 215 nm than without beta-carotene and hypothesized that this was caused by beta-carotene quenching the singlet oxygen photochemically generated by the illumination of all-trans-retinal. Further, in vitro studies with carotenoids were reported by Krinsky and Deneke (1982) who have studied the effect of in liposomes incorporated beta-carotene and canthaxanthin on dye-photosensitized reactions. They concluded that carotenoids functioned by quenching triplet sensitizers, singlet oxygen and radical intermediates. Evidence that, in addition to the singlet oxygen quenching capability of beta-carotene, its potential to react rapidly with peroxyl radicals is important was recently demonstrated by studying the chemistry of the oxidation products of beta-carotene (Kennedy and Liebler, 1991; Handelman et al., 1991b; and Mordi et al., 1991).

Another in vitro experiment (Terao et al., 1989) used the oxidation of methyl-linoleate by 2,2′-azobis(2,4-dimethyl)valeronitrile as a model system for the peroxidation of polyunsaturated fatty acids. The decomposition of this initiator induces a free radical chain oxidation which, via lipid peroxyl radicals, results in the formation of methyllinoleate hydroperoxides. The effect on the rate of their formation of the presence of the carotenoids beta-carotene, zeaxanthin, canthaxanthin or astaxanthin was studied. All carotenoids inhibited the formation of methyl-linoleate peroxides; however, the degree of inhibition appeared to be dependent on the number and kind of oxygen groups present in the carotenoid molecules. Thus beta-carotene without any oxygen group was lowest, followed by zeaxanthin with two hydroxyl groups, which was followed by canthaxanthin with two keto groups. The best inhibitor was astaxanthin having two hydroxyl- and two keto functions. Complementary evidence regarding the inhibition of lipid peroxidation by astaxanthin in vitro was published recently (Kurashige et al., 1990). The iron ion catalyzed lipid peroxidation of rat liver mitochondria by the xanthine oxidase system was investigated. The inhibition of peroxidation by astaxanthin in comparison with that by alpha-tocopherol was assessed by measuring the level of thiobarbituric acid-reactive material. The mean effective concentration of astaxanthin was up to 500 times lower than that of alpha-tocopherol indicating that in this system astaxanthin was by far the better antioxidant. That these data may have some biological relevance was suggested in the same publication by the finding that astaxanthin was more potent than alpha-tocopherol in inhibiting carrageenin-induced paw edema in rats, an often used model system to study inflammatory reactions. Interesting also are observations that antioxidants inhibit lipid peroxidation in bovine rod outer segments (Zimmerman and Keys, 1991) and porcine retinal homogenates (Hiramitsu and Armstrong, 1991), although carotenoids have not been studied in these systems. Palozza and Krinsky (1991), in an experimental set up similar to that of Terao (see above), compared beta-carotene and vitamin E and concluded that one function of beta-carotene could be the protection of endogeneous vitamin E against peroxidation.

Findings in animal experiments

To oppose the hazard by oxidative reactions, elaborate antioxidant defense mechanisms have evolved of which catalase, superoxide dismutase, reduced glutathione, and the nutritional antioxidants selenium, the vitamins C and E and carotenoids are some important constituents. Although it is beyond the scope of this review to cover all ophthalmological implications of the non-carotenoid antioxidants, it should be mentioned that vitamins C and E have been shown to occur in photoreceptor cells, and that lipofuscin in the RPE accumulated to a larger extent in rats which were maintained for 32 weeks on antioxidant-deficient diets (Katz et al., 1978) than in rats on normal diets.

A classical animal experiment was reported by Ham et al. (1984). One macaque monkey was fed 7.5 mg beta-carotene per day for 64 days. The threshold for minimal retinal damage from blue light exposure was measured and when compared with control experiments, a small but definite elevation was found. However, when measured 30 minutes after breathing air with a pO_2 of 316 torr (equivalent to 80% O_2), a 44% elevation of radiant exposure threshold was measured and 30 minutes after breathing air with a pO_2 of 226 torr a 60% rise in threshold was measured. The results indicate that the apparent effects of beta-carotene are most pronounced in the presence of lower oxygen pressures (Burton and Ingold, 1984). Although in this experiment no data were reported on the intraretinal concentration of beta-carotene, another experiment which was carried out in thirteen rats determined average intraretinal beta-carotene concentrations to be 0.4 μg/ml wet tissue after the peritoneal administration of four doses of 35 mg beta-carotene per kg of body weight (Tso et al., 1989). In this experiment photic retinopathy was induced by exposing the animals to continuous green-filtered fluorescent light for 24 hours. Six hours after light injury, there was no difference between the groups in the extent of photic damage as assessed by histological examination. Considerable loss of photoreceptor nuclei was apparent in all groups. However, six to 14 days after light exposure the treated animals appeared to be less affected and although a thinning of their outer nuclear layers could be seen, a clear preservation of photoreceptor elements and RPE was now evident. Thus beta-carotene seemed to have an effect on the long-term consequences, but not on the acute effects of light on the retina in rats. Jaffe and Wood (1988) have also produced photic lesions in the eyes of monkeys. The lesions were targeted to locations adjacent to and overlapping the fovea. The lesions were analyzed histologically and also by fluorescence angiography. It was found that the damage was significantly more pronounced outside of the fovea. An earlier report had mentioned that in monkey retinae exposed to large fields of bright visible light, the central macular area consistently showed less photic damage than the area just surrounding it (Lawwil et al., 1977). This is interesting, as the cones, which prevail in the fovea, are generally much more susceptible to photic damage (Marshall, 1986). One explanation might be that cones in the fovea are better protected by the presence of the macular yellow pigment than cones at the periphery.

Recently it was demonstrated that posterior subcapsular cataracts in Royal College of Surgeons (RCS) rats could be prevented by dietary supplementation with beta-carotene (Hess et al., 1991). RCS rats provide a model of a hereditary dysfunction of the retina, in the course of which the RPE fails to phagocytose the shed apices of the outer segments of the rod photoreceptors. Associated with this condition are posterior subcapsular opacities; the same type of cataract occurs in the aging population. Earlier, evidence was presented that these cataracts were secondary to lipid peroxidation in the degenerating retina whereby photochemical reactions involving rhodopsin were supposed to lead to the generation of cataractogenic products, which would cross the vitreous and cause cataracts at the posterior bend of the lens (Zigler and Hess, 1985). These cataracts can be completely prevented either by rearing the rats in darkness or by feeding beta-carotene or vitamin E. As both regimens can either reduce the formation of singlet oxygen or inactivate it directly by quenching, these results lend support to the etiological involvement of singlet oxygen in this condition of RCS rats. Whether these data are relevant to the human eye, which has a different geometry with a much greater distance between the retina and the lens, remains to be established. That they indeed might be is suggested by recent epidemiological data which seem to indicate that lutein, one of the macular carotenoids, may also have effects at locations other than the macula. A subanalysis of 2000 participants in the Beaver Dam Eye Study, a population-based epidemiological investigation of age-related eye diseases in which dietary and supplemental nutrient intake as well as type and severity of lens opacifications were evaluated, suggested that in males the intake of lutein and some other carotenoids was negatively associated with the severity of nuclear cataract (Mares-Perlmann et al., 1991). One other epidemiological investigation of cataract and vitamin intake showed that the intake of spinach, but not of carrots, had a strong inverse association with the risk of cataracts (Hankinson et al.,

1991). This might again indicate that lutein, which is a major carotenoid of spinach, may be involved in the prevention of cataract. Whether these results can be explained on a similar basis as the findings of Hess et al., discussed above, i.e. by the prevention of the formation of cataractogenic products in the retina, remains to be clarified.

Findings in humans
1. Observational data. In the following, some observational data will be discussed which provide indirect evidence for the potential of the macular carotenoids to prevent light/oxygen-induced damage in the retina. Although these data obviously cannot be a final proof, they give valuable leads as to what direction further research should take. It is known that the operating microscope used during eye surgery can cause photic damage to the retina. Interestingly, as reviewed by Jaffe and Wood (1988), in most such cases the photic lesions are attenuated or absent in the central area of the macula, the fovea, where macular yellow is most concentrated. When compared with younger people, elderly people show no loss of sensitivity of those short-wavelength-sensitive cones which are located within the fovea, whereas cones at other locations are significantly less sensitive (Haegerstrom-Portnoy, 1988).

A number of toxic and degenerative changes in the retina show an annular pattern, called Bull's-eye maculopathy. Thus a circular ring of structural change surrounding the macula can be seen in toxic retinopathy after chloroquine usage, retinitis pigmentosa and other degenerative diseases including age-related macular degeneration. The macula itself does not seem to be affected. Weiter et al. (1988) have tried to find out whether the presence of macular yellow could be one explanation for the peculiar geometric manifestation of this phenomenon. They used fluorescence angiogram fundus pictures from 95 subjects taken in blue light, in which the yellow spot appears as a dark area due to the absorption of blue light by the yellow pigments and compared the diameter of this dark area with the diameter of the annular degeneration. There was a striking match of the diameters ascribed to the yellow spot and those ascribed to the area spared by the degeneration pattern, and in one case the macular yellow and the spared area measured in the same eye actually overlapped completely.

A quite interesting observation was reported by Bornstein (1973). This psychologist investigated the color-naming systems in about 130 different languages with emphasis on primary color names for black and blue. He also evaluated data on color blindness or weakness in the respective cultures. It was observed that 46 of the 130 languages studied semantically did not discern between black and blue and that a significant number of people of these cultures had to be characterized as being blue-blind or blue-weak. Furthermore, the geographical distribution of the majority of these cultures revealed that they all lived in areas close to the equator or at greater altitudes, thus in environments where light intensity is very high. Therefore the phenomenon was speculatively explained by the presence of a dense macular yellow pigmentation, which would be beneficial as a protective measure in these environments, but at the same time would filter out most of the blue light leading to blue blindness. In the absence of chemical data on the composition of macular yellow in different ethnic groups, the scientific validity of the presented analysis is difficult to judge, however, the findings are interesting and show how disciplines very remote from pure ophthalmology have been dealing with the question of the significance of macular yellow.

2. Epidemiological evidence. Nutritional status has been proposed as a potential risk factor in AMD (Tso, 1985) and the first epidemiological investigation of the correlation of nutritional factors and the risk for AMD was a case-control study which compared 26 AMD patients with 23 age- and sex-matched controls (Blumenkranz et al., 1986). The use of vitamin supplements and serum levels of vitamins A, C and E did not differ significantly between the two groups. However, the sample size of this study was small and as a control spouses were used; this may have biased against detecting any dietary effect (Goldberg et al., 1988). Goldberg et al. (1988) analyzed ophthalmological data from the NHANES in the course of which almost 10,000 persons were examined. The diagnosis of AMD was made according to criteria developed by the National Eye Institute. Intake of vitamins A and C were estimated by 24-hour recall and a food-frequency questionnaire for fruits and vegetables rich in vitamins A and C. After adjusting for demographic and medical factors, the frequency of consumption of fruits and vegetables rich in vitamin A appeared to be negatively correlated with the risk of AMD. As most fruits and vegetables rich in preformed vitamin A are also rich in carotenoids, this study provides suggestive evidence for a possible protective function of carotenoids in AMD. Recently first results from another epidemiological study, the Eye Disease Case – Control Study (Sperduto, 1991), were reported. In this five-center study various risk factors for AMD were investigated. Results for 421 cases with neovascular/exudative AMD and 615 controls were presented regarding plasma levels of vitamins C and E, selenium and carotenoids. A logistic regression analysis indicated a statistically significant protective effect for higher carotenoid levels and a higher antioxidant index derived from nutrients with antioxidant potential.

3. Clinical intervention trials. Retinitis pigmentosa (RP) is a hereditary disease which progressively leads to degeneration of the RPE. Although this disease is clearly separate from AMD, there are some similarities, particularly the involvement of the RPE. It was reported that there was subjective improvement of the visual field loss in progressed RP patients when they were treated with high doses of beta-carotene and vitamin E (Heckenlively et al., 1989). Although statistical significance was not attained because of the large variance and a small sample size, in all comparisons the median visual field deterioration rate was better in the treated subjects. More relevant may be a trial reported by Crary (1987). This investigator treated 102 AMD patients for 7 to 12 years with a combination of 500 mg vitamin C, 400 mg vitamin E, 9 mg beta-carotene and 250 mg selenium per day. It was found that this treatment halted or improved the degenerative macular changes in 60% of the patients.

Conclusions

These results of initial attempts towards nutritional intervention, although encouraging, do not prove the efficacy of supplemental vitamins or carotenoids to prevent, retard or ameliorate AMD. Therefore no recommendation as to nutritional supplementation can be given before results from more and better controlled trials are available. The National Eye Institute (NEI) has decided to include in one of its projects on the clinical course and prognosis of AMD and cataracts, a controlled randomized triaGGçtevaluate the effect of a supplementation with antioxidant micronutrients including beta-carotene (Sperduto et al., 1990, 1991). A critical question is, at what time should intervention of any kind be started. For carotenoids to be effective it might be speculated that intervention must start very early in life to ensure the presence of a yellow shield in the macula. As mentioned, lenses of infants and adolescents are colorless and do not have the capacity to absorb the blue light which has been shown to be at least one important factor in the etiology of the complex and multifactorial disease AMD. Also, many other questions remain to be answered. However, from a teleological perspective, the reason nature spends so much effort on concentrating carotenoids within the macula might be not only to correct chromatic aberration, but also to provide one important component of an efficient protection against necessary and also potentially damaging blue light.

Andley, U. P. (1987) Photodamage to the eye. Photochem. Photobiol. 46: 1057–1066.

Appleby, S. J., and Muntz, W. R. A. (1979) Occlusable yellow corneas in tetraodontidae. J. Exp. Biol. 83: 249–259.

Arden, G., and Barker, F. M. (1991) Canthaxanthin and the eye – a critical ocular toxicological assessment. J. Toxicol.-Cut. & Ocular. Toxicol. 10 (1&2): 115–155.

Benolken, R. M., Maude, M. B., and Anderson, R. E. (1976) Photopigments of the lateral eye of limulus. J. Comp. Physiol. 107: 339–347.

Blumenkranz, M. S., Russell, S. R., Robey, M. G., Kott-Blumenkranz, R., and Penneys, N. (1986) Risk factors in age-related maculopathy complicated by choroidal neovascularization. Ophthalmology 96: 552–558.

296

Bone, R. A., Landrum, J. T., and Tarsis, S. L. (1985) Preliminary identification of the human macular pigment. Vision Res. 25: 1531–1535.

Bone, R. A., Landrum, J. T., Fernandez, L., and Tarsis, S. L. (1988) Analysis of the macular pigment by HPLC: retinal distribution and age study. Invest. Ophthalmol. Vis. Sci. 29: 843–849.

Bornstein, M. H. (1973) Color vision and color naming; a psychophysiological hypothesis of cultural difference. Psychol. Bull. 80: 257–285.

Bressler, N. M., Bressler, S. B., and Fine, S. L. (1988) Age-related macular degeneration – a major review. Survey Ophthalmol. 32: 375–412.

Burton, G. W., and Ingold, K. U. (1984) Beta-carotene – an unusual type of lipid antioxidant. Science 224: 569–573.

Cains, A., Bone, R. A., Landrum, J. T., and Zamor, J. (1991) Determination of the absolute configuration of the macular pigment carotenoids. Invest. Ophthalmol. Vis. Sci. (Suppl.) 32: 1009.

Conn, P. F., Schalch, W., and Truscott, T. G. (1991) The singlet oxygen – carotenoid interaction. J. Photochem. Photobiol. B: Biol., 11: 41–47.

Crary, E. J. (1987) Antioxidant treatment of macular degeneration of the aging and macular edema in diabetic retinopathy. South. Med. J. 80: 38.

Daicker, B., Schiedt, K., Adnet, J. J., and Bermond, P. (1987) Canthaxanthin retinopathy – an investigation by light and electron microscopy and physicochemical analysis. Graefe's Arch. Clin. Exp. Ophthalmol. 225: 189–197.

Dayhaw-Barker, P. (1986) Ocular photosensitation. Photochem. Photobiol. 46: 1051–1055.

Deisenhofer, J., and Michel, H. (1989) The photosynthetic reaction center from the purple bacterium Rhodopseudomonas virdis. Chemica Scripta 29: 205–220.

Delmelle, M. (1977) Retinal damage by light: possible implication of singlet oxygen. Biophys. Struct. Mechanism 3: 195–198.

Ferris, F. L. (1983) Senile macular degeneration: review of epidemiologic features. Am. J. Epidemiol. 118: 132–151.

Foote, C. S. (1979) Quenching of singlet oxygen, in: Singlet Oxygen. Wasserman, M. M. & Murray, R. W., eds. Academic Press, New York, pp. 139–171.

Frank, R. N. (1989) Macular degeneration. JAMA 261: 767–768.

Gerster, H. (1991) Review: Antioxidant protection of the ageing macula. Age and Ageing 20: 60–69.

Goldberg, J., Flowerdew G., Smith E., Brody J. A., and Tso, M. O. M. (1988) Factors associated with age-related macular degeneration – Analysis of data from NHANES I. Am. J. Epidemiol. 128: 700–711.

Gross, J. (1987) Pigments in Fruits. Academic Press, New York, pp. 87–182.

Gross, J. (1991) Pigments in Vegetables. Van Nostrand, New York, pp. 85–128.

Gurne, D. H., Tso, M.O.M., Edward, D. P., and Ripps, H. (1991) Antiretinal antibodies in serum of patients with age-related macular degeneration. Ophthalmology 98: 602–607.

Haegerstrom-Portnoy, G. (1988) Short-wavelength-sensitive-cone sensitivity loss with aging: a protective role for macular pigment? J. Opt. Soc. Am. Ser. A. 5: 2140–2144.

Ham, W. T., Mueller, H. A., Ruffolo, J. J., Millen, J. E., Cleary, S. F., Guerry, R. K., and Guerry, D. (1984) Basic mechanisms underlying the production of photochemical lesions in the mammalian retina. Curr. Eye Res. 3: 165–174.

Ham, W. T., and Mueller, H. A. (1989) The photopathology and nature of blue light and near-UV retinal lesions produced by lasers and other optical sources, in: Laser Applications in Medicine and Biology. Wolbarsht, M. L., ed. Plenum, New York, pp. 210.

Handelman, G. J., and Dratz, E. A. (1986) The role of antioxidants in the retina and retinal pigment epithelium and the nature of prooxidant-induced damage. Adv. Free Radical Biol. & Med. 2: 1–89.

Handelman, G. J., Draz, E. A., Reay, C. C., and van Kuijk, F. J. G. M. (1988) Carotenoids in the human macula and whole retina. Invest. Ophthalmol. Vis. Sci. 29: 850–855.

Handelman, G. J., Snodderly, D. M., Krinski, N. I., Russett, M. D., and Adler, A. J. (1991a) Biological control of primate macular pigment. Invest. Opththalmol. Vis. Sci. 32: 257–267.

Handelman, G. J., van Kuijk, F. J. G. M., Chatterjee, A., and Krinsky, N. I. (1991b) Characterization of products formed during the autoxidation of beta-carotene. Free Radical Biology & Medicine 10: 427–437.

Hankinson, S. E., Stampfer, M. J., Seddon, J. M., Colditz, G. A., Rosner, B., Speizer, F. E., and Willett, W. C. (1991) A prospective study of vitamin intake and cataracts. 24th Annual Meeting of the Soc. for Epidemiologic Research, June 11–14, Buffalo, N.Y.

Heckenlively, J. R., Yoser, S. L., and Pearlman, J. T. (1989) Treatment Trial of beta-carotene and vitamin E for Retinitis Pigmentosa (RP). Invest. Ophthalmol. Vis. Sci. (Suppl.) 30: 305.

Hess, H., and Zigler, J. S. (1991) Retina-lens interaction in genesis of cataracts in RCS rats and prevention by dietary supplementation with beta-carotene and vitamin E. Invest. Ophthalmol. Vis. Sci. (Suppl.) 32: 1100.

Hiramitsu, T., and Armstrong, D. (1991) Preventive effect of antioxidants on lipid peroxidation in the retina. Ophthalmic Res. 23: 196–203.

Isler, O. Ed. (1971) Carotenoids. Birkhäuser, Basel.

Jaffe, G. J., and Wood, I. S. (1988) Retinal phototoxicity from the operating microscope: a protective effect by the fovea. Arch. Ophthalmol. 106: 445–446.

Jane, S. D., and Bowmaker, J. K. (1988) Tetrachromatic colour vision in the duck: microspectrophotometry of visual pigments and oil droplets. J. Comp. Physiol. Ser. A 162: 225–235.

Katz, M. L., Stone, W. L., and Dratz, E. A. (1978) Fluorescent pigment accumulation in retinal pigment epithelium of antioxidant-deficient rats. Invest. Ophthalmol. Vis. Sci. 17: 1049–1058.

Kennedy, T. A., and Liebler, D. C. (1991) Peroxyl radical oxidation of beta-carotene: formation of beta-carotene epoxides. Chem. Res. Toxicol. 4: 290–295.

Kini, M. M., Leibowitz, H. M., Colton, T., and Nickerson, R. J. (1978) Prevalence of senile cataract, diabetic retinopathy, senile macular degeneration and open-angle glaucoma in the Framingham Eye Study. Am. J. Ophthalmol. 85: 28–34.

Kirschfeld, K. (1982) Carotenoid pigments: their possible role in protecting against photooxidation in eyes and photoceceptor cells. Proc. R. Soc. Lond. (B) 216: 71–85.

Klein, B. E., and Klein, R. (1982) Cataracts and macular degeneration in older Americans. Arch. Opthalmol. 100: 571–573.

Krebs, W., and Krebs, J. (1991) Primate retina and choroid – Atlas of fine structure in man and monkey. Springer-Verlag, New York, pp. 2–8.

Krinsky, N. I. (1968) The protective function of carotenoid pigments. Photophysiology 3: 123–195.

Krinsky, N. I. (1979) Carotenoid protection against oxidation. Pure & Appl. Chem. 51: 649–660.

Krinsky, N. I. and Deneke, S. M. (1982) Interaction of oxygen and oxy-radicals with carotenoids. J. Natl. Cancer Inst. 69: 205–210.

Krinsky, N. I., Russett, M. D., Handelman, G. J., and Snodderly, D. M. (1990) Structural and geometrical isomers of carotenoid in human plasma. J. Nutr. 120: 1654–1662.

Kurashige, M., Okimasu, E., Inoue, M., and Utsumi, K. (1990) Inhibition of oxidative injury of biologic membranes by Astaxanthin. Physiol. Chem. Phys. & Med. NMR 22: 27–38.

Lawwill, T., Crockett, S., and Currier, G. (1977) Retinal damage secondary to chronic light exposure – thresholds and mechanisms. Docum. Ophthalmol. 44: 379–402.

Lermann, S. (1983) An experimental and clinical evaluation of lens transparency and aging. J. Geront. 38: 293–301.

Liles, M. R., Newsome, D. A., and Oliver, P. D. (1991) Antioxidant enzymes in the aging human retinal pigment epithelium. Arch. Ophthalmol. 109: 1285–1288.

Liu, I. Y., White, L., and LaCroix, A. Z. (1989) The association of age-related macular degeneration and lens opacities in the aged. Am. J. Pub. Health 79: 765–769.

Malinow, M. R., Feeney-Burns, L., Peterson, L. H., Klein, M. L., and Neuringer, M. (1980) Diet-related macular anomalies in monkeys. Invest. Opthalmol. Vis. Sci. 19: 857–863.

Mares-Perlman, J. A., Klein, B. E. K., Klein, R., Ritter, C., Linton, K. L. P., and Luby, M. H. (1991) Relationship between diet and cataract prevalence. Invest. Ophthalmol. Vis. Sci. (Suppl.) 32: 723.

Marshall, J. (1985) Radiation and the ageing eye. Ophthal. Physiol. Opt. 5: 241–263.

di Mascio, P., Kaiser, S., and Sies, H. (1989) Lycopene as the most efficient biological carotenoid singlet oxygen quencher. Arch. Biochem. Biophys. 274: 532–538.

Minke, B., and Kirschfeld, K. (1978) Microspectrophotometric evidence for two photointerconvertible states of visual pigment in the barnacle lateral eye. J. General. Physiol. 71: 37–45.

Mordi, R. C., Walton, J. C., Burton, G. W., Hughes, L., Ingold, K. U., and Lindsay, D. A. (1991) Exploratory study of beta-carotene autoxidation. Tetrahedron Letters 32: 4203–4206.

298

Muntz, W. R. A. (1972) Inert absorbing and reflecting pigments. Sensory Physiology 7: 529–565.

Nussbaum, J. J., Pruett, R. C., and Delori, F. C. (1981) Historic perspectives – Macular yellow pigment – The first 200 years. Retina 1: 296–310.

Ordy, J. M., Brizzee, K. R., Wengenack, T. M., and Dunlap, W. P. (1991) Age-related macular degeneration (AMD) in the retina of the aged rhesus monkey. Invest. Ophthalmol. Vis. Sci. (Suppl.) 32: 1174.

Palozza, P., and Krinsky, N. I. (1991) The inhibition of radical-initiated peroxidation of microsomal lipids by both α-tocopherol and β-carotene. Free Radical Biology & Medicine 11: 407–414.

Sapp, R. J., Christianson, J. S., Maier, L., Studer, K., and Stark, W. S. (1991) Carotenoid replacement therapy in drosophila: recovery of membrane, opsin and visual pigment. Exp. Eye Res. 53: 73–79.

Sickel, W. (1972) Retinal metabolism in dark and light. Handbook of Sensory Physiology 7. pp. 667–727.

Snodderly, D. M., Brown, P. K., Delori, F. C., and Auran, J. D. (1984a) The macular pigment. I. Absorbance spectra, localization, and discrimination from other yellow pigments in primate retinas. Invest. Ophthalmol. Vis. Sci. 25: 660–673.

Snodderly, D. M., Auran, J. D., and Delori, F. C. (1984b) The macular pigment. II. Spatial distribution in primate retinas. Invest. Ophthalmol. Vis. Sci. 25: 674–685.

Snodderly, D. M., Handelmann, G. J., and Adler, A. J. (1991) Distribution of individual macular pigment carotenoids in central retina of Macaque and Squirrel monkey. Invest. Ophthalmol. Vis. Sci. 32: 268–279.

Sperduto, R. D., Ferris, F. L., and Kurinij, N. (1990) Do we have a treatment for age-related cataract or macular degeneration? Arch. Ophthalmol. 108: 1403–1404.

Sperduto, R. (1991) The eye disease case-control study. NEI symposium on Eye Disease Epidemiol., Third National Eye Institute Symposium on Eye Disease Epidemiology – March 25–27, Bethesda, MD.

Stone, W. L., Farnsworth, C. C., and Dratz, E. A. (1979) A reinvestigation of the fatty acid content of bovine, rat and frog outer segments. Exp. Eye Res. 28: 387–397.

Taylor, H. R., Munoz, B., West, S., Bressler, N. M., Bressler, S. B., and Rosenthal, F. S. (1990) Visible light and risk of age-related macular degeneration. Tr. Am. Ophth. Soc. 88: 165–177.

Terao, J. (1989) Antioxidant activity of beta-carotene-related carotenoids in solution. Lipids 24: 659–661.

Truscott, T. G. (1991) personal communication.

Tso, M. O. M. (1985) Pathogenetic factors of aging macular degeneration. Ophthalmology 92: 628–635.

Tso, M. O. M. (1989) Experiments on visual cells by nature and man: in search of treatment for photoreceptor degeneration. Invest. Ophthalmol. Vis. Sci. 30: 2430–2454.

Wald, G. (1945) Human vision and the spectrum. Science 101: 653–658.

Weiter, J. J., Delori, F., and Dorey, C. K. (1988) Central sparing in annular macular degeneration. Am. J. Ophthalmol. 106: 286–292.

Werner, J. S., Steele, V. G., and Pfoff, D. S. (1989) Loss of human photoreceptor sensitivity associated with chronic exposure to radiation. Ophthalmology 96: 1552–1558.

West, S. K., Rosenthal, F. S., Bressler, N. M., Bressler, S. B., Munoz, B., Fine, S. L., and Taylor, H. R. (1989) Exposure to sunlight and other risk factors for age-related macular degeneration. Arch. Ophthalmol. 107: 875–879.

Yoser, S. L., and Heckenlively, J. R. (1989) The appearance of retinal crystals in retinitis pigmentosa patients using beta-carotene. Invest. Ophthalmol. Vis. Sci. (Suppl.) 30: 305.

Young, R. W. (1988) Solar radiation and age-related macular degeneration. Survey Ophthalmol. 32: 252–269.

Yu, N.-T., Cai, M.-Z., Lee, B.-S., Kuck, J. F. R., McFall-Ngai, M., and Horwitz, J. (1991) Resonance raman detection of a carotenoid in the lens of the deep-sea hatchetfish. Exp. Eye Res. 52: 475–479.

Zigler, J. S., and Hess, H. H. (1985) Cataracts in the Royal College of Surgeons Rat: evidence for initiation by lipid peroxidation products. Exp. Eye Res. 41: 67–76.

Zimmermann, W. F., and Keys, S. (1991) Effects of the antioxidants dithiothreitol and vitamin E on phospholipid metabolism in isolated rod outer segments. Exp. Eye Res. 52: 607–612.

Free Radicals and Aging
ed. by I. Emerit & B. Chance
© 1992 Birkhäuser Verlag Basel/Switzerland

Oxidative stress in diabetic retina

Michel Doly[a], Marie-Thérèse Droy-Lefaix[b] and Pierre Braquet[b]

[a]*Laboratoire de Biophysique (Inserm U. 71), Facultés de Médecine et de Pharmacie, BP 38, F-63001 Clermont-Ferrand Cedex, France, and* [b]*IHB/IPSEN Laboratoires de Recherche, 17, Avenue Descartes, F-92350 Le Plessis-Robinson, France*

Summary. The authors describe the alterations usually associated with diabetic retinopathy. They concern the classical thickening of the basal membrane of retinal capillaries and the associated modification of retinal vessel permeability. These alterations correspond to the blood-retinal barrier disruption. The authors then discuss the participation of oxygenated free radicals in the pathogenesis of diabetic retinopathy. They report several experimental studies establishing such a participation and finally describe their own results obtained on a model of retinas isolated from alloxan-induced diabetic rats.
After one month of evolution, the electroretinograms (ERG) recorded on isolated retinas from diabetic rats had an amplitude about 20% lower than the controls, whereas after two months of diabetes, this decrease was about 60%. Under these conditons, the authors tested the protective properties of *Ginkgo biloba* extract (EGb 761) on their model. They observed that in EGb-treated animals (100 mg/kg/day), the ERG had a significantly ($p < 0.001$) greater amplitude than untreated animals after two months of diabetes evolution.
In conclusion, the authors discuss the possible utilization of a free radical scavenger, such as EGb 761, in the prevention of the retinal impairment in diabetes.

Introduction

Retinopathy is one of the most serious consequences of diabetes in man. Its gravity is directly linked to irreversible impairment of visual function.

At the present time, there is only partial comprehension of the mechanism involved in diabetic retinopathy. Several types of histological impairment have been described. Most notably, a thickening of the basal membrane of the capillaries is observed. More recently, several authors have highlighted a disruption in the blood-retinal barrier both in man and in experimental animals. This disruption is accompanied by an increase in capillary permeability that could play a role in the pathogenesis of diabetic retinopathy.

Several recent studies have reported a clear increase in the concentration of oxygenated free radicals in the retinas of diabetic animals (Murata et al., 1981; Cohen, 1984; Nishimura and Kuriyama, 1985a). Moreover, this increase is correlated with an elevation in the rate of membrane lipid peroxidation. Furthermore, the retina which is a tissue very rich in polyunsaturated fatty acids, is particularly sensitive to the reactivity of oxygenated free radicals (Anderson and Sperling, 1971;

Doly et al., 1984). Excessive production of oxygenated radicals may thus explain the specificity of the retinal impairment involved in diabetes.

Retinal alterations in diabetes

Diabetic retinopathy has been an important subject in histological studies on animal models. For some time now (Bloodworth and Moditor, 1985; Tanigushi and Nomura, 1968; Papachristodoulou and Health, 1977), it has been known that the characteristic histological finding in retinopathy is the thickening of the basal membrane of the retinal capillaries. More recent studies, in particular those of Ishibashi et al. (1980), demonstrate that these histological modifications accompany an increase in permeability of certain retinal vessels. Indeed, by injecting horseradish peroxidase (HRP) 30 min before sacrificing animals, a high increase in vesicular transport and the existence of junctional transport between the capillary endothelial cells can be established in rats maintained in a diabetic state for 2 to 6 months. This increase in permeability of the endothelial cells apparently precedes the thickening of the basal membrane. In 1975, Cunha-Vaz et al. introduced the concept of the blood-retinal barrier by analogy with the blood-brain barrier. They noted that one of the early consequences of diabetes is a disruption of the blood-retinal barrier. These results were subsequently confirmed by several workers, such as Ishibashi et al. (1980) on streptozotocin-diabetic rats and by Blair et al. (1984) on a spontaneously diabetic variety of rats. The disruption in the blood-retinal barrier appears to be due to an increase in vesicular transport and a junctional insufficiency between the endothelial cells (Ishibashi et al., 1980). Its mechanism is still unknown, but hyperglycemia and the related accumulation of sorbitol probably play a primary role (Palmberg, 1977). Aldose reductase, an enzyme responsible for the conversion of glucose to sorbitol, is present in the retina and a significant accumulation of sorbitol is observed in diabetic rat retinas. Thus, aldose reductase inhibitors could delay the disruption of the blood-retinal barrier in experimental animals (Palmberg, 1977).

There are fewer studies on the electrophysiological consequences of diabetic retinopathy than there are histological studies on the condition. The first ERG alterations related to diabetes were described at the level of the oscillatory potential (Yonemura et al., 1962). In humans, these wavelets, superimposed on the ascending branch of the b-wave, decrease in amplitude with the evolution of diabetes and finally disappear. Since this oscillatory potential is directly related to the activity of the amacrin cells, functional impairment at this level has been proposed as the consequences of diabetic retinopathy (Yonemura, 1977; Nishimura and Kuriyama, 1985b).

Tamai and Tanaka (1973) have given an account of an electroretino-graphic study in vivo, performed on streptozotocin-diabetic albino rats. They observed that 4 months after the induction of diabetes, there was an average fall in a-wave amplitude of 40% and 37% for b-wave, which they interpreted as being a moderate impairment of the retinal function. More recently, and still concerning streptozotocin-diabetic rats, Tanaka (1981) demonstrated that injection of dextran sulfate ester sodium had the effect of partially preventing the fall in amplitude of a- and b-waves.

Free radicals and diabetic retinopathy

The pathogenesis of diabetic retinopathy is still quite obscure but seems to be closely linked with the formation of oxygenated free radicals. The studies of Murata et al. (1981) performed on alloxan-induced diabetic rats, and by Nishimura and Kuriyama (1985a) on streptozotocin-induced diabetic rats, clearly demonstrate that diabetic retinopathy is accompanied by a high increase in the concentration of oxygenated free radicals in the retina. These radicals react preferentially with unsaturated lipids leading to the formation of lipoperoxide radicals (LO_2^-) inducing an oxidative process which results in membrane lysis (Doly et al., 1984, 1985). Thus Murata et al. (1981) reported a 40% increase in the lipid peroxidation rate in the retina of diabetic rats in comparison with control rats. This attack on the membrane has particularly harmful effects on retinal tissue and might be related to the important decrease of lipid replacement observed on streptozotocin-induced diabetic rat retina (Careaga and Bazan, 1981; Bazan et al., 1985). Indeed, two membranes are implicated in the phototransduction mechanism which generates the electroretino-gram (ERG): the plasma membrane and the saccular membrane of the photoreceptor. These two membranes are particularly rich in polyunsaturated fatty acids (Kagan et al., 1973; Oakley and Pinto, 1981) and so very sensitive to lipoperoxidation. The first one to be impaired and denatured by oxidation seems to be the saccular membrane. In fact, after lipoperoxidation, rhodopsin is more easily extracted (Novikov et al., 1975) which may correspond to a weakening of protein-lipid interaction; moreover, regeneration capacity of rhodopsin decreases after peroxidation (Franswooth and Dratz, 1976). Finally, intracellular records performed in the rod outer segment indicate that lipoperoxidative substances interact essentially on the saccular membrane (Shvedova et al., 1979).

Free radical scavengers and diabetic retinopathy

In order to demonstrate the participation of free radicals in diabetic retinopathy, experiments have been performed on isolated retina of

302

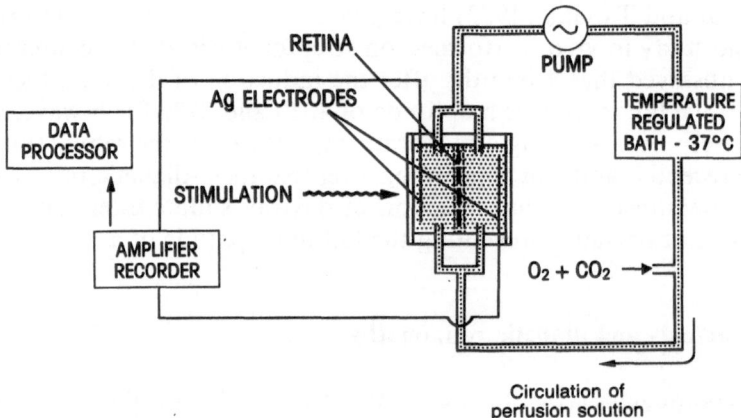

Figure 1. Experimental set-up used to record ERG on retina isolated from albino rats.

alloxan-induced diabetic rats (Doly et al., 1986, 1988). The experimental set-up is shown in Figure 1 and was previously published (Doly et al., 1980).

Experimental diabetes was induced by alloxan injection. The retinal function was assessed by ERG records during the survival of the isolated retina. Alongside the control group, three other groups of rats were examined: the first one comprised animals sacrificed one month after diabetes evolution; the second one was made up of animals sacrificed after two months of diabetes evolution and the third one contained treated animals after two months of diabetes. The treatment consisted in the daily administration of 100 mg/kg per os *Ginkgo biloba* extract (EGb 761) in aqueous solution. This extract is a potent free radical scavenger and exerts a protective effect on the retina against membrane lipid peroxidation (Braquet et al., 1982). An example of ERG is shown in Figure 2, so as the evolution of ERG amplitude during the survival of normal retinas in standard conditions.

We chose to quantify the ERG by measuring the b-wave amplitude; indeed since the latter is generated at the level of the Müller cells (Tomita and Ynagida, 1981), its development involves all the intraretinal nerve connections, and its amplitude thus reflects the quality of the retinal metabolism.

The average survival curve obtained in diabetic animals 1 month after the evolution of diabetes is shown in Figure 3.

It can be observed that during the first hour of survival, the ERG amplitudes recorded in 1-month diabetic rats are comparable to those of the reference curve, whereas during the remainder of the survival period, the ERGs of the diabetic rats are systematically less in amplitude than the reference ERGs. The amplitude of the ERGs obtained on

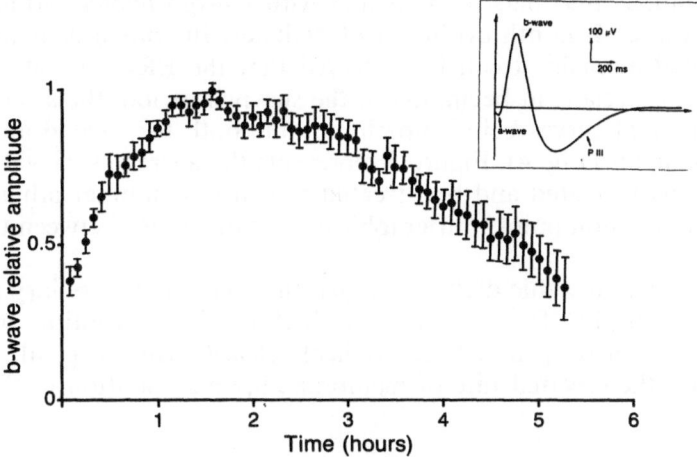

Figure 2. Normal evolution of ERG b-wave amplitude during retina survival in standard conditions. On the upper right an example of electroretinogram (ERG) recorded on isolated retina.

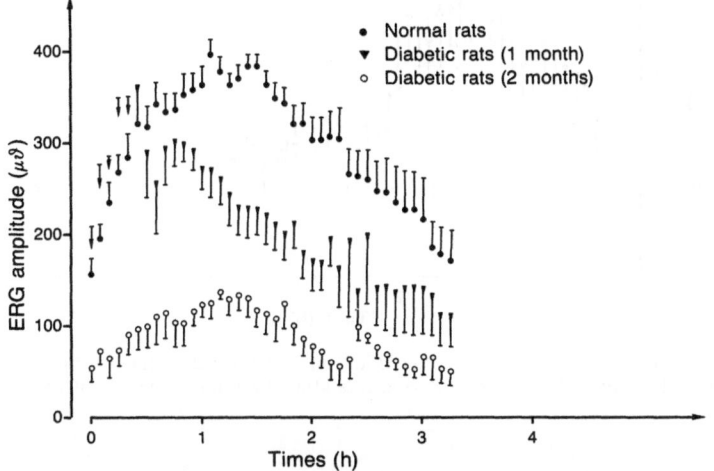

Figure 3. Evolution of ERG amplitude recorded on retina isolated from diabetic rats (1 month and 2 months) compared to ERG recorded on normal rat retina.

2-month diabetic rat retinas are systematically lower than the reference values over the entire survival period (Fig. 3). It can be observed that the decrease in the ERG amplitude is directly correlated with the duration of the evolution of diabetes. At the point of maximum amplitude in the survival curve of the control rats, the decrease in the ERG amplitude is around 20% for 1-month diabetic rats, whereas it is around 60% for 2-month diabetic rats.

304

The animals that had been treated with *Ginkgo biloba* extract were sacrificed after 2-month evolution of diabetes. In comparison with the non-treated animals, it can be observed that the ERG amplitudes are increased, especially at beginning of the survival period, the second half of the survival curves being superimposable both for treated and non-treated animals (Fig. 4). Figure 5 represents the average survival curves obtained with treated and non-treated rats after computer adjustment; this representation points out graphically the difference between the two curves.

On such an alloxanic diabetes model, the increase of capillary permeability responsible for the blood retinal barrier disruption may be correlated with oxygenated free radical release (Murata et al., 1981). Moreover, the elevated rate of membrane lipoperoxidation in the pho-

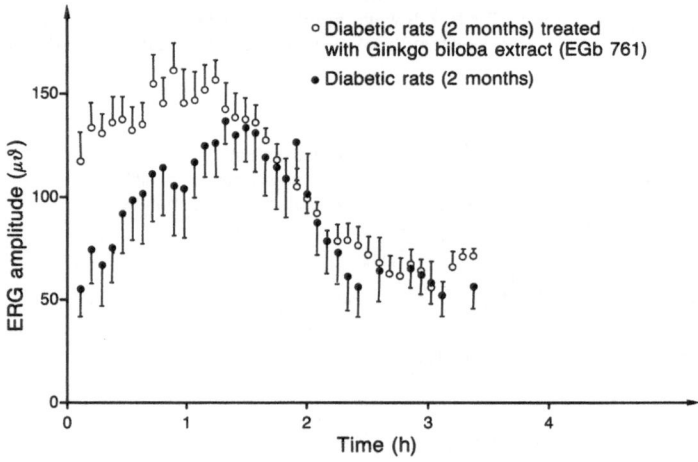

Figure 4. Comparison between ERG recorded on diabetic rats after 2 months of diabetes evolution and on EGb 761 treated diabetic rats after the same evolution of diabetes.

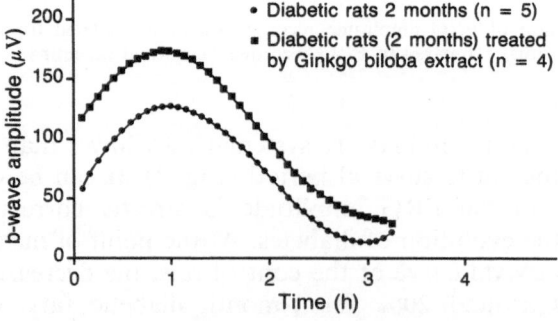

Figure 5. The same curves than on Figure 4 but after computer adjustment.

toreceptors can be responsible for ERG alterations. In a parallel manner, the physiological protective mechanisms (essentially superoxide dismutase and α tocopherol) can also be deficient or overwhelmed (Nishimura and Kuriyamma, 1985b).

In this experiment, ERG effects of *Ginkgo biloba* extract can be correlated both with oxygenated free radical reactivity and production. *Ginkgo biloba* extract is rich in benzopyrones such as heterosides, known for free radical scavenger properties (Doly et al., 1985) and in Ginkgolides A, B and C, (Braquet et al., 1987), which are potent PAF-acether antagonists; PAF is an immunomediator able to amplify eosinophil and leukocyte response and to induce the release of $O_2^{\cdot-}$ and $OH^{\cdot}$ radicals. These radicals may be responsible for retinal membrane impairment and dysfunction in diabetic retinopathy.

Conclusion

In humans, the retina is one of the most aerated organs. The demand for molecular oxygen by the mitochondria of photoreceptors is particularly high. The rod outer segment membrane contains the highest level of polyunsaturated fatty acids of any tissue in the body. In these conditions, the retina appears as a neuro-sensorial tissue particularly sensitive to the reactivity of oxygenated free radicals. The retinal function, which corresponds to photon detection, is essentially based on membrane potential variations. In other words the quality of light perception is markedly linked to the quality of membranes. One of the most deleterious effects of oxygenated radicals is membrane lipoperoxidation which corresponds to a severe disorganization. Several sudies have demonstrated the exaggerated production of free radicals in the retina of diabetic patients. Here, the impairment of retinal function reported during the evolution of diabetes may be closely related to the reactivity of oxygenated radicals. Such experimental data suggest the possible use of free radical scavengers in order to prevent retinal alterations due to diabetes. The preliminary results obtained with *Ginkgo biloba* extract seem to confirm this possibility and demonstrate that free radical scavengers may play a key role in the prevention of diabetic retinopathy.

Anderson, R. E., and Sperling, L. (1971) Lipids of ocular tissues. VII Positional distribution of the fatty acids in the phospholipids of bovine retina rod outer segments. Arch. Biochem. Biophys. 144: 673–677.

Bazan, H. E. P., Careaga, M. M., and Bazan, N. G. (1985) Decreased utilization of [2-^{3}H] glycerol in phospholipid and neutral glyceride biosynthesis in the retina of streptozotocin-diabetic rats. Neurochem. Path. 3: 109.

Blair, N. P., Tso, M. O. M., and Dodge, J. T. (1984) Pathologie studies of the blood-retinal barrier in the spontaneously diabetic BB rat. Invest. Ophthalmol. 25, 302–331.

Bloodworth, J. H. B., and Moditor, D. L. (1985) Ultrastructural aspect of human and canine diabetic retinopathy. Invest. Ophthalmol. 4: 1037–1048.

Braquet, P., Doly, M., Bonhomme, B., and Meyniel, G. (1982) Peroxydation lipidique et activité électrique de la rétine isolée: effect protecteur de l'extrait de Ginkgo biloba, in: Compte-rendus des Journées Internationales du Groupe Polyphénols. Toulouse (France), 29–30 September.

Braquet, P., Touqui, L., Shen, T. Y., and Vargaftig, B. B. (1987) Perspectives in platelet-activating factor research. Pharmacol. Rev. 39: 97–145.

Careaga, M. M., and Bazan, H. E. P. (1981) The rat retina is a useful in vivo model to study membrane lipid synthesis. Rates of biosynthesis of neutral glycerides and phospholipids. Neurochem. Res. 6: 1169–1178.

Cohen, G. (1984) Oxy-radical production in alloxan-induced diabetes: an example of an in vivo metal-catalyzed Haber-Weiss reaction, in: Free Radicals in Molecular Biology. Armstrong D., Sohal, R. S., Cuther, R. G., Slater, T. F. eds. Raven Press, New York, pp. 307–316.

Cunha-Vaz, J., De Abreu, J. R. F., Campos, A. J., and Figo, G. H. (1975) Early breakdown of the blood-retinal barrier in diabetes. Br. J. Ophthalmol. 59: 649–656.

Doly, M., Isabelle, D. B., Vincent, P., Gaillard, G., and Meyniel, G. (1980) Mechanism of the formation of X-ray induced phosphenes. I. Electrophysiological investigations. Rad. Res. 82: 93–105.

Doly, M., Braquet, P., Bonhomme, B., and Meyniel, G. (1984) Effects of lipid peroxidation on the isolated rat retina. Ophthalmic Res. 16: 292–296.

Doly, M., Braquet, P., Droy, M. T., Bonhomme, B., and Vennat, J. C. (1985) Effects des radicaux libres oxygénés sur l'activité électrophysiologique de la rétine isolée de rat. J. Fr. Ophthalmol. 8: 273–277.

Doly, M., Droy-Lefaix, M. T., Bonhomme, B., and Braquet, P. (1986) Effets de l'extrait de Ginkgo biloba sur l'électrophysiologie de la rétine isolée de rat diabétique. Presse Méd. 15: 1480–1483.

Doly, M., Braquet, P., Droy, M. T., et al. (1988) Alteration of electrophysiological function of isolated retina from alloxan-induced diabetic rats: effect of treatment with Ginkgo biloba extract. Neurochem. Pathol. 8: 15–26.

Farnsworth, C. C., and Dratz, E. A. (1976) Oxidative damage of retinal rod outer segment membranes and the role of Vitamin E. Biochim. Biophys. Acta 443: 556–570.

Ishibashi, T., Tanaka, K., and Taniguchi, Y. (1980) Disruption of blood-retinal barrier in experimental diabetic rats: an electron microscopic study. Exp. Eye Res. 30: 401–410.

Kagan, V. E., Shvedova, A. A., Novikov, K. N., and Koslov, Y. P. (1973) Lipid-induced free radical oxidation of membrane lipids in photoreceptors of frog retina. Biochim. Biophys. Acta 330: 76–79.

Murata, R., Nishida, T., Eto, S., and Mukai, N. (1981) Lipid peroxidation in diabetic rat retina. Metab. Pediat. Ophthalmol. 5: 83–87.

Nishimura, C., and Kuriyama, K. (1985a) Alteration of lipid peroxide and endogenous antioxidant contents in retina of streptozotocin-induced diabetic rats: effect of Vitamin A administration. Japan J. Pharmacol. 37: 365–372.

Nishimura, C., and Kuriyama, K. (1985b) Alterations in the retinal dopaminergic neuronal system in rats with streptozotocin-induced diabetes. J. Neurochem. 45: 448–455.

Novikov, K. N., Kagan, V. E., Shvedova, A. A., and Kozlov, Y. P. (1975) Protein-lipid interactions on peroxide oxidation of lipids in the photoreceptor membrane. Biofizika 20: 1039–1042.

Oakley, B., and Pinto, L. H. (1981) $[Ca^{2+}]_i$ modulations of membrane sodium conductance in rod outer segments, in: Cuttent Topics in Membrane and Transport. Muller, ed. Academic Press Inc., New York. p. 405.

Palmberg, P. F. (1977) Diabetic Retinopathy. Diabetes 26: 703–711.

Papachristodoulou, D., and Heath, H. (1977) Ultrastructural alterations during the development of retinopathy in sucrose-fed and streptozotocin-diabetic rats. Exp. Eye Res. 25: 371–384.

Shvedova, A. A., Sidorov, A. S., Novikov, K. N. et al. (1979) Lipid peroxidation and electric activity of the retina. Vision Res. 19: 49–55.

Tamai, A., and Tanaka, K. (1973) The ERG of the streptozotocin-diabetic albino rat. Folia Ophthalmol. Japan. 24: 847–850.

Tanaka, T. (1981) Effects of various agents on the ERG of the streptozotocin-diabetic rat. Folia Ophthalmol. Japan. 72: 1470–1480.

Tanigushi, Y., and Nomura, T. (1968) Fine structure of retinal blood vessels in human diabetics. Acta Soc. Ophthalmol. Japan. 72: 1165–1178.

Tomita, T., and Ynagida, T. (1981) Origins of the ERG waves. Visions Res. 21: 1703–1707.

Yonemura, D., Aoki, T., and Tsuzuki, K. (1962) Electroretinogram in diabetic retinopathy. Arch. Ophthal. 68: 49–54.

Yonemura, D. (1977) An electrophysiological study on activities of neuronal and non-neuronal retinal elements in man with reference to its clinical application. Acta Soc. Ophthalmol. Japan. 81: 1632–1665.

Free Radicals and Aging
ed. by I. Emerit & B. Chance

Active oxygen species, articular inflammation and cartilage damage

Y. Henrotin[a], G. Deby-Dupont[b], C. Deby[b], P. Franchimont[a] and
I. Emerit[c]

[a]Laboratory of Radioimmunoassay, [b]Centre interdisciplinaire de Biochimie normale et
pathologique de l'Oxygène, University Sart-Tilman, Pathology Tower B23 and B35,
B-4000 Liège, Belgium, and [c]University Paris, Centre Biomédical des Cordeliers Free
Radical Research Group, 15 rue de l'Ecole de Médecine, F-75006 Paris, France

Summary. Rheumatoid arthritis and osteoarthritis are age-related diseases, in which degenerative changes (arthrosis) and superimposed inflammatory reactions (arthritis) lead to progressive destruction of the joints. Active oxygen species derived from various sources play a role
in this process, which may be influenced by appropriate treatment with antioxidants and free
radical scavengers.

Introduction

Inflammatory joint disease has been intensively studied with respect
to oxygen free radical involvement both in animal models and humans.
The antiinflammatory effect of the free radical scavenging enzyme,
superoxide dismutase (SOD), had already been used under the trade
name Palosein for joint disease in horses before the enzymatic nature of
this metallo-protein was described by McCord and Fridovich in 1969.
Indeed comparison of the analytical data obtained by these authors for
SOD with those of Huber and Saifer (1977), who had isolated an
antiinflammatory protein from bovine liver in 1965 and developed it for
drug use, allowed the conclusion that the two proteins were identical.
The beneficial effects observed in veterinary medicine were probably the
reason why osteoarthritis and rheumatoid arthritis were among the first
human diseases for which antioxidant therapy was introduced.

Regrettably many studies in the literature on free radical reactions in
inflammatory joint disease do not clearly distinguish between these two
diseases, as defined by the American Rheumatology Association. Both
diseases increase in frequency with age and may therefore be considered
as age-related diseases. The relative contribution of degenerative
changes and reactive inflammatory responses is complex, but one may
suppose that antioxidants ameliorate superimposed inflammation rather
than the underlying degenerative process. In contrast to other age-
related diseases such as cardiovascular disease and cancer, joint disease

is not life-threatening. However it can be disabling and the cause of considerable suffering for the elderly, and it seems worthwhile to evaluate the role of active oxygen species in this process with the hope of amelioration by antioxidant treatment.

Constituents and properties of cartilage

An important function of articular cartilage is to provide an optimal contact of bone surfaces during articular joint movement and to reduce friction to an acceptable level (Broom et al., 1990). Cartilage exerts this function by its mechanical properties which are ascertained by two distinct but interacting components: the proteoglycans and the collagen fibrils. The extracellular matrix of articular cartilage is composed of a network of collagen fibers which are embedded in an hyperhydrated gel, rich in proteoglycans, glycoproteins and other minor constituents (Heineguard et al., 1989). The high concentration of charged sulfate groups in proteoglycans gives a large swelling pressure by Donnan effect and renders the cartilage resistant to the loading imposed by skeletal movement (Kempson, 1980). Type II collagen, the main fibrillar component, is responsible for resistance to tensile stress (Eyre et al., 1987). The structural integrity of human cartilage is maintained by the equilibrium between synthesis and degradation of these matrix constituents (Dingle, 1991). This equilibrium may be disturbed as a consequence of aging (Hammermann et al., 1989) or under various pathologic conditions, which either increase the rate of degradation or decrease the synthesis of these matrix components. Proteinases (Dean, 1991; Shapiro et al., 1991), growth factors (Castor et al., 1987), prostaglandins (Henrotin et al., 1991) and cytokines (Gilman et al., 1987; Towle et al., 1987; Yaron et al., 1989; Martel-Pelletier et al., 1991; Shimei et al., 1991) are involved in this complicated interplay, in which not only inflammatory cells, but also synovium and chondrocytes themselves take part (Steinberg 1991). We will show in the following that active oxygen species are also implicated.

Sources of active oxygen species

Release by inflammatory cells

Besides proteases and inflammatory mediators, stimulated neutrophils release reactive oxygen species into the surrounding medium with obvious potentially adverse effects on adjacent tissues as well as the necessary adverse effects on invading micro-organisms. Neutrophils are known to accumulate readily in many pathophysiological conditions, as for example the inflamed joint. The oxygen species known to be released by neutrophils (through the action of a plasma bound NADPH oxidase) are O_2^- and H_2O_2 (formed by dismutation of O_2^-) and HOCl (produced by the action of myeloperoxidase with H_2O_2) (Babior, 1978; Klebanoff, 1978). The formation of hydroxyl radicals by interaction of O_2^- and H_2O_2 is postulated through the metal-catalyzed Haber-Weiss reaction or Fenton-type reaction (Halliwell, 1985; Gutteridge, 1987; Buettner, 1991). The role of iron as a catalyzor of these reactions appears to be generally accepted, while a certain controversy exists as to the bioavailability of "free iron". It is noteworthy that the synovial fluid from rheumatoid arthritis patients has higher concentrations of catalytic iron salts than that of healthy persons (Blake et al., 1984). Also ferritin, from which iron can be mobilized, was found to be increased (Biemond et al., 1986).

Generation of active oxygen by xanthine oxidase

If hypoxia alone may be inhibitory for oxygen free radical production, temporary ischemia followed by reoxygenation, in contrast, can be a source of oxygen free radicals generated via the xanthine oxidase pathways. This mechanism has been proposed by others for ischemia-reperfusion injury of the gut (Granger et al., 1986) and the myocardium (McCord et al., 1985). Adenosin triphosphate is progressively degraded during ischemia to yield hypoxanthine, which then acts as a substrate for xanthine oxidase. This enzyme is derived from its dehydrogenase form by calcium-dependent proteases during ischemia. Upon reperfusion, molecular oxygen becomes available, and all participants are present for the reaction producing O_2^-. Recently a group of British investigators (Blake et al., 1989) suggested the possibility of an ischemia-reperfusion phenomenon in degenerative joint disease. During exercise, the pressure in the joint may reach critical levels leading to the occlusion of superficial synovial capillaries and consequently to tissue hypoxia. When rest allows the reperfusion of blood, active oxygen species are formed according to the above-mentioned mechanism. The presence of xanthine dehydrogenase activity in human synovium at low but detectable levels (Allen et al., 1987) and of increased concentrations of hypoxanthine, xanthine and urate in synovium and plasma of patients with rheumatoid arthritis and osteoarthritis (Gubjorsson et al., 1991) are in favor of this hypothesis. Formation of oxygen free radicals by chondrocytes, exposed to anoxia-reoxygenation in vitro (Henrotin et al., 1991), is further evidence in favor of such a mechanism and will be discussed in more detail in the following paragraph.

Generation of active oxygen species by chondrocytes

Recent studies suggest that rabbit, mouse or bovine chondrocytes are potent sources of oxygen free radicals. It could be shown that human and bovine immunoglobulin induces superoxide production by bovine chondrocytes (Cooke et al., 1991). Normal chondrocytes from rabbit cartilage released hydrogen peroxide in greater amounts than pulmonary macrophages (Tiku et al., 1988; 1990). Low levels of spontaneous hydrogen peroxide production were increased by phorbol myristate acetate (PMA), but not by concanavalin A (Robertson et al., 1988). However when chondrocytes were primed with IFN gamma or TNF, concavalin A stimulation resulted in higher H_2O_2 values than in controls (Tiku et al., 1988). Other authors have confirmed the role of cytokines in this process (Iwasaki et al., 1988; Matsubara et al., 1988).

The ability of human chondrocytes to produce O_2^-, H_2O_2 and $OH^{\cdot}$ radicals could be shown by Henrotin et al. (1992) using freshly isolated chondrocytes according to previously developed methods (Franchimont et al., 1989; Bassler et al., 1991). For each sample, taken from the knee immediately after death or during total prothesis placing, the severity of cartilage pitting was recorded according to the scale of Moskowitz et al. (1991). Chondrocytes were stimulated with PMA or by a period of anoxia followed by reoxygenation.

Superoxide production was demonstrated on chondrocytes in culture using formazan blue formation from nitroblue tetrazolium. Staining was partially suppressed by pretreatment of cells with SOD.

Hydrogen peroxide production, as measured with a fluorimetric assay using dichlorofluorescin (Keston et al., 1965), was increased over control values by PMA and even more by anoxia-reoxygenation.

Hydroxyl radical generation was demonstrated with a chromatographic method (Weiss et al., 1978), based on ethylene detection. In contrast to H_2O_2 production, PMA was as efficient as anoxia-reoxygenation for $OH^{\cdot}$ production. The combination of both stimulants yielded higher values than each stimulant alone. Anoxia lasting 4 h yielded higher values for OH radical production than anoxia lasting 12 h. The effects of PMA stimulation were significantly decreased by catalase (5 mM), demonstrating the role of H_2O_2 in the generation of hydroxyl radicals.

In contrast to a report by Tiku et al. (1990), release of active oxygen species was observed not only with chondrocytes in adherence, but also with cells maintained in suspension by stirring. Active oxygen species were detected inside the cells as well as in the supernatants. Therefore one may assume that chondrocytes can participate actively in cartilage destruction via the intermediacy of oxygen-derived free radicals.

Oxygen free radical-related damage to macromolecules

Synovial fluid

Hyaluronic acid is the dominant macromolecule in synovial fluid accounting for its viscosity. Its depolymerization is accompanied by loss of viscosity. This could be demonstrated in vitro and in vivo after exposure to various free radical generating systems. McCord (1974) reported that enzymatically generated superoxide radical depolarized purified hyaluronic acid from bovine synovial fluid by reacting with hydrogen peroxide to produce the hydroxyl radical. More recently, the reactivity of free radicals with hyaluronic acid was studied with the pulse radiolysis technique by Myint et al. (1987). Superoxide production by phagocytosing polymorphonuclear leukocytes was suggested to be responsible in vivo for synovial fluid degradation in inflamed joints (Greenwald et al., 1986; Nurcombe, 1991; Saari, 1991). The end product of the degradation process in vitro is a polymer of about 10,000 Da. In agreement herewith, synovial fluid from inflamed joints contains hyaluronate of reduced molecular weight (McNeil et al., 1985).

Cartilage

Oxygen free radicals may induce damage in all matrix components. Many in vitro studies have investigated the degradation of cartilaginous tissue slices by active oxygen species generated with the xanthine-xanthine oxidase system. Damage was supposed to be due to direct attack of proteoglycans and collagen fibers by free radicals or to occur indirectly via free radical mediated activation of latent collagenase (Burkhardt et al., 1986). Because of the combined protection given by SOD and catalase, these authors proposed the OH$^{\cdot}$ radical as the ultimate damaging species. Kowanko et al. (1989), on the other hand, suggested that hypochlorous acid from neutrophils attacks the cartilage together with lysosomal enzymes.

Proteoglycans form aggregates with hyaluronic acid. They are important for the mechanical properties of cartilage. Effects of oxygen-derived reactive species on proteoglycan synthesis (Schalwijk et al., 1985) and proteoglycan-hyaluronate aggregates were studied by Bates et al. (1984; 1985). When human neonatal cartilage is exposed to H_2O_2 proteoglycan monomers lose their ability to interact with hyaluronic acid. This is due to fragmentation of the link protein and hyaluronic acid cleavage (Roberts et al., 1987; 1989). On the other hand, the side chains of chondroitin sulfate and keratan sulfate can be a trap for free radicals before either protein or carbohydrate are attacked.

Also the damaging effect of active oxygen species on collagen was intensively studied. Collagen in solution can be degraded by exposure to ozone or Fenton-type reactions, suggesting the role of hydroxyl radicals in this process (Curran et al., 1984). Collagen from calf skin was degraded by oxyradicals generated using the xanthine-xanthine oxidase system (Monboisse et al., 1984). These authors, using pulse and gamma radiolysis, concluded that superoxide was responsible for the degradation of collagen (Monboisse et al., 1988). Peptides smaller than alpha chains were excised from the collagen. The cleavage occurred in the helical region, as concluded from the 4-hydroxyproline content specific for this part of the molecule. Release of hydroxyproline, concomitant with uronic acid, was also observed by Burkhardt et al. (1986). The degradation of the fibrils, initiated via free radical mechanisms, is completed by non-specific protease digestion.

The action of active oxygen species on collagen is treated in more detail in another chapter of this book.

Lipids

Hydroxyl radicals react with unsaturated fatty acids of membrane lipids to initiate chain reactions, resulting in the generation of other, more long-lived radical species and in a variety of lipid peroxidation products. An increase in malondialdehyde and substances with conjugated diene structure is observed in plasma and synovial fluid of arthritis patients. Patients with rheumatoid arthritis exhale increased amounts of pentane. The pentane exhaled is directly proportional to the disease activity (Humad et al., 1986).

The most important unsaturated fatty acid in membranes is arachidonic acid, which is released by phospholipase activity and metabolized via the cyclo- and the lipoxygenase pathway to prostaglandins and leukotrienes. Phospholipase A2 activity was found to be increased in monocytes, serum and synovial fluid of patients with rheumatoid arthritis and osteoarthritis (Bomalaski et al., 1986; Pruzanski et al., 1985). Lipid peroxidation was also observed in rat adjuvant arthritis and inhibited by superoxide dismutase (Yoshikawa et al., 1985).

DNA

Oxidative damage to DNA results in strand breakage and base damage, histologically expressed as chromosome breakage and sister chromatid exchange (Emerit et al., 1982). An increase in structural chromosome aberrations and sister chromatid exchanges has been reported for patients with rheumatoid arthritis by several laboratories

(Emerit et al., 1974; Kong-oo Goh and Jacox, 1982; Vincent et al., 1986). The chromosome damage seen in dividing blood cells of patients is the consequence of circulating DNA-damaging material, called clastogenic factor (CF) (Emerit, 1982). CF may be isolated from patients' plasma, synovial fluid and from cell culture supernatants. When media of lymphocyte cultures from patients with rheumatoid arthritis were examined for clastogenic activity, addition of increasing numbers of monocytes to the culture system resulted in increasing clastogenic activity. Addition of neutrophils did not considerably influence the production of CF. No CF was found in culture media, when monocytes were removed from the mononuclear cells by adherence.

In patients with osteoarthritis, CF was not present in plasma or cell cultures, but the synovial fluid from inflamed osteoarthritic knees was clastogenic.

Similar CF can be observed in other chronic inflammatory diseases or after irradiation, where its formation is also related to superoxide radical generation (Emerit, 1982; 1986; 1988). CF can be induced in vitro by exposure of blood cultures to superoxide generated with PMA or with a xanthine-xanthine-oxidase system (Khan and Emerit, 1985; Emerit et al., 1985). This is prevented by simultaneous treatment of the cells with superoxide dismutase. Not only CF formation, but also CF action is mediated via superoxide, since the clastogenic effect on normal cells can be prevented by SOD.

Besides their clastogenic effect, CF stimulates superoxide production by neutrophils and monocytes. This is also inhibitable by SOD (Emerit et al., 1990). The fact that CF stimulates superoxide production and is formed via superoxide results in a vicious circle, which may be responsible for the persisting prooxidant state.

Preliminary results of biochemical studies indicate that CF contains clastogenic lipid peroxidation products such as 4-hydroxynonenal (Emerit et al., 1991), but also clastogenic nucleotides such as inosine di- and triphosphate (Auclair et al., 1990), which may be derived from nucleotid pool imbalances.

Oxygen toxicity defense mechanisms

The defense mechanisms against oxygen toxicity consist in an enzymatic and a nonenzymatic detoxication. The possibility of insufficient defense mechanisms was investigated with controversial results. SOD levels were found to be deficient in leukocytes of patients with juvenile arthritis (Rister et al., 1978). All antioxidant enzymes were decreased in erythrocytes from rheumatoid arthritis according to Imadaya et al. (1988). Other authors did not find significant differences for erythrocyte

SOD between patients and controls (Scudder and Dormandy, 1976; Pasquier et al., 1984); nor did they observe increased SOD activity in neutrophils and monocytes of these patients (Youssef and Baron, 1983).

The results were also discordant for the synovial fluid of rheumatoid arthritis patients (Igari et al., 1982; Biemond et al., 1984). In contrast to SOD, catalase and glutathione peroxidase were found to be increased. Blake et al. (1984) noted only trivial amounts of SOD and catalase.

The reasons for the observed discrepancies may be multiple and related to choice of patients and influence of treatment.

While many studies deal with antioxidant enzymes in red and white blood cells, very little is known about chondrocytes. According to Baker et al. (1988), chondrocytes contain catalase and glutathione peroxidase. During our investigations mentioned above, chondrocytes homogenized by Potter treatment removed 70% of the H_2O_2 to which they had been exposed, while the corresponding cultures supernatants removed only 35%. Catalase activity could be inhibited by 94% with 3-aminotriazole and by 100% with azide.

Among the nonenzymatic defenses, antioxidant vitamin levels (C and E) were significantly lower in sera of rheumatoid arthritis patients (Situnayake et al., 1991). In addition, selenium was reduced in serum of severe cases of rheumatoid arthritis (Tarp et al., 1985). According to O'Dell et al. (1991), selenium levels were lower in patients with active disease than in those with quiescent disease. Interestingly, serum selenium concentrations in patients with osteoarthritis were significantly lower than those from seropositive RA and controls. This could not be explained by the age of the respective study groups.

Imbalances in antioxidant levels seem to be the consequence rather than the cause of the prooxidant state in arthritis patients.

Antioxidant therapy

When the antioxidant defense system is overwhelmed, oxygen free radicals induce tissue damage. This represents the rationale for antioxidant treatment.

As already mentioned, the beneficial effects of SOD in joint disease of horses (Huber and Saifer, 1977) may have determined the choice of rheumatoid arthritis and osteoarthritis as the first human diseases to be treated with this antioxidant enzyme. The first pilot study with bovine Cu-Zn SOD was published by Lund-Olesen and Menander in 1974. They injected 2-3 mg SOD intraarticularly into the synovial cleft of patients suffering from active osteoarthritis of the knee joint. After three injections, 16 out of 19 patients improved, and the remission appeared to last for about 3 months after cessation of treatment. The results of this open uncontrolled study were in the following confirmed by randomized, double-blind trials conducted separately on groups of patients from Scandinavia, Great Britain and Central Europe (for review see Flohé, 1988). Similar doses (2–4 mg) of SOD or placebo were injected intraarticularly at weekly or biweekly intervals. The drug was found to be superior to placebo for clinically relevant parameters like pain, functional improvement in terms of walking ability and climbing stairs, joint circumference etc. In another study SOD was compared to intraarticular corticoid injection (Gammer and Broback, 1984). While all patients continuously improved during the treatment period (1 injection every second week up to the sixth week), the corticoid group deteriorated rapidly after cessation of treatment, whereas the therapeutic effect of SOD persisted until the end of the follow up period (up to 6 months). It was concluded from these trials that SOD ameliorates the symptoms of osteoarthritis when applied locally. Furthermore, a recent study conducted by McIlwain et al. (1989) yielded satisfactory results in osteoarthritic patients receiving injections into the affected joint. Again there were few clinical differences in the beginning of the treatment, but in later periods, after cessation of the injections, clinical efficiency became apparent.

As expected, no difference in the radiological appearance of placebo- and drug-treated joints was seen. SOD acts on the superimposed inflammation rather than on the underlying degenerative process of osteoarthritis. However, inflammation-dependent tissue destruction may be influenced and therefore progress of the disease may be retarded.

In contrast to osteoarthritis, the results for rheumatoid arthritis were less encouraging. With regard to the general character of the disease process, systemic treatment with intramuscular injection of 8 mg, 3 or 4 times weekly was chosen. The therapeutic benefits were small and reached statistical significance only for certain parameters (Menander-Huber, 1980). They could not be confirmed by others (Camus et al., 1980; Flohé, 1988; Greenwald, 1991).

Prolonged amelioration obtained by local injection of SOD into the inflamed joint has been observed in more than one study. Action mechanisms other than the classical notion that exogenous SOD dismutates extracellularly released superoxide have been discussed in recent years (Michelson et al., 1987; Vaille et al., 1990). Since there was no correlation between the degree of anti-inflammatory activity and the level of circulating exogenous SOD, the action mechanism was thought to be indirect and independent of the maintenance of high extracellular concentrations. These authors speculated that part of the injected SOD remains attached to semispecific sites on external cell membranes.

Although the safety of SOD has been stressed, a risk for immunological complications cannot be ruled out, in particular for long-term treatments. Therefore the development of so-called SOD mimicks, restricted to the active site of the enzyme without the complete protein, seemed to be of interest. Copper salicylate alleviated symptoms in patients with rheumatoid arthritis (Sorenson, 1976). D-Penicillamine, a drug supposed to have SOD-like properties, similar to those of copper-penicillamine (Younes and Weser, 1979), was found to be beneficial in a series of a hundred patients (Camus et al., 1974). Other action mechanisms have been proposed for this drug, such as improvement of the glutathione peroxidase/reductase system by increasing the reduced glutathione pool within the cell (Lipsky et al., 1989).

Among the antioxidant vitamins, alpha- tocopherol given at a dose of 400 IU/day over a period of 6 weeks allowed to reduce the intake of non-steroidal antiinflammatory drugs (Schmidt and Bayer, 1990). NSAIDS are at present the most commonly used treatment for the symptomatic relief of arthritis patients. They are inhibitors of cyclo- and lipoxygenases, but some have also free radical scavenging properties or inhibit superoxide production by neutrophils (Kaplan et al., 1984, Biemond et al., 1986, Minta et al., 1986, Montecucco et al. 1990). Also steroids (Goldstein et al. 1976) and gold (Davies et al., 1985; Parente et al., 1986) may suppress superoxide production by neutrophils.

In contrast to these reports concerning neutrophils, Hurst et al. (1984) reported that the use of disease remittive drugs such as corticoids, penicillamine and gold is associated with enhanced superoxide radical generation by monocytes of rheumatoid arthritis patients.

Contradictory reports exist also for selenium. Patients with rheumatoid arthritis treated with selenium-enriched yeast (256 μg of selenium per day) for 6 months showed no improvement despite normalization of their selenium status. On the other hand, supplementation with selenium selenite (160 μg/day) (Munthe et al., 1980) or selenomethionine (100 μg/day) (Meltzer et al., 1985) resulted in amelioration in 40% of the patients. No improvement was observed in a group of patients with osteoarthritis (Hill et al., 1990), who have even lower selenium levels than rheumatoid arthritis patients (see above).

Because of the promoting role of iron for hydroxyl radical generation, desferroxamine, a drug used in iron overload, was tried as a treatment for rheumatoid diseases, but proved to have toxic side effects (Blake et al., 1985).

Conclusion

The data reported in this review provide a strong rationale for the therapeutic use of antioxidants. The variety of drugs proposed for the treatment of rheumatoid arthritis and osteoarthritis indicates that there is no perfect solution. Most anti-inflammatory drugs have side effects, which become more and more important with duration of treatment. Since articular pain often starts before the age of sixty, the treatment of

318

these patients is a long-term strategy. Antioxidants such as vitamin E known for its good tolerance, or SOD known for its sustained effect after intraarticular injection, may prolong the pain-free intervals and reduce the need for analgetics.

Allen, R. E., Outhwaite, J., Morris, C. J., and Blake, D. R. (1987) Xanthine oxidoreductase is present in human synovium. Ann. Rheum. Dis. 46: 843–45.

Auclair, C., Gouyette, A., Levy, A., and Emerit, I. (1990) Clastogenic inosine nucleotides as components of the chromosome breakage factor in scleroderma. Arch. Biochem. Biophys. 278: 238–244.

Babior, B. M. (1978) Oxygen-dependent microbial killing by phagocytes. N. Engl. J. Med. 298: 203–209.

Baker, M. S., Feigan, J., and Lowther, D. A. (1988) Chondrocyte anti-oxydant defences: the role of catalase and glutathione peroxidase in protection against H_2O_2 dependent inhibition of proteoglycan biosynthesis. J. Rheumatol. 15: 4: 670–677.

Baker, M. S., Feigan, J., and Lowther, D. (1989) The mechanism of chondrocyte hydrogen peroxide damage. Depletion of intracellular ATP due to suppression of glycolysis caused by oxidation of glyceraldehyde-3-phosphate deshydrogenase. J. Rheumatol. 16: 7–14.

Bassleer, C., Gysen, P., Foidart, J. M., Bassleer, R., and Franchimont, P. (1986) Human chondrocytes in tridimensional culture. In Vitro 22: 115–120.

Bates, E. J., Harper, G. S., Lowther, D. A., et al. (1984) Effect of oxygen derived reactive species on cartilage proteoglycan-hyaluronate aggregates. Biochem. Int. 8: 629–637.

Bates, E. J., Johnson, C. C., and Lowther, D. A. (1985) Inhibition of proteoglycan synthesis by hydrogen peroxide in cultured bovine articular cartilage. Biochem. Biophys. Acta 838: 221–228.

Biemond, P., Swaak, A. J., and Koster, J. F. (1984) Protective factors against hydrogen peroxide in rheumatoid arthritis synovial fluid. Arthr. Rheum. 27: 760–765.

Biemond, P., Swaak, A. J., Elijk, H. G., et al. (1986) Intraarticular ferritin bound iron in rheumatoid arthritis. Arthr. Rheum. 29: 1187–1193.

Blake, D. R., Gallagher, P. J., Potter, A. R., et al. (1984) The effect of synovial iron on the progression of rheumatoid disease: an histologic assessment of patient with early rheumatoid synoviotis. Arth. Rheum. 27: 495–501.

Blake, D. R., Winyard, P., and Lunec, J. (1985) Cerebral and ocular toxicity induced by desferrioxamine. Quart. J. Med. 56: 345–355.

Blake, D. R., Unsworth, J., Outhwaite, J. M., Morris, C. J., Merry, P., Kido, B. L., Ballard, R., Gray, L., and Lunec, J. (1989) Hypoxic-reperfusion injury in the inflamed human joint. Lancet Feb 11: 290–293.

Bomalaski, J. S., Hirata, F., and Clark, M. (1986) Aspirin inhibits phospholipase c. Biochem. Biophys. Res. Commun. 139: 115–121.

Broom, N. D., and Silyn-Roberts, H. C. (1990) Collagen-collagen versus proteoglycan interactions in the determination of cartilage strength. Arthr. Rheum. 33: 1512–1517.

Buettner, G. R., and Chamulitrat, W. (1990) The catalytic activity of iron in synovial fluid as monitored by the ascorbate free radical. Free Radical Biol. Med. 8: 55–56.

Burkhardt, H., Schwingel M., Meninger, H., McCartney, H. W., and Tschesche, H. (1986) Oxygen radicals as effectors of cartilage destruction. Direct degradative effect on matrix components and indirect action via activation of latent collagenase from polymorphonuclear leukocytes. Arth. Rheum. 29, 3: 379–387.

Camus, J. P., Crouzet, J., Benichou, Ch., and Lièvre, J. A. (1974) Cent cas de polyarthrite rhumatoide traités par la D-penicillamine. Ann. Int. Med. 125: 9–18.

Camus, J. P., Emerit, I., Michelson, A. M., Prier, A., Koeger, A. C., and Merlet, C. (1980) Superoxyde dismutase et polyarthrite rhumatoide. Rev. Rhumat. 47: 489–492.

Castor, C. W., Ragsdale, C. G., Cabral, A. R., Smith, E. M., and Aaron B. P. (1987) Anabolic and catabolic responses of human articular cells to growth factors. J. Rheumatol. (suppl. 14) 14: 67–69.

Cooke, D., Saura, S., Uno, R., and Scuddamore, A. (1991) Interaction of immunoglobulins and chondrocytes. J. Rheumatol. (suppl. 27) 18: 114–116.

Curran, S. F., Amusoro, M. A., Goldstein, B. D., et al. (1984) Degradation of soluble collagen by ozone or hydroxyl radicals. FEBS Lett. 176: 155–160.

Davis, P., Johnston, C., Bertouch, J., et al. (1987) Depressed superoxide radical generation by neutrophils from patients with rheumatoid arthritis and neutropenia: correlation with neutrophil reactive IgG. Ann. Rheum. Dis. 46: 51–54.

Dean, D. D. (1991) Proteinase-mediated cartilage degradation in osteoarthritis. Sem. Arthr. Rheum. 20, 6 (suppl. 2): 2–11.

Dingle, J. T. (1991) Cartilage maintenance in osteoarthritis: interaction of cytokines, NSAID and prostaglandins in articular cartilage damage and repair. J. Rheumatol. (suppl. 28) 18: 30–37.

Emerit, I., Feingold, J., Camus, J. P., and Housset, E. (1974) Etude chromosomique des maladies du collagène. Ann. Génét. 17: 251–256.

Emerit, I. (1982) Chromosome breakage factors: origin and possible significance. Progr. Mutat. Res. 4: 61–74.

Emerit, I., Kech, M., Levy, A., Feingold, J., and Michelson, A. M. (1982) Activated oxygen species at the origin of chromosome breakage and sister-chromatid exchanges. Mutat. Res. 103: 165–177.

Emerit, I., Khan, S. H., and Cerutti, P. (1985) Treatment of lymphocyte cultures with a hypoxanthine- xanthine oxidase system induces the formation of transferable clastogenic material. Free Rad. Biol. Med. 1: 51–57.

Emerit, I. (1986) Oxygen-derived free radicals and DNA damage in autoimmune diseases, in: Free Radicals, Aging and Degenerative Diseases, pp. 327–334. Ed. J. E. Johnson. Alan R. Liss Inc., New York.

Emerit, I. (1988) Increased oxyradical formation in connective tissue disease, in: Cellular Antioxidant Defense Mechanisms, pp. 112–121. Ed. C. K. Chow. CRC Press, Boca Raton.

Emerit, I., Levy, A., and Camus, J. P. (1988) Monocyte-derived clastogenic factor in rheumatoid arthritis. Free Rad. Biol. Med. 6: 245–250.

Emerit, I., Levy, A., and Khan, S. H. (1990) Superoxide generation by clastogenic factors, in: Free Radicals, Lipoproteins, and Membrane Lipids, pp. 99–104. Eds Crastes de Paulet et al. Plenum Press, New York.

Emerit, I., Khan, S. H., and Esterbauer, H. (1991) Hydroxynonenal, a component of clastogenic factors? Free Rad. Biol. Med. 10: 371–377.

Eyre, D. R., Apon, S., Wu, J. J., Ericsson, L. H., et al. (1987) Collagen type IX: evidence for covalent linkages to type II collagen in cartilage. FEBS Lett. 220: 337–341.

Flohé, L. (1988) Superoxide dismutase for therapeutic use: Clinical experience, dead ends and hopes. Mol. Cell. Biochem. 84: 123–131.

Franchimont, P., Bassleer, C., and Henrotin, Y. (1989) Effects of hormones and drugs on cartilage repair. J. Rheumatol. 18: 5–9.

Franchimont, P., Bassleer, C., Henrotin, Y., Gysen, P., and Bassleer, R. (1989) Effects of human and salmon calcitonin on human articular chondrocytes cultivated in clusters. J. Clin. Endocr. Met. 69, 2: 259–266.

Gammer, W., and Broback L. G. (1984) Clinical comparison of orgotein and methylprednisolone acetate in the treatment of osteoarthrosis of the knee joint. Scand. J. Rheumatol. 13: 108–112.

Gilman, S. C. (1987) Activation of rabbit articular chondrocytes by recombinant human cytokines. J. Rheumatol. 14: 1002–7.

Goldstein, I. M., Roos, D., Weissman, G., et al. (1976) Influence of corticosteroids on human polymorphonuclear leukocyte function in vitro. Inflammation 1: 301–315.

Granger, D. N., Hollworth, M. E., and Parks, D. A. (1986) Ischemia-reperfusion injury: role of oxygen-derived free radicals. Acta Physiol. Scand. 548: 47–63.

Greenwald, R. A., and Moak, S. A. (1986) Degradation of the hyaluronic acid by polymorphonuclear leukocytes. Inflammation 10: 15–30.

Greenwald, R. A. (1991) Oxygen radicals, inflammation and arthritis: pathological considerations and implications for treatment. Sem. Arthr. Rheum. 20, 4: 219–240.

Gubjornsson, B., Zak, A., Niklasson, F., and Hallgren, R. (1991) Hypoxanthine, xanthine and urate in synovial fluid from patients with inflammatory arthritis. Ann. Rheum. Dis. 50: 669–672.

Gutteridge, J. C. (1987) Bleomycin detectable iron in knee joint synovial fluid arthritis patients and its relationship to the extracellular oxidant activities of ceruloplasmin, transferrin and lactoferrin. Biochem. J. 245: 415–421.

320

Halliwell, B., Gutteridge, J. M. C., and Blake, D. C. (1985) Metal ions and oxygen radical reactions in human inflammatory joint disease. Phil. Trans. R. Soc. Lond. B 311: 659–671.

Hammerman, D. (1989) The biology of osteoarthritis. N. Engl. J. Med. 320: 1322–30.

Heineguard, D., and Oldberg, A. (1989) Structure and biology of cartilage and bone matrix noncollagenous macromolecules. FASEB J. 3: 2042–2051.

Henrotin, Y., Bassleer, C., Reginster, J. Y., Bassleer, R., and Franchimont, P. (1989) Effects of Etodolac on human chrondrocytes cultivated in three dimensional culture. Clin. Rheumatol. (suppl. 1) 8: 36–42.

Henrotin, Y., Deby-Dupont, G., Deby, C., Debruin, M., Lamy, M., and Franchimont, P. (1992) Production of active oxygen species by isolated human chondrocytes. British J. Rheum. (in press).

Hill, J., Birdha, D. (1990) Failure of selenium ACE to improve osteoarthritis. Br. J. Rheumatol. 29: 211–213.

Huber, W., and Saifer, M. G. P. (1977) Orgotein, the drug version of bovine Cu-Zn superoxide dismutase, in: Superoxide and Superoxide Dismutase, pp. 517–536. Eds A. M. Michelson et al. Academic Press, London.

Humad, S., Zarling, M., Clapper, M., and Skosey, J. L. (1986) Breath pentane excretion as a marker of disease activity in rheumatoid arthritis. Free Rad. Res. Commun. 5: 101–106.

Hurst, N. P., Bessac, B., and Nuki, G. (1984) Monocyte superoxide anion production in rheumatoid arthritis: preliminary evidence for enhanced rates of superoxide production by monocytes from patients receiving penicillanine, sodium aurothionalate and corticosteroids. Ann. Rheum. Dis. 43: 27–33.

Igari, T., Kaneda, H., Horiuchi, S., and Ono, S. A. (1982) A remarkable increase of superoxide dismutase activity in synovial fluid of patients with rheumatoid arthritis. Clin. Orthop. 162: 282–287.

Imadaya, A., Terasawa, K., Tosa, R., et al. (1988) Erythrocyte antioxydant enzymes are reduced in patients with rheumatoid arthritis. J. Rheumatol. 15: 1968–1974.

Iwasaki, Y., Matsubara, T., and Hirohata, K. (1988) A mechanism of cartilage destruction in immunologically-mediated inflammation: increased superoxide anion release from chondrocytes in response to interleukin-1 and interferons. Orthop. Trans. 12: 438–444.

Kaplan, H. I., Edelson, H., Korchak, H. I., Given, W., Abramson, S., and Weissman, G. (1984) Effects of non-steroidal anti-inflammatory agents on human neutrophil functions in vitro and in vivo. Biochem. Pharmacol. 33: 371–378.

Kempson, G. E. (1980) The mechanical properties of articular cartilage, in: The Joints in Synovial Fluid, pp. 215–290. Ed. L. Sokoloff. Pitman, London.

Keston, A. S., and Brandt, R. (1965) The fluorimetric analysis of ultramicro quantities of hydrogen peroxide. Analyt. Biochem. 11: 1–5.

Khan, S. H., and Emerit, I. (1985) Lipid peroxidation product and clastogenic material in culture media of human leukocytes exposed to the tumor promoter phorbol myristate acetate. Free Rad. Biol. Med. 1: 443–449.

Klebanoff, S. J., and Clark, R. A. (1978) The Neutrophil: Function and Chemical Disorders. North-Holland, Amsterdam.

Kong-oo Goh, and Jacox, R. F. (1982) Chromosomes in rheumatoid arthritis. Ann. Rheum. Dis. 41: 644–646.

Kowanko, I. C., Bates, E. J., and Ferrante, A. (1989) Mechanisms of human neutrophil mediated cartilage damage in vitro: the role of lysosomial enzymes, hydrogen peroxide and hypochlorous acid. Immunol. Cell. Biol. 67: 321–329.

Lipsky, D. E. (1989) Mechanisms of action of slow acting drugs in rheumatoid arthritis. Clin. Exp. Rheumatol. 7/S-3: 177–180.

Lund-Olesen, K., and Menander, K. B. (1974) Orgotein, a new anti-inflammatory metalloprotein drug: preliminary evaluation of clinical efficacy and safety in degenerative joint disease. Curr. Ther. Res. 16: 706–717.

Martel-Pelletier, J., Zafarullah, M., Kodama, S., and Pelletier, J. P. (1991) In vitro effects of interleukin-1 on the synthesis of metallo-proteases, TIMP, plasminogen activators and inhibitors in human articular cartilage. J. Rheumatol. (suppl. 27) 18: 80–84.

Matsubara, T., Chin, T., and Hirohata, K. (1988) Mechanism of cartilage destruction in immunologically mediated inflammation increase superoxide anion release from chondrocytes in response to interleukin-1 and interferon gamma. Arthr. Rheum. 31, 4: 533.

McCord, J. M., and Fridovich, I. (1969) Superoxide dismutase, an enzymic function for erythrocuprein (Hemocuprein). J. Biol. Chem. 244: 6049–6055.

McCord, J. M. (1974) Free radicals and inflammation: protection of synovial fluid by superoxide dismutase. Science 185: 529–531.

McCord, J. M. (1985) Oxygen-derived free radicals in post-ischemic tissue injury. New Engl. J. Med. 312: 159–161.

McNeil, J. D., Wiebkino, W., Betts, W. H., et al. (1985) Depolymerisation products of hyaluronic acid after exposure to oxygen derived free radicals. Ann. Rheum. Dis. 44: 780–789.

McIlwain, H., Silverfield, J. C., Cheatum, D. E., et al. (1989) Intra-articular orgotein in osteoarthritis of the knee: a placebo controlled efficacy, safety and dosage comparison. Am. J. Med. 87: 295–300.

Meltzer, H. M., and Myskja, A. (1985) Selenium supplementation to rheumatics. Who responds? Abstracts of the Proceedings "Metabolisms of trace elements related to human disease", p. 36. Symposium Loen, Norway.

Menander-Huber, K. (1980) Double-blind controlled clinical trials in man with bovine copper-zinc superoxide dismutase (Orgotein); in: Biological and Clinical Aspects of Super-oxide and Superoxide Dismutase. pp. 408–423. Eds W. H. Bannister, and J. V. Bannister. Elsevier, New York.

Michelson, A. M., Puget, K., and Jadot, G. (1986) Antiinflammatory activity of superoxide dismutases: comparison of enzymes from different sources in different models in rats: mechanism of action. Free Rad. Res. Commun. 2: 43–56.

Minta, J. O., and William, M. P. (1986) Interactions of antirheumatic drugs with the superoxide generation system of activated human polymorphonuclear leukocytes. J. Rheumatol. 13: 498–504.

Monboisse, J. C., Braquet, P., and Borel, J. P. (1984) Oxygen-free radicals as mediators of collagen breakage. Agents Actions 15: 1–2.

Monboisse, J. C., Gardes-Albert, M., Randoux, A., et al. (1988) Collagen degradation by superoxide anion in pulse and gamma radiolysis. Biochem. Biophys. Acta 965: 29–35.

Montecucco, C., Mazcone, A., Pasotti, D., and Fratino, P. (1990): Effect of piroxicam therapy on granulocyte function elastase concentration in peripheral blood and synovial fluid in rheumatoid arthritis patients. Inflammation 26: 152–158.

Moskowitz, R. W., Reese, J. H., Young, G. R. et al. (1980) The effects of Rumalon, a glycosaminoglycan peptide complex in a partial menisectomy model of osteoarthritis in rabbits. J. Rheumatol. 18, 2: 205–209.

Munthe, E., Aaseth, J., and Jellun, E. (1986) Trace elements and rheumatoid arthritis. Acta Pharmacol. Toxicol. 59: 365–373.

Myint, P., Deeble, D. J., and Beaumont, P. C. (1987) The reactivity of various free radicals with hyaluronic acid: a steady state and pulse radiolysis studies. Biochem. Biophys. Acta 925: 194–199.

Nurcombe, H. L., Buchnall, R. C., and Edwards, S. W. (1991) Neutrophils isolated from the synovial fluid of patients rheumatoid arthritis: priming and activation in vivo. Ann. Rheum. Dis. 50: 147–153.

Nurcombe, H. L., Bucknall, R. C., and Edwards, S. W. (1991) Activation of the neutrophil myeloperoxidase -H202 system by synovial fluid isolated from patients with rheumatoid arthritis. Ann. Rheum. Dis. 50: 237–242.

O'Dell, J. R., Lemley-Guillespie, S., Palmer, W. R., Weaver, A. L., Moore, G. F., and Klassen L. W. (1991) Serum selenium concentrations in rheumatoid arthritis. Ann. Rheum. Dis. 50: 376–78.

Palmer, D. G., Hogg, N., and Revell, P. A. (1986) Lymphocytes, polymorphonuclear leukocytes, macrophages and platelets in synovium involved by rheumatoid arthritis. A study with mono-clonal antibodies. Pathology 18: 431–437.

Pasquier, C., Masch, P. S., Raichvar, G. D., et al. (1984) Manganese containing superoxide dismutase deficiency in polymorphonuclear leukocytes of adults with rheumatoid arthritis. Inflammation 8: 27–32.

Pruzanski, W., Vadas, P., Stefanski, E., and Urowitz, M. B. (1985) Phospholipase A2 activity in sera and synovial fluid in rheumatoid arthritis and osteoarthritis. Its possible role as a proinflammatory enzyme. J. Rheumatol. 12: 211–217.

Rister, M., and Bauermeister, K. (1982) Superoxidedismutase and superoxide radical release in rheumatoid arthritis. Klin. Wochenschr. 60: 561–565.

Roberts, C. R., Mort, J. S., and Roughley, P. J. (1987) Treatment of cartilage proteoglycan aggregate with hydrogen peroxide. Biochem. J. 247: 349–357.

Roberts, C. R., Roughley, P. J., and Mort, J. S. (1989) Degradation of human proteoglycan aggregate induced by hydrogen peroxide. Protein fragmentation, amino acid modification and hyaluronic acid cleavage. Biochem. J. 259: 805–811.

Robertson, F. M., Liesch, J. B., Tiku, K., and Tiku, M. L. (1988) Production of reactive oxygen intermediates by rabbit articular chondrocytes. Arthritis and Rheum. (suppl. 4) 31: 984.

Saari, H. (1991) Oxygen derived free radicals and synovial fluid hyaluronate. Ann. Rheum. Dis. 50: 389–392.

Schalwijk, J., Vandenberg, W. B., Vandeputte, L., and Joosten, L. A. B. (1985) Hydrogen peroxide suppresses the proteoglycan synthesis of intact articular cartilage. J. Rheumatol. 12: 205–210.

Schmidt, K. H., and Bayer, W. (1990) Efficacy of vitamin E as a drug in inflammatory joint disease, in: Antioxidants in Therapy and Preventive Medicine, pp. 147–150. Eds I. Emerit et al. Plenum Press, New York.

Scudder, P., Stocks, J., and Dormandy, T. L. (1976) The relationship between erythrocyte superoxide dismutase activity and erythrocyte copper levels in normal subjects and in patients with rheumatoid arthritis. Clin. Chim. Acta 69: 397–403.

Shapiro, S. M., Campbell, E. J., Sewior, R. M., and Welgus, H. G. (1991) Proteinase secreted by human mononuclear phagocytes. J. Rheumatol. (suppl. 27) 18: 95–98.

Shimnei, M., Masuda, K., Kikuchi, T., Shimomura, Y., and Okada, Y. (1991) Production of cytokines by chondrocytes and its role in proteoglycan degradation. J. Rheumatol. (suppl. 27) 18: 89–91.

Situnayake, R. D., Thurnham, D. I., Kootathep, S., Chirico, S., Lunec, J., Davis, M., and McCornkey, B. (1991) Chain breaking antioxidant status in rheumatoid arthritis: clinical and laboratory correlates. Ann. Rheum. Dis. 50: 81–86.

Sorenson, J. R. (1976) Copper chelates as possible active forms of the antiarthritic agents. J. Med. Chem. 19: 135–148.

Steinberg, J. J., and Sledge, C. B. (1991) Chondrocyte mediated cartilage: regulation by prostaglandin E2, cyclic AMP and interferon α. J. Rheumatol. (suppl. 27) 18: 63–65.

Tarp, V., Overvad, K., Hansen, J. C., and Thorling, E. B. (1985) Low selenium level in severe rheumatoid arthritis. Scand. J. Rheumatol. 14: 97–101.

Tiku, M. L., Liesch, J. B., and Robertson, F. M. (1988) Chondrocytes produce oxygen radicals after treatment with concanavalin A, gamma interferon or tumor necrosis factor. Arthr. Rheumat. 131, 4 (suppl.): 569.

Tiku, M. L., Liesch, J. B. and Robertson, F. M. (1990) Production of hydrogen peroxide by rabbit articular chondrocytes. J. Immunol. 145: 690–696.

Towle, C. A., Trice, M. E., Ollivierre, F., et al. (1987) Regulation of cartilage remodeling by IL-1: evidence for autocrine synthesis of IL-1 by chondrocytes. J. Rheumatol. (suppl. 14) 14: 11–13.

Vaille, A., Jadot, G., and Elizagaray, A. (1990) Antiinflammatory activity of various superoxide dismutases on polyarthritis in the Lewis rat. Biochem. Pharmacol. 39: 247–255.

Vincent, G., Croquette, M. F., Houvenagel, E., and Lelore, E. (1986) Anomalies chromosomiques au cours de la polyarthrite rhumatoide. Taux de cassures et recherche du facteur cassant. Rev. Rhumat. 11: 625–630.

Vincent, F., Brun, H., Clain, E., Ronot, X., and Adolphe, M. (1989) Effects of oxygen-free radicals on proliferation kinetics of cultured rabbit articular chondrocytes. J. Cell. Phys. 141: 262–266.

Weiss, S. J., Rustagi, P. K., and Buglio, A. F. (1978) Human granulocyte generation of hydroxyl radical. J. Exp. Med. 147: 316–322.

Yaron, I., Meyer, F. A., Dayer, J. M., et al. (1989) Some recombinant human cytokines stimulate glycosaminoglycan synthesis in human synovial fibroblast cultures and inhibit it in human articular cartilage cultures. Arthr. Rheum. 32: 173–180.

Yoshikawa, T., Tanaka, H., and Kondo, M. (1985) The increase of lipid peroxidation in rat adjuvant arthritis and its inhibition by superoxide dismutase. Biochem. Med. 33: 320–325.

Younes, M., and Weser, U. (1977) Superoxide dismutase activity of copper-penicillamine: possible involvement of Cu(I) stabilized sulphur radical. Biochem. Biophys. Res. Commun. 78: 1247–1251.

Yousef, A., and Baroun D. (1983) Leukocyte superoxide dismutase in rheumatoid arthritis. Ann. Rheum. Dis. 42: 558–562.

Free Radicals and Aging
ed. by I. Emerit & B. Chance

Oxidative damage to collagen

J. C. Monboisse and J. P. Borel

Lab. Biochemistry, CNRS URA 610, UFR Medicine, Univ. Reims Champagne-Ardenne, 51 rue Cognacq Jay, F-51095 Reims Cedex, France

Summary. Extracellular matrix molecules, such as collagens, are good targets for oxygen free radicals. Collagen is the only protein susceptible to fragmentation by superoxide anion as demonstrated by the liberation of small 4-hydroxyproline-containing-peptides. It seems likely that hydroxyl radicals in the presence of oxygen cleave collagen into small peptides, and the cleavage seems to be specific to proline or 4-hydroxyproline residues. Hydroxyl radicals in the absence of oxygen or hypochlorous acid do not induce fragmentation of collagen molecules, but they trigger a polymerization of collagen through the formation of new cross-links such as dityrosine or disulfure bridges. Morever, these cross-links can not explain the totality of high molecular weight components generated under these experimental conditions, and the nature of new cross-links induced by hydroxyl radicals or hypochlorous acid remains unclear.

Introduction

Oxygen free radicals are involved in a number of pathological situations such as inflammation, diabetes, ocular diseases, renal diseases, skin diseases and fibrosis. They have also been implicated in aging (Halliwell, 1987). In vivo, they are usually generated during several electron transport chain reactions. In pathophysiological situations, particularly during the postischemic reperfusion, the enzyme xanthine dehydrogenase is converted to xanthine oxidase and generates high amounts of superoxide anions. Oxygen free radicals are also generated by phagocytes through the NADPH-oxidase system and are essential for the killing of bacteria or foreign substances (Morel et al., 1991). They can also be generated by the increase of intracellular oxidant formation after adding toxins (such as alloxan, paraquat or adriamycin) or after irradiation. In the particular case of diabetes mellitus, oxygen free radicals are produced during the reaction of glucose with proteins (Wolff and Dean, 1987) or by the reaction of glycated proteins with oxygen (Gillery et al., 1988).

The extracellular matrix molecules are good targets for oxygen free radicals and mainly collagen that is degraded during the first stages of inflammation by the action of both oxygen free radicals and proteinases. In turn, collagen I activates polymorphonuclear neutrophils to produce oxygen free radicals (Monboisse et al., 1987; Monboisse et al., 1990; Monboisse et al., 1991).

Collagen

Collagens constitute a large family of proteins comprising some 20 species located in the extracellular matrix whose important functions include mechanical solidity, resistance of connective tissues and cohesion of organs, support of cell adhesion, induction of cell division and differentiation during growth of the organism or wound repair (Van der Rest and Garrone, 1990; Borel, 1991).

The various types of collagen share structural and metabolic properties: they are all based on the polymerization of an elementary unit which has the form of a rod containing three polypeptide chains folded together in a triple helix in all or part of the molecule. The polypeptide sequence contains a glycine at every third residue; proline is abundant and the polypeptides also contain residues of several amino acids nearly specific to collagen; 4-hydroxyproline, 3-hydroxyproline and hydroxylysine, all formed by post-translational reactions.

Degradation of collagen by superoxide anion

Superoxide anion is not very reactive. Proteins such as serum albumin or various enzymes are not degraded by superoxide anion (Schuessler and Schilling, 1984; Wolff and Dean, 1986; Davies, 1987). Moreover, proteins incubated with superoxide anion generating systems or with hydrogen peroxide alone are more sensitive to the proteolysis by non-specific proteinases (Fligiel et al., 1984; Davies et al., 1987). Unlike these proteins, collagen is the only protein found to be sensitive to superoxide anion. Greenwald and Moy (1979) demonstrated that the incubation of acid soluble collagen with superoxide anion generated by the xanthine oxidase-hypoxanthine system prevented the formation of fibrils by this collagen.

In our laboratory, we also demonstrated that superoxide anions generated by the xanthine oxidase-hypoxanthine system were able to degrade acid soluble collagen. This degradation concerned only a small percent (about 5%) of the total amount of collagen, and was characterized as the liberation of small 4-hydroxyproline-containing-peptides into the incubation medium, suggesting cleavages in the helical part of the collagen molecule (Monboisse et al., 1983; Monboisse et al., 1984).

The collagen degradation induced by superoxide anions was quantitatively studied in gamma radiolysis experiments (Monboisse et al., 1988). Suspensions of delipidated acid soluble type I collagen were incubated with superoxide anions generated by gamma irradiation of a 50-mM phosphate buffer pH 7.4 in the presence of 100 mM sodium formate to trap hydroxyl radicals. At the end of the irradiation period, the liberation of small 4-hydroxyproline-containing-peptides was chosen as a marker of the collagen fragmentation. Under these experimental conditions, irradiation of collagen suspensions induced a liberation of small 4-hydroxyproline-containing-peptides. The amount of these peptides increased linearly with the irradiation doses, and for a same irradiation dose, also increased linearly with the collagen concentration in the range of concentrations used (0 to 3.3×10^{-6} M). Superoxide dismutase inhibited the collagen degradation, but catalase did not. In addition, the presence of chelating agents such as 0.05 mM desferrioxamine or 0.5 mM DTPA in the incubation medium did not inhibit collagen degradation. These results confirmed the ability of superoxide anions to cleave collagen by themselves and ruled out the hypothesis of the formation of hydroxyl radicals near the collagen molecule.

On the other hand, pulse radiolysis experiments showed that delipidated type I collagen was able to trap superoxide anion with a kinetic constant $k(O_2^- + \text{collagen}) = 4.8 \times 10^6 \text{ mol}^{-1}.1.\text{s}^{-1}$ (Monboisse et al., 1988).

Action of hydroxyl radical OH· on collagen

The action of OH· radicals on collagen is quite different in the presence or in the absence of oxygen.

In the presence of oxygen, hydroxyl radicals generated by gamma radiolysis of a 50 mM phosphate buffer pH 7.4 without sodium formate, induced a liberation of small 4-hydroxyproline-containing-peptides into the incubation medium. The amount of these peptides increased linearly with the irradiation dose and with the collagen concentration (Monboisse, 1989).

The analysis by reverse phase hplc of collagen breakdown-products gave different patterns of peptides obtained by the action of O_2^- or OH· radicals (Monboisse, unpublished results). The amino acid composition of the peptides obtained by the action of OH· radicals in the presence of oxygen showed a significant decrease in the amount of 4-hydroxyproline and proline residues and an increase of aspartic acid and glutamic acid residues (Monboisse, 1989). These results were in good agreement with those obtained by Wolff et al. (1986) for the degradation of serum albumin by OH· radicals in the presence of oxygen. These authors demonstrated a decrease of proline content and an increase of glutamic acid content. They explained the degradation of albumin by an oxidation of the proline residue and the formation of a new N-glutamic acid residue (Dean et al., 1989).

When irradiations of collagen were performed in the absence of oxygen, under controlled atmosphere of nitrogen protoxide (N_2O), we did not measure a significant degradation of collagen expressed as the liberation of 4-hydroxyproline-containing-peptides into the incubation medium. We observed a polymerization of collagen in the incubation cuvette (Monboisse, 1989). Several authors obtained identical results with collagen (Bailey et al., 1964; Boguta and Dancewicz, 1983; Fujimori, 1988) or other proteins (Girotti et al., 1986). They explained the formation of high molecular weight components by the generation of dityrosine residues giving a specific fluorescence to modified proteins.

Action of hypochlorous acid on collagen

When acid-soluble collagen is exposed to hypochlorous acid generated by the myeloperoxidase-Cl^--H_2O_2 system of neutrophils, there are large changes in the structural features of the collagen molecule. The measurement of 4-hydroxyproline-containing-peptides in the incubation medium did not show a significant increase in the amount of these peptides suggesting that hypochlorous acid did not cleave collagen.

The analysis by SDS-PAGE, under non-reducing conditions, of collagen incubated with hypochlorous acid demonstrated a large decrease in the content of $\alpha1(I)$ and $\alpha2(I)$ chains. In addition, we could measure an increase in high molecular weight components resulting of the formation of cross-links of unknown nature. Vissers and Winterbourn (1991) obtained similar results by incubating human plasma fibronectin under identical experimental conditions. These authors explained the formation of these high molecular weight components by the generation of dityrosine residues. Moreover, the collagen I molecule contains only three tyrosine residues per α chain, and this low amount of tyrosine could not explain the totality of cross-links generated by hypochlorous acid onto collagen.

Conclusion

These different results demonstrate that collagen is a good target for oxygen free radicals. Cleavage of many proteins by hydroxyl radicals has been already demonstrated, but collagen appears to be the only protein that is cleaved by superoxide anion O_2^-. The susceptibility of collagen to superoxide anions may be due to its special triple-helical structure. The specificity of the cleavage induced by O_2^- is under study. The separation by hplc of the peptides generated by the action of O_2^- and the analysis of their amino acid sequences should provide some evidence.

Oxygen free radicals induce fragmentation of collagen or cross-linking of collagen. These opposite effects may reflect different pathophysiological situations. Fragmentation of collagen may occur during the degradative processes of inflammation, while cross-linking of collagen may reflect what happens in aging.

Bailey, A. J., Rhodes, D. N., and Cater, C. W. (1964) Irradiation-induced crosslinking of collagen. Radiation Res. 22: 606–621.

Boguta, G., and Dancewicz, A. M. (1983) Radiolytic and enzymatic dimerization of tyrosyl residues in insulin, ribonuclease, papain and collagen. Int. J. Radiat. Biol. 43: 249–265.

Borel, J. P. (1991) Les collagènes. L'Eurobiologiste 15: 247–271.

Davies, K. J. A. (1987) Protein damage and degradation by oxygen radicals. I-General aspects. J. Biol. Chem. 262: 9895–9901.

Davies, K. J. A., Lin, S. W., and Pacifici, R. E. (1987) Protein damage and degradation by oxygen radicals. IV. Degradation of denatured protein. J. Biol. Chem. 262: 9914–9920.

Dean, R. T., Wolff, S. P., and McElligott, M. A. (1989) Histidine and proline are important sites for free radical damage to proteins. Free Rad. Res. Commun. 7: 97–103.

Fligiel, S. E. G., Lee, E. C., McCoy, J. P., Johnson, K. J., and Varani, J. (1984) Protein degradation following treatment with hydrogen peroxide. Am. J. Pathol. 115: 418–425.

Fujimori, E. (1988) Cross-linking of collagen CNBr peptides by ozone and UV light. FEBS Lett. 235: 98–102.

Gillery, P., Monboisse, J. C., Maquart, F. X., and Borel, J. P. (1988) Glycation of proteins as a source of superoxide. Diab. Metab. 14: 25–30.

Girotti, A. W., Thomas, J. P., and Jordan, J. E. (1986) Xanthine-oxidase-catalyzed crosslinking of cell membranes proteins. Archs. Biochem. Biophys. 251: 639–653.

Greenwald, R. A., and Moy, W. W. (1979) Inhibition of collagen gelation by action of the superoxide radical. Arth. Rheum. 22: 251–259.

Halliwell, B. (1987) Oxidants and human disease: some new concepts. FASEB J. 1: 358–364.

Monboisse, J. C. (1989) Contribution à l'étude des interactions entre collagène et radicaux libres oxygénés. Thesis Univ. Reims.

Monboisse, J. C., Bellon, G., Dufer, J., Randoux, A., and Borel, J. P. (1987) Collagen activates superoxide anion production by human polymorphonuclear neutrophils. Biochem. J. 246: 599–603.

Monboisse, J. C., Bellon, G., Randoux, A., Dufer, J., and Borel, J. P. (1990) Activation of human neutrophils by type I collagen. Requirement of two different sequences. Biochem. J. 270: 459–462.

Monboisse, J. C., Braquet, P., Randoux, A., and Borel, J. P. (1983) Non enzymatic degradation of acid soluble collagen by superoxide anion: protective effect of flavonoids. Biochem. Pharmacol. 32: 53–58.

Monboisse, J. C., Gardès-Albert, M., Randoux, A., Borel, J. P., and Ferrandini, C. (1988) Collagen degradation by superoxide anion in pulse and gamma radiolysis. Biochim. Biophys. Acta, 965: 29–35.

Monboisse, J. C., Garnotel, R., Randoux, A., Dufer, J., and Borel, J. P. (1991) Adhesion of human neutrophils to and activation by type I collagen involving a $\beta 2$ integrin. J. Leukoc. Biol. 50: 373–380.

Monboisse, J. C., Poulin, G., Braquet, P., Randoux, A., Ferradini, C., and Borel, J. P. (1984) Effects of oxy radicals on several types of collagen. Int. J. Tissue React. 6: 385–390.

Morel, F., Doussière, J., and Vignais, P. (1991) The superoxide-generating oxidase of phagocytic cells. Physiological, molecular and pathological aspects. Eur. J. Biochem. 201: 523–546.

Schuessler, H., and Schilling, K. (1984) Oxygen effect in the radiolysis of proteins – Part 2 – Bovine serum albumin. Int. J. Radiat. Biol. 45: 267–281.

Van der Rest, M., and Garrone, R. (1990) Collagens as multidomain proteins. Biochimie 72: 473–484.

Vissers, M. C. M., and Winterbourn, C. C. (1991) Oxidative damage to fibronectin – I – The effects of the neutrophil myeloperoxidase system and HOCl. Archs. Biochem. Biophys. 285: 53–59.

Wolff, S. P., and Dean, R. T. (1986) Fragmentation of proteins by free radicals and its effect on their susceptibility to enzymic hydrolysis. Biochem. J. 234: 399–403.

Wolff, S. P., and Dean, R. T. (1987) Glucose autoxidation and protein modification. The potenteial role of "autoxidative glycosylation" in diabetes. Biochem. J. 245: 243–250.

Wolff, S. P., Garner, A., and Dean, R. T. (1986) Free radicals, lipids and protein degradation. TIBS 11: 27–31.

Free Radicals and Aging
ed. by I. Emerit & B. Chance

Free radicals and aging of the skin

Ingrid Emerit

Free Radical Research Group, Centre de Recherches Biomédicales des Cordeliers, University of Paris VI, 15 rue de l'Ecole de Médecine, Paris, France

Summary. Cutaneous aging is the result of genetically determined or intrinsic aging superimposed by degenerative changes due to actinic irradiation, also called photoaging. The manifestations of cutaneous aging, as it relates to the perception of age, is caused by ultraviolet light, in particular in those parts of the body exposed daily to solar radiation. Free radical generation in the skin by UV light and from other sources, such as cellular infiltrations or the xanthine oxidase reaction, may be detected by direct and indirect methods. The decrease in antioxidant enzymes and small molecular weight antioxidants such as glutathione, vitamine E and ubiquinone upon exposure to UV light is an indication that the pro-antioxidant balance can be overwhelmed by acute or chronic photo-oxidative stress. Antioxidant supplementation is therefore a means for prevention or at least retardation of premature cutaneous aging.

Introduction

The effect of aging on the skin is a neglected issue in most textbooks of gerontology. This may reflect sparsity of knowledge, but also lack of interest. Compared to age-related diseases of the cardiovascular and central nervous system, age-related disorders of the skin are in general benign, and even the most common skin cancers, such as squamous and basal cell carcinoma, usually do not kill the patient. Nevertheless significant discomfort and suffering may result from the pathology associated with aging skin.

Oxygen-derived free radicals are generated in the skin from various sources and by various mechanisms, as in any other organ. In addition, the skin is exposed to free radical generating environmental agents such as air pollutants and solar radiation. A considerable body of circumstantial evidence has been amassed that free radicals are responsible for at least part of the degenerative changes leading to cutaneous aging (Black, 1987).

Intrinsic versus extrinsic aging

The innate or intrinsic aging process also called chronologic aging, has to be distinguished from changes resulting from actinic damage, which is mainly due to UV light and therefore called photoaging. Both processes are superimposed on sun-exposed parts of the body. Clinical, morphological and biochemical characteristics of both processes are different, and photoaging does not only represent an acceleration of the innate aging process. The following macroscopic and microscopic differences have been described by various authors (Gilchrest et al., 1983; Kligman, 1989; Montagna et al., 1989):

- macroscopically, intrinsic aging results in fine wrinkling, thinning and laxity of the skin, while photoaged skin displays a telangiectatic, leathery, dry, nodular surface with deep wrinkles, accentuated skin furrows, sags and bags. In addition photoaged skin shows irregularities in pigmentation, actinic keratoses, as well as a variety of benign or premalignant growths.
- microscopically, the dominant change in photoaging is the hyperplasia of elastic tissue in the dermis, which may lead to complete disorganization described as solar elastosis. In contrast hereto, intrinsic aging induces only a modest increase in the number and thickness of the elastic fibers. Collagen undergoes only minor changes in intrinsic aging, while photoaging results in loss of collagen bundles, a decrease in mature collagen and an increase in type III collagen. The ground substance of the dermal matrix, which is composed of proteoglycans and hyaluronic acid, increases greatly in photoaged skin, while in protected skin the ground substance decreases with age.

Free radicals are involved in both aging processes. For practical reasons, free radical related damage to the skin has been primarily studied on photodamaged skin.

Actinic damage and photoaging

The detrimental effects of UV radiation (220–400 nm) on biological organisms have been studied for many years. On the skin, erythema and sunburn are acute effects, while aging and cancer are the consequences of chronic exposure.

The solar spectrum is a complex band of radiation including UVC (220–290 nm wavelength), UVB (290–320 nm) and UVA (320–400 nm), as well as visible light and infrared radiation. The biologically most deleterious wavelength is UVC, which is fortunately blocked by the ozone layer in the stratosphere. UVA and UVB penetrate to the surface of the earth and reach our body. UVA is about 1000 times less erythemogenic on the skin than UVB, but sunlight contains about 100 times more UVA than UVB. While UVB effects are principally limited to the epidermis, UVA penetrates more deeply and has been shown to induce mononuclear and polymorphonuclear cell infiltrations, as well as endothelial cell damage in the microvasculature of the dermis (Rosario et al., 1979). It has been claimed that UVA plays a far more important role in contributing to the harmful effects of sun exposure than previously suspected (NIH Conference Statement May, 1989). Its responsability for photocarcinogenesis is probably less than that of UVB. With respect to aging, UVA is even more dangerous than UVB.

Free radical formation

By melanin and its precursors

Already 30 years ago, free radicals were detected in the skin by electron paramagnetic resonance (EPR) spectroscopy at low temperature (Norrins, 1962) and electron spin resonance (ESR) techniques (Stratton and Pathak, 1968). A stable ESR signal was observed in pigmented, but not in fair skin. Also skin samples from albino mice and guinea pigs gave no such intrinsic free radical signal, which appeared to be related to the presence of melanin. Previous in vitro studies on melanins from human hair had ascribed the free radical property of melanins to a semiquinone form of these pigments (Mason et al., 1960). It was hypothesized that melanin was able by means of its capacity for oxidation and reduction and its stable free radical state to protect a tissue against reducing or oxidizing conditions, which might otherwise set free reactive free radicals within living cells. In keeping with epidemiologic evidence, melanin was considered to act as a photoprotective agent. This was confirmed in the following (see next paragraph), but it was also recognized that melanins may be a source of free radicals with damaging effects.

Melanins are pigments produced by melanocytes, which are located in the basal layer of the epidermis. As a first effect of exposure to sun, immediate pigment darkening of the skin occurs followed by new melanin formation 2–3 days after the initial exposure. This new melanin formation implies several steps including synthesis of the enzyme tyrosinase, formation of a cell organelle, the melanosome, polymerization of tyrosine into melanin and transfer of the melanized melanosome into the keratinocytes of the epidermis.

The biosynthesis of eumelanin involves the enzymatic oxidation of tyrosine to 3,4-dihydroxy-phenylanalin (DOPA) and to dopaquinone via a semiquinone radical intermediate (Chedekel, 1982). Pheomelanin is formed by a branching of the eumelanin pathway by the interaction with cysteine, glutathione or related sulfhydryl compounds. From red-haired sun-sensitive to dark-haired sun-resistant individuals, pheomelanins and eumelanins are mixed in variable proportions. UV irradiation enhances the ESR signals of eumelanin and pheomelanin, indicating an increase in free radical generation.

Oxygen is consumed during the photooxidation of catechols to their respective semiquinones. Superoxide and hydrogen peroxide are formed as products of oxygen reduction (Sarna and Sealy, 1984; Kalyanaraman et al., 1984) and may ultimately give rise to hydroxyl radicals in pigments containing trace metals (Koryotowski et al., 1987). Recently generation of superoxide during the enzymatic action of tyrosinase was confirmed by Koga et al. (1992), using chemiluminescence. The complete quenching of light emission by superoxide dismutase indicated that singlet oxygen production was negligible in this system.

Radicals may also be produced during photolysis of melanogenic metabolites such as cysteinyldopas and dihydroxyindoles, which are very unstable in presence of UV light of >300 nm wavelengths. ESR spin trapping, laser flash photolysis and pulse radiolysis techniques indicated formation of hydrated electrons, hydrogen atoms, hydroxy-, semiquinone-, aryl thiyl and alanyl radicals (Koch and Chedekel, 1987).

Damage resulting from these active oxygen intermediates may partly counteract the protective action of these pigments. Pheomelanins are particularly prone to photooxidation, and it has been suggested that the high incidence of skin cancer among red-haired individuals is not only due to the lower shielding effect resulting from smaller amounts of total melanin, but also to the larger proportion of pheomelanin in their skin. The ESR signals from melanins are enhanced in the presence of photosensitizers such as rose bengal and porphyrins, and this enhancement is again more important for red than for black hair (Persad et al., 1983; Sealy et al., 1984). It was suggested that the photosensitizing effects of other compounds e.g. chlorpromazine, chloroquine etc., which also bind to melanins, may depend on the quantity and type of melanin present in the skin of an individual.

By photosensitization reactions

When light of appropriate wavelength is absorbed by a sensitizer molecule, this molecule passes through its first excited singlet state to a more long-lived triplet state. By either hydrogen abstraction or electron transfer, radicals or radical ions are formed, which in presence of oxygen may produce secondary oxygen-derived free radicals. In so-called photodynamic reactions, direct interaction with molecular oxygen leads mainly to singlet oxygen by energy transfer. However, the electron transfer can occur in either direction and may lead also to an oxidized sensitizer and the superoxide anion radical (Lee and Rodgers, 1987; Foote, 1991).

The production of active oxygen species (superoxide and singlet oxygen) was studied by Pathak and Carraro (1984) in psoralen-sensitized reactions. Psoralens are used in combination with UVA light in the treatment of psoriasis and other skin disorders. In our laboratory, we used PUVA-induced skin reactions for testing the protective effects of topical SOD applications on mouse skin. Psoralens are also components of certain cosmetic preparations for skin tanning, despite their mutagenic and carcinogenic effects. Photosensitized skin reactions may be responsible for permanent tissue damage and cutaneous aging, given the increasing number of exogenous sensitizers such as drugs, cosmetics, air pollutants, etc.

The role of endogenous photosensitizing metabolites has been stressed by Cunningham et al. (1985). Using the cytochrome C assay, these authors could show that superoxide is generated by exposure of riboflavin to the light from a solar simulator. The quantities produced were a function of riboflavin concentration and fluence. Similar results were observed with monochromatic light of different wavelength acting on various other photosensitizers such as NADPH, NADH and the nucleosiders 2-thiouracil and 4-thiouridine. These studies indicated that the quantum yields for superoxide production by riboflavin were highest at 290 nm, but that considerable quantities were also produced at 405 nm and in the visible range. If the deleterious effects of UVB have been attributed principally to its direct action on DNA, these studies show that UVB may also create damage via indirect action like UVA. The latter may be particularly dangerous with respect to cutaneous aging via these action mechanisms. Exposure to UVA is increasing because of growing use of UVA sun-tanning parlors and of UVB sunscreens, which allow for more prolonged sun exposure and increase selective exposure to UVA. Again individuals with fair skin are particularly at risk.

From other sources

Not only UV light, but also ionizing radiation produces free radicals in skin. Invasion of the skin by pathogenic bacteria, viruses and fungi attracts neutrophils and macrophages to inflamed areas, which through their respiratory burst release active oxygen species in their environment. Also exposure to UV light is followed by dermal inflammatory infiltration, resulting in additional free radical production with prolonged damaging effects. Liberation of arachidonic acid from membranes and initiation of the AA cascade is another source of active oxygen in inflammatory conditions.

Free radical generation by an xanthine oxidase reaction was supposed to occur in thermal burns (Till et al., 1989). Xanthine dehydrogenase and oxidase were found in murine epidermis and conversion from the D to the O-form of this enzyme was correlated with undergoing differentiation in keratinocytes. (Reiners and Rupp, 1989).

These sources of free radicals contribute probably less to cutaneous aging than the permanent exposure to light and photosensitizing agents.

Defenses of the skin against free radical attack

Since the majority of free radicals is formed as a consequence of light exposure, the shielding by melanin pigments represents an important defense. Melanin acts as a filter in the skin by absorbing the radiant energy of ultraviolet and visible light. In addition melanin may act as a scavenger of oxygen radicals, as already mentioned above (Sealy et al., 1980; Geremia et al., 1984; Ranadive and Menon, 1986). Superoxide generated via a xanthine oxidase reaction is efficiently scavenged by dopamelanin (Goodschild et al., 1981). Comparative studies with SOD showed that this effect is weak: equivalent scavenging was observed with 66 μg/ml dopamelanin and 0.05 μg/ml superoxide dismutase (Tomita and Tayani, 1986). Since melanin is stored in cytoplasmic vesicles, it may not be available for radical scavenging outside the melanosome, i.e. it would not protect against cell membrane or cytoplasmic oxidant insults.

Another particularity of skin is the protection given by cystin-rich proteins and keratin present in the stratum corneum (Yoshino et al., 1981; Tezuka and Takahashi, 1986).

In addition to these defense systems, skin is protected by antioxidant enzymes and various small molecular weight antioxidants. Superoxide dismutase (SOD) levels were found to be lower in skin than in heart, kidney or liver (Carraro and Pathak, 1988). Comparative studies showed that fibroblasts contain higher levels of SOD, catalase and glutathione peroxidase than keratinocytes and melanocytes, the latter having the lowest activities (Yohn et al., 1991). In agreement herewith, human epidermal melanocytes were more easily damaged by hydrogen peroxide than keratinocytes (Muglia et al., 1986). The catalase activity of keratinocytes increases as stages of differentiation increase (Reiners et al., 1988). Keratinocytes in the outermost layers of the epidermis may require the highest antioxidant defense capability in order to shield the basal layers from oxidant stress. It has been hypothesized that the antioxidant enzymes are low in melanocytes, since these cells principally rely on their protection by pigments. However no difference in antioxidant activity levels was noted between cell cultures from Black or Caucasian skin.

Besides the antioxidant enzymes SOD, catalase and GSH peroxidase, another enzyme may play a major role for free radical reduction processes in the epidermis. The FAD-containing enzyme thioredoxin reductase acts as a free radical scavenger at the outer surface of human keratinocytes. It utilizes NADPH as electron donor and catalyzes the reduction of oxygen radicals to peroxide ions, which are then reduced to water (Schallreuther et al., 1986). It is also supposed to trap nitroxide-type radicals at the surface of the skin (Schallreuter and Wood, 1986).

Most major free radicals arising in cells are likely to react with cellular reducing agents like ascorbate and thiols, of which glutathione is most abundant. It is evenly distributed throughout the viable and non-viable cell layers of the epidermis. In its protective roles glutathione may act as a radical scavenger per se or as a cofactor for protective enzymes such as glutathione peroxidase and glutathione transferases (Connor and Wheeler, 1987).

In addition to these hydrophilic antioxidants, the skin contains lipophilic antioxidants, including alpha-tocopherol (vitamin E) and ubiquinone/ubiquinol.

Enzymic and non-enzymic antioxidants decrease significantly after exposure to light. In UVB-exposed skin, inhibition of superoxide dismutase, glutathione peroxidase and glutathione reductase was observed (Maisuradze, 1987; Fuchs et al., 1989). A single exposure of mouse skin to UVB irradiation resulted in a significant decrease of SOD activity 24–48 h after exposure, returning to normal levels after 72 h (Miyachi, 1987). Similar observations were reported by Pence and Naylor (1990) after exposure of hairless mice to UVB ($0.09 \, J/cm^2$). Catalase was decreased by 12 h after UV irradiation and remained depressed for up to 72 h. Also exposure to UVA and visible light resulted in an impairment of the enzymic antioxidant defense system (Fuchs et al., 1989; Punnonen et al., 1991). In addition, these authors observed a tendential, but not significant decrease in cutaneous ascorbic acid, tocopherol and ubiquinone levels.

Glutathione depletion was observed as a consequence of ultraviolet radiation by Connor and Wheeler (1987). UVB irradiation of hairless mouse skin caused rapid transient fluctuations in the epidermal and dermal glutathione level and the relative amount present as the oxidized form. UVA irradiation depleted epidermal and dermal glutathione for several hours, but required much higher doses than UVB. However in presence of the photosensitizer psoralen, prolonged depletion of glutathione levels was observed, which were dependent on the psoralen concentration and lasted several days for the highest doses. Even UVA doses of $1 \, J/cm^2$, which did not influence cutaneous glutathione levels, resulted in considerable depletion in presence of the photosensitizer. There was failure to recover with the highest doses ($50 \, mg/kg$ 8-methoxypsoralen and $5 \, J/cm^2$ UVA), as indicated by the bulk of residual glutathione assayed as GSSG. A decrease in glutathione peroxidase levels occurred after the highest PUVA regimes, accompanied by a fall in glutathione reductase levels.

Consequences

When the antioxidant defense system is overwhelmed, active oxygen species induce tissue damage by attacking macromolecules, such as lipids, nucleic acids, proteins, proteoglycans and hyaluronic acid. Histologically this is expressed in the epidermis by loss of normal maturation from the basal to the granular cell layer and presence of dyskeratotic cells. These so-called sunburn cells represent an example of apoptosis (Weedon et al., 1979), and cell death is probably due to DNA fragmentation (Young, 1987). The protective effect of SOD indicates involvement of superoxide or secondarily generated hydroxyl radicals at the origin of the DNA damage (Danno et al., 1984; Danno and Horio, 1987). Sunburn cells are observed after exposure to UVB, rarely after UVA (Rosario et al., 1979). Another histologic marker for DNA damage is perinuclear halo formation. This phenomenon has been observed in electron microscope studies after exposure to UVB. In addition to perinuclear halos, the keratinocytes showed clumping of tonofilaments, cytoplasmic vacuolization and nuclear degeneration (Wilgram et al., 1970). Perinuclear halos can be induced also by exposure of cells to UVA light (365 nm) in presence of a photosensitizer (riboflavin). Pretreatment of the cells with SOD prevented halo formation (Emerit et al., 1981). Active oxygen species are also involved in damage to the immunocompetent Langerhans cells, which lose their surface markers upon exposure to UVB light. This could be prevented by SOD, but not by catalase, mannitol or histidine (Horio and Okamoto, 1987).

In the dermis, the most striking findings are due to free radical induced damage to endothelial cells of capillaries and venules, resulting in edema formation and extravasation of mono- and polymorphonuclear cells (Rosario et al., 1979; Margolis et al., 1989). According to Connor and Wheeler (1987), edema formation occurs in mouse skin, when glutathione depletion reaches 50%. This is the case after 10 min with a UVB dose of 90 m J/cm^2. The time course for glutathione depletion is considerably longer and more protracted with UVA and PUVA. Correspondingly, edema formation is delayed under these conditions.

The lesions induced in endothelial cells result in necrosis and thickening of the vessel wall. The consequence is impaired nutrition of the dermis. Free radical damage to the fiber network, by fragmentation and abnormal cross-linking, modifies the mechanical properties of the skin. Collagen is primarily responsible for the tensile strength, whereas the intact elastic fiber network provides resilience and elasticity. Also hyaluronic acid and proteoglycans, which form a kind of hyperhydrated gel in the extracellular matrix, may be damaged by oxygen-derived free radicals. This is well-known from studies on free radical induced injury in connective tissue and has been discussed more in detail in the chapters on age-related changes in cartilage (Henrotin et al., this volume) and collagen (Monboisse and Borel, this volume).

While the changes in the dermis, which are in great part due to free radical reactions induced by UVA light in presence of sensitizing agents, play an important role in cutaneous aging, free radical reactions in the epidermis are responsible for photocarcinogenesis, due to mutagenesis in the actively dividing cells of the basal layer.

Free radicals are not only creating damage at the site of their generation. Lipid peroxidation of membranes may result in more long-lived peroxides, which may diffuse in the skin and even enter the blood stream (Yagi, 1987). Exogenously formed lipid peroxides, if applied to the skin, may induce necrosis and bleeding (Tanaka, 1979). Generation of lipid peroxides during storage of cosmetic creams may lead to irritation of the skin (Tanaka et al., 1986). Increased levels of lipid peroxidation products were already reported by Meffert in 1976 in chronically sun-exposed human skin and with advancing age. Ogura et al. (1987) could show that topical application of SOD diminished the lipid peroxide content in the UV-exposed skin.

Prevention of premature aging by antioxidants

Generation of free radicals in the skin has been demonstrated with direct and indirect means, as exposed above. The decrease of enzymic and non-enzymic antioxidants in the skin after exposure to light, despite its transient character, indicates that the defenses may be insufficient. Chronic oxidant stress, as it occurs in modern life due to excessive sun exposure and increasing air pollution, may justify supplementation with antioxidants in order to retard the cutaneous aging process.

Animal studies have shown that dietary antioxidant supplementation may result in marked reduction of number and severity of UV-induced skin carcinomas (Black et al., 1985). High levels of dietary lipids with unsaturated fatty acids increased tumor formation, which could be inhibited by antioxidants. The protective effect of beta-carotene and other carotenoids (canthaxanthine, phytoene) was reported (Mathews-Roth, 1986). These findings suggest that appropriate nutrition may also be beneficial for retardation of cutaneous aging.

Systemic application of SOD influenced sunburn cell formation. A single injection of 10–30 units/g body weight given either just before or less than 15 min after exposure to UVB proved to be efficient (Danno et al., 1984). Depletion of Langerhans cells in mouse skin was reduced by 20–200 μg of SOD injected intradermally (Horio and Okamoto, 1987). The UV-induced decrease in SOD levels could be avoided by 100 μg of liposomal SOD, injected intraperitoneally to mice, 2 h before UV irradiation (Miyachi et al., 1986).

Topical application of SOD (0.3 g of 0.05% liposomal SOD) reduced the lipid peroxide content of mouse skin, if applied twice, 24 and 18 h before UV-exposure (Ogura et al., 1986). Protection with topically administered SOD was also obtained in psoralen-mediated photosensitized reactions on guinea pig skin (Pathak and Joshi, 1984).

Work of our laboratory confirmed the protective effect of topical SOD application on PUVA-treated mouse skin. Two mouse strains were used for this study: epilated New Zealand Black mice with black coat color and pigmented skin and hairless mice of the albino type. Both sides of the back were painted with 5- or 8-methoxypsoralen at a concentration of 0.05%. A total dose of 4 J/cm^2 of UVA light was given with a mineralight having an output of 3.3 $mWatt/cm^2$ at a distance of 20 cm. During the 20 min of irradiation, the mouse was kept in a restrained position under light chloroform anesthesia. The PUVA-induced inflammatory skin reaction was evaluated by inspection and by measurement of skinfold thickness with a cutimeter. The response was delayed in agreement with data from the literature on PUVA-treated human skin (Rosario et al., 1979). Skinfold thickness increased from the second day on and reached a maximum after 5 days. The reaction was more important in the albino hairless mice than in pigmented NZB mice. Psoralen or UVA alone did not induce skin reactions, even at a 10 times higher fluency of 40 J/cm^2.

The PUVA-induced skin reactions could be regularly prevented by application of a gel containing native SOD from bovine erythrocytes at a concentration of 0.01-0.1% w/v. The SOD gel was applied on the right side of the back, while the left side received only the vehicle or gel prepared with heat-inactived SOD, such that each animal was its own control. The skin reaction was partially inhibited by a single application 1 h before UVA irradiation. The protection became more and more complete with increasing number of pretreatments. Application of the SOD gel as a posttreatment resulted only in a small protective effect (Alaoui and Emerit, in preparation).

Since photosensitized reactions between psoralens and UVA are supposed to occur in the dermis, the protective effects observed by Pathak and Joshi (1984) and by ourselves indicate that SOD can penetrate the skin in its native form. Whether it enters through the stratum corneum or through hair follicles needs further study.

Also tocopherols are able to inhibit PUVA-induced erythema, as shown by Potapenko et al. (1984). Interestingly the protective effect was maximal at concentrations of 2.5×10^{-10} to 5×10^{-9} mol/cm^2, while the antioxidant action decreased at higher doses. In our NZB mouse model, the maximal protection was observed at a concentration of 1%, while 0.1 and 5% concentrations were equivalent (Emerit et al., in preparation). On the other hand, concentration-dependent protection was observed in hairless mice exposed to UVB, where edema formation was more efficiently inhibited with D-alpha tocopherolacetate at 5% than at 1% concentration (Moller et al., 1987). Dermal penetration and systemic distribution of radioactive vitamin E in human skin grafted onto nude mice showed that vitamin E penetrates through stratum corneum and hair follicles, that it is found after 4 h in the epidermis and upper portion of the dermis, but that after 24 h the maximum is detected in the lower dermis, which acts as a barrier or reservoir for this highly lipophilic antioxidant (Klain, 1989).

Vitamins A and C are often introduced in skin care products in combination with vitamin E, which is supposed to be regenerated by vitamin C. Addition of ascorbate is however a double-edged sword, since it becomes a prooxidant in the presence of trace metals.

The provitamin beta-carotene and other singlet oxygen quenchers such as DABCO and NaN_3 were found to be protective in photosensitized skin reactions (Pathak and Carraro, 1986).

Glutathione, the major low molecular weight antioxidant in living tissues cannot penetrate the skin and is also unstable. However other sulfhydryl compounds such as lipoic acid, thiazolodine-4-carboxylic acid and dithiothreitol are more easily absorbed and may be beneficial in topical application because of their disulfide-reducing capabilities. Anisyldithiolthione is an efficient singlet oxygen quencher and is thought to protect tissues against oxidant stress by maintaining GSH levels (Tissié et al., 1990).

Also flavonoid aglycones, members of a ubiquitous class of plant phenols, have often been proposed to act as antioxidants and radical scavenging agents. Structure-activity relationships, which would allow to make a choice among the more than 300 known flavonoids, have been studied by Bors et al. (1990). Optimal antioxidant capacity of a given substance depends on a high reactivity with radicals of different origin and a relatively high stability of the radical 'intermediate' to prevent participation in chain reactions. Similar speculations were made for synthetic phenolic antioxidants, among which butylated hydroxyanisole and hydroxytoluene (BHA and BHT) have been in use for a long time for prolongation of the shelf-life of lipids in skin care products (Kahl et al., 1990).

Conclusions

Cutaneous aging is in great part due to chronic light exposure. Controlled exposure and use of sunscreens may efficiently reduce aging due to actinic damage. However, both solutions prevent or at least diminish suntan. Oral intake and topical application of antioxidants diminish UVL-induced free radical tissue injury and represent a third solution not influencing tanning.

By the age of 15, the skin of most Caucasians already shows a certain degree of deterioration of the elastic fiber network characteristic of actinic damage. Therefore prophylactic antioxidant administration has to start early. It should be pursued throughout the year, since even small fluencies of UVA light may generate free radicals in the presence of the numerous exogenous and endogenous sensitizers. Also, most antioxidants appear to be more protective before than after exposure to UV irradiation. Only regular use can therefore provide optimal conditions for protection.

Black, H. S., Lenger, W. A., Gerguis, J., and Thornby, J. I. (1985) Relation of antioxidants and level of dietary lipid to epidermal lipid peroxidation and ultraviolet carcinogenesis. Cancer Res. 45: 6254–6259.

Black, H. S. (1987) Potential involvement of free radical reactions in ultraviolet light-mediated cutaneous damage. Photochem. Photobiol. 46: 213–221.

Bors, W., Heller, W., Michel, C., and Saran, M. (1990) Radical chemistry of flavonoid antioxidants, in Antioxidants in Therapy and Preventive Medicine, pp. 165–170. Eds I. Emerit et al. Plenum Press, New York.

Carraro, C., and Pathak, M. A. (1988) Characterization of superoxide dismutase from mammalian skin epidermis. J. Invest. Dermatol. 90: 31–36.

Chedekel, M. R. (1982) Photochemistry and photobiology of epidermal melanins. Photochem. Photobiol. 35: 881–885.

Connor, M. J., and Wheeler, L. A. (1987) Depletion of cutaneous glutathione by ultraviolet radiation. Photochem. Photobiol. 46: 239–245.

Cunningham, M. L., Krinsky, N. I., Giovanazzi, S. M., and Peak, M. J. (1985) Superoxide anion is generated from cellular metabolites by solar radiation and its components. Free Rad. Biol. Med. 1: 381–386.

Danno, K., Horio, T., Takigawa, M., and Imamura, S. (1984) Role of oxygen intermediates in UV-induced epidermal injury. J. Invest. Dermatol. 83: 166–168.

Danno, K., and Horio, T. (1987) Sunburn cell: Factors involved in its formation. Photochem. Photobiol. 45: 683: 690.

Emerit, I., Michelson, A. M., Martin, E., and Emerit, J. (1981) Perinuclear halo formation as an indication of phototoxic effects. Dermatologica 163: 295–299.

Foote, C. S. (1991) Definition of type I and type II photosensitized oxidation. Photochem. photobiol. 54: 659.

Fuchs, J., Huflejt, M. E., Rothfuss, L. M., Wilson, D. S., Carcamo, G., and Packer, L. (1989) Acute effects of near ultraviolet and visible light on the cutaneous antioxidant defense system. Photochem. Photobiol. 50: 739–744.

Fuchs, J., Huflejt, M. E., Rothfuss, L. M., Wilson, D. S., Carcamo, G., and Packer L. (1989) Impairment of enzymic and nonenzymic antioxidants in skin by UVB irradiation. J. Invest. Dermatol. 93: 767–773.

Geremia, E. C., Corsano, C., Bonomo, R., Giardinelli, R., Vanella A., and Sichel, G. (1984) Eumelanins as free radical trap and superoxide dismutase activities in amphibia. Comp. Biochem. Physiol. 79 B: 67–69.

Gilchrest, B. A., Szabo, G., Flynn, E., and Goldwyne, R. M. (1983) Chronologic and actinically induced aging in human skin. J. Invest. Dermatol. 80: 81–85.

Goodchild, N. T., and Kwock, L. (1981) Melanin: A possible cellular superoxide scavenger, in Oxygen and Oxygen Radicals in Chemistry and Biology, pp. 645–648. Eds M. A. J. Rodgers and E. L. Powers. Academic Press, New York.

Horio, T., and Okamoto, H. (1987) Oxygen intermediates are involved in ultraviolet radiation-induced damage of Langerhans cells. J. Invest. Dermatol. 88: 699–702.

Kahl, R., Weinke, S., and Kappus, H. (1990) Comparison of antioxidant and prooxidant activity of various synthetic antioxidants, in: Antioxidants in Therapy and Preventive Medicine, pp. 283–290. Eds I. Emerit et al. Plenum Press, New York.

340

Kalyanaraman, B., Korytowski, W., Pilas, B., and Sarna, T. (1986) Photoinduced generation of hydrogen peroxide and hydroxyl radical in eumelanins. Photochem. Photobiol. 43s: 27.

Klain, G. J. (1989) Dermal penetration and systemic distribution of 14 C-labeled vitamin E in human skin grafted athymic nude mice. Internat. J. Nutri. Res. 59: 333–337.

Kligman, L. H. (1989) Skin changes in photoaging: Characteristics, prevention and repair, in: Aging and the Skin, pp. 331–346. Eds A. K. Balin and A. M. Kligman. Raven Press, New York.

Koch, W. H., and Chedekel, M. R. (1987) Photochemistry and photobiology of melanogenic metabolites: Formation of free radicals. Photochem. Photobiol. 46: 229–238.

Koga, S., Nakano, M., and Tero-Kubota, S. (1991) Generation of superoxide during the enzymatic action of tyrosinase. Arch. Biochem. Biophys. 292: 570–575.

Korotowski, W., Pilas, B., Sarna, T., and Kalyanaraman, B. (1987) Photoinduced generation of hydrogen peroxide and hydroxyl radicals in melanins. Photochem. Photobiol. 45: 185–190.

Lee, P. C. C., and Rodgers M. A. J. (1987) Laser flash photokinetic studies of rose bengal sensitized photodynamic interactions of nucleotides and DNA. Photochem. Photobiol. 45: 79–96.

Maizuradse, V. N., Platonov, A. G., Gudz, T. I., Goncharenko, E. N., and Kudriashov, I. U. B. (1987) Effect of ultraviolet rays on lipid peroxidation and various factors of its regulation in the rat skin. Biol. Nauki 5: 31–35.

Margolis R. J., Sherwood, M., Maytum, D. J., Granstein, R. D., Weinstock M. A., Parrish, J. A., and Gange, R. W. (1989) Longwave UV radiation (UVA, 320–400 nm)-induced tan protects human skin against further UVA injury. J. Invest. Dermatol. 93: 713–718.

Mason, H. S., Ingram, D. J. E., and Allen, B. (1960) The free radical property of melanins. Arch. Biochem. Biophys. 86: 225–230.

Mason, H. S., Kalyanaraman, B., Trainer, B. E., and Eling, T. E. (1980): A carbon-centered free radical intermediate in the prostaglandin synthetase oxidation of arachidonic acid: Spin trapping and oxygen uptake studies. J. Biol. Chem. 255: 5019–5020.

Mathews-Roth, M. M. (1986) Carotinoids quench evolution of excited species in epidermis exposed to UVB (290–320 nm) Light. Photochem. Photobiol. 43: 91–93.

Meffert, H., Diezel, W., and Sonnichsen, N. (1976) Stable lipid peroxidation products in human skin: Detection, ultraviolet light-induced increase, pathogenic importance. Experientia 32: 1397–1398.

Miyachi, Y., Imamura, S., and Niwa, Y. (1987) Decreased skin superoxide dismutase activity by a single exposure of ultraviolet radiation is reduced by liposomal superoxide dismutase pretreatment. J. Invest. Dermatol. 89: 111–112.

Moller, H., Potokar, M., and Wallat, S. (1988) Vitamin E as a cosmetic agent. Henkel Referate 24/Int. ed, pp. 91–95.

Montagna, W., Kirchner, S., and Carlisle, K. (1989) Histology of sun-damaged human skin. J. Am. Acad. Dermatol. 21: 907–918.

Muglia, J. J., Tonnesen, M. G., Osborn, R. L., and Norris, D. R. (1986) Ineffective antioxidant defense in melanocytes. J. Invest. Dermatol. 86: 496. (Abstr.)

Norrins, A. L. (1962) Free radical formation in the skin following exposure to ultraviolet light. J. Invest. Dermatol. 39: 445–448.

Ogura, R., Sugiyama, M., Sakanashi, T., and Hidaka, T. (1987) Role of oxygen in lipid peroxide of skin exposed to ultraviolet light, in: The Biological Role of Reactive Oxygen Species in Skin, pp. 55–63. Eds O. Hayashi et al. Elsevier, New York.

Pathak, M. A., and Stratton, K. (1968) Free radicals in human skin before and after exposure to light. Arch. Biochem. Biophys. 123: 468–476.

Pathak, M. A., and Joshi, P. C. (1984) Production of active oxygen species by psoralens and ultraviolet radiation (320–400 nm). Biochem. Biophys. Acta 798: 115–128.

Pathak, M. A., and Carraro, C. (1987) Reactive oxygen species in cutaneous photosensitized reactions in porphyrias and PUVA photochemotherapy and in melanin pigmentation, in: The Biological Role of Reactive Oxygen Species in Skin, pp. 75–94. Eds O. Hayashi et al. Elsevier, New York.

Pence, B. C., and Naylor, M. F. (1990) Effects of single-dose ultraviolet radiation on skin superoxide dismutase, catalase, and xanthine oxidase in hairless mice. J. Invest. Dermatol. 95: 213–216.

Persad, S., Menon, I. A., and Haberman, H. F. (1983) Comparison of the effects of UV-visible irradiation of melanins and melanin-hematoporphyrin complexes from human black and red hair. Photochem. Photobiol. 73: 63–68.

Potapenko, A. Y., Abijev, G. A., Pistsov, M. Y. Roshchupkin, D. I., Vladimirov, Y. A., Pliquett, F., Ermolayev, A. V., Sarycheva, I. K., and Evstigneeva, R. P. (1984) PUVA-induced erythema and changes in mechanoelectrical properties of skin. Inhibition by tocopherols. Arch. Dermatol. Res. 276: 12–16.

Punnonen, K., Jansen, C. T., Puntala, A., and Ahotupa, M. (1991) Effects of in vitro UVA irradiation and PUVA treatment on membrane fatty acids and activities of antioxidant enzymes in human keratinocytes. J. Invest. Dermatol. 96: 255–259.

Ranadive, N. S., and Menon, I. A. (1986) Role of reactive oxygen species from melanins in photoinduced cutaneous inflammations. Pathol. Immunopathol. Res. 5: 118–139.

Reiners, J. J., Hale, M. A., and Cantu, A. R. (1988) Distribution of catalase and its modulation by 12-O-tetradecanoylphorbol-13-myristate in murine dermis and subpopulations of keratinocytes differing in their stages of differentiation. Carcinogenesis 9: 1259–1263.

Reiners, J. J., and Rupp, T. (1989) Conversion of xanthine dehydrogenase to xanthine oxidase occurs during keratinozyte differentiation: Modulation by 12-O-tetradecanoylphorbol-13-acetate. J. Invest. Dermatol. 93: 132–135.

Rosario, R., Mark, G. J., Parrish, J. A. and Martin, C. (1979) Histological changes produced in skin by equally erythemogenic doses of UVA, UVB, UVC and UVA with psoralens. Brit. J. Dermatol. 101: 299–308.

Roshchupkin, D. I., Pistsov, M. Y., and Potapenko, A. Y. (1979) Inhibition of ultraviolet light-induced erythema by antioxidants. Arch. Dermatol. Res. 266: 91–94.

Sarna, T., and Sealy, R. C. (1984): Photoinduced oxygen consumption in melanin systems. Action spectra and quantum yields for eumelanin and synthetic melanin. Photochem. Photobiol. 39: 69–74.

Schallreuter, K. U., Pittelkow, M. R., and Wood, J. M. (1986) Free radical reduction by thioredoxin reductase at the surface of the epidermis. J. Invest. Dermatol. 87: 728–732.

Schallreuter, K. U., and Wood, J. M. (1989) Free radical reduction in the human epidermis. Free Rad. Biol. Med. 6: 519–532.

Sealy, R. C., Sarna, T., Wanner, E. J., and Reszka, K. (1984) Photosensitization of melanin: an electron spin resonance study of sensitized radical production and oxygen consumption. Photochem. Photobiol. 40: 453–460.

Tanaka, T. (1979) Skin damage and its prevention from lipoperoxide. Vitamins 53: 577–586.

Tanaka, T., and Hayakawa, R. (1986) Lipid peroxides in cosmetic products and their effect to irritate the skin. J. Clin. Biochem. Nutr. 1: 200–206.

Tezuka, T., and Takahashi, M. (1987) Stratum corneum membrane proteins in newborn rat as scavengers of lipid peroxide, in: The Biological Role of Reactive Oxygen Species in Skin, pp. 125–134. Eds O. Hayaishi et al. Elsevier, New York.

Tissié, G., Latour, E., Coquelet, C., and Bonne, C. (1990) Singlet oxygen-induced damage to rat lenses in vitro. Protection by anisyldithiothione, in: Antioxidants in Therapy and Preventive Medicine, pp. 529–532. Eds I. Emerit et al. Plenum Press, New York.

Till, G. O., Guilds, L. S., Mahrougui, M., Friedl, H. P., Trentz, O., and Ward, P. A. (1989) Role of Xanthine Oxidase in Thermal Injury of Skin. Am. J. Pathol. 135: 195–202.

Tomita, Y., and Tagami, H. (1987) The scavenging and filter effect of melanin against superoxide anion produced by photoirradiation, in: The Biological Role of Reactive Oxygen Species in Skin, pp. 95–100. Eds O. Hayashi et al. Elsevier, New York.

Weedon, D., Searle, J., and Kerr, J. F. (1979) Apoptosis. Its nature and implications for dermatopathology. Am. J. Dermatopathol. 1: 133–144.

Wilgram, G. F., Kidd, R. L., Krawezyk, W. S., and Cole, P. L. (1970) Sunburn effect on keratinosomes. Arch. Dermatol. 101: 505–519.

Wood, J. M., and Schallreuter, K. U. (1991) Studies on the reactions of human tyrosinase, superoxide anion, hydrogen peroxide and thiols. Biochim. Biophys. Acta 1074: 378–385.

Yagi, K. (1987) Lipid peroxides in the skin, in: The Biological Role of Reactive Oxygen Species in Skin, pp. 109–116. Eds O. Hayashi et al. Elsevier, New York.

Yohn, J. J., Norris, D. A., Yrastorza, D. G., Buno, I. J., Leff, J. A., Hake, S. S., and Repine, J. E. (1991) Disparate antioxidant enzyme activities in cultured human cutaneous fibroblasts, keratinocytes, and melanocytes. J. Invest. Dermatol. 97: 405–409.

Yoshino, K., Matsuo, I., and Ohkido, M. (1981) Antioxidation effects of human epidermal keratin on skin surface lipid peroxides. Jpn. J. Dermatol. 91: 1175–1179.

Young, A. R. (1987) The sunburn cell. Photodermatology 4: 127–134.

Free Radicals and Aging
ed. by I. Emerit & B. Chance
© 1992 Birkhäuser Verlag Basel/Switzerland

Modulators of free radical activity in diabetes mellitus: Role of ascorbic acid

Alan J. Sinclair[a], Joseph Lunec[b], Alan J. Girling[c] and
Anthony H. Barnett[d]

[a]University Department of Geriatric Medicine, University of Wales College of Medicine,
Cardiff CF2 1SZ; [b]Wolfson Laboratories; [c]Department of Mathematics and Statistics, and
[d]Department of Medicine, University of Birmingham B15 2TT, UK

Summary. Free radical mechanisms are increasingly being implicated in the pathogenesis of tissue damage in diabetes. Various sources of free radicals may modulate oxidative stress in diabetes, including non-enzymatic glycosylation of proteins and monosaccharide autooxidation, polyol pathway activity, indirect production of free radicals through cell damage from other causes, and reduced antioxidant reserve. Ascorbic acid, which may be a principal modulator of free radical activity in diabetes, is shown to be consumed, presumably through free radical scavenging, thus preserving levels of other antioxidants such as glutathione.

Introduction

Diabetes mellitus is a complex syndrome (or syndromes) of hyperglycaemia in association with both vascular and metabolic abnormalities. The vascular lesion consists of (1) microangiopathy, characterized by thickening of capillary basement membranes resulting in increased vascular permeability, which is clinically manifested as diabetic retinopathy and/or nephropathy, and (2) macroangiopathy, which consists of atheromatous involvement of large blood vessels, morphologically similar to non-diabetic atheroma, but tending to occur earlier and be more extensive.

The view that microangiopathy may have a free radical aetiology is supported by evidence such as increased lipid peroxidation in diabetes (Sato et al., 1979; Nishigaki et al., 1981), reduced antioxidant reserve in patients with diabetes (Illing et al., 1951; Karpen et al., 1985), and from studies showing a direct cytotoxic effect of free radicals and their metabolic end-products (lipod hydroperoxides) on vascular endothelial cells (Sacks et al., 1978; Ward et al., 1978). Figure 1 illustrates the complex interrelationships between factors likely to be involved in the pathogenesis of diabetic microangiopathy. A summary of the studies linking free radical mechanisms to diabetes is given in Figure 2. It has recently been postulated that free radicals may mediate part of the beta cell cytotoxicity of interleukin 1 (IL-1) (Nerup et al., 1988), a peptide

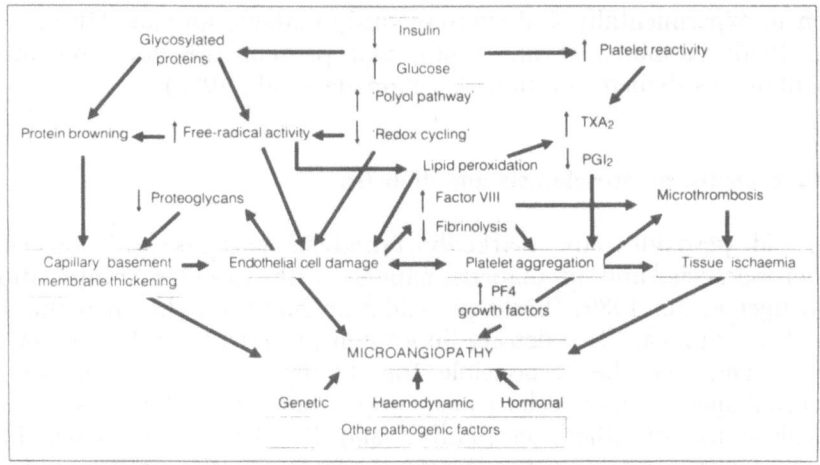

Figure 1. Pathogenesis of diabetic microangiopathy: possible mechanisms (with kind permission of the Editor of "Diabetic Medicine").

(a) **Increased activity suggested by:**

    1. Elevation of : Plasma diene conjugates

                        Serum TBA reactivity

                        Fluorescent proteins (IgG/alb)

    2. Reduction of :

                        Red cell superoxide dismutase

                        Hepatic/renal catalase

                        white cell glutathione (GSH)

(b) **Antioxidants (AOS) and the beta-cell**

    1. AOS protect against alloxan and streptozocin-induced beta-cell damage.

    2. AOS prevent diabetes development in spontaneously-diabetic animals.

Figure 2. Free radicals and diabetes.

hormone produced by macrophages, monocytes and other cell types (Dinarello et al., 1986), which may have a pathogenetic role in type 1 (insulin-dependent) diabetes. The idea for an association between IL-1 and free radicals comes from studies which show that IL-1 induces free radical formation in eosinophilis and endothelial cells (Matsubara et al., 1986; Pincus et al., 1986), and that beta cells are known to be very sensitive to free radical damage since they have a dramatically reduced antioxidant reserve (Malaisse et al., 1982; Asayama et al., 1986). This hypothesis suggests a therapeutic role for antioxidants in protecting beta cells, and this view is also supported by evidence showing that antioxidants protect culture-conditioned islet allografts from destruction in experimentally and spontaneously diabetic animals (Bradley et al., 1986; Nomikos et al., 1986), and prevent the development of spontaneous diabetes in animals (Yamada et al., 1982).

Free radicals, atherosclerosis and diabetes

Lipid peroxides are markedly raised in patients with extensive atherosclerosis, and in diabetic patients with vascular complications (Stringer et al., 1989). Lipid peroxidation has been shown to modify by chemical means low density lipoproteins (LDL) (Steinbrecher et al., 1984), and may be responsible for the in vivo transformation of macrophages to foam cells which are characteristic of the early fatty streak lesion of atheroma (Traber and Kayden, 1980). Thus, lipid peroxidation may be a fundamental process in the pathogenesis of diabetic and non-diabetic vascular disease (Betteridge, 1989).

Modulation of free radical activity in diabetes

There are various potential sources of free radical activity which may exert a modulatory effect on the level of oxidative stress in diabetes:

(a) Non-enzymatic glycosylation, ageing and free radicals

Based on biochemical and physiological similarities, diabetes mellitus has been proposed as a suitable model of premature ageing (Cerami, 1986). Glycosylation of protein (by an Amadori pathway) appears to increase its susceptibility to hydroxyl radical attack. Glucose autooxidation (catalysed by transition metals) may be the ultimate source of free radicals in this situation since some studies have shown that fragmentation of protein by glucose is inhibitable by metal chelating agents and free radical scavengers such as sorbitol and catalase (Hunt and Wolff, 1991). By a series of dehydrations and rearrangements, Amadori products can form a family of related fluorescent adducts called advanced glycosylation end-products (AGEs) (Brownlee et al., 1988). These AGEs form slowly and irreversibly over time on long-lived proteins, cell surfaces and nucleic acids, and are believed to contribute to tissue damage by protein-protein cross-linking. Examples of this damage include cataractous opacification of the human lens, and increased rigidity of collagen. Accumulation of AGEs increases with age and in patients with diabetes (Monnier et al., 1984). In a recent study, serum concentrations of AGEs were markedly elevated in patients with diabetic nephropathy (Makita et al., 1991).

AGE-specific receptors on macrophages and monocytes are thought to be involved with the expression of several growth factors such as tumor necrosis factor (TNF), interleukin-1, and insulin-like growth factor (Vlassara et al., 1988); this raises the possibility that these interactions may be of importance in the pathogenesis of atherosclerosis during ageing, and in patients with diabetes (Kirstein et al., 1990).

(b) The polyol (sorbitol) pathway

This pathway, which comprises the rate-limiting enzyme aldose reductase (AR), converts glucose to sorbitol and has been implicated in the pathogenesis of diabetic complications (Fig. 3) (Cogan, 1984). AR is widely distributed in mammalian tissues, which are susceptible to diabetic complications such as the peripheral nerve, retina, lens of the eye or renal glomerulus. With hyperglycaemia and increased flux through the pathway, intracellular sorbitol accumulates and promotes functional and structural changes in tissues through several mechanisms, including exerting a powerful osmotic effect and depleting myoinositol, which is an important component of membrane phospholipids and determinant of neuronal adenosine triphosphatase (ATPase) (Green et al., 1987). Increased polyol pathway activity results in a depletion of both NADPH and NAD^+ with consequent major alterations in cellular redox potentials. This leads to a reduction in the antioxidant, glutathione, rendering cells less likely to cope with increased oxidative stress. Inhibitors of AR have been demonstrated to prevent diabetic complications in animal models, providing the rationale for current clinical trials of these agents (Collier and Small, 1991).

(c) Free radicals and secondary cell damage in diabetes

Tissue damage, which is associated with inactivation of antioxidants and release of metal ions which are potent catalysts of free radical

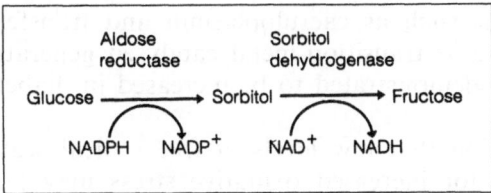

Figure 3. The polyol pathway. NADPH = reduced nicotinamide adenine dinucleotide phosphate. NADH = reduced NAD.

reactions, can itself lead to lipid peroxidation (Halliwell and Gutteridge, 1984). This raises the possibility that the level of lipid peroxidation in diabetes may reflect several contributory mechanisms, namely, increased free radical activity as a primary event in diabetes; unchecked free radical activity due to reduced antioxidant defences; tissue damage from the diabetic process itself; and tissue damage from other causes. We have already discussed the possible role of free radicals in atherogenesis. It is not unreasonable to speculate that vascular wall damage at all stages from the earliest intimal lesion to ulceration and plaque formation results in a superimposed flux of increased free radical reactions. Essential hypertension which rises steadily with age in the population often co-exists with diabetes and may be another source of free radical activity. It has been demonstrated that the vascular lesions of hypertension can be produced by free radical reactions (Selwign, 1983). Further support for a role for free radicals in the aetiology of hypertension comes from the recent Kuopio Ischaemic Heart Risk Factor Study in Finnish men (Salonen et al., 1988) which showed that a marked elevation of blood presssure was associated with low levels of both plasma ascorbic acid and serum selenium concentrations. However, it is as valid an argument as in the case of atheroma, that blood vessel damage in hypertension (which is a combination of hyperplastic arteriosclerosis and hyaline sclerosis) may initiate further free radical reactions.

(d) Reduced antioxidant reserve in diabetes

Antioxidant defences appear to be compromised in patients with diabetes. Glutathione has been found to be decreased in erythrocytes (Seltzer, 1957), plasma (Awadalla et al., 1978; Chari et al., 1984), and in platelets (Thomas et al., 1985). These are not universal findings since other investigators have found GSH to be unchanged in diabetes (Som et al., 1981; Banerjee, 1982). Further evidence of increased oxidative stress in diabetes is implied from studies demonstrating reduced vitamin E (Karpen et al., 1984), uric acid (Herman et al., 1976), and reduced activity of specific free radical scavenging enzymes such as superoxide dismutase (Arai et al., 1987; Collier et al., 1989), and catalase (Matkovics et al., 1982). In addition, iron-oxidising and iron-binding proteins such as caeruloplasmin and transferrin, whose role may be to decrease transition-metal catalysed generation of free radicals, have been demonstrated to be increased in diabetes (Jones et al., 1988).

The assumption that low levels of free radical scavengers is prima fascie evidence for increased oxidative stress may not be valid. For example, such findings may represent an adaptation to reduced oxidative stress. This latter conclusion is supported by a recent study demonstrating decreased diene conjugates in diabetic patients (Collier et al., 1988); however, in this latter study, a non-peroxide isomer of linoleic acid (PL-9, 11-LA) was measured, and its validity as a marker of free radical activity has been questioned (Sinclair et al., 1989). Also, most studies are unable to account for the possible contributions of type and duration of diabetes, presence or absence of complications and glycaemic control on the measured concentrations of antioxidants.

Ascorbic acid and diabetes

Reduced levels and altered metabolic turnover of ascorbic acid (AA, vitamin C) have been reported in several tissues in experimentally induced diabetes (Yew, 1983; Rikans, 1981; Yue et al., 1989), and in diabetic patients (Som et al., 1981). There are reports that high-dose vitamin C regimens are associated with reversal of early signs of retinopathy (Crary and McCarty, 1984), and normalisation of capillary strength in diabetes mellitus (Cox and Butterfield, 1975). In the elderly non-diabetic population, AA is often deficient and may be correctable by supplementation (Neale et al., 1988) – this problem might be of particular significance in elderly diabetic patients who are reported to show more rapid progress of some diabetic complications (Caird, 1982).

Ascorbic acid functions as an important component of cellular defence against oxygen toxicity and lipid peroxidation caused by free radical mechanisms (Procter and Reynold, 1984; Levine, 1986). During scavenging of free radicals it is converted to dehydroascorbic acid, DHA, in serum and in the mitochondrial fractions of various tissues (Siperstein et al., 1968). It is also a co-factor in the biosynthesis and post-translational modification of collagen due to interaction with proline hydroxylase (Barnes, 1976; McLennan et al., 1988), and has been demonstrated to affect a variety of biochemical processes as indicated in Table 1. These observations suggest a possible close interrelationship between ascorbic acid and pathways known to be influenced by the diabetic process.

We have recently demonstrated that serum ascorbate concentrations in elderly diabetic patients are markeldy reduced in comparison with age-matched control subjects, and this reduction is more pronounced in

Table 1. Biochemical roles of ascorbic acid.

Free radical scavenging
Collagen biosynthesis
Leucocyte function/chemotaxis
Platelet activation
Prostaglandin metabolism
Polyol pathway

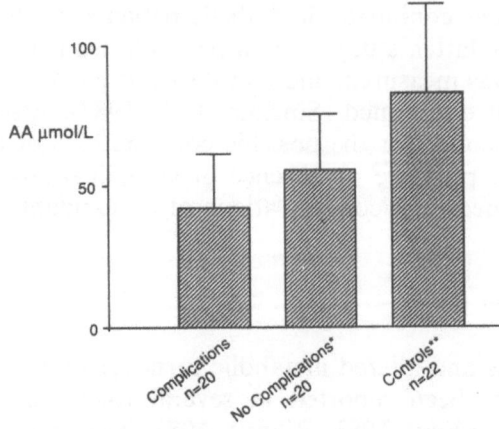

* p<0.01, compared with complications group
** p<0.001, compared with complications group

Figure 4. Baseline ascorbic acid concentrations in diabetic patients and controls.

those patients with diabetic retinopathy (Fig. 4) (Sinclair et al., 1991). Supplementation of the diet for six weeks with ascorbic acid, 1 gram daily by mouth, was able to correct the deficiency but highlighted a disturbance of ascorbate metabolism in the diabetic patients whose serum concentrations fell despite continued supplementation (Fig. 5). In this study, we also measured three other 'markers' of free radical activity. The results are given in Table 2, which shows that there were

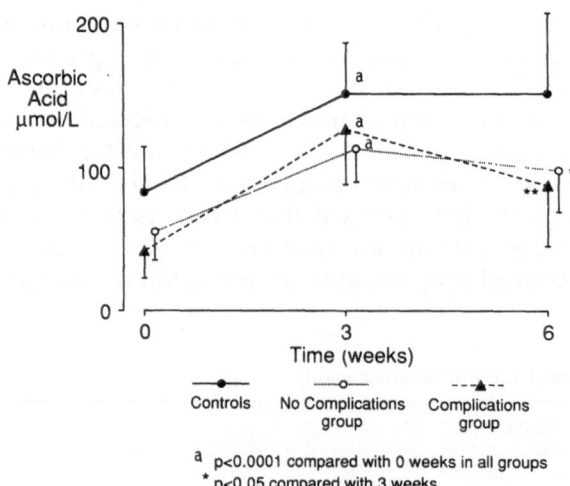

a p<0.0001 compared with 0 weeks in all groups
* p<0.05 compared with 3 weeks
** p<0.01 compared with 3 weeks

Figure 5. Ascorbate concentrations in diabetic patients and controls during ascorbate supplementation.

Table 2. Values for glutathione (GSH), diene conjugates (DC) and thiobarbituric acid (TBA) reactivity in diabetic patients and controls at baseline and during ascorbate supplementation.

Group/Time (wks)		GSH ug/ml		DC OD u/ml		TBA nmol MDA	
Complications	0	276.8	(33.0)	0.41	(0.20)	2.04	(0.82
	3	285.4	(35.8)	0.41	(0.19)	2.11	(0.88)
	6	288.0	(31.6)	0.44	(0.25)	2.13	(0.81)
No complications	0	280.1	(66.2)	0.36	(0.18)	2.36	(0.92)
	3	279.9	(51.4)	0.40	(0.21)	2.12	(0.7)
	6	280.2	(50.3)	0.38	(0.21)	2.53	(1.25) *
Controls	0	270.3	(41.4)	0.33	(0.10)	2.22	(0.95)
	3	293.2	(45.8)	0.36	(0.18)	2.06	(0.77)
	6	273.2	(45.1)	0.33	(0.09)	1.66	(0.61) *

* p < 0.05, compared with 3 and 6 weeks for log-transformed data.

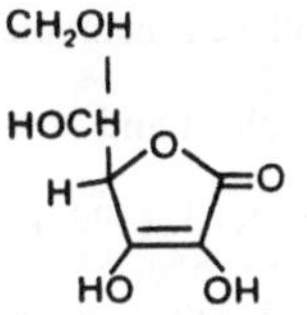

Figure 6. Molecular structures of L-ascorbic acid and D-glucose.

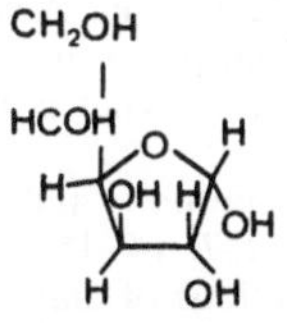

no major differences between the patients and controls at baseline or during antioxidant supplementation for any of the 'markers' (Sinclair et al., 1992).

The finding of reduced ascorbate in patients with diabetes is consistent with previous reports, although for the first time oxidative stress is shown to be significantly higher in those with microangiopathy. Consumption of ascorbate is likely to be at the expenses of other antioxidants (such as vitamin E or glutathione), and this might account for similar values for free radical markers in patients and controls. An

350

alternative explanation is that free radical activity is not increased in diabetes, and that ascorbate loss is due to other mechanisms. For example, studies using human lymphocytes suggest that competitive inhibition between glucose and ascorbic which have structural similarities (Fig. 6), for transport across cell membranes may be responsible for part of the reduction of ascorbate in patients with diabetes (Davis et al., 1983).

Although the biological importance of these observations remains unclear, we feel that further investigation from longer term controlled trials of vitamin C supplementation in patients with diabetes are required.

Arai, K., Iizuka, S., Tada, Y., Oikawa, K., and Taniguchi, N. (1987) Increase in the glycosylated form of erythrocyte Cu–Zn–SOD in diabetes and close association of non-enzymic glycosylation with enzyme activity. Biochim. Biophys. Acta. 924: 292–296.

Asayama, K., Kooy, N. W., and Burr, I. M. (1986) Effect of vitamin E deficiency and selenium deficiency on insulin secretory reserve and free radical scavenging systems in islets: decrease of islet manganosuperoxide dismutase. J. Lab. Clin. Med. 107: 459–464.

Awadalla, R., El-Dessoukey, E. A., Doss, H., and Klalifa, K. (1978) Blood-reduced glutathione, serum caeruloplasmin and mineral changes in juvenile diabetes. Z. Ernahrungsweiss. 17: 72–78.

Banjerjee, A. (1982) Blood dehydroascorbic acid and diabetes mellitus in human beings. Ann. Clin. Biochem. 19: 65–70.

Barnes, M. J. (1976) Function of ascorbic acid in collagen metabolism. Ann. N. Y. Acad. Sci. 258: 264–275.

Betteridge, D. J. (1989) Diabetes, Lipoprotein metabolism and atherosclerosis. Br. Med. Bull. 45: 285–311.

Bradley, B., Prowse, S. J., Bauling, P., and Lafferty, K. J. (1986) Desferrioxamine treatment prevents chronic islet allograft damage. Diabetes 35: 550–555.

Brownlee, M., Cerami, A., and Vlassara, H. (1988) Advanced glycosylation end products in tissue and the biochemical basis of diabetic complications. N. Engl. J. Med 318: 1315–1321.

Caird, F. I. (1982) Complications of diabetes in old age, in: Advanced Geriatric Medicine. Evans, J. G. and Caird, F. I. eds. Pitman, London, pp. 3–9.

Cerami, A. (1986) Ageing of proteins and nuclei acids. What is the role of glucose? Trends Biol. Sci. 11: 311–314.

Chari, S. W., Nath, N., and Rathi, A. B. (1984) Glutathione and its redox system in diabetic polymorphonuclear leucocytes. Am. J. Med. Sci. 287: 14–15.

Cogan, D. G. (1984) Aldose reductase and complications of diabetes. Ann. Intern. Med. 101: 82–91.

Collier, A., Jackson, M., Dawkes, R. M., Bell, D., and Clarke, B. F. (1988) Reduced free radical activity detected by decreased diene conjugates in insulin-dependent diabetic patients. Diabetic Med. 5: 747–749.

Collier, A., Wilson, R., Bradley, H., Thomson, J. A., and Small, M. (1989) Free radical activity in type 2 diabetes. Diabetic Med. 7: 27–30.

Collier, A., and Small, M. (1991) The role of the polypol pathway in diabetes mellitus. Br. J. Hosp. Med. 45: 38–40.

Cox. B. D., and Butterfield, W. J. H. (1975) Vitamin C supplements and diabetic cutaneous capillary fragility. Br. Med. J. 3: 205.

Crary, E. J., and McCarty, M. F. (1984) Potential clinical applications for high dose nutritional antioxidants. Med. Hypothesis. 13: 77–98.

Davis, K. A., Lee, W. Y. L., and Labbe, R. F. (1983) Energy dependent transport of ascorbic acid into lymphocytes. Fed Proc. 42: 2011.

Dinarello, C. A. (1986) Interleuken-1:amino acid sequences, multiple biological activities and comparison with tumor necrosis factor (cachetin). Year Immunol. 2: 68–89.

Greene, D. A., Lattimer, S. A., and Sima, A. A. F. (1987) Sorbitol, phosphoinositides, and sodium-potassium-ATPase in the pathogenesis of diabetic complications. N. Engl. J. Med. 316: 599–606.

Halliewll, B., and Gutteridge, J. M. C. (1984) Lipid peroxidation, oxygen radicals, cell damage, and antioxidant therapy. Lancet. i: 1396–1397.

Herman, J. B., Medalie, J. H., and Goldbourt, U. (1976) Diabetes, prediabetes and uricaemia. Diabetolagra 12: 47–52.

Hunt, J. V., and Wolff, S. P. (1991) Oxidative glycation and free radical production: a causal mechanism of diabetic complications. Free Rad. Res. Comms. 12/13: 115–123.

Illing, E. K., Gray, C. H., and Lawrence, R. D. (1951) Blood glutathione and non-glucose substances in diabetes. Biochem. J. 48: 637–640.

Jones, A. F., Winkles, J. W., Jennings, P. E., Florkowski, C. M., Lunec, J., and Barnett, A. H. (1988) Serum antioxidant activity in diabetes mellitus. Diabetes Res. 7: 89–92.

Karpen, C. W., Cataland, S., and O'Dorisio, T. M. (1985) Production of 12 HETE and vitamin E status in platelets from type 1 human diabetic subjects. Diabetes 34: 526–531.

Kirstein, M., Brett, J., Radoff, S., Ogawa, S., Stern, D., and Vlassara, H. (1990) Advanced protein glycosylation induces transendothelial human monocyte chemotaxis and secretion of platelet-derived growth factor: role in vascular disease of diabetes and ageing. Proc. Natl. Acad. Sci. 87: 9010–9014.

Levine, M. (1986) New concepts in the biology and biochemistry of ascorbic acid. N. Eng. J. Med. 314: 892–902.

Makita, Z., Radoff, S., Rayfield, E. J., Yang, Z., Skolnik, E., Delaney, V., Friedman, E. A., Cerami, A., and Vlassara, H. Advanced glycosylation end products in patients with diabetic nephropathy. New Engl. J. Med. 325: 836–842.

Malaisse, W. J. (1982) Alloxan toxicity to the pancreatic beta-cell: a new hypothesis. Biochem. Pharmacol. 31: 3527–3534.

Matkovics, B., Varga, S., and Seabo, L. (1982) The effect of diabetes on the activity of the peroxide metabolising enzymes. Horm. Metab. Res. 14: 77–79.

Matsubara, T., and Ziff, M. (1986) Increased superoxide anion from human endothelial cells in response to cytokines, J. Immunol. 137: 3295–3298.

McLennan, S., Yue, D. K., Fisher, E., Capogreco, C., Heffernan, S., Ross, G. R., and Turtle, J. R. (1988) Deficiency of ascorbic acid in experimental diabetes: relationship with collagen and polyol pathway abnormalities. Diabetes 37: 359–361.

Monnier, V. M., Kohn, R. R., and Cerami, A. (1984) Accelerated age-related browning of human collagen in diabetes mellitus. Proc. Natl. Acad. Sci. 81: 583–587.

Neale, R. J., Lim, H., Turner, J., and Freeman, J. R. (1988) The excretion of large vitamin C loads in young and elderly subjects: an ascorbic acid tolerance test. Age and Ageing 17: 35–41.

Nerup, J., Mandrup-Poulson, T., Molvig, J., Helqvist, S., Wogensen, L., and Egeberg, J. (1988) Mechanisms of pancreatic beta-cell destruction in type 1 diabetes. Diabetes Care 11: 16–23.

Nishigaki, I., Hagihara, M., Tsunekawa, H., Maseki, M., and Yagi, K. (1981) Lipid peroxide levels of serum lipoprotein fractions of diabetic patients. Biochem. Med. 25: 373–378.

Nomikos, I. N., Prowse, S. J., Carotenuto, P., and Lafferty, K. J. (1986) Combined treatment with nicotinamide and desferrioxàmine prevents islet allograft destruction in NOD mice. Diabetes 35: 1302–1304.

Pincus, S. H., Whitcomb, E. A., and Dinarello, C. A. (1986) Interaction of IL-1 and TPA in modulation of eosinophil function. J. Immunol. 137: 3509–3514.

Procter, P. H., and Reynolds, E. S. (1984) Free radicals and disease in man. Physiol. Chem. Phys. 16: 175–195.

Rikans, L. E. (1981) Effect of alloxan diabetes on rat ascorbic acid. Horm. Metab. Res. 13: 123.

Sacks, T., Moldow, C. F., Craddock, P. R., Bowers, T. K., and Jacob, H. S. (1978) Oxygen radicals mediate endothelial cell damage by complement-stimulated granulocytes. J. Clin. Invest. 61: 1161–1167.

Salonen, J. T., Salonen, R., Ihanainen, M., Parviainen, M., Seppanen, R., Kantola, M., Seppanen, K., and Rauramaa, R. (1988) Blood pressure, dietary fats, and antioxidants. Am. J. Clin. Nutr. 48: 1226–1232.

Sato, Y., Hotta, N., Sakamoto, N., Matsuoka, S., Ohishi, N., and Yagi, K. (1979) Lipid peroxide level in plasma of diabetic patients. Biochem. Med. 21: 104–107.

352

Seltzer, H. S. (1957) Blood glutathione in mild diabetes mellitus before treatment and during sulphonylurea-induced hypoglycaemia. Proc. Soc. Exp. Biol. Med. 95: 74–76.

Selwign, A. P. (1983) The cardiovascular system and radiation. Lancet 2: 152–154.

Siperstein, M. D., Unger, R. H., and Madison, L. L. (1968) Studies of muscle capillary basement membranes in normal subjects, diabetic and pre-diabetic patients. J. Clin. Invest. 47: 1973–1999.

Som, S., Basu, S., Mukherjee, D., Deb. S., Choudary, R., Mukherjee, S., Chatterjee, S. N., and Chatterjee, I. B. (1981) Ascorbic acid metabolism in diabetes mellitus. Metabolism 30: 572–577.

Sinclair, A. J., Lunec, J., and Barnett, A. H. (1989). Diene conjugates and microangiopathy. Diabetic Med. 6: 458.

Sinclair, A. J., Girling, A. J., Gray, L., Le Guen, C., Lunec, J., and Barnett, A. H. (1991) Disturbed handling of ascorbic acid in diabetic patients with and without microangiopathy during high dose ascorbate supplementation. Diabetologia 34: 171–175.

Sinclair, A. J., Girling, A. J., Gray, L., Lunec, J., and Barnett, A. H. (1992) An investigation of the relationship between free radical activity and vitamin C metabolism in elderly diabetic subjects with retinopathy. Gerontology (in press).

Steinbrecher, U. P., Parthasarathy, S., Leake, D. S., Witztum, J. L., and Steinberg, D. (1984) Modification of low density lipoprotein by endothelial cells involves lipid peroxidation and degradation of low density lipoprotein phospholipids. Proc. Natl. Acad. Sci. 81: 3883–3887.

Stringer, M. D., Gorog, P. G., Freeman, A., and Kakkar, V. V. (1989) Lipid peroxides and atherosclerosis. Br. Med. J. 298: 281–284.

Thomas, G., Skrinska, V., Lucas, F. V., and Schumacher, O. P. (1985) Platelet glutathione and thromboxane synthesis in diabetes. Diabetes 34: 951–954.

Trabser, M. G., and Kayden, H. J. (1980) Low density lipoprotein receptor activity in human monocyte-derived macrophages and its relation to atheromatous lesions. Proc. Natl Acad. Sci. 77: 5466–5470.

Vlassara, H., Brownlee, M., Manogue, K., Dinarello, C., and Pasagian, A. (1988) Cachectin/TNF and IL-1 induced by glucose-modified proteins: role in normal tissue remodelling. Science 204: 1546–1548.

Ward, P. A., Till, G. O., Kunkel, R., and Beauchamp, G. (1983) Evidence for role of hydroxyl radical in complement and neutrophil-dependent tissue injury. J. Clin. Invest. 72: 789–801.

Yamada, K., Nonaka, K., Hanafusa, T., Miyazaki, A., Toyoshima, H., and Tarui, S. (1982) Preventive and therapeutic effects of large-dose nicotinamide injections on diabetes associated with insulitis: an observation in nonobese diabetic (NOD) mice. Diabetes 31: 749–753.

Yew, M. S. (1983) Effect of streptozocin diabetes on tissue absorbic acid and dehydroascorbic acid. Horm. Metab. Res. 15: 158.

Yue, D. K., McLennan, S., Fischer, E., Heffernan, S., Capogreco, C., Ross, G. R., and Turtle, J. R. (1989) Ascorbic acid metabolism and the polyol pathway. Diabetes 38: 257–261.

Free Radicals and Aging
ed. by I. Emerit & B. Chance
© 1992 Birkhäuser Verlag Basel/Switzerland

Possible role of free radicals in the chronic inflammation of the gut

Toshikazu Yoshikawa, Shuji Takahashi and Motoharu Kondo

First Department of Medicine, Kyoto Prefectural University of Medicine, Kamigyo-ku, Kyoto 620 Japan

Summary. Recent studies have demonstrated that intracolonic administration of trinitrobenzenesulfonic acid (TNB) dissolved in ethanol produces chronic colitis in rats, and that this model shares many features of human inflammatory bowel disease (IBD), particularly Crohn's disease. We investigated the role of free radicals in the pathogenesis of this colitis model. In the early stage of this colitis, antioxidant enzymes (such as superoxide dismutase, glutathione peroxidase) and an antioxidant, α-tocopherol, were significantly decreased with the severity of colonic damage. Mn-SOD at a dose of 50000 U/kg attenuated this colitis when preadministered subcutaneously one hour before the induction of colitis. These results suggest that oxygen-derived free radicals may play an important role in this colitis.

Introduction

In recent years, oxygen-derived free radicals have been shown to be closely involved in many experimental and clinical diseases, and the possibilities of prevention or treatment of these disorders by enhancing the antioxidant capacity of the host has been intensively investigated. Previous studies of our department have indicated that impaired antioxidative activity may be a factor in the development of gastric mucosal injuries. Indeed these are more frequent in elderly persons, whose gastric mucosa has significantly lower α-tocopherol levels than that of young people. In this study we extended our investigations to chronic inflammatory bowel disease, also a frequent disorder in the elderly.

The role of oxygen-derived free radicals was studied in the early stage of experimentally induced colitis. Recent studies have demonstrated that intracolonic administration of 2,4,6-trinitrobenzenesulfonic acid (TNB) dissolved in ethanol results in severe, transmural, granulomatous inflammation of the distal colon, which evolves for several weeks in rats. This model shares many of the histopathological and clinical features of human inflammatory bowel disease (IBD), particularly Crohn's disease, and may provide an opportunity to elucidate the pathophysiology of IBD in humans.

Induction of colitis

Colonic inflammation was induced using the technique described by Morris et al. (1989). Briefly, after a 48-h fast, male Wistar rats (190–210 g) were lightly anesthetized with ether, and a rubber catheter (OD, 2 mm) was inserted into the lumen of the colon via the anus. The catheter was advanced so that the tip was 8 cm proximal to the anus. TNB (Wako Pure Chemical Industry, Osaka, Japan) dissolved in 50% ethanol (120 mg/ml) was injected into the colon through the catheter (total volume of 0.25 ml per rat). After injection, the anus was occluded with a clip for 60 min. Initially, a dose-response study was performed in which TNB was administered at doses of 10, 20, or 30 mg per rat in the 50% ethanol vehicle. And another dose-response study was also performed in which ethanol was administered at concentrations of 10, 20, or 50%, containing 30 mg of TNB per rat. The rats were killed one week after the enema. Based on the results of these study, the rats in experimental group were administered 30 mg TNB in 0.25 ml of 50% ethanol (TNB/ethanol group). In control experiments, rats received 0.25 ml of either 50% ethanol alone (ethanol group), or 30 mg of TNB in 0.9% saline (TNB/saline group).

Assessment of colonic inflammation and damage

At various times after intracolonic administration, several rats from each treatment group were sacrificed by exsanguination via the abdominal aorta. The distal colon was removed, opened by a longitudinal incision, and immediately examined both macroscopically, and microscopically. The macroscopic lesions were scored on a 0–5 scale according to the damage score by Morris et al. (1989). Colonic myeloperoxidase (MPO) activity and wet weight of the distal 8 cm of the colon were determined as indices of inflammation. The tissue levels of MPO activity were determined using the technique described by Krawisz et al. (1984).

Measurement of α-tocopherol and antioxidant enzymes

The levels of α-tocopherol in serum and colonic mucosa were determined by the method of Abe (1975). The superoxide dismutase (SOD) activity in colonic mucosa was assayed by the method of Nakano et al. (1990), and the glutathione peroxidase (GSH·Px) activity assayed by the method of Gunzler et al. (1974).

Effects of treatment with radical scavengers

Cu,Zn-SOD (recombinant human SOD, Nippon Kayaku Co., Ltd., Tokyo) or Mn-SOD (Unitika Ltd., Osaka, Japan) was administered subcutaneously at doses of 90000 U/kg, or 50000 U/kg respectively one hour before the intracolonic injection. The rats were sacrificed 7 days after administration of TNB, the colonic damage was scored and the wet weight of distal colon was measured.

Results

The severity of colonic damage induced by TNB/ethanol increased both with the TNB amount administered and with the ethanol concentration (Figs. 1 and 2). With doses of TNB of 10–30 mg, damage score and colonic wet weight increased in a dose-related manner. There was a significant difference between the group with 30 mg TNB in 50% ethanol and the group with 50% ethanol alone. But there was no significant difference between the group with 30 mg TNB in 10 or 20% ethanol and the group with 50% ethanol alone, and the TNB/saline group produced no colonic lesions.

All of the animals that received 30 mg of TNB in 50% ethanol developed areas of grossly visible bowel wall thickening, inflammation, and ulcers. Such lesions were most severe one or two weeks after the induction of colitis, and persisted for up to 6 weeks (Fig. 3). In the rats

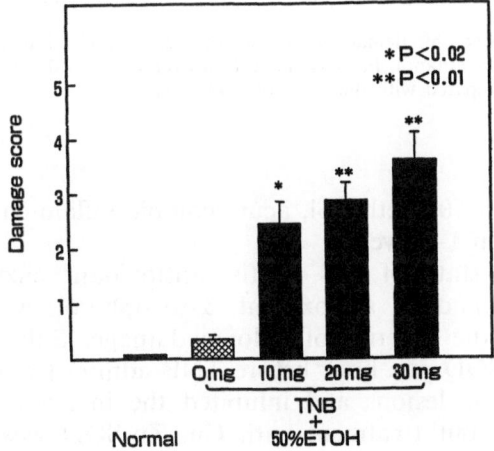

Figure 1. Effect of amount of TNB dissolved in 50% ethanol on the colonic damage score 7 days after enema. Each value indicates the mean $\pm$ SE of 6–8 rats. *$p < 0.02$, **$p < 0.01$ when compared with the value of control group.

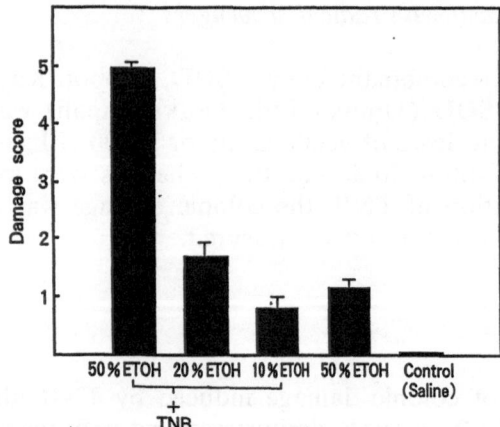

Figure 2. Effect of ethanol concentration on the damage score of colitis 7 days after enema. Each value indicates the mean ± SE of 6–8 rats.

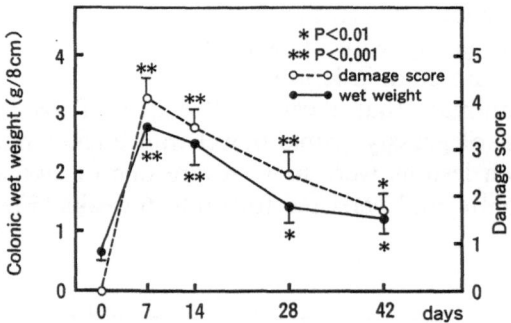

Figure 3. Changes of colonic damage score and wet weight after administration of 50 mg of TNB dissolved in 50% ethanol. Each value indicates the mean ± SE of 6–8 rats. *$p < 0.01$, **$p < 0.001$ when compared with the value of control group.

treated only with 50% ethanol, acute colonic inflammation did occur, but was healed in 1–2 weeks.

At the early state of this colitis, antioxidant enzymes (such as SOD, GSH·Px) and an antioxidant, α-tocopherol, were significantly decreased with the severity of colonic damage. Subcutaneous treatment with Mn-SOD one hour before TNB administration significantly attenuated colonic lesions and inhibited the increase of colonic wet weight (Fig. 4), but treatment with Cu, Zn-SOD was inadequate in inhibiting the colonic lesions (data not shown). Preadministration of Mn-SOD did not attenuate colonic damage induced by 50% ethanol alone.

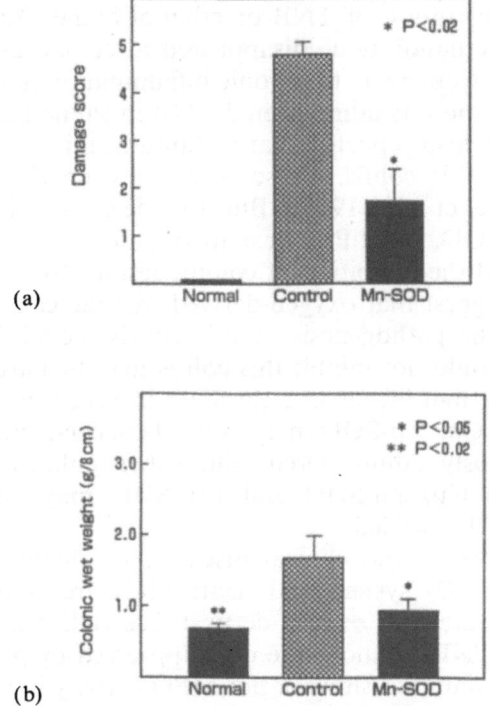

(a)

(b)

Figure 4. Effect of Mn-SOD (50000 U/kg, s.c.) on the increase of colonic damge score (a) and wet weight (b) 7 days after enema of 50 mg TNB dissolved in 50% ethanol. Each value indicates the mean ± SE of 6 rats.

Discussion

Despite the clinical evidence (Shiratora et al., 1989; Suematsu et al., 1987) supporting a role of oxygen-derived free radicals in IBD, the demonstration of this role has been made difficult by the paucity of reproducible and histopathologically relevant animal models of IBD. But this TNB model designed by Morris et al. (1989), a simple, reproducible, and inexpensive model, has many similarities with human IBD, particularly Crohn's disease. It has been suggested by Little et al. (1966) that TNB, when ethanol acts as a mucosal barrier breaker, binds easily to a substance of high molecular weight (such as cell-surface protein) in colonic tissue, resulting in subsequent immunological responses, but the precise mechanisms by which TNB dissolved in ethanol caused chronic inflammation in the colon are not apparent.

In this study, we have demonstrated that intracolonic administration of TNB dissolved in ethanol causes long-lasting ulceration and inflammation of the distal colon in rats. This chronic colitis model was not

358

produced by administration of TNB or ethanol alone. Acute mucosal lesions induced by ethanol alone disappeared after one or two weeks, and the progression from acute to chronic inflammation never occurred. On the other hand, the rats administered TNB in saline had no inflammatory lesions. It is also reported that intramural injection of TNB in saline can produce this colitis. These results are consistent with the hypothesis of Little et al. (1966). But the decrease of antioxidant enzymes (such as SOD, GSH·Px), or α-tocopherol in colonic tissue at the acute phase, and the inhibition of colonic lesions by preadministration of Mn-SOD suggest that oxygen-derived free radicals may play an important role in the pathogenesis of this colitis model. The reasons why Cu, Zn-SOD could not inhibit this colitis may be partly explained by the fact that the half-life of Cu, Zn-SOD in serum is shorter than that of Mn-SOD. Cu, Zn-SOD may have beneficial effects on this colitis, if continuously administered. But other unknown biological differences between Cu, Zn-SOD and Mn-SOD may be important. Further studies will be needed.

In many experimental and clinical diseases, the hypoxanthine-xanthine oxidase (HX-XO) system and neutrophils are considered very important as the sources of oxygen-derived free radicals. The colonic lesions induced by TNB/ethanol were not suppressed by preadministration of allopurinol (data not shown), but MPO activity in colonic tissue significantly increased and the lesions were attenuated in neutropenic rats (data not shown). These results suggest that neutrophils may be more important than the HX-XO system as a source of free radicals in this model.

Our data suggest that oxygen-derived free radicals, especially derived from neutrophils, may play an important role in the onset of chronic colitis induced by TNB dissolved in ethanol.

Abe, K., Yoguchi, Y., and Katsui, G. J. (1975) J. Butr. Sci. Vitaminol. 21: 183–188.
Gunzler, W. A., Kremers, H., and Flohe, L. (1974) An improved coupled test procedure for glutathione peroxidase (EC 1.11.1.9.) in blood. Z. Klin. Chem. Klin. Biochem. 12: 444–448.
Krawisz, J. E., Sharon, P., and Stenson, W. F. (1984): Quantitative assay for acute intestinal inflammation based on myeloperoxidase activity. Gastroenterology 87: 1344–1350.
Little, J. R. and Eison, H. N. (1966) Preparation and characterization of antibodies specific for the 2,4,6-trinitrophenyl group. Biochem. 5: 3385–3395.
Morris, G. P., Beck, P. L., Herridge, M. S., Depew, W. T., Szewczuk, M. R., and Wallace, J. L. (1989) Hapten-induced model of chronic inflammation and ulceration in the rat colon. Gastroenterology 96: 795–803.
Nakano, M., Kimura, H., Hara, M., Kuroiwa, M., Kato, M., Totsune, K., and Yoshikawa, T. (1990) A highly sensitive method for determining both Mn- and Cu-Zn superoxide dismutase activities in tissues and blood cells. Analyt. Biochem. 187: 277–280.
Shiratora, Y., Aoki, S., Takada, H., Kiriyama, H., Ohto, K., Hai, K., Teraoka, H., Matano, S., Matsumoto, K., and Kamii, K. (1989) Oxygen-derived free radical generating capacity of polymorphonuclear cells in patients with ulcerative colitis. Digestion 44: 163–171.
Suematsu, M., Suzuki, M., Kitahora, T., Miura, S., Suzuki, K., Hibi, T., Watanabe, M., Nagata, H., Asakura, H., and Tsuchiya, M. (1987) J. Clin. Lab. Immunol. 24: 125–128.

Free Radicals and Aging
ed. by I. Emerit & B. Chance
© 1992 Birkhäuser Verlag Basel/Switzerland

Age-related variations of enzymatic defenses against free radicals and peroxides

Yves Artur[a], Bernard Herbeth[a, b], Laila Guémouri[a], Edith Lecomte[a], Claude Jeandel[c] and Gérard Siest[a]

[a]*Centre de Médecine Préventive, 2 avenue du Doyen J. Parisot, F-54500 Vandoeuvre-lès-Nancy, France;* [b]*INSERM U115, Faculté de Médecine, BP 184, F-54505 Vandoeuvre-lès-Nancy, France; and* [c]*Service de Médecine Interne B, Centre Hospitalier Régional Universitàire, Avenue de Bourgogne, F-54500 Vandoeuvre-lès-Nancy, France*

Summary. Superoxide dismutase, glutathione peroxidase and catalase are the three main enzymatic systems of defense of the organism against free radicals and peroxides. A survey of the literature shows that no general tendency of evolution of these systems in aging emerges, even if some recent studies in humans demonstrate the existence of a concomitant decrease in most of the antioxidant enzymes in blood of the elderly. The study of the antioxidant systems and their interrelations in the elderly represents a large field of future investigations.

Introduction

The free radical theory of aging postulates that free radical reactions contribute in a large way to the process of aging (Harman, 1988). At least, as recalled by Pryor (1987), oxygen-derived reactive chemical species are involved in the etiology and development of diseases, especially chronic diseases that shorten life below the maximum life span possible for the organism.

The biological effects of free radicals and peroxides are controlled in vivo by a large number of antioxidative defense mechanisms: Vitamins E and C, carotenoids, metabolites such as glutathione and uric acid, and antioxydant enzymes. Among these enzymes, superoxide dismutase (SOD; EC 1.15.1.1.) catalyzes dismutation of the superoxide anion (O_2^-) into hydrogen peroxide (H_2O_2), catalase (CAT; EC 1.11.1.6.) detoxifies H_2O_2, and glutathione peroxidase (GPX; EC 1.11.1.9.) both detoxifies H_2O_2 and converts lipid hydroperoxides into nontoxic alcohols. Logic suggests that tissue damage could occur when production of free radicals and peroxides is excessive or when the defense mechanisms are deficient. Still very little is known concerning a possible age-related variation in this balance between attack and defense. One major reason for this is the obvious lack of reliable biological markers for evaluating oxidative stress in vivo (Hotz et al., 1987). Conversely, literature provides information concerning evolution of some tissular protection systems in aging. In the following paragraphs we review the data concerning age-related variations of SOD, GPX and CAT in animals and humans.

Age-related variations of antioxidant enzymes in animals

Table 1 summarizes the main published results concerning age-related variations of antioxidant enzymes in animal tissues, mostly obtained in rodents. Data from the various authors appear somewhat conflicting,

Table 1. Age-related variations of antioxidant enzymes in animal tissues

Authors	Animals, age	SOD	CAT	GPX
Pinto and Bartley (1969)	Rats, 18 m vs 4 m			↑:L
Kellog and Fridovich (1976)	Rats, 26 m vs 8 m	Cu-Zn SOD = :B, L Mn-SOD = :B Mn-SOD ↓:L		
Hazelton and Lang (1985)	Mouses, 36 m vs 10 m			↓:B, H, L
Del Maestro and McDonald (1989)	Rats, 10 m vs 3 m	Cu-Zn SOD ↑:B Mn-SOD = :B	↑B	↓:B
Laganière and Yu (1989)	Rats, 24 m vs 12 m		↓:L	↓:L
Laganière and Yu (1989)	Rats, 24 m vs 4 m		↓:L	↑:L
Cand and Verdeti (1989)	Rats, 24 m vs 4 m	= :B, H ↓:K, L	↑:H ↓:B, K, L	= :B, H ↓:K, L
Sohal (1989)	Flies, 3 weeks vs less than 1 week	↓:Hg	↓:Hg	
Vertechy et al. (1989)	Rats, 27 m vs 4 m	Cu-Zn SOD ↑:H Cu-Zn SOD = :SM Mn-SOD ↓:H Mn-SOD ↑:SM	= :H ↑:SM	↑:H ↑:SM
Semsei et al. (1989)	Rats, 29 m vs 6 m	Cu-Zn SOD ↓:L	↓:L	
Sohal et al. (1990)	Rats, Mouses pigs, cows	↑:B, H, L	↑:B, H, L	↑:B ↓:H, L
Mote et al. (1990)	Rats, 28 m vs 4 m	Cu-Zn SOD ↓:L	↓:L	
Ji et al. (1990)	Rats, 26 m vs 4 m	Cu-Zn SOD ↑:SM Cu-Zn SOD ↓:L Mn-SOD ↑:L, SM	= :SM ↑:L	↑ = L, SM
Rao et al. (1990a)	Rats, 26 m vs 6 m	Cu-Zn SOD ↓: B, G, H, K, L	= :G ↓:B, H, K, L	= :B, G, H, L ↓:K
Rao et al. (1990b)	Rats, 28 m vs 4 m	Cu-Zn SOD ↓:L	↓:L	= :L
Rao et al. (1990b)	Rats, 21 m vs 4 m			↑:L

SOD: Superoxide dismutase; CAT: Catalase; GPX: Glutathione peroxidase; B: Brain; G: Gut; H: Heart; Hg: Homogenate; K: Kidney; L: Liver; SM: Skeletal muscle; m: Month; vs: Versus; = : No variation*; ↑: Increase*; ↓: Decrease*; *in the old animals, in comparison with the young animals.

and it is quite impossible to draw general conclusions or even some tendencies. However it must be emphasized that the variations of each enzyme in aging depend on the organ examined. Similarly, tissue activities of two isoenzymes of SOD, which are known to have different cellular and subcellular localizations (Marklund, 1984), seem to vary in different ways in aging.

Age-related variations of antioxidant enzymes in humans

Data relevant to the effects of aging on human antioxidant enzymes are scarce, especially for tissues other than blood. A paper by Gutman et al. (1987) reports an absence of age-related variation of SOD in human skin fibroblasts. Other data, which concern the erythrocyte and plasma enzymes, are presented in Table II.

Our group has particularly studied the biological variability of SOD, GPX and CAT in human blood (Guémouri et al., 1991a). These enzyme activities were measured in plasma (P) and erthrocytes (E) of 1836 apparently healthy subjects, aged 4–97 years, recruited in the Center for Preventive Medicine of Vandoeuvre-lès-Nancy (France) or in a convalescent home of the Nancy suburbs. Multiple-regression analysis showed that age has a statistically significant effect on all antioxidant enzyme activities. This effect is clearly illustrated in Figure 1. E- and P-SOD are stable in subjects less than 65 years of age and slightly decrease in the elderly (Fig. 1a, b). In comparison with adults aged 20 to 30 years, the average decrase in the elderly reaches -13.5% (Not Significant, NS) and -21 to -22% ($p < 0.05$) in men and women, respectively, for E-SOD, and -13% (NS) in men and -7.5 to 9.5% (NS) in women for P-SOD. E- and P-GPX activities seem to increase until the age of 18 years, P-GPX being particularly low in adolescents between 10 and 14 years of age (Fig. 1c, d). They are stable in adults under 65 years of age and decrease thereafter. In comparison with subjects between 20 and 30 years old, the average decrease in individuals older than 65 years is -29.5% ($p < 0.05$) in men and -21 to -26% ($p < 0.05$) in women for E-GPX, and -12% (NS) in men and -30 to 34.4% ($p < 0.01$) in women for P-GPX. Variations of E-CAT with age are similar to those of E-SOD. E-CAT slightly decreases after 65 years: -14.5% (NS) in men and -4.4 to 11.5% (NS) in women, in comparison with adults 20 to 30 years of age (Fig. 1e). These data are in agreement with those we obtained previously in more restricted populations (Perrin et al., 1990) and with results from Schäfer and Thorling (1990) concerning E-GPX. Our data on CAT and GPX are inconsistent with those of other authors (Table 2). This may be explained by large differences in the type and size of the populations studied; moreover, in the absence of standardized methods for the determination of the antioxidant enzyme activities,

Table 2. Age-related variations of antioxidant enzymes in human blood

Authors	Populations	Erythrocyte			Plasma	
		SOD	CAT	GPX	SOD	GPX
Jozwiak and Jasnowska (1985)	29 subjects 65–80 y.o. vs 34 subjects 25–50 y.o.	↓	↑	↑		↑
Campbell et al. (1989)	28 subjects ≥65 y.o. vs 75 subjects <65 y.o.			=		
Neve et al. (1989)	10 subjects ≥80 y.o. vs 145 subjects 20–79 y.o.			=		=
Schäfer and Thorling (1990)	30 subjects ≥65 y.o. vs 25 subjects <65 y.o.			↓		
Perrin et al. (1990)	25 subjects 74–98 y.o. vs 41 subjects 49–63 y.o.	↓	↓	↓	=	
Guémouri et al. (1991a)	54 subjects 65–97 y.o. vs 356 subjects 20–30 y.o.	↓	↓	↓	=	↓

SOD: Superoxide dismutase; CAT: Catalase; GPX: Glutathione peroxidase; vs: Versus; =: No variation*; ↑: Increase*; ↓: Decrease*; y.o.: years old; *in the elderly, in comparison with younger subjects

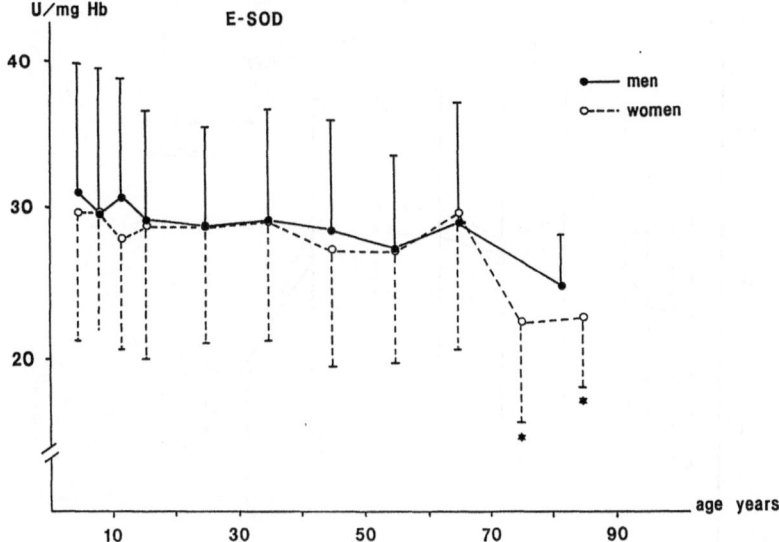

Figure 1a

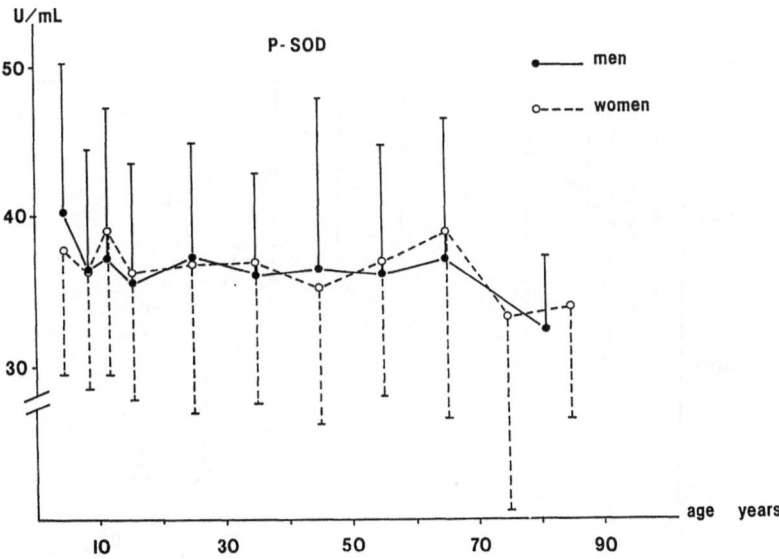

Figure 1b

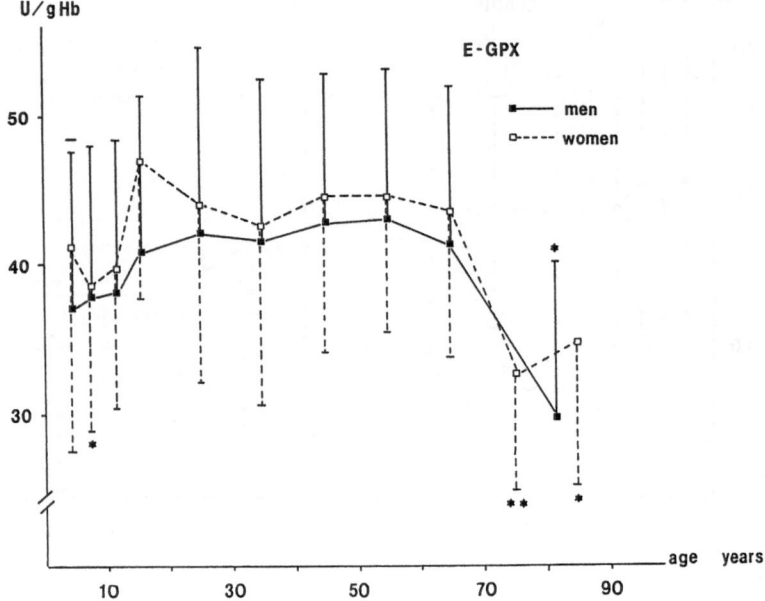

Figure 1c

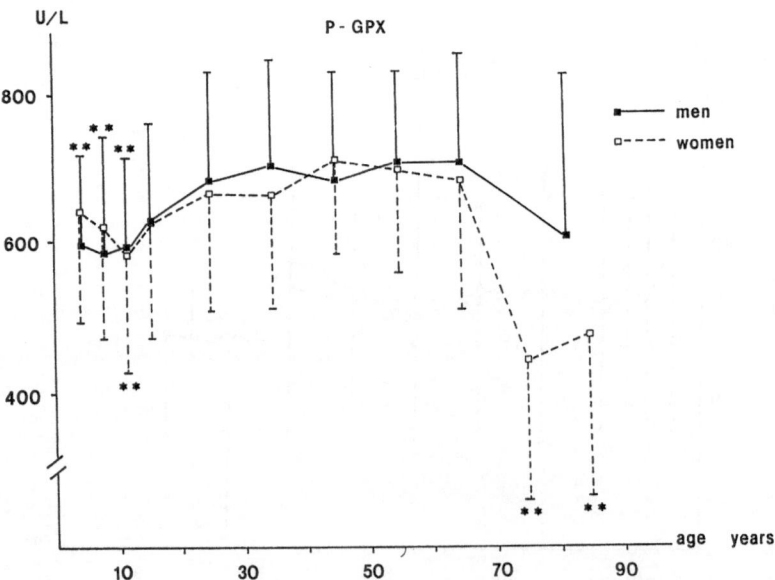

Figure 1d

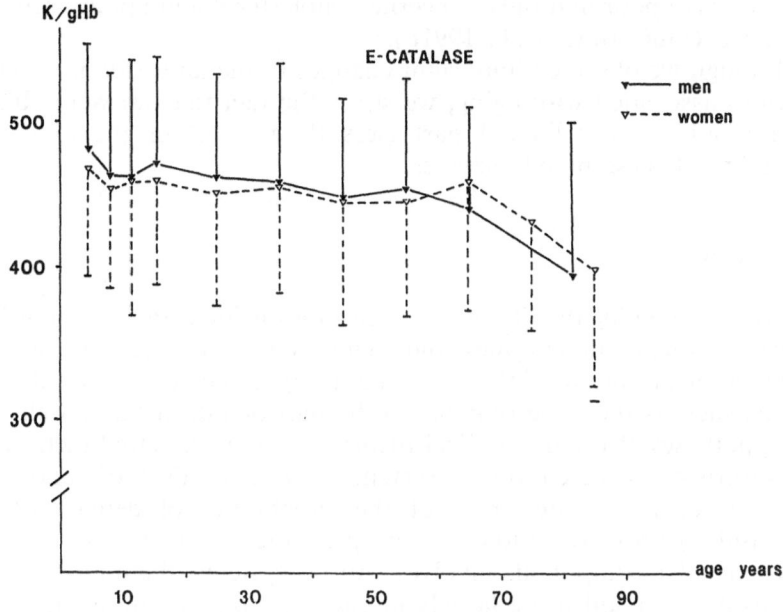

Figure 1e

Figure 1. Variations in the means of the plasma (P) and erythrocyte (E) antioxidant enzyme activities as a function of age for men (● ■ ▼) and women (○ □ ▽). Vertical lines correspond to SDs in the various age subgroups of population. For each subgroup mean activity was compared with that of subjects 20 to 30 years of age; statistical significance of variations (Tukey studentized range method) are marked as follows: $^-p < 0.10$; $*p < 0.05$; $**p < 0.01$. The figures were drawn from data presented as scatter plots in Guémouri et al. (1991a). *a*, Age-related variations of E-SOD (● ○); *b*, Age-related variations of P-SOD (● ○); *c*, Age-related variations of E-GPX (■ □); *d*, Age-related variations of P-GPX (■ □); *e*, Age-related variations of E-CAT (▼ ▽).

differences in the assay conditions (e.g., types of substrates used and precision of measurements) may also affect the results and their degree of significance. These arguments are also valid for explaining the large heterogeneity of the results concerning the animal tissues (Table 1).

Beside these effects of aging, in the same study (Guémouri et al., 1990a), we found no evidence that gender (except for E-GPX), weight, blood pressure of menopause influence the enzyme activities. Variations related to smoking and alcohol consumption are slight and concern only P-SOD and P-GPX, respectively. Conversely, intake of some drugs, such as anti-inflammatory agents, antidepressants and thyroid hormones, modifies activity of some of the three enzymes. E-SOD positively correlates with P-SOD ($r = 0.216$, $p < 0.01$) and E-CAT ($r = 0.123$, $p < 0.001$), and E-GPX with P-GPX ($r = 0.218$, $p < 0.001$), but we found no statistically significant correlation between GPX and SOD or CAT. The correlations of the antioxidant enzymes with serum

366

analyses were poor and only concerned cholesterol and apolipoproteins A1 and B (Guémouri et al., 1991b).

Although we observed important changes in the antioxidant enzymes activities associated with aging, we stress the fact that we were able to explain only a very limited part (less than 10%) of the biological variability of these blood enzymes.

Conclusion

After a survey of the literature, it becomes evident that relationships between antioxidant enzymes and aging remain unclear. No general tendency of evolution of these systems in aging emerges, even if some results, such as those we obtained in human blood, fit fairly well with the hypotheses of the free radical theory of aging. In actual fact, this is not surprising, since enzymatic systems such as SOD, GPX and CAT represent only a limited part of the capabilities of defense of the organism against free radicals and peroxides. Elimination of these highly reactive compounds involves various systems, the multiple interactions of them still being largely unknown, especially in aging.

Acknowledgments. The authors thank the Caisse Nationale d'Assurance Maladie des Travailleurs Salariés (Paris) for its financial support and Mrs M. Cafaxe for secretarial help.

Campbell, D., Bunker, V. W., Thomas, A. J., and Clayton, B. E. (1989) Selenium and vitamin E status of healthy and institutionalized elderly subjects: Analysis of plasma, erthyrocytes and platelets. Br. J. Nutr. 62: 221–227.

Cand, F., and Verdeti, J. (1989). Superoxide dismutase, glutathione peroxidase, catalase and lipid peroxidation in the major organs in the aging rats. Free Radical Biol. Med. 7: 59–63.

Del Maestro, R., and McDonald, W. (1987) Distribution of superoxide dismutase, glutathione peroxidase and catalase in developing rat brain. Mech. Aging Dev. 41: 29–38.

Guémouri, L., Artur, Y., Herbeth, B., Jeandel, C., Cuny, G., and Siest, G. (1991a) Biological variability of superoxide dismutase, glutathione peroxidase, and catalase in blood. Clin. Chem. 37: 1932–1937.

Guémouri, L., Herbeth, B., Artur, Y., Jeandel, C., and Siest, G. (1991b) Weak relationships between blood antioxidant enzymes and serum lipids or apolipoproteins. J. Int. Med. 229: 297–299.

Gutman, R. L., Cohen, M. R., McAmis, W., Ramchand, C. N., and Sailer, V. (1987) Free radical scavenging systems and the effect of peroxide damage in aged human skin fibroblasts. Exp. Gerontol. 22: 373–378.

Harman, D. (1988) Free radicals in aging. Mol. Cell. Biochem. 84: 155–161.

Hazelton, G. A., and Lang, C. A. (1985) Glutathione peroxidase and reductase activities in the aging mouse. Mech. Aging Dev. 29: 71–81.

Hotz, P., Hoet, P., and Lauwerys, R. (1987) Peroxydation lipidique en pathologie humaine: Evaluation des données de la littérature. Path. Biol. 35: 1067–1073.

Ji, L. L., Dillon, D., and Wu, E. (1990) Alteration of antioxidant enzymes with aging in rat skeletal muscle and liver. Am. J. Physiol. 258: R918–R923.

Jozwiak, Z., and Jasnowska, B. (1985) Changes in oxygen-metabolizing enzymes and lipid peroxidation in human erythrocytes as a function of age of donor. Mech. Aging Dev. 32: 77–83.

Kellog, E. W., and Fridovich, I. (1976) Superoxide dismutase in the rat and mouse as a function of age and longevity. J. Gerontol. 31: 405–408.

Laganière, S., and Yu, B. P. (1989) Effect of chronic food restriction in aging rats. II. Liver cytosolic antioxidants and related enzymes. Mech. Aging Dev. 48: 221–230.

Marklund, S. L. (1984) Extracellular-superoxide dismutase and other superoxide dismutase isoenzymes in tissues from nine mammalian species. Biochem. J. 222: 649–655.

Mote, P. L., Grizzle, J. M., Walford, R. I., and Spindler, S. R. (1990) Age-related down regulation of hepatic cytochrome P1-450, P3-450, catalase and CuZn-superoxide dismutase RNA. Mech. Aging Dev. 53: 101–110.

Neve, J., Vertongen, F., Peretz, A., and Carpentier, Y. A. (1989) Valeurs usuelles du sélénium et de la glutathion peroxidase dans une population belge. Ann. Biol. Clin. 47: 138–143.

Perrin, R., Briançon, S., Jeandel, C., Artur, Y., Minn, A., Penin, F., and Siest, G. (1990) Blood activity of Cu/Zn superoxide dismutase, glutathione peroxidase and catalase in Alzheimer's disease: A case-control study. Gerontology 36: 306–313.

Pinto, R. E., and Bertley, W. (1969) The effect of age and sex on glutathione reductase and glutathione peroxidase activities and on aerobic glutathione oxidation in rat liver homogenates. Biochem. J. 112: 109–115.

Pryor, W. A. (1987) The free-radical theory of aging revisited: A critique and a suggested disease – specific theory, in: Modern Biological Theories of Aging, H. R. Warner, R. N. Butler, R. L. Sprott and E. L. Schneider, pp. 89–112. Eds Raven Press Publ., New York.

Rao, C., Xia, E., and Richardson, A. (1990a) Effect of age on the expression of antioxidant enzymes in male Fischer F 344 rats. Mech. Aging Dev. 53: 49–60.

Rao, G., Xia, E., Nadakavukaren, M. J., and Richardson, A. (1990b) Effect of dietary restriction on the age-dependent changes in the expression of antioxidant enzymes in rat liver. J. Nutr. 120: 602–609.

Schäfer, L., and Thorling, E. B. (1990) Lipid peroxidation and antioxidant supplementation in old age. Scand. J. Clin. Lab. Invest. 50: 69–75.

Semsei, I., Rao, G., and Richardson, A. (1989) Changes in the expression of superoxide dismutase and catalase as a function of age and dietary restriction. Biochem. Biophys. Res. Commun. 164: 620–625.

Sohal, R. S. (1989) Oxidative stress hypothesis of aging, in: Advances in Myochemistry, 2nd edn, pp. 21–34. Ed. G. Benzi. John Libbey Publ., London.

Sohal, R. S., Sohal, B. H., and Brunk V. T. (1990) Relationship between antioxidant defenses and longevity in different mammalian species. Mech. Aging Dev. 53: 217–227.

Vertechy, M., Cooper, M. B., Ghirardi, O., and Ramacci, M. T. (1989). Antioxidant enzyme activities in heart and skeletal muscle of rats of different ages. Exp. Gerontol. 24: 211–218.

Free Radicals and Aging
ed. by I. Emerit & B. Chance
© 1992 Birkhäuser Verlag Basel/Switzerland

Antioxidant status (selenium, vitamins A and E) and aging

M. Simonoff, C. Sergeant, N. Garnier, P. Moretto, Y. Llabador, G. Simonoff and C. Conri

Centre d'Etudes Nucléaires de Bordeaux-Gradignan URA 451 du CNRS, F-33175 Gradignan Cedex, France, and Médecine Interne et Cardiologie, Hôpital St André 1, Rue Jean Burguet, F-33075 Bordeaux Cedex, France

Summary. Antioxidant status can be evaluated by blood selenium, vitamins A and E. The level of selenium was determined in whole blood, erythrocytes and plasma of 170 French people (70–95 years old) healthy and with intercurrent illness, by using PIXE (proton-induced X-ray emission analysis). These results are discussed with other values from the literature. Plasma levels of vitamins A and E have been measured by HPLC. All data were compared with those obtained for younger subjects. Healthy elderly people residing in a geriatric home received selenium supplements during 1 month. The influence of this supplementation brought to light a deficiency for this trace element. The correlation of aging and nutritional requirements with immune function, heart and cancer death rate is presented and discussed.

Introduction

Aging is a physiological process. Nevertheless, two factors play a major role in aging: 1) the genetically determined aging process affecting all individuals of a species, independent of the environment and 2) accidental aging, which is due to damaging environmental effects, e.g. diseases and/or free radical reactions. It has been suggested that free radical formation and lipid peroxidation with consequent damage are important factors in the aging process (Harman, 1981). Accelerated aging has been observed in diseases in which there is an increased free radical flux (Down's syndrome, Systemic Lupus erythematodes). Some of the changes and diseases associated with aging may be explained on the basis of the free radical theory: increased frequency of tumors, atherosclerosis and hypertension immune dysfunction, changes in the central nervous system (Halliwell and Gutteridge, 1985). In most of Europe, life expectancy is around 70 years and in the developed countries this value is increasing owing to the high level of medical care and public health.

The potential life expectancy of some species shows a close correlation with the animal's basal metabolism (O_2 uptake per unit body weight per day) (Cutler, 1984). It has been maintained by several authors that O_2^- leakage occurring during the function of the mitochondrial respiratory chain, the amount of which is proportional to the basal metabolism, may be a primary factor in aging owing to its slow cumulative damaging effect on mitochondria or to its action on regulator genes. Thus, according to Cutler, aging is in fact a process of dysdifferentiation resulting from free radical action. This explains the observation that potential life expectancy can be prolonged with caloric restriction decreasing the basal metabolism or by decreasing the surface temperature of the body.

Although the causative role of free radical reactions in aging has not been sufficiently proved, their involvement in the process must be taken into consideration by all modern theories proposed to explain aging (Feher et al., 1987). The increasing effect on average life span of antioxidants has been proved in several animal experiments. It has been shown that administration of antioxidants and lipid peroxidation decreasing diets with little copper, little unsaturated fatty acids and large amounts of antioxidants (selenium, vitamin E) and essential nutrients increases the average life expectancy of the animals.

Selenium is a constituent of the enzyme selenium-dependent glutathione peroxidase which prevents lipid peroxidation, and acts synergetically in the cytoplasm and mitochondrion the way liposoluble antioxidant vitamins E and A act in membranes. Complementary antioxidant systems, including superoxide dismutase, catalyse, ascorbate, β carotene and urate provide an effective defense against oxidative damage.

Aging may change nutrient intake, increase the need for specific ones and interfere with their absorption, storage and utilization. Elderly persons are prone to disorders; many have to take drugs which may render them vulnerable to nutritional deficiencies. Changes occurring in antioxidant protection in humans with aging are not very well established and more information is needed.

So, at the Bordeaux Center of Nuclear Research we measured whole blood, plasma and erythrocyte selenium in 170 elderly people (healthy or hospitalized with intercurrent illness), 70 to 95 years old, and compared these values to the younger French healthy population (30–50 years) that we had measured previously (Simonoff et al., 1988). Fifteen of them, receiving only a mixed diet in a geriatric home, were given selenium supplements and compared with younger adults receiving the same type of selenium supplementation.

For a broader view between aging and antioxidant status we also measured plasma retinol and α tocopherol. Our results concerning selenium and vitamin E and A status in elderly people will be discussed and reviewed with others from the literature.

Selenium

Selenium is considered to be an essential trace element for animals (Schwarz and Foltz, 1957) and for humans (Van Riij et al., 1979; Keshan Disease Group, 1979; Young, 1981). It appears to function mainly as a protective agent against the toxic effects of hydroperoxides. Indeed, selenium is a constituent of Se-dependent glutathione peroxidase, which is found in various tissues such as erythrocytes and liver.

Glutathione peroxidase catalyzes the removal of hydrogen peroxide and lipid peroxides which are the products of reactive oxygen species, especially of the superoxide anion (O_2^-) and the hydroxyl radical ($OH^·$).

Low blood levels of selenium have been observed in numerous pathological conditions (Simonoff et al., 1988; Neve et al., 1985; Shamberger, 1986; Virtamo and Huttunen, 1985), but the significance of this finding is not clear. Only severe deficits appear to provoke observable symptoms, particularly myopathies and/or cardiomyopathies in geographical zones which are poor in selenium and in some isolated cases. However, an absolute or relative deficit of selenium has been implicated in the development of cancers (Combs, 1985), cardiovascular diseases (Oster et al., 1990; Salonen et al., 1982). Low levels of selenium are regularly and significantly correlated with the severity of hepatic lesions (biological and/or histological) observed in liver disease and alcoholism (Dworkin et al., 1985; Korpela et al., 1985). Reports about the association between low selenium status and risk of coronary heart disease, stroke and cancer, though controversal, clearly point to the importance of this trace element.

Blood selenium levels are influenced by numerous factors (geography, nutrition, age, tobacco, alcohol, health status, etc) (Simonoff et al., 1991). Many degenerative diseases, common in the elderly, may be associated with lipid peroxidation increasing the requirements of antioxidants (Halliwell and Gutteridge, 1985). Thus, the elderly may particularly require adequate antioxidant protection but relatively little is known about tissue or circulating levels of selenium.

Blood and tissue levels of selenium are related to dietary intake. Few studies have investigated this status in the elderly compared with younger people (Vatassery et al., 1983; Vandewoude and Vandewoude, 1987). These studies are contradictory because most of them did not extend the range examined to include a significant number of subjects over 70 years of age. In an attempt to elucidate the effect of old age and of underlying disease we determined blood selenium (plasma, whole blood then erythrocytes) in apparently healthy individuals aged 70–95 years and in hospitalized patients suffering from cancers or cardiovascular diseases and compared these values to those of a younger population in the same area (Bordeaux). In addition, the effect of chronic diseases in an institutionalized elderly population was assessed. Moreover, it seemed very important to know the tendency for elderly people (70–95 years) and the result of a supplementation compared to younger populations.

Subjects

Blood was taken by venipuncture (8 AM, fasting) in vacuum-containing tubes (Becton-Dickinson) on Li-Heparin. Subjects were from a hospital or geriatric home:

- 145 elderly people (in the region of Bordeaux), over 70 years old (mean age ± S.D.:79 ± 6, range from 70 to 95 years), hospitalized in an internal medicine ward. This population comprised patients as well as apparently healthy elderly subjects considered as control.
- 25 elderly people living in a geriatric home.

• *Elderly control group from the hospital (normal diet)*
This group consisted of 112 subjects considered as healthy (41 men and 71 women), all showing normal results in clinical and biological examinations in spite of the short hospitalization for fever, painful rheumatism, various troubles of the digestive system like diarrhea. Some of the participants were suffering from cerebrovascular troubles or senility.

• *Elderly patients from the hospital*
33 patients were classified as follows in two groups according to the International Classification of Disease (9th revised edition)
  - Cancers: 21 patients
  - Cardiovascular diseases (hypertension, valvular or ischemic cardiopathy): 12 patients

• *Elderly from a geriatric home (mixed diet)*
25 subjects (8 men and 17 women). This group of long-term geriatrics (many years at the home) consisted of apparently healthy subjects receiving only mixed intake. Almost all of them suffered from cerebrovascular disease, Parkinson's, Alzheimer's disease, and were receiving various medications, but none had been diagnosed as having a renal, gastrointestinal or malignant disease.

Selenium measurements

Blood selenium was determined by proton induced X-ray emission (PIXE) after selenium preconcentration as described elsewhere (Simonoff et al., 1988). The limit of detection was ~1 ng/ml. The coefficient of variation was <10%. Total blood selenium and plasma selenium were both measured and red blood cell selenium calculated using the hematocrit.

The results are expressed in ng/ml (1 μmol = 79 ng).

Blood selenium levels

Table 1 and Figures 1–2 recapitulate: number, age ± standard deviation, plasma selenium values ± SEM for all the subjects. The results for the three groups can be summarized as follows:

Elderly control group but hospitalized:
Plasma selenium: 52 ng/ml for men, 60 ng/ml for women (difference non significant)

Geriatric group eating a mixed diet:
Plasma selenium: 63 ng/ml for men, 68 ng/ml for women (difference non significant)

Patients group (cardiovascular diseases or cancers):
Plasma selenium levels (Fig. 2) were significantly lower than in the two healthy control groups (Fig. 1) ranging from 43 to 55 ng/ml. Whole blood and erythrocyte selenium values can be seen in Table 2 for the discussion with other values from the literature. Few studies published until now, with regard to the effect of age on the Se status, have investigated subjects over the age of 60. In our study, plasma selenium levels were significantly lower in the elderly population around 80 years old (52 to 68 ng/ml) as compared to the younger one (in France around 80 ng/ml for subjects between 25 and 60 years of age).

Table 1. Subjects and plasma selenium

	Age ± SD	Number	Plasma selenium ± SEM ng/ml
Healthy normal diet (112)	F 79.6 ± 5.9	71	60 ± 2.5 (16 → 138)
	M 78.3 ± 6.6	41	52 ± 2.9 (20 → 117)
Healthy mixed diet (25)	F 84 ± 9	17	68 ± 2.2 (62 → 80)
	M 82 ± 7	8	63 ± 3.7 (53 → 68)
Cancers (21)	F 77.7 ± 5.8	7	55 ± 6.4
	M 76.5 ± 5.9	14	43 ± 2.3
Cardiovascular disease (12)	F 78 ± 2.5	6	52 ± 2.8
	M 79.1 ± 4.1	6	52 ± 3.8

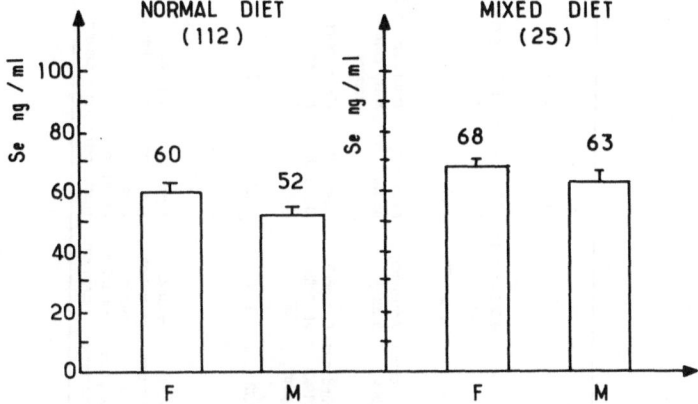

Figure 1. Plasma selenium for healthy elderly.

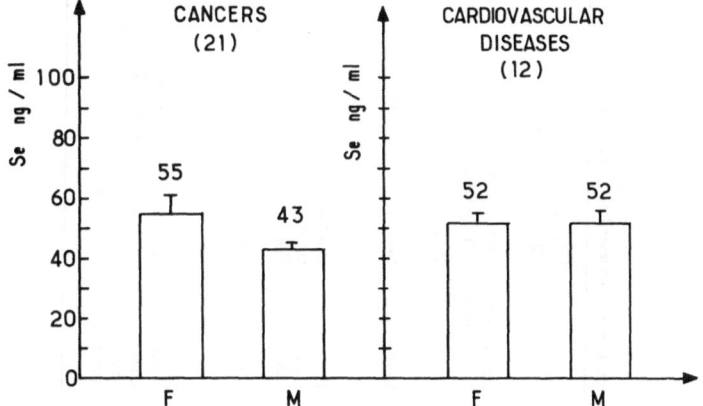

Figure 2. Plasma selenium for elderly patients.

Table 2 summarizes whole blood, plasma or serum and erythrocyte selenium levels expressed in μg/l for several age groups, situations and authors. Healthy elderly people have significantly lower plasma and whole blood levels of selenium when compared with subjects younger than 65. Campbell et al. (1989) found that selenium in platelets was unaffected by age and that there was no difference in concentration between healthy and institutionalized elderly subjects. This result agrees also with those of Kasperek et al. (1982) and Menzel et al. (1983) suggesting a special role for selenium in platelets and a difference in tendency to keep this trace element deficient, even when concentration in plasma or erythrocytes is low. Consequently, the measurement of

Table 2. Selenium in whole blood, plasma or serum and erythrocytes expressed in µg/l for several ages, situations and authors

Whole blood	86	100	72	63	138	129	105	123	130	108	97	79
Plasma or serum	66	83	59	50	116	112	91	97	73	95	81	64
Erythrocytes	115	125	65	58	0.494 µg/g Hb (165 calculated)	0.453 µg/g Hb (151 calculated)		161	200	0.382 µg/g Hb 126	0.374 µg/g Hb 118	0.337 µg/g Hb 99
Mean age (years)	79	35	68	79	18–55	55–85	40	20–59	60–99	<65	>65	>65
Clinical situation	Healthy	Healthy	Malignancy	Malignancy or cardiovascular diseases	Healthy	Healthy	Healthy	Healthy	Apparently healthy	Apparently healthy	Apparently healthy	Institutionalized with chronic diseases
Authors	Simonoff et al., 1991 France	Simonoff et al., 1991 France	Simonoff et al., 1988 France	Simonoff et al., 1991 France	Lloyd et al., 1983 U.K.	Lloyd et al., 1983 U.K.	Girre et al., 1990 France	Verlinden et al., 1983 Belgium	Verlinden et al., 1983 Belgium	Campbell et al., 1989 U.K.	Campbell et al., 1989 U.K.	Campbell et al., 1989 U.K.
Ratio whole blood/ plasma or serum	1.33	1.20	1.22	1.26	1.20	1.15	1.15	1.27	1.78	1.14	1.20	1.23
Technique	PIXE	PIXE	PIXE	PIXE	Hydride generation and atomic absorption	Hydride generation and atomic absorption	Electrothermal atomic absorption spectrometry with Zeeman effect	Hydride generation atomic absorption spectrometry	Hydride generation atomic absorption spectrometry	Hydride generation atomic absorption spectrometry	Hydride generation atomic absorption spectrometry	Hydride generation atomic absorption spectrometry

selenium in platelets is not a useful indicator in assessing selenium status as suggested by others (Neve et al., 1985).

The tissue level of selenium is generally related to dietary intake: plasma would be a short-term indicator and erythrocytes longer-term indicators. In our study, the 25, well-fed, elderly French, over 75 years old, residing in a retirement home in the region of Bordeaux, and receiving only a mixed diet, had significantly higher whole blood and plasma selenium levels than those of subjects of the same age suffering from malignancy or severe cardiovascular diseases. The evaluation of composite diets gave an average daily selenium intake around 35 μg. Daily diets comprised approximately 80 g of meat or fish, 400 g of cereals and bread, 250 g of fruits and vegetables, 100 g of dairy products and 1 liter of beverages. Fruits and vegetables are poor in selenium. Fish, meat and cereals are the main sources (Simonoff et al., 1988). Assuming that healthy young people consume approximately twice as much meat, this difference is sufficient to explain the lower selenium content in plasma and erythrocytes of elderly people.

- The average plasma selenium values in the group of elderly healthy people in our study was 66 ng/ml and thus significantly lower than the value of 83 ng/ml for young healthy adults.
- The average whole blood selenium, 86 ng/ml for elderly, was lower than for young adults (100 ng/ml).

Many authors have detected a trend for serum or plasma selenium values (Verlinden et al., 1983; Miller et al., 1983; Morisi et al., 1989) and whole blood selenium values (Campbell et al., 1989; Lloyd et al., 1983) to diminish with age. There is a significant correlation between reduced plasma selenium and the intake of selenium and/or protein in the elderly. The mean calculated erythrocyte selenium concentration was 115 ng/ml in the elderly population, a little less, but comparable to the values 125 ng/ml in younger adults. The same tendency and order of magnitude were observed by Campbell et al. (1989). These results suggest that even with a reduced dietary selenium intake the erythrocyte long-term indicator is not depleted.

Selenium and cardiovascular disease

Among the main causes of arteriosclerotic diseases, two of them (cholesterol and hypertension) are related to the diet (Puska et al., 1986). But other dietary factors can contribute to these diseases: inadequate intake of trace elements and vitamins. Selenium, copper, zinc, lithium, vanadium, chromium and magnesium are inversely associated with cardiovascular diseases (Huttunen and Virtamo, 1986). In the last ten years attention has been specially focused on selenium for which there are several concordant indications of its relation to heart diseases coming from animal studies, epidemiological and clinical studies.

In selenium deficiency an accumulation of lipid peroxides in the heart may occur, damage the cell membrane and lead to an improved calcium transport with an uncontrolled calcium accumulation in the cell. The result may be an activation of phospholipids rather than an enhanced formation of arachidonic acid, shifting prostaglandin synthesis from prostacyclin to thromboxan. These deviations enhance blood pressure and platelet aggregability.

In animal experiments, selenium protects against cardiotoxic drugs (adriamycin), cardiotoxic xenobiotics and viral infections (Coxsackie B4 viruses) that affect the heart.

In humans, ecological and epidemiological results have been reported that show a relationship between low serum selenium and cardiovascular disease (Oster and Prellwitz, 1990) or cardiomyopathy and an increased cardiovascular risk when selenium status is low. Reeves et al. (1989) described a young black woman with Crohn's disease in whom a congestive cardiomyopathy developed and was subsequently reversed following administration of selenium. For some authors (Schaeffer et al., 1990) the level of selenium does not seem to be a risk factor for coronary artery disease.

In clinical studies done in Germany (Oster et al., 1987), Finland (Salonen et al., 1982), Sweden (Moore et al., 1984) and USA, subnormal serum selenium and whole blood selenium concentrations have been reported in patients with acute myocardial infarction, and coronary arteriosclerosis.

In Germany, young adult patients with coronary heart disease were found to have serum and whole blood selenium lower than healthy:

Patients $55 \pm 15 \, \mu g/l$ Controls $78 \pm 11 \, \mu g/l$ for serum selenium (Oster et al., 1987)

Patients $76 \pm 15 \, \mu g/l$ Controls $93 \pm 18 \, \mu g/l$ for whole blood (Oster et al., 1987)

These values for young adult patients suffering from cardiovascular diseases show the same tendency as in our study for elderly people (mean 79 years old), but the selenium deficiency of our patients and controls is even more pronounced:

Patients $52 \pm 12 \, \mu g/l$ Controls $66 \pm 12 \, \mu g/l$ plasma selenium (Simonoff et al., 1991)

Patients $63 \pm 11 \, \mu g/l$ Controls $86 \pm 13 \, \mu g/l$ whole blood (Simonoff et al., 1991)

In conclusion, if a selenium deficiency may be considered as a risk factor for cardiovascular disease, the threshold seems relatively independent of age. However, the deficiency increases with age: the elderly patients erythrocytes containing only 58 μg selenium/l as compared to 115 μg selenium/l for apparently healthy subjects of the same age (79 years old).

Selenium and cancer

Evidence of selenium enzymes having specific anticarcinogenic action by destroying reactive oxygen radical has been reviewed (Shamberger, 1986).

Protection of livestock and humans from cancer by selenium is indicated by epidemiological relationships (Salonen et al., 1984; Salonen et al., 1985; Stead et al., 1985; Willett et al., 1983) and by experimental studies of selenium and known carcinogens in the development of specific cancer cell lines.

The effects of selenium (Na_2SeO_3) on aflotoxin B1 (AFB1) induced hepatic neoplasia in the rat permit to conclude that Se has an inhibitory effect on the initiation and promotion stages of AFB1 induced preneoplasic foci and nodules. Selenium also prevents progression of these nodules (Lei et al., 1990).

The study of Helzlsouer et al. (1989) assesses a role for selenium in the prevention of bladder cancer. The anticarcinogenic activity of selenium in animal models is well established. In general, selenite is more effective than selenomethionine in inhibiting the development of chemically induced tumors (Thompson et al., 1984; Ip and Hayes, 1989) as well as the growth of implanted neoplasic cells. Both selenite and selenomethionine must be metabolized to exert their activities (Ip, 1988; Ip and Ganther, 1988). Arsenite reduces the effectiveness of selenite in chemoprevention. A substantial prophylactic efficacity of methylated forms of selenium have been shown, suggesting that partially methylated forms of selenium may be directly involved in anticarcinogenic action of selenium (Ip and Ganther, 1990).

For some researchers, the low selenium status of cancer patients is more a consequence than a cause of their illness. This seems particularly true for gastrointestinal cancer. Low plasma selenium may be related to insufficient intake, urinary loss, redistribution, insufficient levels of carrier proteins.

In a previous work (Simonoff et al., 1988) we measured serum selenium for a group of 46 patients (mean age 68 years old) suffering from different types of cancer. The mean serum selenium value was 59 ± 4 μg/l, not very different from that obtained for elderly patients (50 ± 7 μg/l, 77 years old). It does not seem that age plays a role on the selenium status in cancer but much more the situation of the primary tumor, evolution, distant metastasis, multiple recurrences, denutrition.

Selenium and longevity

Selenium is related to heart and cancer death rates, the two largest factors in human longevity. Longevity in China has been increased by selenium supplementation in areas of selenium deficiency. According to the U.S. National Research Council, those states in the USA with the highest longevity rates largely follow the pattern of crop Se content. The geochemical abundance of selenium is high in marine sediments from which soils formed in the central states of Montana-Dakota to New Mexico-Texas where ischemic heart death rates are lowest.

Glacial disturbance of landscapes in the northeastern quadrant of the USA with a high leaching rate leads to low availability in soil selenium and to relatively high ischemic heart death rates (Jackson, 1988). Besides China, New Zealand and Finland stand out among the low average blood selenium areas. In Finland, heart and cancer death rates are sensitive to small changes in blood selenium levels, as shown by Salonen et al. (1985). In New Zealand blood selenium levels (20–140 μg/l) extend from the lower levels in China to the low-midrange in USA. In New Zealand cancer death rate is as high as in the 'low Se' region of USA, for example New Hampshire (Schrauzer, 1979).

The fact that heart and cancer death rate varies geographically in China and the USA is taken by Cairns (1985) as an indication for possible prevention. The adoption of a favorable lifestyle (no smoking) and proper mineral and vitamin nutrition according to the amounts and balanced ratio set up by US National Academy of Sciences Recommended Daily Allowances 10th Ed, Washington (1989) including available supplements as needed would be beneficial in increasing longevity.

Selenium supplementation

Substantial research indicates that dietary selenium is reflected by blood selenium level. Supplementation shows us whether a subject is more or less deficient. During one month,

* 20 subjects (30 years old) received selenium-enriched yeast (120 μg Se/day). The plasma selenium increase was 16%.
* 6 subjects (45 years old) received sodium selenite (120 μg Se/day). The plasma selenium increase was 18%.
* 10 elderly people (82 years old) received either enriched yeast or selenite (120 μg Se/day). We observed a very much more important plasma selenium increase: 100% and 80%, respectively. All these results are summarized in Table 3 and Figure 3.

It is interesting to compare such a rapid plasma selenium level increase for elderly people in France (not seen for younger people) with

Table 3. Selenium supplementation

Age	Numbers	Sex ratio F/M	Plasma selenium ng/ml		Increase (%)	Supplementation
			t_0	4 weeks		
30 ± 5	20	13/7	83	96	16%	Yeast 120 μg Se/day
45 ± 3	6	3/3	77	91	18%	Na_2SeO_3 120 μg Se/day
80 ± 7	6	3/3	69	138	100%	Yeast 120 μg Se/day
84 ± 6	4	2/2	64	115	80%	Selenite 120 μg/day
Adults* China	10	10/0	27.3	59.9	122%	Na_2SeO_3 150 μg Se/day

*Bioavailability of selenium to residents in a low selenium area of China. Luo et al., Am. Jour. Clin. Nut. 42 (1985) 439–448.

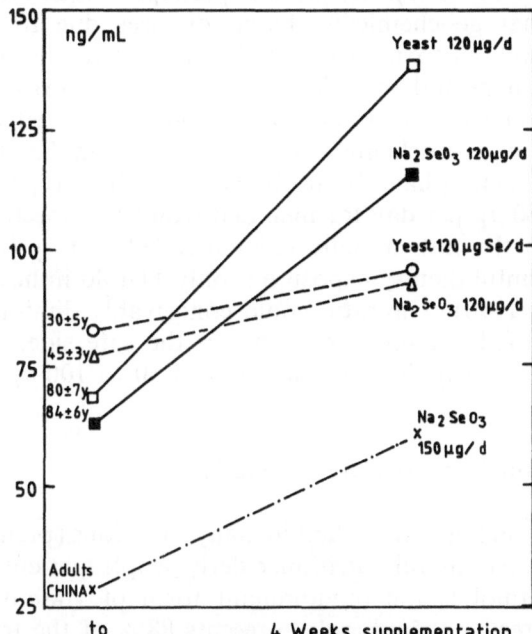

Figure 3. Plasma selenium after four weeks of supplementation.

the observation of Luo et al. (1985) where 10 adults from a low selenium area of China receiving 150 μg Se/day had a plasma selenium increase of 122%. These data agree also with those reported by Steiner et al. (1982) indicating that selenium increases promptly and significantly in plasma when people of low selenium status are supplemented with selenite or selenium-enriched yeast.

Conclusion

The results of our study suggest that the reduced selenium intake of elderly people diminished the plasma level, if compared to younger people, but that the erythrocytes do not seem to be affected in the absence of severe intercurrent illness. Nevertheless, the global selenium status and homeostasia must be affected by this subdeficient intake since the elderly people's responsiveness is quite different from that of the younger subjects. These results suggest that most elderly people would benefit from a selenium supplementation or an increased selenium intake from the richest sources like eggs and fish.

In France, the mean intake is between $45-50$ μg Se/day for a young adult and around $30-35$ μg Se/day for elderly people consecutively to a reduced animal protein intake and calories. A dietary selenium recommendation of 70 and 55 μg/day has been suggested for men and women, respectively (RDA, 1989). Selenium is perhaps unique among the trace elements in that geochemically based diseases due to both element deficiency as well as toxic levels of the element in the diet have been observed in humans and animals under naturally occurring environmental conditions (Levander, 1986). Deleterious clinical effects (fingernail deformations) and biochemical changes (shifts in Se distribution in blood compartments) have been observed in Chinese persons consuming 900 and 750 μg per day for men and women, respectively. Thus the factor between the nutritionally recommended dietary intake and the potentially harmful dietary exposure is only 11-fold in healthy individuals (Levander, 1990). The ratio could conceivably diminish in the case of sickly or ill-fed persons. So to be on the safe side, elderly people should not receive supplements that exceed 50 to 100 μg per day.

Liposoluble antioxidant vitamins A and E

Vitamins A and E are important in many situations (premature infants, malabsorption, parenteral nutrition, elderly people, patients with cancer).

All trans retinol is the predominant form of vitamin A in serum ($>95\%$), whereas α tocopherol represents 88% of the total vitamin E pool of serum.

Currently, the nutritional status with regard to retinol and α tocopherol is assessed by measurement of these vitamins in plasma.

Determination of retinol and α tocopherol in serum or plasma by HPLC

Several procedures for quantitating retinol and α tocopherol separately or simultaneously have been published. Nevertheless, the measurement of serum vitamins A and E by HPLC is very convenient: fast, reliable, simultaneous measurement of the two vitamins, small serum sampling, no interferences to which calorimetric assays are often subject with β carotene or phytofluene.

The HPLC method that we used is a modification of that published by Castagnani and Bieri (1983). After addition of internal standards (retinyl acetate and α tocopheryl acetate) to serum or plasma, deproteinization is realized with acetonitrile. Lipids and liposoluble vitamins are extracted with hexane. This solvent is evaporated under a stream of nitrogen, at room temperature, then the lipidic residue is dissolved in ethanol. A portion of this solution is injected onto a C18-reversed phase chromatographic column. Our chromatography column is a reversed phase Adsorbosphere HS C18 (20%) $3\text{-}\mu m$ particle size (4, 6 mm ID $\times$ 100 mm), coupled with a guard column reversed phase Adsorbosphere HS C18 (20%) $3\text{-}\mu m$ particle size (4, 6 mm ID 10 mm) from ALLTECH associates, Inc. Absorbances of the vitamins and internal standards are measured at 285 nm. Retinol and α tocopherol are quantified from standard curves of peak surface ratios versus weight ratios for each vitamins.

Either plasma or serum can be used as samples. Direct exposure to natural light is to be avoided by protecting glass tubes with aluminium foil. Blood samples are collected after an overnight fast. As a rule, after venipuncture they are immediately centrifuged, then plasma kept frozen at $-40°C$, or in the refrigerator ($4°C$) for some days.

Stability of vitamins A and E in serum or plasma

The stability of vitamins A and E in serum under various conditions has been very well documented using the HPLC techniques. In the case of Driskell et al. (1982) samples from a single donor human serum pool were stored at $25°C$ (1 day), $4°C$ (4 weeks), $-20°C$ (16 months) and $-70°C$ (16 months). Samples were assayed at intervals during these periods of storage. Vitamin A (coefficient variation 3.6%) and E (c.v. 4.5%) were found to be stable. The values were constant within the limits of the precision of the assay and the chromatographic peaks were clean with no shoulders or other indication of breakdown of the vitamins. Vitamins A and E in serum were found to be stable to freezing and thawing (seventeen freezing and thawing cycles over a period of five weeks).

Our results for vitamins A and E in plasma

Figures 4, 5 and 6 and Table 4 give the mean retinol and α tocopherol in plasma, for healthy subjects in function of age.

382

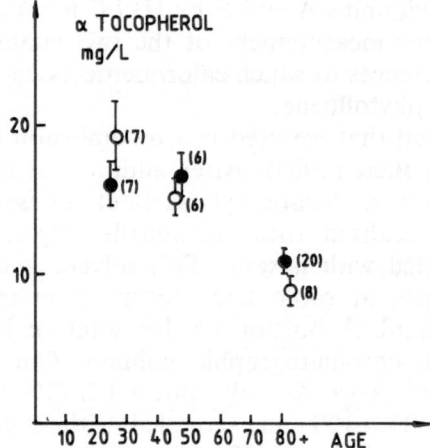

Figure 4. ● → ♂ – ○ → ♀ – Mean serum alpha-tocopherol in different age for healthy subjects (number is shown in parentheses).

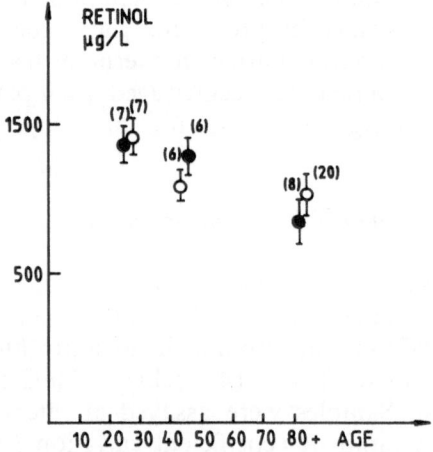

Figure 5. ● → ♂ – ○ → ♀ – Mean retinol in different age groups for healthy subjects (number is shown in parentheses).

For healthy subjects in our different age groups (25–45 and 82 years old), the vitamin levels diminish with age. Our dietary survey revealed a significantly reduced intake of α tocopherol and retinol for elderly people compared to younger people. As in the Finnish study by Knekt et al. (1988), the plasma α tocopherol levels were positively correlated with plasma retinol and plasma selenium for both sexes. There was no

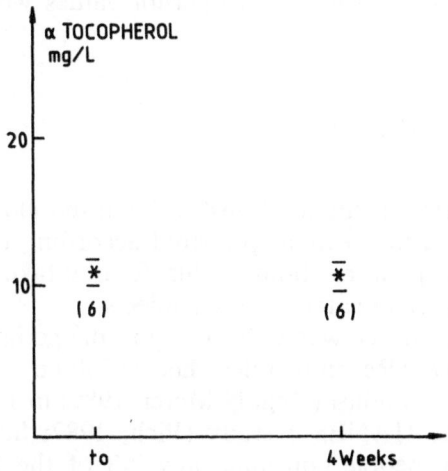

Figure 6. Plasma α-tocopherol after elderly's supplementation (UI/Day: 8) during 4 weeks.

Table 4. Mean retinol and α tocopherol for healthy subjects in different age groups

Age	Sex	Number	Retinol ± SEM μg/l	α tocopherol ± SEM mg/l
27 ± 2	F	7	971 ± 69	19.3 ± 1.7
25 ± 2	M	7	941 ± 45	16.1 ± 1.2
45 ± 3	F	6	808 ± 64	15.6 ± 1.7
47 ± 3	M	6	906 ± 54	16.7 ± 1.8
84 ± 9	F	20	756 ± 59	9.3 ± 0.9
82 ± 7	M	8	681 ± 94	11.4 ± 1.0

Table 5

	Geriatric home RDI*			
	Men	Women	Men	Women
Energy (MJ) and [Cal]	(9)	[2150]	(9.0)	(7.5)
Protein (g) and [Cal]	(78)	[320]	(75)	(61)
Fat (g) and [Cal]	(72)	[660]	(71)	(58)
Carbohydrate (g) and [Cal]	(320)	[1200]	(300)	(300)
Selenium μg/day	35–42	50	70	50
Retinol [mg equ]	1.2	1.0	1.0	
Tocopherol [mg equ]	14	12	12	

evidence of avitaminose since α tocoperhol values were > 5 mg/l and retinol > 100 μg/l.

Food intake for elderly subjects

The mean nutritional intake of institutionalized elderlies was calculated on the basis of menu plans prepared according to food tables by the kitchen of the geriatric home. This food intake was mixed but generally not completely eaten by the subjects.

Vitamin A and E intake was satisfactory in the geriatric home and in agreement with RDI (Recommended dietary intake). Food intake was very similar to other studies (Regöly Merei, 1989) in Europe (Table 5).

According to the HANES I study (Pich, 1987; Suter et al., 1987) 42–65% of elderly people consume only 2/3 of the RDA values for vitamin A but the serum retinol level is low only in 0.3% of them. Intake of margarine, vegetable oils, and green vegetables predicts the serum level of α tocopherol. In our examinations vitamin A and E status of institutionalized elderlies, as shown by retinol and tocopherol intake and serum levels, are satisfactory.

1) Vitamin E (α tocopherol)

Vitamin E is the term used for eight naturally occurring essential fat soluble nutrients. Four have the tocol structure with a saturated phytyl C_{16} side chain (α, β, γ, δ tocopherol) and four the tocotrienol structure with three double bonds in the phytyl side chain (α, β, γ, δ tocotrienol). Alpha tocopherol has the highest biological activity.

The side chain facilitates the incorporation and retention of vitamin E in biomembranes. Vitamin E is the most important lipid soluble antioxidant. It quenches free radicals and acts as a terminator of lipid peroxidation (Liebler et al., 1986; McCay, 1985). The vitamin E radical then formed is fairly stable and may be reduced back to vitamin E by ascorbate and gluthathione.

In humans, vitamin E deficiency may produce diminished erythrocyte life span, neurological dysfunction, myopathies. Alterations of vitamin E status have been associated with the development of certain forms of cancer (Wald et al., 1984; Willett et al., 1984; Menkes et al., 1986; Trickler et al., 1987; Knekt et al., 1988), cardiovascular diseases (Gey et al., 1987; Bieri et al., 1983) and impaired immune response (Panush et al., 1985; Tanaka et al., 1979; Boxer, 1986).

Research on vitamin E requirements for healthy adults has not been conclusive (Van Gossum et al., 1988). Vitamin E requirements may vary more than fivefold, depending on dietary intake and/or tissue composition from previous dietary habits. Animal research has demonstrated that a high polyinsaturated fatty acid intake increases the vitamin E requirement (Horwitt, 1986).

Epidemiological studies indicate a lower incidence of infectious disease in elderly with high plasma tocopherol concentrations (Chevance et al., 1985).

Vitamin E and immune function

When vitamin E is absent from the diet, all circulating cells derived from the common hemopoietic stem cell precursor present membrane changes. Red blood cells from vitamin E deficient animals and humans lyse in the presence of hydrogen peroxide; platelets are more adhesive (Steiner et al., 1982) and produce more thromboxane than platelets from vitamin E replete animals; neutrophils have an increased level of peroxidized lipid in the membranes and have depressed chemotaxis and phagocytosis.

Vitamin E acts as a major fat-soluble antioxidant. It is an essential constituent of all cellular membranes, outer cellular membranes and internal mitochondrial and nuclear membranes (Molennaar et al., 1984). The immune response initiates at the cell membrane level (Roitt, 1984) and is important in maintaining the lymphocyte membrane fluidity which is indispensable for proliferative responses. Vitamine E content of lymphocytes and mononuclear cells is about ten times greater than that found in red blood cells and platelets.

Vitamin E deficient animals have severely depressed immune responses (Bendich et al., 1986). Macrophage membranes are altered, interleukin-2 production is lowered and production of prostaglandins altered (Meydani et al., 1986). In animals, high doses of vitamin E increase resistance to infection and decrease morbidity and mortality. In agreement with these findings epidemiological studies have shown a significant association between high plasma vitamin E level and a lower incidence of infections in healthy adults over the age of 60 (Chevance et al., 1984, 1985).

Aging is associated with altered regulation of the immune system (Spallholz, 1990). Age-related functional changes have been characterized for both humoral and cell-mediated immune responses. T-cells have the major alterations (Makinodan et al., 1981; Goodwin et al., 1982). Cooperation between monocytes and lymphocytes is essential in antigen recognition, lymphocyte differentiation, antibody production; moreover, macrophages synthesize IL_1 which induces the production of IL_2 by the activated T-cells.

High concentrations of arachidonic acid are observed in the phospholipid membranes of macrophages. These cells, upon stimulation, release up to half the arachidonic acid content to form prostaglandin (PG), hydroxeicosatetraenoic acid (HETE) and leukotrienes (LT). PG_2 and other oxydative metabolites of activated macrophages (H_2O_2) suppress lymphocyte proliferation and lymphokine synthesis. Increased PGE_2 production by macrophages has been reported for aged rats and mice (Meydani et al., 1986).

Short-term vitamin E supplementation improves immune responsiveness in healthy elderly individuals appearing to be mediated by a decrease in PGE_2 and/or other lipid peroxidation products (Meydani et al., 1990).

Plasma level of vitamin E

The vitamin E status in humans can be assessed by several methods; one of the most common is the measurement of α tocopherol concentrations in plasma.

The various factors contributing to the observed interindividual variation in serum α-tocopherol levels in humans remain largely unknown. It is thought that diet is an important contributor to this variation but other actors such as absorption efficiency, rate of utilization and excretion rate of tissue mobilization may also be involved. The vitamin E status in men and women aged 60–90 years was measured in different countries.

The coefficient conversions for dl α tocopherol can be summarized as follows:

– 1 mg is equivalent to 1.10 IU
– 1 μmole/l is equivalent to 0.430 mg/l or 0.473 IU/l

The distribution of plasma-free tocopherol level in Slovenia (Pokorn et al., 1990) ranged from 10 μmol/l to 69.9 μmol/l for men and 20 μmol/l to 64.9 μmol/l for women (median 40–44.9 for both groups). In an aged Finnish male population the mean serum vitamin E was about 10 mg/l, and no significant differences were found between age groups from 65 to 85 years (Kivela et al., 1989). The plasma α tocopherol concentration, before vitamin E supplementation, was about 26 μmol/l for an American population (60–75 years). The concentrations in the vitamin E-supplemented group (800 mg α tocopherol acetate) increased threefold after one month (Meydani et al., 1990). In Hungary, institutionalized elderly people (60–94 years) have serum tocopherol values around 27 μmol/l and non-institutionalized people of the same age a mean in the range of 40–45 μmol/l. Campbell et al.

(1989) studied the status of healthy and institutionalized elderly subjects in England. The healthy subjects' plasma levels of vitamin E increase with age to 60 years and decreased above 80 years. The vitamin E status in the institutionalized elderly was reduced compared with the healthy elderly (18.8 μmol/l in plasma compared to 26.2 μmol/l). Other works (Vandewoude and Vandewoude, 1987) also demonstrated a fall in plasma vitamin E levels in subjects over 70–80 years of age. The results of a study done by Campbell et al. (1989) suggest that aging per se has little effect on vitamin E status and that recurrent illness and reduced food intake are the most important factors in the reduced status of vitamin E in the elderly.

Vitamin E intake and supplementation

The importance of tocopherol in human nutrition has been recognized for several decades. Vegetable oils and oil products are among the major contributors of dietary tocopherols, either directly (salad oil) or indirectly as ingredients in prepared or processed foods. Safflower and sunflower oils are the richest sources of α tocopherol. Alteration of tocopherol content may be observed during oil production, processing and storage time. Oil products such as mayonnaise, salad dressing, margarines generally contain high concentrations of tocopherols. Animal fats are generally two or three orders of magnitude less concentrated in tocopherols than vegetable oils. Many fish and fish oils are also reasonably good sources of tocopherols, mostly α tocopherols (Kinsella, 1987). Salmon and cod liver-oils contain up to 200 mg tocopherol/kg oil. Fish oil supplements usually contain added tocopherol to increase the stability of these unsaturated oil concentrates.

The current recommended daily intake of vitamin E is 10 international units (IU) for adult men and 8 for adult women. Precise estimates of the daily intake of tocopherols by individuals are difficult to obtain particularly if the intake is calculated from food composition tables.

In healthy adults, vitamin E supplementation with 800 mg dl α tocopherol acetate (Meydani et al., 1990) produces a more significant reduction in PGE_2 (than in young subjects) and in plasma lipid peroxides (TBAR). But as noticed by Herbert et al. (1991), the enhancement in cellular immunity observed with vitamin E supplementation in healthy elderly subjects may be harmful in situations involving chronic autoimmune diseases.

Other nutrients, for instance zinc, are known to have a beneficial effect on immune response in animals. Nevertheless, a recent double-blind placebo-controlled study by Bogden et al. (1988) showed no significant effect of zinc supplementation on immune response in elderly subjects.

In conclusion, concording results suggest that many elderly individuals might benefit from a supplemented intake of vitamin E. But public health recommendations involve long-term studies to establish whether the immuno-stimulation is due to pharmacological or physiological properties of vitamin E.

2) Vitamin A (Retinol)

Vitamin A is a general term used for the β-ionone derivatives with a structure and a biological activity similar to the basic molecule trans retinol or retinol. The side chain is C9 with four conjugated double bounds and two methyl groups.

The term 'provitamin A' is used for all the carotenoïdes with a biological activity comparable to vitamin A. There are about 50 but the most important is β-carotene. These have 2 cycles separated by a C18 chain and 9 conjugated double bounds.

Retinoïds are synthetic derivatives of vitamin A. They have a cyclic group, a polyene chain and a polar group; but each part can be modified to obtain more efficient new molecules, especially as concerns the biological effects in dermatology and cancerology.

Food contains derivatives of vitamin A (generally as retinyl esters) and provitamins (mostly β-carotene). Retinyl esters are hydrolyzed; provitamins A are mostly hydrolyzed, too, but only a small part of β-carotene circulates under this form.

The functions of vitamin A are various and needed for: promotion of growth, reproduction, vision, differentiation of epithelia, myelization processes, immune processes, presumed inhibitory effect on cancerogenesis.

Plasma levels of vitamin A with age

Serum vitamin A concentration is an indicator of dietary intake of all forms of vitamin A. Unbound vitamin A is stored in the liver. The mobilization of vitamin A is dependent on the total amount stored in the liver (Varma et al., 1972). Serum vitamin A concentration does not reflect short-term dietary intake, but a long-term intake, in contrast to serum selenium which indicates short-term selenium status.

Mean vitamin A concentrations did not differ between age groups (65–84 years) in the study of Kivelä et al. (1989). This finding is consistent with many others as long as old but healthy groups are considered. In the Kivalä study only 1% of the aged population has a severe vitamin A deficiency (below 300 μg/l) and 4% a mild deficiency (300–399 μg/l). Subclinical or clinical deficiencies in vitamin A are rare among the elderly in western countries (Smith et al., 1984). However abnormally low serum vitamin A values have been reported in every third patient admitted to a hospital in the UK.

Vitamin A and cancer

Two epidemiological studies (Wald et al., 1980; Kark et al., 1981) have reported that low serum retinol levels are associated with an increased risk of cancer. Another study did not find low serum retinol to be associated with an increased cancer risk (Willett et al., 1984). In some epidemiological studies it seemed that the risk of cancer in human beings was correlated inversely both with the level of retinol in blood and with the dietary intake of β-carotene (Peto et al., 1981; Shekelle et al., 1981).

The mechanism of the effect of dietary carotene is not completely known. β-carotene is highly effective in quenching singlet oxygen, and this may serve to inhibit the transformation of cells under a number of circumstances. Since supplemental β-carotene can be ingested for long periods virtually without risk of toxicity (except innocuous skin pigmentation) it would seem reasonable to advise patients and the public to consume a balanced diet containing an adequate supply of both β-carotene and vitamin A.

Recently, retinoïds received increased attention for their anticarcinogenic action. They are used in the prevention and treatment of squamous cell epithelial carcinoma, because they inhibit metaplasia of the squamous epithelium. Oral doses produce regression of bladder papilloma and oral and laryngeal leukoplakia.

Vitamin A: Sources, requirements

The major natural sources of vitamin A in the diet are:
- provitamins A contained in plant carotenoïd pigment such as β-carotene (carrot, parsley, spinach, melon, apricot, cabbage)
- and the long chain retinyl esters found in animal tissues (cod liver oil, animal livers, eggs, butter, cheese).

In the United States the Recommended daily Dietary Allowance of vitamin A has been set at 800 and 1000 retinol equivalents for adult women and men, respectively (RDA 1981).

1 retinol equivalent = 1 μg of retinol = 6 μg of β-carotene = 3.3 IU

In France the advised requirements (of the same order) are given in the articles of Dupin (1981), Mareschi et al. (1984) and Apfelbaum et al. (1989).

This recommended intake is above the probably true requirement. US nutrition surveys (the Ten-State Nutrition Survey and the Health and Nutrition Examination Survey [HANES]) have not found vitamin A deficiency to be a serious public health problem. However, a considerable portion of the population in the United States and Canada was found to have serum vitamin A levels in the range classified as a moderate deficiency risk (100 to 200 μg/l in the United States). In addition, post mortem studies have found reduced liver reserves of vitamin A in a significant number of people. These observations may reflect a marginal vitamin A nutritional status in this country or unknown metabolic abnormalities in a population with adequate nutrition (Goodman, 1984).

Recently, hypervitaminosis A is becoming a clinical problem of increasing frequency in the United States and other developed Western countries because of increasing self-medication and overprescription. Chronic intake of high doses of vitamin A (10 to 20 times the recommended daily allowance) can lead to variable signs and symptoms: headache, nausea, hepatomegaly and hepatic injury.

True deficiencies in vitamins are rare, except in some underdeveloped countries where malnutrition is endemic. Nevertheless, the existence of marginal nutritional deficiency has been confirmed by several nutritional enquiries realized in many industrialized countries (France, United States, Canada, Germany, United Kingdom, Switzerland), as a consequence of physiological or pathological situations or a way of life with special food selection and food intake. In all these situations (smokers, vegetarians, excessive alcohol consumption, hypocaloric diet), elderly people are exposed to a vitamin or trace element sub-deficiency or deficiency.

Conclusion on vitamins A and E

Low vitamin A and E antioxidant status can contribute to heart disease or cancer, the two largest factors in human longevity.

Reports on the association of low serum vitamin A, vitamin E and/or selenium status with the risk of coronary heart disease, stroke (Virtamo et al., 1985) and cancer (Salonen et al., 1985), although controversial clearly point out the importance of these nutrients.

Recently, Riemersma et al. (1991) studied the relation between the risk of angina pectoris and plasma concentration of vitamin A, E and carotene. Vitamin E was independently and inversely correlated to the risk of angina after adjustment for age, smoking habit, blood pressure, lipids and relative weight. These findings suggest that some populations with a high incidence of coronary heart disease (C.H.D.) should supplement their diets with more cereals, vitamin E rich oils, vegetables and fruits. How do low plasma and (presumably) tissues levels of natural antioxidants relate to C.H.D.? An increased tendency to peroxidation of polyunsatured fatty acids resulting from a reduction of antioxidants favors thrombosis. Oxidative modification of LDL (Esterbauer et al., 1987) increases atherogenesis, thrombosis and myocardial ischemic damage. Formation of foam cells from monocytes/macrophages is favored when LDL is in the oxidized and modified form (Steinberg, 1989). Antioxidants are more easily destroyed by food processing than polyunsaturated fatty acids (Esterbauer et al., 1987). For prevention of C.H.D. and increase of life expectancy, it seems important to have a food antioxidant status as approximated by vitamin A and E plasma levels.

The protective effect of vitamin A and β-carotene (highly effective in quenching singlet oxygen) against cancer increased our interest in these nutrients. Nutritional and specific physiological status (digestive tract) may influence metabolism and requirements of vitamins A and E. From reviewed findings and/or existing RDA's (RDI's) it seems desirable to have a daily diet ratio of vitamin A to vitamin E in the range 1:5–10 (if expressed in μmole of retinol and tocopherol). Some recent works on vitamins A and E confirm the concept of their interactions (Berger, 1989). Most research devoted to vitamin A and E interactions were carried out on animals. But to what extent could these results be valid also for humans?

In most recommended dietary allowances (RDA-RDI) there is no special suggestion for maintaining specific or special amounts of vitamins A and E or observing a strict relationship expressed in weight. Table 6 shows the results concerning the mean recommended dietary intake for vitamin A and E (RDI, 1983). The mean for vitamin A and E is obtained for 41 and 12 countries, respectively.

In case of vitamin deficiency (indicated by plasma or serum level) especially in disease states, vitamins A and E usually occur at the same (percentage of) low level, when compared to normal. However in elderly

Table 6

Subjects	Vitamin E		Vitamin A	
	mg/day	μmole/day	μg/day	μmole/day
Child	5.5	12.8	372	1.3
Adult woman	10.3	23.9	848	3.0
Adult man	12.0	27.9	910	3.2
Elderly	12.0	27.9	910	3.2

subjects deficiency in vitamin E is more significant than in vitamin A. In our measurements, the decrease in vitamin E in institutionalized elderlies is more pronounced than that of vitamin A as compared to the values of healthy 40-year-old subjects.

In conclusion concording results suggest that many elderly individuals might benefit from vitamin E supplementation. The supposed protective effect of vitamin A against cancer must be considered with caution in order to avoid toxic hypervitaminosis A from self-medication and overprescription, which is becoming a clinical problem of increasing frequency in some countries.

Acknowledgments. The authors wish to thank the geriatric home's staff and also M. F. Fraigneau, C. Chouard, C. Hamon, for technical assistance, N. Cachau for help with manuscript, the Ligue Nationale contre le Cancer and Herbaxt Laboratories, Torcy France, for financial support.

Apfelbaum, M., Forrat, C., and Nillus, P. (1989) Diététique et Nutrition, 2e éd. Masson, Paris.

Bauernfeind, J. C. (1980) Tocopherols in foods, in: Vitamin E: A Comprehensive Treatise. Machlin, L. J., ed. Dekker, New York.

Bendich, A., Gabriel, E., and Machlin L. J. (1986) Dietary vitamin E requirement for optimum immune responses in rats. J. Nutr. 116: 675–681.

Berger, S. (1989) Interactions between vitamin A and vitamin E and their influence in nutrition. ROCZN, PZH 40: 114–124.

Bieri, J. G., Corash, L., and Hubbard, V. S. (1983) Medical uses of vitamin E. N. Engl. J. Med. 308: 1063–1071.

Bodgen, J. D., Oleska, J. M., Lavenhar, M. A. et al. (1988). Zinc and immuno-competence in elderly people: effects of zinc supplementation for 3 months. Am. J. Clin. Nutr. 48: 655–663.

Boxer, L. A. (1986) Regulation of phagocyte function by α-tocopherol. Proc. Nutr. Soc. 45: 333–344.

Bunker, V. W., Lawson, M. S., Stansfield, M. F., and Clayton, B. E. (1988) Selenium balance studies in apparently healthy and housebound elderly people eating self-selected diets. Br. J. Nutr. 59: 171–180.

Cairns, J. (1985) Scien. Am. 253/5: 51.

Castignani, G. L., and Bieri, J. G. (1983) Simultaneous determination of retinol and α-tocopherol in serum or plasma by liquid chromatography. Clin. Chem. 29/4: 708–712.

Campbell, D., Bunker, V. W., Thomas, A. J., and Clayton, B. E. (1989) Selenium and vitamin E status of healthy and institutionalized elderly subjects: analysis of plasma, erythrocytes and platelets. Br. J. Nutr. 62: 221–227.

Chevance, M., Brubacher, G., Herbeth, B., Vernhes, G., Mikstacki, T., Dete, F., Fournier, C., and Janot, C. (1984) Immunological and nutritional status among the elderly, in: Lymphoid Cell Functions in Aging. de Wech A. L., ed. Eurage Rijswick. pp. 231–237.

Chevance, M., Brubacher, G., Herbeth, B., et al. (1985). Immunological and nutritional status among the elderly, in: Nutrition, Immunity, and Illness in the Elderly. Chandra, R. K., ed. Pergamon Press, New York. pp. 137–142.

Combs, G. F. (1985) Can dietary selenium modify cancer risk? Nutr. Rev. 43: 325–331.

Cutler, R. G. (1984) Antioxidants, aging and longevity, in: Free Radicals in Biology, vol. 6. Pryor, W. A., ed. Academic, New York. pp. 371–428.

Driskell, W. J., Neese, J. W., Bryant, C. C., and Bashor, M. M. (1982) Measurement of vitamin A and vitamin E in human serum by high-performance liquid chromatography. J. Chromat. 231: 439–444.

Dupin, H. (1981) Apports nutritionnels conseillés pour la population française. CNRS-CNERMA, Technique et Documentation, Paris 101.

Dworkin, B., Rosenthal, W. S., Jankowski, R. A., Gordon, G. G., and Haldae, D. (1985) Low blood selenium levels in alcoholics with and without advanced liver disease. Correlation with clinical and nutritional status. Digest. Dis. Sci. 30: 838–844.

Esterbauer, H., Jurgens, G., Quehenberger, O., and Koller, E. (1987) Autooxidation of human low density lipoprotein: loss of polyinsaturated fatty acids and vitamin E and generation of aldehydes. J. Lipid. Res. 28: 495–509.

Feher, J., Csomos, G., and Vereckei, A. (1987) Free radical reactions in Medicine. Springer-Verlag, Berlin/Heidelberg/New York/London/Paris/Tokyo.

Gey, K. F., Brubacher, G. B., and Stähelin, H. B. (1987) Plasma levels of antioxidant vitamins in relation to ischemic heart disease and cancer. Am. J. Clin. Nutr. 45: 1368–1377.

Girre, C., Hispard, P., Therond, P., Guedj, S., Bourdon, R., and Dally, S. (1990) Effect of abstinence from alcohol on the depression of glutathione peroxidase activity and Selenium and Vitamin E levels in chronic alcoholic patients. Alcohol Clin. Exp. Res. 14/6: 909–912.

Goodman D. (1984) Vitamin A and retinoids in health and disease. New J. Med. 310/16: 1023–1031.

Goodwin, J. S., Searles, R. P., and Tung, K. S. K. (1982) Immunological response of a healthy elderly population. Clin. Exp. Immunol. 48: 403–410.

Halliwell, B., and Gutteridge, J. M. (1985) Free radicals in Biology and Medicine. Clarendon Press, Oxford.

Harman, D. (1981) The aging process. Proc. Nat. Acad. Sci. USA 78: 7124–7128.

Harman, D. (1984) The free radical theory of aging, in: Free Radicals in Biology, vol. 5, Pryor, W. A. ed. New York, Academic Press. pp. 255–273.

Harman, D. (1982) Nutritional implication of the free radical theory of aging. J. Am. Col. Nutr. 1: 27–34.

Helzlsouer, K. J., Comstock, G. W., Morris, J. S. (1989) Selenium lycopene, α-tocopherol, β-carotene, retinol and subsequent bladder cancer. Cancer Res. 49: 6144–6148.

Herbert, V. (1991) Vitamin E supplementation of elderly people. Am. J. Clin. Nutr. 53: 976.

Horwitt, M. K. (1986) Interpretations of requirements for thiamin, riboflavin, niacin tryptophan and vitamin E plus comments on balance studies and vitamin B6. Am. J. Clin. Nutr. 44: 973–985.

Huttunen, J. K., and Virtamo, J. (1986) Diet and prevention of coronary heart disease and cancer. Halgren, B. ed. Raven Press.

Ip, C. (1988) Differential effect of dietary methionine on the biopotency of selenome-thionine and selenite in cancer chemoprevention. J. Natl. Cancer Inst. 80: 258–262.

Ip, C., and Ganther, H. (1988) Efficacy of trimethylselenomium versus selenite in cancer chemoprevention and its modulation by arsenite. Carcinogenesis (Lond.) 9: 1481–1484.

Ip, C, and Hayes, C. (1989) Tissue selenium levels in selenium supplemented rats and their relevance in mammary cancer protection. Carcinogenesis (Lond.) 10: 921–925.

Ip, C., and Ganther, H. E. (1990) Activity of methylated forms of selenium in cancer prevention. Cancer Res. 50: 1206–1211.

Jackson, M. L. (1988) Selenium: geochemical distribution and associations with human heart and cancer death rate and longevity in China and the United States. Biol. Trace Elem. Res. 15.

Janghorbani, M., Rockway, S., Mooers, C. S., Roberts, E. M., Ting, B. T., and Siltrin, M. D. (1990) Effect of chronic selenite supplementation on selenium excretion and organ accumulation in rats. J. Nutr. 120/3: 274–279.

Kark, J. D., Smith, A. H., Switzer, B. R., and Hames, C. G. (1981) Serum vitamin A (retinol) and cancer incidence in Evans Country, Georgia. JNCI 66: 7–16.

Kasperek, K., Lombeck, I., Kiem, J., Iyengar, G. V., Wang, Y. X., Feinendegen, L. E., and Bremer, H. J. (1982) Platelet selenium in children with normal and low selenium intake. Biol. Trace Element Res. 4: 29–34.

Keshan Disease Research Group (1979) Epidemiology studies on the etiologic relationship of selenium and Keshan disease. Chin. Med. J. 92: 477–482.

Kinsella, J. E. (1987), in: Seafoods and Fish Oils in Human Health and Disease. Dekker, New York. pp. 186–189.

Kivela, S. L., Menapaa, P., Nissinen, A., Alfthan, G., Punsar, S., Enlund, H., and Puska, P. (1989) Vitamin A, vitamin E and Selenium status in an aged Finnish male population. Int. J. Vit. Nutr. Res. 59: 373–380.

Knekt, P., Aromaa, A., Maatela, J., et al. (1988a) Serum vitamin E, serum selenium and the risk of gastro-intestinal cancer. Int. J. Cancer 42: 846–850.

394

Knekt, P. (1988b) Serum vitamin E level and risk of female cancers. Int. J. Epidemiol. 17: 281–286.

Knekt, P., Aromaa, A., Maatela, J., Aaran, R. K., Nikkari, T., Hakama, M., Hakulinen, T., Peto, R., Saxen, E., and Teppo, L. (1988c) Serum vitamin E and risk of cancer among Finnish men during a 10-year follow-up. Am. J. Epidemiol. 127: 28–41.

Kok, F. J., De Bruijn, A. M., Vermeeren, R., et al. (1987) Serum selenium, vitamins, antioxidants and cardiovascular mortality: a 9-year follow up study in the Netherlands. Am. J. Clin. Nutr. 45: 462–468.

Korpela, H., Kumpulainen, J., Luoma, P. V., Arranto, A. J., and Sotaniemi, E. (1985) Decreased serum selenium in alcoholics as related to liver structure and function. Am. J. Clin. Nutr. 42: 147–151.

Lei, D. N., Wang, L. Q., Ruebner, B. H., Hsieh, D. P., Wu, B. F., Zhu, C. R., and Du, M. J. (1990) Effect of selenium on aflatoxin hepatocarcinogenesis in the rat. Biomed. Environ. Sci. 3/1: 65–80.

Lemoine, A., Le Devehat, C., and Herbeth, B. (1984) Statut vitaminique: étude d'un groupe d'adultes français. Bull. Acad. Nat. Med. 168: 522–530.

Lemoine, A., Le Devehat, C., Herbeth, B., Bourgeay-Causse, M., Delacoux, E., Mareschi, J. P., Martin, J. Y., Miravet, L., Potier De Courcy, G., and Zittoun, J. (1986) Vitamin status in three groups of French adults. Controls, obese subjects, alcohol drinkers. Ann. Nutr. Meta. 30, Suppl. 1: 1–94.

Levander, O. A. (1986) Selenium in: Trace Elements in Humans and Animal Nutrition, vol. 2. Mertz, W., ed. Academic Press, Orlando FL. pp. 209–279.

Levander, O. A. (1990) Selenium: Essentiality versus toxicity in man. J. Environ. Geochem. Health 12(suppl.): 11–19.

Liebler, D. C., Kling, D. S., and Reed, D. J. (1986) Antioxidant protection of phospholipid bilayers by α-tocopherol. J. Biol. Chem. 261: 12114–12119.

Luo, X., Wei, H., Yang, C., Xing, J., Liu, X., Qiao, C., Feng, Y., Liu, J., Liu, Y., Wu, Q., Liu, X., Guo, J., Stoecker, B. J., Spallholz, J. E., and Yang, S. P. (1985) Bioavailibility of selenium to residents in a low selenium area of China. Am. J. of Clin. Nutr. 42: 439–448.

Makinodan, T. (1981) Cellular basis of immunologic aging, in: Biological Mechanisms in Aging. Schimke, R. T., ed. US Department of Health and Human Services, Bethesda, MD. pp. 488–500.

Mareschi, J., Cousin, F., De la Villeon, B., and Brubacher, G. B. (1984) Valeur calorique de l'alimentation et couverture des apports nutritionnels conseillés en vitamines de l'homme adulte. Ann. Nutr. Metab. 28: 11–23.

McCay, P. B. (1985) Vitamin E: Interactions with free radicals and ascorbate. Annu. Rev. Nutr. 5: 323–340.

Menkes, M. S., Comstock, G. W., Vuilleumier, J. P., Helsing, K. J., Rider, A. A., and Brookmeyer, R. (1986) Serum β-carotene, vitamins A and E, selenium, and the risk of lung cancer. N. Engl. J. Med. 315: 1250–1254.

Menzel, H., Steiner, G., Lombeck, I., and Ohnesorge, F. K. (1983) Glutathione peroxidase and glutathione S-transferase activity of platelets. Eur. J. Paed. 140: 244–247.

Meydani, S. N., Meydani, M., Verdon, C. R., Shapiro, A. A., Blumberg, J. B., and Hayes, K. C. (1986) Vitamin E supplementation suppresses prostaglandin E_2 synthesis and enhances the immune response of aged mice. Mec. Aging Dev. 34: 191–201.

Meydani, S. N., Barklund, M. P., Liu, S., Meydiani, M., Miller, R. A., et al. (1990) Vitamin E supplementation inhances cell mediated immunity in healthy elderly subjects. Am. J. Clin. Nutr. 52: 557–563.

Miller, L., Mills, B. J., Blotcky, A. J., and Lindeman, R. D. (1983) Red blood cell and serum selenium concentrations as influenced by age and selected diseases. J. Amer. College of Nutrition 4: 331–334.

Miyamoto, H., Araya, Y., Ito, M., et al. (1987) Serum selenium and vitamin E concentrations in families of lung cancer patients. Cancer 60: 1159–1162.

Molennaar, I., Hulstaert, C. E., and Hardonk, M. J. (1984) Role in function and ultra-structure of cellular membranes, in: Vitamin E: A Comprehensive Treatise. Machlin, L. J., ed. Marcel Dekker, New York, pp. 372–389.

Morisi, G., Patriarca, M., Marano, G., Giampaoli, S., and Taggi, F. (1989) Age and sex specific reference serum selenium levels estimated for the Italian population. Ann. Ist. Super Sanita 25/3: 393–403.

Moore, J. A., Noira, R., and Wells, I. C. (1984) Clin. Chem. 307: 1171.

Munnich, A., Ogier, H., Saudubray, J. M., et al. (1985) Les vitamines: Aspects métaboliques, génétiques, nutritionnels et thérapeutiques. Masson, Paris.

Nagel, J. E., Chopra, R. K., Chrest, F. J., et al. (1988) Decreased proliferation, interleukin 2 synthesis and interleukin 2 receptor expression are accompanied by decreased mRNA expression in phytohemagglutinin-stimulated cells from elderly donors. J. Clin. Invest. 81: 1096–1101.

Neve, J., Vertongen, F., and Molle, L. (1985) Selenium deficiency. Clin. Endocrinol. Metabol. 14: 629–656.

Oster, O., Prellwitz, W., Luley, C., Meinertz, T., Geibel, A., and Kasper, W. (1987) Medicine and Biology, vol. 4. Brätter, P. and Schramel, P., eds. Walter de Gruyter.

Oster, O., and Prellwitz, W. (1990) Selenium and cardiovascular disease. Biol. Trace Elem. Res. 24: 91– .

Packer, L. (1991) Protective role of vitamin E in biological systems. Am. J. Clin. Nutr. 53: 1050S–1055S.

Panush, R. S., and Delafuente, J. C. (1985) Vitamins and immunocompetence. World Rev. Nutr. Diet. 45: 97–132.

Parker, R. S. (1989) Dietary and biochemical aspects of vitamin E. Adv. Food Nutr. Res. 33: 157–232.

Paul, A. A., and Southgate, D. A. T. (1987) MacCance and Widdowson's Composition of Foods. Elsevier/North Holland Biomedical Press, Amsterdam.

Peto, R., Doll, R., Buckley, J. D., and Sporn, M. B. (1981) Can dietary beta-carotene materially reduce human cancer rates? Nature 290: 201–208.

Pich, S. M. (1987) J. Nutr. 117: 636–640.

Pokorn, D., Accetto, B., and Prevorenik, A. (1990) Vitamin E status in men and women aged 60–90 years. Acta Med. Iug. 44, No 3: 223–232.

Puska, P., Pietinen, P., Nissinen, A., Salonen, J., Tuomilehto, J., and Ehmholm, C. (1986) Diet and prevention of coronary heart disease and cancer. Halgren, B., ed. Raven Press.

Recommended Dietary Allowances (1980) 9° édition, National Academy of Sciences, Washington, p. 186.

Recommended dietary intakes around the world (1983) Nutrition Abstracts and Reviews 53: 1075.

Recommended Dietary Allowances (1989) 10° édition, National Academy of Sciences, Washington.

Reeves, W. C., Marcuard, S. P., Willis, S. E., and Movahed, A. (1989) Reversible cardiomyopathy due to selenium deficiency. JPEN J. Parenter Enteral. Nutr. 13/6: 663–665.

Regöly Merei, A., Laszlofi, M., Zajkos, G., Vertes, L., Gergely, A., Antal, M., and Biro, G. (1989) From vitamin A and E of institutionalized elderlies. Die Nahrung 33: 695–697.

Riemersma, R. A., Oliver, M., Elton, R. A., et al. (1990) Plasma antioxidants and coronary heart disease: Vitamins C and E, and Selenium. Eur. J. Clin. Nutr. 44: 143–150.

Riemersma, R. A., Wood, D. A., Macintyre, C. C. A., Elton, R. A., and Oliver, M. F. (1991) Risk of angina pectoris and plasma concentrations of vitamins A, C and E and carotene. The Lancet 337: 1.

Roitt, I. M. (1984) Essential Immunology. Blackwell Scientific Publications, London.

Salonen, J. T., Alfthan, G., Huttenen, J. K., Pikkarainen, J., and Puska, P. (1982) Association between cardiovascular death and myocardial infarction and serum selenium in matched-pair longitudinal study. Lancet 2: 175–179.

Salonen, J. T., Alfthan, G., Huttunen, J. K., and Puska, P. (1984) Association between serum selenium and the risk of cancer. Am. J. Epidemiol. 3: 342–347.

Salonen, J. T., Salonen, R., Lappetelainen, R., et al. (1985) Risk of cancer in relation to serum concentrations of selenium and vitamins A and E: matched case-control analysis of prospective data. Br. Med. J. 290: 417–420.

Schaffer, M., Kronberger, E., Eber, B., Wawschinek, O., Dusleaq, J., Friehs, I., Gogg, R., Petek, W., and Klein, W. (1990) Serum selenium levels in distinct manifestations of coronary artery disease. Acta Med. Austriaca. 17/1: 22–24.

Schrauzer, G. N. (1979), in: Advances in Nutritional Research, vol. 2. Draper, H. H., ed. Plenum, New York. pp. 219–244.

Schwarz, K., and Foltz, C. M. (1957) Selenium as an integral part of factor 3 against dietary necrotic liver degeneration. J. Am. Chem. Sov. 79: 3292–3293.

Shamberger, R. J. (1986) Medical implications of selenium biochemistry. Trace Elem. Med. 3: 105–111.

396

Shekelle, R. B., Lepper, M., Liu, S., et al. (1981) Dietary vitamin A and risk of cancer in the Western Electric study. Lancet 2: 1185–1190.

Simonoff, M., Hamon, C., Moretto, P., Llabador, Y., and Simonoff, G. (1988a) High sensibility Pixe determination of selenium in food and biological samples using a pre-concentration technique. Nucl. Ins. Meth. B31: 442–448.

Simonoff, M., Hamon, C., Moretto, P., Llabador, Y., and Simonoff, G. (1988b) Selenium in Foods in France. J. Food Comp. Anal. 1: 295–302.

Simonoff, M., Conri, C., Fleury, B., Berdeu, B., Moretto, P., Ducloux, G., and Llabador, Y. (1988c) Serum and erythrocyte selenium in normal and pathological states in France. Trace Elem. Med. 5/2: 64–69.

Simonoff, M., and Simonoff, G. (1991) Le Sélénium et la vie. Masson, Paris.

Smith, J. L., Wickiser, A. A., Korth, L. L., Grandjean, A. C., Schaeffer, A. E. (1984) J. Am. Coll. Nutr. 3: 13–25.

Sohal, R. S., Farmer, K. J., Allen, R. G., and Cohen, N. R. (1983) Effect of age on oxygen consumption, superoxide dismutase, catalase, glutathione, inorganic perioxides and chloroform soluble antioxidants in the adult male housefly, Musca Domestica. Mech. Ageing Der. 24: 185–195.

Spallholtz, J. E. (1990) Selenium and glutathione peroxidase: essential nutrient and antioxidant component of the immune system. Adv. Exp. Med. Biol. 262: 145–158.

Stead, R. J., Hinks, L. J., Hodson, M. E., Redington, A. N., Clayton, B. E., and Batten, J. C. (1985) Selenium deficiency and possible increased risk of carcinoma in adults with cystic fibrosis. Lancet 2: 862–863.

Steinberg, D., Parthasarathy, S., Carew, T. E., Khoo, J. C., and Witzum, J. L. (1989) Beyond cholesterol, modifications of low-density lipoprotein that increase its atherogenicity. N. Engl. J. Med. 320: 915–924.

Steiner, G., Menzel, H., Lombeck, I., Ohnesorge, F. K., and Bermer, H. J. (1982a) Plasma glutathione peroxidase after Se supplementation in patients with reduced Se state. Eur. J. Pediatr. 138: 138–140.

Steiner, M., and Mower, R. (1982b) Mechanism of action of vitamin E on platelet functions. Ann. N. T. Acad. Sci. 393: 289–299.

Suter, P. M., and Russel, R. M. (1987) Am. J. Clin. Nutr. 45: 501–512.

Tanaka, J., Fuiwara, H., and Torisu, M. (1979) Vitamin E and immune response. Immunology 38: 727–734.

Thomas, A. J., Bunker, V. W., Hinks, L. J., Sodha, N., Mullee, M. A., and Clayton, B. E. (1988) Energy, protein, zinc and copper status of twenty-one elderly in patients: analyzed dietary intake and biochemical indices. Br. J. Nutr. 59: 181–191.

Thompson, H. J., Meeker, L. D., and Kokoska, S. (1984) Effect of an inorganic and organic form of dietary selenium on the promotional stage of mammary carcinogenesis in the rat. Cancer Res. 44: 2803–2806.

Tolonen, M., Halme, M., and Sarna, S. (1985) Vitamin E and selenium supplementation in geriatric patients. Biol. Trace Elem. Res. 7: 161–168.

Tolonen, M., Sarna, S., Halme, M., et al. (1988) Antioxidant supplementation decreases TBA reactants in serum of elderly. Biol. Trace Elem. Res. 17: 221–228.

Trickler, D., and Shklar, G. (1987) Prevention by vitamin E of experimental oral carcino-genesis. JNCI 78: 1615–1619.

U.S. Department of Agriculture (1979). Composition of Foods Fats and Oils. Agriculture Handbook No 8–4. U.S. Government Printing Office. Washington, D.C.

U.S. Department of Agriculture (1985). Nationwide Food Consumption Survey. Continuing Survey of Food Intakes by Individuals. Human Nutrition Information Service. Nutrition Monitoring Division. CSFII Report (85).

U.S. National Academy of Sciences: Aging and the Geochemical Environment. Washington, D.C. (1981).

Vandewoude, M. F. J., and Vandewoude, M. G. (1987) Vitamin E status in a normal population: the influence of age. J. Amer. Colleg of Nutr. 6: 307–311.

Van Gossum, A., Shariff, R., Lemoyne, M., et al. (1988) Increased lipid peroxidation after lipid infusion as measured by breath pentane out. Am. J. Clin. Nutr. 48: 1394–1399.

Van Riij, A. M., Thomson, C. D., McKenzie, J. M., and Robinson, M. F. (1979) Selenium deficiency in total parenteral nutrition. Am. J. Clin. Nutr. 32: 2076–2085.

Varma, R. N., and Beaton, G. H. (1972) Can. J. Physiol. Pharmacol. 50: 1026–1037.

Vatassery, G. T., Johnson, G. J., and Krerowski, A. M. (1983) Changes in vitamin E concentrations in human plasma and platelets with age. J. Amer. College of Nutrition 4: 369–375.

Verlinden, M., Van Sprungel, M., Van der Auwera, J. C., and Eylenbosch, W. J. (1983) The selenium status of Belgian population groups. II. Newborns, children and the aged. Biol. Trace Element Research 5: 103–113.

Virtamo, J., and Huttunen, J. K. (1985) Selenium in human disease. Ann. Clin. Res. 17: 87–89.

Wald, N., Idle, M., Boreham, J., and Bailey, A. (1980) Low serum-vitamin A and subsequent risk of cancer: preliminary results of a prospective study. Lancet 2: 813–815.

Wald, N. J., Boreham, J., Hayward, J. L., and Bulbrook, R. D. (1984) Plasma retinol, β-carotene and vitamin E levels in relation to the future risk of breast cancer. Br. J. Cancer 49: 321–324.

Willett, W. C., Morris, J. C., Pressel, S., Taylor, J. O., Polk, B. F., Stampfer, M. J., Rosner, B., Schneider, K., and Hames, C. G. (1983) Prediagnostic serum selenium and risk of cancer. Lancet 2: 130–134.

Willett, W. C., Polk, B. F., Underwood, B. A., Stampfer, M. J., Pressel, S., Rosner, B., Taylor, J. O., Schneider, K., and Hames, C. G. (1984) Relation of serum vitamins A and E and carotenoids to the risk of cancer. N. Engl. J. Med. 310: 430–434.

Younger, V. R. (1981) Selenium: a case for its essentiality in man. N. Engl. J. Med. 304: 1228–1230.

Free Radicals and Aging
ed. by I. Emerit & B. Chance
© 1992 Birkhäuser Verlag Basel/Switzerland

Inverse correlation between essential antioxidants in plasma and subsequent risk to develop cancer, ischemic heart disease and stroke respectively: 12-year follow-up of the Prospective Basel Study

Monika Eichholzer[a], Hannes B. Stähelin[a] and K. Fred Gey[b]

[a]*Geriatric Clinic Kantonsspital, CH-4031 Basel and* [b]*Vitamin Substudy WHO/MONICA Project, Institute of Biochemistry and Molecular Biology, University of Bern, CH-3000 Bern, Switzerland*

Summary. There is accumulating evidence that free radicals may contribute to various diseases such as cancer or cardiovascular disease. Possible health hazards can to some extent be prevented by the body's multilevel defense system against free radicals, which comprises, besides others, antioxidant vitamins. The 12-year mortality follow-up of 2,974 participants of the Basel Study allowed to test the hypothesis that low antioxidant vitamin plasma concentrations (vitamin A, C, E and carotene) were associated with increased death from cancer of various sites and death from atherosclerosis such as ischemic heart disease and stroke, respectively. For the analysis 204 cancer cases, 132 fatalities from ischemic heart disease (IHD) and 31 deaths from cerebral vascular disease were available.

Cancer mortality. Overall mortality from cancer was associated with low mean plasma levels of carotene adjusted for cholesterol ($p < 0.01$) and of vitamin C ($p < 0.01$). Bronchus and stomach cancers were associated with a low mean plasma carotene level ($p < 0.01$). Subjects with subsequent stomach cancer had also lower mean vitamin C and lipid-adjusted vitmain A levels than survivors ($p < 0.05$). Calculating the relative risk with exclusion of mortality during the first two years of follow-up, low plasma carotene was associated with an increased risk for bronchus cancer (RR 1.8, $p < 0.05$), and the small number of stomach cancer cases (RR 2.95, $p < 0.05$) low plasma levels of carotene and vitamin A with all cancer types (RR 2.47, $p < 0.01$), and low plasma retinol in older subjects (> 60 years) with lung cancer (RR 2.17, $p < 0.05$). Studies in other cohorts with a poor vitamin E status revealed an increased risk of subsequent cancer at low vitamin E levels as well. It is concluded that low plasma levels of all major essential antioxidants are associated with an increased risk of subsequent cancer mortality.

Cardio-vascular mortality. Plasma carotene concentration below quartile 1 was associated with an increased risk for IHD (RR 1.53, $p = 0.02$). The same was true for low levels of both carotene and vitamin C (RR $= 1.96$, $p = 0.022$). The risk of cerebrovascular death was elevated in subjects with low carotene in the presence of low vitamin C plasma concentration (RR 4.17, $p < 0.01$). These data confirm and extend recent findings on an inverse correlation of β-carotene and vitamin C respectively to CVD. In the Basel population (with relatively low IHD mortality) the variation of plasma vitamin E levels occurred at high levels, all presumably above the critical threshold, which revealed correlation to CVD. In summary, a poor plasma level of any essential antioxidant can be associated with an increased risk of CVD. The evidence is sufficient to justify intervention trials to test the actual CVD-preventive potentials of a combination of all major essential antioxidants, in order to optimize the total antioxidant defense potential.

Introduction

Free radicals are highly reactive chemical species since they have unpaired electrons in their outer orbitals. The human body is exposed to free radicals that are continuously formed in the regular endogenous metabolism or that have external sources such as ionizing radiation or smoking. Free radicals are capable of damaging all cellular components, i.e. lipids (peroxidation of polyunsaturated fatty acids in organelles and plasma membranes), proteins (denaturation and inactivation of enzymes, respectively), carbohydrates (polysaccharide depolymerization), nucleic acids (causing mutation and inhibition of protein nucleotide, and fatty acid sythesis) (Southorn and Powis, 1988, part I).

There is accumulating evidence that free radicals may contribute to various diseases such as cancer or cardiovascular disease. Carcinogenesis is thought to occur in stages. In the initiation stage of carcinogenesis, a physical, chemical, or biological agent directly causes an irreversible alteration in the molecular structure of DNA of the cell. In the subsequent promotion stage, tumor-promoting radicals usually do not react directly with the genetic material but alter the expression of the genes that regulate cell differentiation and growth. Oxygen free radicals (hyperbaric oxygen, superoxide anion radical, and certain organic peroxides) may act mainly as tumor promoters although they can also be weak carcinogens (Southorn and Powis, 1988, part I). It has also been postulated that certain promoters of carcinogenesis act by generation of oxygen radicals and resultant lipid peroxidation. Lipid peroxidation cross-links proteins and affects all aspects of cell organization, including membrane and surface structures, and the mitotic apparatus (Ames, 1983).

In relation to atherosclerosis, recent studies suggest that oxygen free radicals might be involved in the development of atherosclerotic plaques (Gey, 1986; Steinberg and Witztum, 1990). In the subendothelial space the low-density lipoproteins might undergo oxidation by oxygen free radicals released from arterial endothelial and smooth muscle cells. Peroxidatively modified low-density lipoproteins are known to be toxic to all arterial cells, to attract macrophages from the blood and to convert macrophages into lipid-laden foam cells which can colonize as fatty streaks (Southorn and Powis, 1988, part II).

Possible health hazards by these oxygen species can to some extent be prevented by the body's multilevel defense system against free radicals, which comprises enzymes, endogenous nonessential antioxidants and essential antioxidants. Among the latter antioxidant vitamins E, C, A and carotenoids are of particular interest. Vitamin E, being fat soluble, is found in all cell membranes. Here it may prevent the destructive non-enzymatic oxidation of polyunsaturated fatty acids by molecular oxygen, and inhibit thereby, e.g. the production of promoters in carcinogenesis or the modification of low-density lipoproteins (Passmore and Eastwood, 1986; Bertram et al., 1987). The antioxidant role of the water-soluble vitamin C is based on the one hand on the reaction with aqueous radicals, including peroxyl radicals, similar to the chain-breaking reaction of the liposoluble vitamin E, and on the other hand on its capacity to regenerate and spare vitamin E (Sies, 1989). Beta-carotene is considered as an efficient quencher of singlet molecular oxygen (Sies, 1989). The antioxidant vitamins may act by various mechanisms in carcinogenesis since they can also modify the activity of several crucial enzymes (e.g. lipoxygenases, protein kinase C), either by regulation of the "peroxide tone" and/or as specific ligands (Gey, 1992). Vitamins C and E may play an additional role by preventing nitrosamins formation in the stomach and by reducing fecal mutagenicity (Tannenbaum, 1983; Dion et al., 1982). The thiyl radical scavenger vitamin A (D'Aquino et al., 1989) may have an impact on the promotional phase of carcinogenesis and certainly suppresses the expression of precancerous cells (metaplasia) and cancer cells (Olson, 1990).

For humans, most direct evidence of an effect of a suboptimal status of antioxidant vitamins on the risk of cancer and ischemic heart disease is based on epidemiological studies.

The Western Electric study (Shekelle et al., 1981) was the first of a series of epidemiologic studies in which the intake of carotene was related to the incidence of lung cancer. The corresponding fact that low plasma levels of carotene can predict an increased risk of bronchus cancer was confirmed later by several studies (Stähelin et al., 1984; Menkes et al., 1986; Nomura et al., 1985 etc.) Wald et al. (1980) initially reported an increased risk of all cancers among persons with low levels of absolute serum retinol levels but in a further follow-up did not support these findings (Wald et al., 1986).

Only a few prospective studies have been conducted that address vitamin E and cancer risk of specific sites, and the results remain controversial (e.g. Knekt et al., 1988; Menkes et al., 1986; Willett et al., 1984; Nomura et al., 1985).

The overall evidence from surveys concerned with vitamin C is suggestive of a role of this vitamin in the prevention of cancer of the digestive tract, especially the esophagus and the stomach (Bertram et al., 1987).

The risk of IHD is predicted by classical risk factors such as hypercholesterolemia, hypertension, and smoking only to roughly 50% (Gey, 1989); thus dietary factors such as antioxidant vitamins may play a complementary role. Randomized blood samples and coronary heart disease mortality in standardized European study populations (Vitamin Substudy of the WHO/MONICA Project) show that low plasma levels of essential antioxidants, with the rank order alpha-tocopherol > vitamins A,C, are associated with high coronary heart disease

(Gey 1986; Gey et al., 1991). Some within-population studies have also shown an inverse relation between consumption of antioxidant vitamins and cardiovascular mortality. A time trend analysis of data from Israel has suggested an inverse association between vitamin A intake and coronary heart disease rates (Palgi, 1981). In the USA, industrial vitamin C production is inversely related to CHD mortality over the past 20 years (Ginter, 1979). Regional standardized mortality ratios for CHD in the UK relate between the consumption of fresh fruit and green vegetables and mortality from all cerebrovascular disease, independent of social class (Acheson and Williams, 1983).

This working hypothesis is in accordance with recent data obtained in individuals:

- among CHD patients the angiographically evaluated degree of coronary occlusion was inversely related to blood leukocyte ascorbate levels among both smokers and nonsmokers (Ramirez and Flowers, 1980);
- in an area of poor antioxidant status but common values of plasma cholesterol and blood pressure (Edinburgh/Scotland) low levels of either vitamin E and vitamin C were associated with an increased risk of early, previously undiagnosed angina pectoris. Thereby, low levels of vitamin C were at least in part due to cigarette smoking (Riemersma et al., 1991). However, in a case-control study in Finland on hypercholesterolemic patients with both early and treated advanced CHD no corresponding correlation was found for cholesterol-adjusted vitamin E and for vitamin C (Salonen et al., 1988). This discrepancy could be due to markedly smaller variation of antioxidant levels in Finland (Gey, 1992) or to predominance of other, i.e. antioxidant-unrelated risk factors in Finland; but it could also be due to selecting only hypercholesterolemic subjects;
- in the US Physicians Study the supplementation of β-carotene for 5 years was reported to reduce coronary events in males (Gaziano et al. 1990);
- in the Prospective US Nurses Study the highest intake of essential antioxidants correlated after the 8-year follow-up with a lower risk of IHD, i.e. the relative risk was 0.66 for high vitamin E intake, 0.70 for high vitamin A consumption and 0.78 for high β-carotene intake (Manson et al., 1991); similar trends were found for stroke (W. C. Willett, personal communication, 1990).

The 12-year mortality follow-up of the Basel Study allowed further to test the hypothesis that low antioxidant vitamin plasma concentrations (vitamin A, C, E and carotene) are associated with an increased death rate from cancer of various sites as well as from atherosclerosis such as ischemic heart disease and stroke, respectively. For the analysis 204 cancer cases, 132 fatalities from ischemic heart disease and 31 deaths from cerebral vascular disease were available.

Methods

Study population and follow-up

The Basel Study was started in 1960 (Basel Study I) to elucidate risk factors for cardiovascular diseases and is described in detail elsewhere (Widmer et al., 1981; Zinniker et al., 1978; Stähelin et al., 1983). Over 6,000 healthy subjects (4,858 men and 1,461 women) working in the three major pharmaceutical companies of Basel volunteered to participate. The male study group can be considered as representative of the average urban working Swiss male population. Follow-up examinations were carried out in 1965–66 (Basel Study II: 3,796 men and 829 women) and in 1971–73 (Basel Study III: 3,528 men and 688 women). At the 1971–1973 examination cycle, plasma specimens were collected in 2,974 men for the measurement of essential antioxidants.

The determination of the plasma vitamin A, C, E and carotene (approximately 80% beta-carotene and 20% alpha-carotene) levels from 1971–73 provided the primary baseline information for the present study. Other relevant risk factors such as age, smoking, blood pressure, weight, diabetes mellitus, serum lipid levels and alcohol were also assessed in 1971–73 and taken into consideration for the analysis.

The last mortality follow-up was carried out in 1985 (Stähelin et al., 1988; Stähelin et al., 1991). Death certificates of the Swiss Federal Office of Statistics in Bern (Switzerland) were used to identify causes of death. The autopsy rate was 60% and allowed validation of the death certificates. A total of 553 men died during the 12-year follow-up, including as already mentioned 204 cancer cases and 132 fatalities from ischemic heart disease (ICD-8 codes 410–414) and 31 deaths from vascular disease (ICD-8 codes 430–438). The cancer cases were grouped into bronchus cancer ($n = 68$, ICD-8 code 162), stomach cancer ($n = 20$, ICD-8 code 151), gastrointestinal ($n = 37$, ICD-8 codes 151 and 153) and 'all cancers' ($n = 204$). Because it was not possible to identify persons who were known to have cancer at the time, blood was drawn, additional cancer analyses were carried out excluding men who died from cancer within the first two years of the follow-up.

Laboratory analysis

Plasma vitamin levels were – in contrast to most other studies on this topic – measured immediately after blood samples were drawn. Because

all fat soluble vitamin plasma levels were correlated with plasma cholesterol and triglycerides respectively, vitamin A and E were – according
to the strongest correlations – adjusted to the sum of cholesterol
(220 mg/dl) and triglycerides (110 mg/dl) and carotene for cholesterol
(220 mg/dl) (Gey and Puska, 1989).

Statistical analysis

Group means of death cases and survivors – adjusted for smoking
and age – were compared. Relative risks (RR) were calculated after
adjustment of explanatory variables by using the Cox proportional
hazard regression model (Cox, D. R., 1972). Low vitamin levels were
defined as lower than first quartile (Q 1) except for vitamin C where a
biological threshold value of marginal vitamin C deficiency, i.e.
22.7 μmol/l, is already available (Gey et al., 1987). In fact, the cut-off
levels of Q 1 were 30.02 μmol/l for lipid-standardized α-tocopherol,
2.45 μmol/l for lipid-adjusted retinol and 0.23 μmol/l for cholesterol-adjusted carotene.

For more details about statistical methods see previous publication
(Stähelin et al., 1991).

Results and discussion

Cancer mortality

In the present 12-year follow-up a total of 553 men (out of 2,974 men
with complete data set) died. Cancer was the major cause of death, i.e.
in 204 men (37%), including 68 with bronchus cancer (12%) and 37
(7%) with a gastrointestinal cancer (20 with stomach cancer and 17
with large bowel cancer excluding cancer of the rectum).

Descriptive analysis (Fig. 1). Comparing mean values of plasma
vitamins from cancer cases with those of the survivors, the overall
mortality from cancer was associated with low mean plasma levels
of carotene ($p < 0.01$) and of vitamin C ($p < 0.01$). Bronchus and
stomach cancers were associated with a low mean plasma carotene
level ($p < 0.01$). Subjects with subsequent stomach cancers also had
lower mean vitamin C and vitamin A levels than did survivors
($p < 0.05$).

Risk analysis (Fig. 2). Calculating the relative risk of low vitamin
levels compared with higher ones, using the Cox model with exclusion
of mortality during the first 2 years of follow-up, low plasma carotene
(below quartile 1) was associated with a significantly increased risk for
bronchus cancer ($RR = 1.8$, $p < 0.05$), low plasma levels of carotene

404

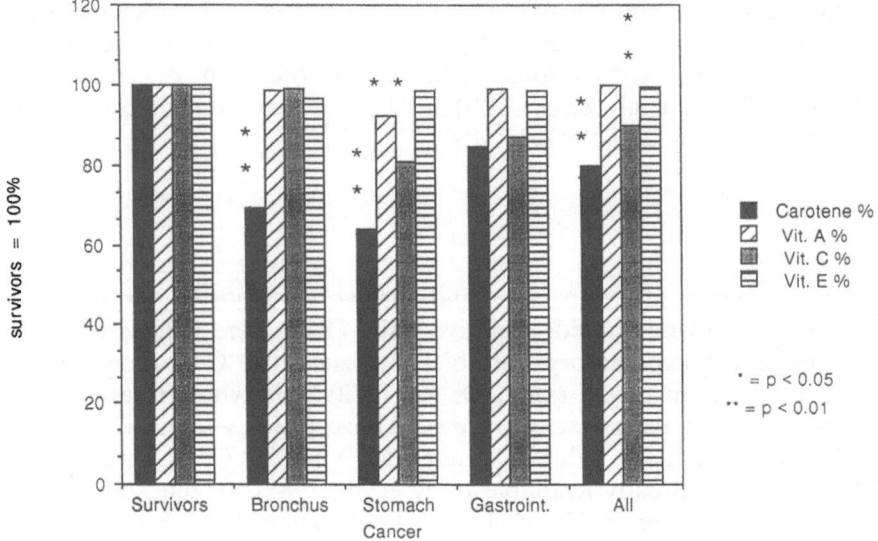

Figure 1. Comparison of various cancer sites regarding levels of mean plasma vitamin concentrations at base line (1971/73) in the 12-year follow-up of the Prospective Basel Study (survivors = 100%).

and vitamin A with all cancers (RR = 2.47, $p < 0.01$), and low plasma retinol in older subjects (over 60 years of age) with lung cancer (RR = 2.17, $p < 0.05$). Low levels of vitamin C increased the risk of stomach cancer (RR = 2.38) and gastronintestinal cancer (RR = 2.46) in older subjects, but only significantly with the inclusion of the first 2 years of follow-up.

Thus, the most prominent finding of our cancer analysis was the increased risk of cancer in cases with low plasma carotene levels. The relative risk was significantly higher for bronchus cancer and the small group of cases with stomach cancer. The major finding of the present 12-year follow-up is in line with evaluations of the Prospective Basel Study after 5 and 7 years respectively (Stähelin et al., 1984; Gey et al., 1987). In consequence, the Basel Study in which decay of carotene in stored samples was prevented by immediate plasma assay extends and consolidates associations of other studies, i.e. of low plasma levels of carotene with an increased risk of lung cancer in the major recent studies of the blood bank design (Menkes et al., 1986; Nomura et al., 1985) as well as with dietary intake (Ziegler 1991). A puzzling result is the increased risk of low lipid-standardized vitamin A for subsequent lung cancer (RR 2.17, $p < 0.05$ in subjects over 60) and the likely synergism of carotene and lipid-standardized vitamin A in all cancer cases with low plasma level of both carotene and vitamin A (RR of all cancer, 2.42, $p < 0.01$). The results for low plasma retinol levels are

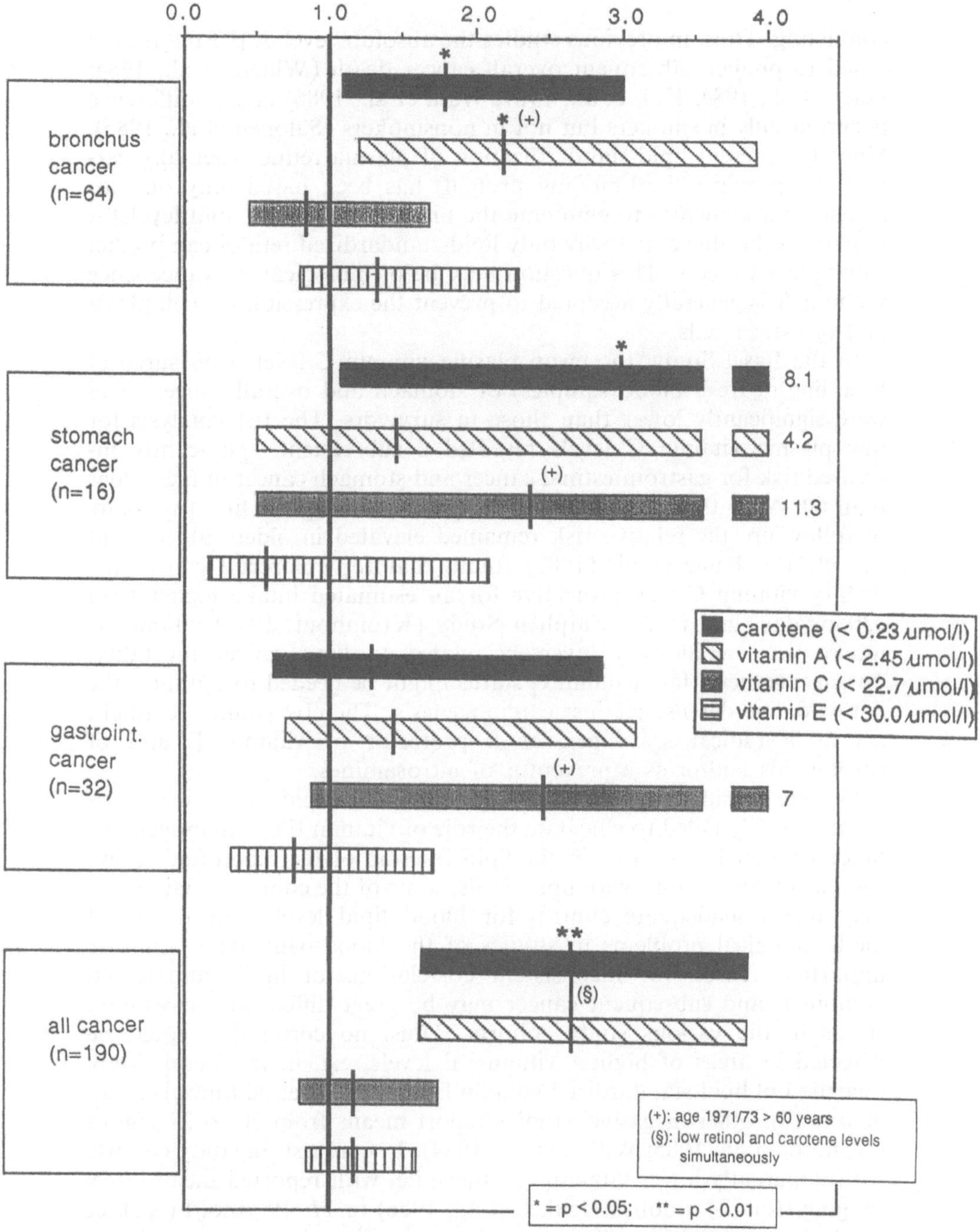

Figure 2. Relative risk adjusted for age, smoking and lipids and 95%-confidence intervals for all cancer groups and all measured vitamins (first two years excluded) in the 12-year Prospective Basel Study.

conflicting. Thus, in previous studies the absolute level of plasma retinol failed to predict subsequent overall cancer death (Willett et al., 1984; Peleg et al., 1984; Kok et al., 1987; Wald et al., 1986) or any difference occurred only in smokers but not in nonsmokers (Salonen et al., 1985). Since the present lipid-standardization of plasma retinol (actually carried by specific retinol-binding protein) has been based only on biostatistical arguments (to eliminate the bias of concurrent lipid levels) it remains to be elucidated why only lipid-standardized retinol can predict subsequent cancers. This question may be of biological relevance since vitamin A is generally accepted to prevent the expression of metaplasia and of cancer cells.

In the Basel Study, the mean plasma vitamin C levels (measured at base line in fresh blood samples) of stomach and overall cancer cases were significantly lower than those in survivors. The risk analysis for low plasma vitamin C levels revealed a subsequent significantly increased risk for gastrointestinal cancer and stomach cancer in men older than 60. After the exclusion of the mortality during the first two years of follow-up, the relative risk remained elevated in older subjects but less clearly. Kune et al. (1987) found that for colorectal cancer the dietary vitamin C was protective for an estimated intake higher than 230 mg/day, and in the Zutphen Study (Kromhout, 1987) vitamin C intake was significantly inversely related to lung cancer mortality. Taken together a fair vitamin C status might be needed to optimize the antioxidative defense against carcinogenesis. Thereby vitamin C might act as a radical scavenger, as a protector of vitamin E and of carotenoids and/or as a preventor of nitrosamines.

Despite considerable theoretic interest, previous epidemiologic investigations clearly failed to elucidate the role of vitamin E in carcinogenesis. Since vitamin E is carried in the lipid fraction of the blood and shows significant correlations with lipid levels, some of the controversial results may reflect inadequate control for blood lipid levels and/or special methodological problems in studies of the blood-bank type. Another important reason for inconsistent correlations of lipid-standardized vitamin E and subsequent cancer may be great differences in vitamin status in the various study cohorts. Thus, no correlation could be detected in areas of highest vitamin E levels, e.g. in the Basel Study quartile 1 of lipid-standardized vitamin E was as high as 30 μmol/l (mean 36 μmol/l), other 'negative' studies report means from 30 to 27 μmol/l (Nomura et al., 1985; Willett et al., 1984). In contrast, in study cohorts with apparently lower vitamin E values, i.e. with reported mean levels ranging from 24 μmol/l (Menkes et al., 1986) to 17–19 μmol/l (Kok et al., 1987; Knekt et al., 1988) there was a significantly increased relative risk for low alpha-tocopherol for cancer of all sites. Furthermore, Willett et al. (1983) and Salonen et al. (1985) found that low plasma levels of vitamins E and A may enhance the risk associated with low levels of selenium and pointed out that results based on assessment of any of these nutrients alone may be misleading. Taken together, the available data may suggest that low plasma α-tocopherol might increase the risk of carcinogenesis only below a threshold level of about 27–28 μmol/l.

In conclusion, the results of the 12-year follow-up of the Basel Study provides further evidence for an increased risk of bronchus and overall cancer mortality in subjects with low plasma carotene and vitamin A levels, and for an increased risk of stomach and gastrointestinal cancers at low plasma vitamin C levels. Antioxidant vitamins seem, in our study, to be effective mainly in people older than 60. This could be in connection with cohort effects or with the inherent relation between aging and carcinogenesis. Significant correlations between vitamin E and cancer in study cohorts with a poor status – in contrast to lacking correlations in areas with high antioxidant status – might suggest a risk below an 'optimal' threshold. In general, our epidemiological results provide further support for the hypothesis that essential antioxidants could be involved in the etiology of certain cancers.

Cardio-vascular mortality

The 1985 mortality analysis allowed to test the hypothesis that low antioxidant vitamin plasma concentrations were associated with increased risk of atherosclerosis such as ischemic heart disease or cerebrovascular disease. A total of 132 fatalities from ischemic heart disease (24% of the study cohort) and 31 deaths from cerebral vascular disease (6%) were recorded, i.e. somewhat less than from cancer. The CVD mortality of male Swiss is well known to be of the lowest order in Europe.

Plasma carotene concentrations below quartile 1 were associated with a significantly higher risk for ischemic heart disease (RR = 1.53, $p = 0.02$) (Table 1). The same was true for low levels of both carotene and vitamin C (RR = 1.96, $p = 0.022$).

The risk of cerebrovascular death was significantly elevated in subjects who had both low carotene and low vitamin C plasma concentrations (RR = 4.17, $p < 0.01$) (Table 1). For carotene or vitamin C alone the correlations lacked statistical significance.

In the Basel Study, as already mentioned, the average plasma vitamin E levels are rather high in comparison to other populations so far studied and even the lowest quartile had vitamin E levels above the presumable 'protective' threshold of approximately 27–28 μmol/l (Gey et al., 1991; Riemersma et al., 1991). A prospective study in the Netherlands (Kok et al., 1987a) failed also to reveal a significant

Table 1. 12-year mortality from Ischemic Heart Disease and Stroke in relation to the plasma status of essential antioxidants at base line of the 1973–85 Propsective Basel Study

Cause of death	Antioxidant (lowest quartile)	Relative risk	95% Cl	p
IHD ($n = 132$)	Low carotene, (<0.23 μmol/l)	1.53	1.07–2.20	0.024
	Low carotene and low vitamin C	1.96	1.10–3.50	0.022
Stroke ($n = 31$)	Low carotene and low vitamin C	4.17	1.68–10.33	<0.01

Netherlands (Kok et al., 1987a) failed also to reveal a significant correlation between plasma vitamin E and CHD, presumably mainly for reasons similar to those of the Basel Study (α-tocopherol/cholesterol ratio ≥ 5, i.e. above the 'protective' threshold) (Gey 1992). Thus the lack of inverse correlation does not contradict the results of other studies.

The new data on the increased risk of low carotene levels for subsequent IHD as well as low levels of both carotene and vitamin C on subsequent IHD and stroke in the prospective Basel Study is an intriguing extension of previous reports from Harvard Medical School (Gaziano et al., 1990; Manson et al., 1991).

The present data are consistent with the hypothesis of antioxidant protection against peroxidative lipoprotein modification (Steinberg and Witztum, 1990; Esterbauer et al., 1992) but do not exclude corresponding effects of essential antioxidants on immunological responses as well as cellular regulations in endothelial cells, macrophages, smooth muscle cells and platelets (Gey 1986; Gey 1992).

Epidemiological studies, even in case of consistency of various types, are limited to the suggestion of preventive potentials of all major essential antioxiants regarding cancer, IHD and/or stroke. Their actual potentials can, however, conclusively be tested only by randomized, double-blind intervention studies with specific supplementation of essential antioxidants in high-risk groups of poor antioxidant status. Presently more than 20 intervention studies (mostly initiated by the US National Cancer Institute) are already under way to evaluate the cancer preventive potentials of β-carotene alone or in combination with vitamin E or A (and/or C). Because of the epidemiological correlations and many mechanistic links of all major antioxidants regarding the counteraction of aggressive oxygen species it seems reasonable simultaneously to optimize the status of all essential antioxidants. Future intervention studies testing the prevention of CVD by essential antioxidants seem equally justified since the epidemiological data are still incomplete but fairly consistent.

409

Acheson, R. M., and Williams, D. R. R. (1983) Does the consumption of fruit and vegetables protect against stroke? Lancet i: 1191–1193.

Ames, B. N. (1983) Dietary Carcinogens and Anticarcinogens. Science 221: 1256–1264.

Bertram, J. S., Kolonel, L. N., and Meyskens, F. L. (1987) Rationale and strategies for chemoprevention of cancer in humans. Cancer Res. 47: 3012–3031.

Cox, D. R. (1972) Regression models and life tables. J. Stat. Soc. B. 34: 187–220.

D'Aquino, M., Dunster, C., and Wilson, R. (1989) Vitamin A and glutathione-mediated free radical damage: competing reactions with polyunsaturated fatty acids and vitamin C. Biochem. Biophys. Res. Comm. 161: 1199–1203.

Dion, P. W., Bright-See, E. B., Smith, C. C., and Bruce, W. R. (1982) The effect of dietary ascorbic acid and alpha-tocopherol on fecal mutagenicity. Mutat. Res. 102: 27–37.

Esterbauer, H., Puhl, H., Waeg, G., Krebs, A., and Dieber-Rotheneder, M. (1992) Oxidation of lipoproteins and atherosclerosis, in: Vitamin E: Biochemistry and Clinical Applications. Packer, E. and Fuchs, J. eds. Dekker, New York. in press.

Gaziano, J. M., Manson, J. E., Ridker, P. M., Buring, J. E., and Hennekens, C. H. (1990) Beta carotene therapy for chronic stable angina. Circ. 82, Suppl. III-201: Abstr. 0796.

Gey, K. F. (1986) On the antioxidant hypothesis with regard to arteriosclerosis. Bibl. Nutr. Dieta 35: 53–91.

Gey, K. F., Brubacher, G. B., and Stähelin, H. B. (1987) Plasma levels of antioxidant vitamins in relation to ischemic heart disease and cancer. Am. J. Clin. Nutr. 45: 1368–1377.

Gey, K. F. (1989) Inverse correlation of vitamin E and ischemic heart disease, in: Elevated Dosages of Vitamins. Walter, P., Stähelin, H. and Brubacher, G. eds. Hans Huber, Toronto/Lewiston NY/Bern/Stuttgart.

Gey, K. F., and Puska, P. (1989) Plasma vitamins E and A inversely related to mortality from ischemic heart disease in cross-cultural epidemiology. Ann. N. Y. Acad. Sci. 570: 268–282.

Gey, K. F. (1992) Vitamin E and other essential antioxidants regarding coronary heart disease: risk assessment studies, in: Vitamin E: Biochemistry and Clinical Applications. Packer, L. and Fuchs, J. eds. Dekker, New York. in press.

Ginter, E. (1979) Decline of coronary mortality in United States and vitamin C. Am. J. Clin. Nutr. 32: 511–512.

Knekt, P., Aromaa, A., Maatela, J., Aaran, R. K., Nikkari, T., Hakama, M., Hakulinen, T., Peto, R., Saxén, E., and Teppo, L. (1988) Serum vitamin E and risk of cancer among Finnish men during a 10-year follow-up. Am. J. Epidemiol. 127: 28–41.

Knox, E. G. (1973) Ischemic heart disease mortality and dietary intake of calcium. Lancet i: 1465–1467.

Kok, F. J., van Duijn, C. M., Hofman, A., Vermeeren, R., de Bruijn, A. M., and Valkenburg, H. A. (1987) Micronutrients and the risk of lung cancer. N. Engl. J. Med. 316: 1416.

Kok, F. J., de Bruijn, A. M., Vermeeren, R., Hofman, A., van Laar, A., de Bruin, M., Hermus, R. J. J., and Valkenburg, H. A. (1987a) Serum selenium, vitamin antioxidants, and cardiovascular mortality: a 9-year follow-up study in the Netherlands. (1987a) Am. J. Clin. Nutr. 45: 462–468.

Kromhout, D. (1987) Essential micronutrients in relation to carcinogenesis. Am. J. Clin. Nutr. 45: 1361–1367.

Kune, S., Kune, G. A., and Watson, L. F. (1987) Case-control study of dietary etiological factors: the Melbourne colorectal cancer study. Nutr. Cancer 9: 21–42.

Manson, J. E., Stampfer, M. J., Willett, W. C., Colditz, G. S., Rosner, B., Speizer, F. E., and Hennekens, C. H. (1991) A prospective study of antioxidant vitamins and incidence of coronary heart disease in women. Circ. 84, Suppl. 11–546: Abstr. 2168.

Menkes, M. S., Comstock, G. W., Vuilleumier, J. P., Helsing, K. J., Rider, A. A., and Brookmeyer, R. (1986) Serum beta-carotene, vitamins A and E, selenium, and the risk of lung cancer. N. Engl. J. Med. 315: 1250–1254.

Nomura, A. M. J., Stemmermann, G. N., Heilbrunn, L. K., Salkeld, R. M., and Vuilleumier, J. P. (1985) Serum vitamin levels and the risk of cancer of specific sites in men of Japanese ancestry in Hawaii. Cancer Res. 45: 2369–2372.

Olson, J. A. (1990) Vitamin A, in: Handbook of Vitamins. Machlin, L. J. ed. Dekker, New York & Basel, pp. 1–57.

410

Palgi, A. (1981) Association between dietary changes and mortality rates; Israel 1949 to 1977; a trend free regression model. Am. J. Clin. Nutr. 34: 1569–1583.

Passmore, R., and Eastwood, M. A. (1986) Human Nutrition and Dietetics. Davidson, L. S. P., and Passmore, R. eds. Churchill Livingstone, Edinburgh/London/Melbourne/New York.

Peleg, I., Heyden, S., Knowles, M., and Hames, C. G. (1984) Serum retinol and risk of subsequent cancer: extension of the Evans County, Georgia Study. J. Natl. Cancer Inst. 73: 1455–1458.

Ramirez, J., and Flowers, N. C. (1980) Leucocyte ascorbic acid and its relationship to coronary heart disease in man. Am. J. Clin. Nutr. 33: 2079–2087.

Riemersma, R. A., Wood, D. A., Macintyre, C. C. A., Elton, R. A., Gey, K. F., and Oliver, M. F. (1991) Risk of angina pectoris and plasma concentrations of vitamins A, C, and E and carotene. Lancet 337: 1–5.

Salonen, J. T., Salonen, R., Lappetelainen, R., Mäenpää, P. H., Alfthan, G., and Puska, P. (1985) Risk of cancer in relation to serum concentrations of selenium and vitamins A and E: matched case-control analysis of prospective data. Br. Med. J. 290: 417–420.

Salonen, J. T., Salonen, R., Seppänen, K., Kantola, M., Parviainen, M., Alfthan, G., Mäenpää, P. H., Taskinen, E., and Rauramaa, R. (1988) Relationship of serum selenium and antioxidants to plasma lipoproteins, platelet aggregability and prevalent ischemic heart disease in eastern Finnish men. Atherosclerosis 70: 155–160.

Shekelle, R. B., Lepper, M., Liu, S., Maliza, C., Raynor, W. J., and Rossof, A. H. (1981) Dietary vitamin A and risk of cancer in the Western Electric Study. Lancet ii: 1185–90.

Sies, H. (1989) Relationship between free radicals and vitamins: an overview, in: Elevated Dosages of Vitamins. Walter, P., Stähelin, H., and Brubacher, G. eds. Hans Huber, Toronto/Lewiston NY/Bern/Stuttgart.

Southorn, P. A., and Powis, G. (1988) Free Radicals in Medicine. I. Chemical Nature and Biologic Reactions. Mayo Clin. Proc. 63: 381–389.

Southorn, P. A., and Powis, G. (1988) Free Radicals in Medicine. II. Involvement in Human Disease. Mayo Clin. Proc. 63: 390–408.

Stähelin, H. B., Buess, E. H., Brubacher, G., and Widmer, L. K. (1983) Karzinom-Mortalität and Ernährung (Resultate der propektiven Basler Studie). Prakt. Arzt 37: 1625–1643.

Stähelin, H. B., Rösel, F., Buess, E., and Brubacher, G. (1984) Cancer, vitamins, and plasma lipids: prospective Basel Study. J. Natl. Cancer Inst. 73: 1463–1468.

Stähelin, H. B., Thurneysen, J., Buess, E., Rösel, F., Eichholzer-Helbling, M., Torhorst, J., and Widmer, L. K. (1988) Todesfälle und Todesursachen im 20-Jahres-Follow-up der Basler Studie. Schweiz. Med. Wochenschr. 118: 1039–1047.

Stähelin, H. B., Gey, K. F., Eichholzer, M., Lüdin, E., Bernasconi, F., Thurneysen, J., and Brubacher, G. (1991) Plasma antioxidant vitamins and subsequent cancer mortality in the 12-year follow-up of the prospective Basal Study. Am. J. Epidemiol. 133: 766–775.

Steinberg, D., and Witztum, J. L. (1990) Lipoproteins and Atherosclerosis. Current concepts. J. Am. Med. Assoc. 264: 3047–3052.

Tannenbaum, S. R. (1983) N-Nitroso compounds: a perspective on human exposure. Lancet i: 629–632.

Wald, N. J., Idle, M., Boreham, J., and Bailey, A. (1980) Low serum-vitamin A and subsequent risk of cancer. Lancet ii: 813–819.

Wald, N., Boreham, J., and Bailey, A. (1986) Serum retinol and subsequent risk of cancer. Br. J. Cancer 54: 957–961.

Widmer, L. K., Stähelin, H. B., Nissen, C., and da Silva, A. (1981) Basler Studie: Venen-, Arterienkrankheiten; koronare Hertzkrankheit bei Berufstätigen. Hans Huber, Bern, Switzerland.

Willett, W. C., Morris, J. S., Pressel, S., Taylor, J. O., Polk, B. F., Stampfer, M. J., Rosner, B., Schneider, K., and Hames, C. G. (1983) Prediagnostic serum selenium and risk of cancer. Lancet ii: 130–134.

Willett, W. C., Polk, B. F., Underwood, B. A., Stampfer, M. J., Pressel, S., Rosner, B., Taylor, J. O., Schneider, K., and Hames, C. G. (1984) Relation of serum vitamins A and E and carotenoids to the risk of cancer. N. Engl. J. Med. 310: 430–434.

Ziegler, R. G. (1991) Vegetables, fruits, and carotenoids and the risk of cancer. Am. J. Clin. Nutr. 53 (suppl.): 251S-259S.

Zinniker, O., Stähelin, H. B., and Widmer, L. K. (1978) Todesfälle und Todesursachen in der Basler Studie. Schweiz. Med. Wochenschr. 108: 869–874.

Free Radicals and Aging
ed. by I. Emerit & B. Chance
© 1992 Birkhäuser Verlag Basel/Switzerland

Vitamin E requirement in relation to dietary fish oil and oxidative stress in elderly

Mohsen Meydani

Antioxidant Research Laboratory, USDA-Human Nutrition Research Center on Aging at Tufts University, 711 Washington Street, Boston, MA 02111, USA

Summary. A growing body of evidence shows that oxygen radicals and other products of free radical reactions are involved in aging and age-related degenerative diseases. Recent studies have suggested that fish oils (FO) have a potentially beneficial effect on age-associated diseases. Consumption of FO may increase requirement for vitamin E, especially under conditions where oxidative stress is increased. Vitamin E requirement increases wtih increased intake of dietary polyunsaturated fatty acids (PUFA). This relationship may be exaggerated in elderly subjects. Our studies, as well as those of others, have shown that plasma lipid peroxides are significantly higher in older subjects compared to young subjects. Thus, in conditions where the percentage of highly unsaturated fatty acid increases in the membrane, older subjects may be more susceptible to oxidative damage. In a series of human studies, we found that older women, receiving FO supplements for 3 months exhibited a greater increase in plasma PUFA compared to young subjects. By substituting membrane fatty acids with the potentially unstable (n-3) fatty acids of FO, older subjects were found to be at greater risk of oxidative stress than young subjects. In addition, when exposed to eccentric exercise-induced oxidative stress, older men, receiving vitamin E supplements for 48 days, exhibited significantly lower levels of lipid peroxides in urine compared to placebo control. These data indicate that older subjects are more susceptible to oxidative stress and may benefit from the antioxidant protection provided by vitamin E.

Introduction

Animal and human studies suggest that the formation of oxygen free radicals and lipid peroxides occur as continuous biological processes in which a number of defensive enzymatic and non-enzymatic systems are involved. Free radicals, i.e., atoms and molecules with an unpaired electron can damage molecules having important roles in cellular homeostasis, resulting in a total loss of cellular funtion crucial to the survival of organism. According to the free radical theory of aging, the aging process and degenerative changes associated with this process may occur due to accumulative effect of the random free radical reactions that are continuously occurring at the cellular level (Harman, 1980; Samarajski et al., 1968).

A growing body of evidence indicates that oxygen-derived species such as O_2^-, $HO^\cdot$, H_2O_2, ..., and other products of free radical and lipid peroxidation reactions play an important role in the pathogenesis of

412

several age-associated disorders (Halliwell, 1987; Meydani et al., 1990). During the reduction of oxygen which occurs in the respiratory chain in the inner mitochondrial membrane, superoxide radicals are formed. In normal, balanced homeostatic conditions, cells are continuously exposed to free radicals. These radicals have the capacity to trigger chain reactions and therefore, produce membrane lipid peroxidation (Harman, 1984). However, in oxidative stress conditions, higher levels of oxygen radicals are produced, exceeding homeostatic antioxidant protection of cells, and thereby resulting in peroxidation of biological membranes (Davies et al., 1982; Dillard et al., 1978). Lipid peroxidation of the biological membranes can lead to loss of integrity of membrane and dysfunction of receptors and membrane-bound enzymes. Lipid peroxidation events also release reactive free radicals and toxic aldehydes which can then completely inactivate enzymes and other cell components (Bowles et al., 1991). There are multiple cellular enzymatic and non-enzymatic antioxidant defense systems in cells which function to protect the membranes and other cell organelles from the deleterious effects of free radical reactions.

Vitamin E is the major fat-soluble, chain breaking antioxidant in biological membranes. It protects membrane polyunsaturated fatty acids (PUFA) from lipid peroxidation (Chow, 1991). Early studies have demonstrated that the requirement for this antioxidant increases with increased consumption of PUFA (Witting, 1974). The ability of vitamin E to protect membranes from oxidative damage is dependent on the magnitude and duration of oxidative stress. Fatty acid composition of a diet influences the fatty acid composition of membrane phospholipids (Witting, 1974). An increased requirement for vitamin E has been suggested in cases of high intake of PUFA (Draper, 1980). In addition, a high intake of PUFA may interfere with vitamin E absorption from the intestine (Gallo-Torres, 1980; Leka et al., 1989).

A relative increase in mono- and poly-unsaturated fatty acids has been recommended by the U.S. National Cholesterol Education Panel as an approach for reducing the risk of cardiovascular disease among Americans. Experimental and epidemiological data suggest that the level and the type of dietary fat play a significant role in the incidence and pathogenesis of age-associated degenerative diseases such as atherosclerosis and coronary artery disease. Several population based studies have recommended increasing physical fitness and physical exercise to attain a variety of beneficial health effects, for both young and elderly people. In a series of human studies which involved fish oil supplementation and used eccentric exercise to induce oxidative stress, the potential of dietary vitamin E supplementation in young and older subjects was investigated.

Fish oil and oxidative stress

Fish and fish oil concentrates containing long chain (n-3) fatty acids have been indicated to protect humans against coronary artery disease (Herold and Kinsella, 1986; Glomset, 1985). Recently, accumulated evidence from animal experiments and human trials have widely promoted (n-3) fatty acids of fish oils in the prevention and treatment of various pathological conditions, including rheumatoid arthritis, psoriasis, cancer and other inflammatory diseases (Kinsella et al., 1990). With renewed interest in fish oil as a preventative measure and treatment of various conditions, its potentially harmful effects have been overlooked. Without adequate antioxidant protection, the substitution of membrane fatty acids with highly oxidizable (n-3) fatty acids of fish oil, i.e. eicosapentaenoic acid (EPA) with 5 double bonds and docosahexaenoic acid (DHA) with 6 double bonds, may potentiate peroxidation of cellular membranes (Odeley and Watson, 1991). Fish oil capsules currently available on the market contain variable levels of EPA and DHA. In order to prevent the oxidation of EPA and DHA in fish oil products and extend capsules shelf-life, manufacturers add vitamin E in a variety of forms. The level of α-tocopherol contained in fish oil products was measured in our laboratory (Chee et al., 1990) and ranged from 180 μg to 2240 μg per g of fish oil; certain products available on the market also contain γ-tocopherol which has 1/10 the biological activity of α-tocopherol.

In animal studies, Meydani et al. (1987) reported that in both young and old mice supplemented with different levels of vitamin E, fish oil fed animals had significantly lower α-tocopherol concentrations in plasma and tissues compared to mice fed corn oil or coconut oil. This reduced status of vitamin E in fish oil fed animals might be due to decreased absorption of vitamin E and/or its increased demand and utilization by other tissues. We have found that percent lymphatic appearance of intact [14]-C-α-tocopherol orally administered in fish oil to rats was significantly ($p < 0.05$) lower that when administered in either corn oil or olive oil (fish oil: $19.9 \pm 16.3\%$, corn oil: $41.7 \pm 22.4\%$, and olive oil: $49.4 \pm 24.8\%$) (Leka et al., 1989). The absorption of α-tocopherol by the intestine was reduced and a significant portion of the administered dose decomposed before lymphatic absorption. Therefore, consumption of meals comprised of fish or fish oil prodcuts, may necessitate increased intake of vitamin E, at least in part to compensate for the partial destruction and reduction in absorption of α-tocopherol from the gut.

Oxidative damage to tissues has been suggested as the bases for the pathology of several human diseases (Halliwell, 1987). Long-term fish oil supplementation without adequate antioxidant protection, may result in in vivo peroxidation of (n-3) fatty acid and thereby contribute to

414

Table 1. Change of plasma fatty acids ratio, α-tocopherol and lipid peroxides following fish oil supplementation

	PUFA/SFA	α-Tocopherol/ (EPA + DHA)	MDA[a]
Young (22–35 y)	+10.1%**	−72%**	+29.4%*
Older (51–71 y)	+18.6%***	−79%***	+63.1%*

[a]% change after 2 months of fish oil supplementation. Significant change from pre-supplementation level, *$p < 0.05$, **$p < 0.001$, ***$p < 0.0001$.

the onset and/or progression of some of age-associated diseases. We investigated the potential change of plasma lipid peroxides of 15 young (22–35 years) and 10 healthy older (51–71 years) women taking 6 capsules of fish oil with their meals daily for a three-month period (Meydani et al., 1991). The fish oil capsules used in the study provided 1.68 g of EPA, 0.72 g of DHA, 0.6 g of other fatty acids and 6 IU of vitamin E per day. After three months of supplementation, older women exhibited a significantly higher increase in plasma EPA and DHA and showed a dramatic increase in PUFA to saturated fatty acids (SFA) ratio compared to young women. Although plasma vitamin E levels did not change, fish oil supplementation significantly decreased the plasma vitamin E/(EPA + DHA) and significantly increased the plasma level of lipid peroxides (Table 1). Increase in plasma lipid peroxides due to fish oil supplementation was significantly higher in the older women compared to young women. Even though the long-term supplementation with fish oil in this study was found to be beneficial in reducing plasma total triglyceride, without providing adequate antioxidant protection by vitamin E, an alteration in fatty acid composition of plasma, and most likely of other tissues to more highly unsaturated fatty acids with a higher potential for peroxidation, could interfere with the homeostatic balance between antioxidants and prooxidants.

Exercise and oxidative stress

Animal and human studies have shown that unaccustomed exercise as well as strenuous and exhaustive exercise can induce oxidative damage and cause muscle injury. Exercise influences oxidative metabolism and produces reactive oxygen species which appear to play a key role in changing the membrane fatty acid composition, permeability and leakage of enzymes, all of which eventually cause the injury to muscle membrane. There is evidence indicating that the generation of free radicals and lipid peroxides increases with the increased respiration during exercise (Gohil et al., 1986; Ji et al., 1990; Sumida et al., 1989).

This increase may last for quite some time following an unaccustomed exercise. In vitamin E deficient animals, exercise increases susceptibility to free radical damage and results in premature exhaustion, greater fragility of lysosomal membrane, and marked depression of muscle mitochondrial respiratory control (Davies et al., 1982; Gohil et al. 1986; Quintanilha et al., 1982). Bowles et al. (1991) reported that a single bout of exercise at 70% of VO_{2max} in untrained rats resulted in 30% decline in vitamin E content of the quadricep muscle. Davis et al. (1982) have shown that exhaustive exercise in animals causes lipid peroxidation and mitochondrial damage.

The limited human studies conducted to date indicate that vitamin E supplementation reduces oxidative stress and lipid peroxidation events following exercise. This implies that vitamin E requirements may increase with exercise (Pincemail et al., 1988; Simon-Schnass and Pabst, 1988). In order to investigate the protective role of vitamin E supplementation against exercise-induced oxidative stress, nine young (22–29 years) and 12 older (55–74 years) healthy men were studied following supplementation with either 800 IU of vitamin E or placebo for 48 days. Volunteers were then subjected to a 45-min bout of eccentric exercise, running downhill on a treadmill at 75% of maximum of their heart rate (Meydani et al., 1991). After 48 days, the concentration of plasma α-tocopherol increased (56% in young, 70% in older subjects) and γ-tocopherol decreased (99% in young, 60% in older subjects). Using an eccentric exercise model in which the muscle is contracted while lengthening, showed a decrease of α- and γ-tocopherol levels in the muscle biopsies obtained from the vastus lateralis muscle of young subjects (Table 2). The change observed in the vitamin E status of muscle following exercise was accompanied with 148% increase in total muscle lipid conjugated dienes placebo groups compared to a 38% increase in the subjects who received vitamin E supplements (Table 2). Therefore, significant decreases in both α- and γ-tocopherol and concomitant increase of lipid conjugated dienes in muscle was demonstrated immediately following exercise. This indicated that acute exercise increases active oxygen radicals species and lipid peroxidation.

Table 2. Change in vitamin E and lipid peroxidation index in muscle biopsy obtained from young subjects following eccentric exercise

	Placebo	Vitamin E-supplemented
α-Tocopherol	−15%*	−23%*
γ-Tocopherol	−20%	−29%
Conjugated dienes	+145%	+38%

*significant ($p < 0.05$) decrease from pre-exercise, both groups combined.

416

Furthermore, these results imply that during this process, vitamin E is utilized in the muscle to counteract deleterious action of free radicals and prevent oxidative injury to membrane lipids. Similar results have been reported in exercised animals (Gohil et al., 1986, 1987; Allessio et al., 1988; Allessio et al., 1988). The protective effect of vitamin E was further substantiated by measuring 24 h urinary TBA-adducts as an index of whole body response to oxidative stress (Draper et al., 1984; Ekstrom et al., 1986; Wu et al., 1990). The protective effect of vitamin E extended up to 12 days post exercise (Meydani et al., 1991) at which time muscle protein injury and remodeling was evident by excretion of 3-methylhistidin (Cannon et al., 1991). Vitamin E supplemented subjects excreted less TBA-adducts as compared to placebo groups ($p < 0.05$). Percent increase in urinary TBA-adducts at 12 days post-exercise was 32.7 ± 13.3 and 16.9 ± 22.4 in young and older vitamin E supplemented subjects, respectively; whereas percent increase of urinary TBA-adducts in placebo groups was 58.7 ± 24.5 and 76.8 ± 34.9 in young and older subjects, respectively.

Conclusion

Epidemiological studies reveal an inverse association between antioxidant nutrient intake and incidence of various age-associated disorders (Gey et al., 1991; Esterbauser et al., 1991). Increased levels of dietary antioxidants significantly reduce the tissue level of lipid peroxides in animal and humans (Pubelle et al., 1982; Meydani et al., 1985; Meydani et al.,, 1988; Meydani et al., 1986; Chavance et al., 1984; Wartanowicz et al., 1984; Lemoyne et al., 1987). The level of plasma lipid peroxides increases with age (Pubelle et al., 1982) which may be due, in part, to diminished status of dietary antioxidants. Low intake and/or low plasma antioxidant levels have been observed in older adults (Campbell et al., 1989). Factors such as inadequate antioxidant protection, long-term exposure to pro-oxidant toxicants, pollutants, and drugs, high intake of specific nutrients such as iron, copper, long-term intake of fish oil concentrates and performing unaccustomed and acute exercise, can increase the potential for cellular and tissue lipid peroxidation and damage. These environmental conditions can contribute significantly to the acceleration of the aging process and age-associated degenerative disorders. Therefore, higher than recommended levels of dietary antioxidants such as vitamin E might be beneficial in older adults.

Allessio, H. M., and Goldfarb, A. H. (1988) Lipid peroxidation and scavenger enzymes during exercise: adaptive response to training. J. Appl. Physiol. 64: 1333–1336.
Allessio, H. M., Goldfarb, A. H., and Cutler, R. G. (1988) MDA content increases in fast- and slow-twitch skeletal muscle with intensity of exercise in rat. Cell Physiol. 24: C874–C877.

Bowles, D. K., Torgan, C. E., Ebner, S., Kehrer, J. P., Ivy, J. L., and Starnes, J. W. (1991) Effects of acute, submaximal exercise on skeletal muscle vitamin E. Free Rad. Res. Comms. 14: 139–143.

Campbell, D., Bunkder, V. W., Yhomas, A. J., and Clayton, B. E. (1989) Selenium and vitamin E status of healthy and institutionalized elderly subjects: analysis of plasma, erythrocytes and platelets. Br. J. Nutr. 62: 221–227.

Cannon, J. G., Meydani, S. N., Fielding, R. A., Fiatarone, M. A., Meydani, M., Farhang-mehr, M., Orencole, S. F., Blumberg, J. B., and Evans, W. (1991) The acute phase response in exercise. II. Associations between vitamin E, cytokines and muscle proteolysis. Am J. Physiol. 29: R1235–R1240.

Chee, K. M., Gong, J. X., Good Ress, D. M., Meydani, M., Ausman, L., Johnson, J., Siguel, E. N., and Schaefer, E. J. (1990) Fatty acid content of marine oil capsules. Lipids 25: 523–528.

Chevance, M., Brubacher, G., Herbeth, B., Vernbes, G., Mikstacki, T., Dete, F., Fournier, C., and Janot, C. (1984) Immunological and nutritional status among the elderly, in: Lymphoid Cell Functions in Aging. DeWick, Al, ed. Eurage, Paris, pp. 231–237.

Chow, C. K. (1991) Vitamin E and oxidative stress. Free Radic. Biol. Med. 11: 215–22.

Davies, K. J. A., Quintanilha, A. T., Brooks, G. A., and Packer, L. (1982) Free radicals and tissue damage produced by exercise. Biochem. Biophys. Res. Commun. 107: 1198–1205.

Dillard, C. H., Litov, R. E., Savin, W. M., Dumelin, E. E., and Tappel, A. L. (1978) Effect of exercise, vitamin E and ozone on pulmonary function and lipid peroxidation. J. Appl. Physiol.: Respirat. Environ. Exer. Physiol. 45: 927–932.

Draper, H. H. (1980) Nutrient interrelationships, in: Vitamin E, A Comprehensive Treatise. Machlin, L. J., ed. Marcel Dekker, New York, pp. 272–288.

Draper, H. H., Polensek, L., Hadley, M., and McGirr, L. G. (1984) Urinary malonaldehyde as an indicator of lipid peroxidation in the rat and in the tissues. Lipids 19: 836–843.

Ekstrom, T., Stahl, A., Sigvardsson, K., and Hogberg, J. (1986) Lipid peroxidation in vivo monitored as ethane exhalation and malondialdehyde excretion in urine after oral administration of chloroform. Acta Pharmacol. Toxicol. 58: 289–296.

Esterbauser, H., Dieber-Robtheneder, M., Strieg, G., and Waeg, G. (1991) Role of vitamin E in preventing the oxidation of low-density lipoprotein. Am. J. Clin. Nutr. 53: 314S–21S.

Gallo-Torres, H. E. (1980) Absorption, in: Vitamin E, A Comprehensive Treatise. Machlin, L. J., ed. Marcel Dekker, New York, pp. 170–267.

Glomset, J. A. (1985) Fish, fatty acids, and human health. N. Engl. J. Med. 312: 1253–1255.

Gey, K. F., Puska, P., Jordan, P., and Moser, U. K. (1991) Inverse correlation between plasma vitamin E and mortality from ischemic heart disease in cross-cultural epidemiology. Am. J. Clin. Nutr. 53: 326S–34S.

Gohil, K., Rothfuss, L., Lang, J., and Packer, L. (1987) Effect of exercise training on tissue vitamin E and ubiquinone content. J. Appl. Physiol. 63: 1638–1641.

Gohil, K., Packer, L., deLumen, B., Brooks, G. A., and Terblanche, G. E. (1986) Vitamin E deficiency and vitamin C supplements: exercise and mitochondrial oxidation. J. Appl. Physiol. 60: 1986–1991.

Halliwell, B. (1987) Oxidants and human disease: some new concepts. FASEB J. 1: 358–364.

Harman, D. (1980) Free radical theory of aging: beneficial effect of antioxidant on the life span of male NZB mice; role of free radical reactions in the deterioration of the immune system with aging and in the pathogenesis of systemic lupus erythrematous. Age 3: 64–73.

Harman, D. (1984) Free radicals and the origination, evolution and present status of free radical theory of aging, in: Free Radicals in Molecular Biology, Aging and Disease. Armstrong, D., Cutler, R. G., Sohal, R. S., and Slater, T. F., eds. Ravenswood Press, New York, pp. 1–12.

Herold, P. M., and Kinsella, J. E. (1986) Fish oil consumption and decreased risk of cardiovascular disease: a comparison of findings from animal and human feeding trials. Am. J. Clin. Nutr. 43: 566–598.

Ji, L. L., Dillon, D., and Wu, E. (1990) Alteration of antioxidant enzymes with aging in rat skeletal muscle and liver. Am. J. Physiol. 258: R918–R923.

Kinsella, J. E., Lokesh, B., Stone, R. A. (1990) Dietary n-3 polyunsaturated fatty acids and amelioration of cardiovascular disease: possible mechanisms. Am. J. Clin. Nutr. 52: 1–28.

Leka, L. S., Remaly, K. M., Bizinkauskas, P. A., and Meydani, M. (1989) Effects of fish oil on intestinal absorption of vitamin E in the rat. FASEB J. 3: A951.

418

Lemoyne, M., Van Gossen, A., Kurian, R., Ostro, M., Axler, J., and Jeejeebhoy, K. N. (1987) Breath pentane analysis as an index of lipid peroxidation: a functional test of vitamin E status. Am. J. Clin. Nutr. 46: 267–272.

Meydani, S. N., Shapiro, A. C., Meydani, M., Macauley, J. B., and Blumberg, J. B. (1987) Effect of age and dietary fat (fish, corn and coconut oils) on tocopherol status of C57BL/6Nia mice. Lipids 22: 345–350.

Meydani, S. N., Barklund, M. P., Liu, S., Meydani, M., Miller, R. A., Cannon, J. G., Morrow, F. D., Rocklin, R., and Blumberg, J. B. (1990) Vitamin E supplementation enhances cell-mediated immunity in healthy elderly subjects. Am. J. Clin. Nutr. 52: 557–563.

Meydani, M., Verdon, C. P., and Blumberg, J. B. (1985) Effect of vitamin E, selenium and age on lipid peroxidation events in rate cerebrum. Nutr. Res. 5: 1227–1236.

Meydani, M., Macauley, J. B., and Blumberg, J. B. (1988) Effect of dietary vitamin E and selenium on susceptibility of brain regions to lipid peroxidation. Lipids 23: 405–409.

Meydani, M., Natiello, F., Goldin, B., Free, N., Woods, M., Schaefer, E., Blumberg, J. B., and Gorbach, S. L. (1991) Effect of long-term fish oil supplementation on vitamin E status and lipid peroxidation in women. J. Nutr. 121: 484–491.

Meydani, M., Evans, W., Handelman, G., Biddle, L., Fielding, R. A., Meydani, S. N., Orencole, S. F., Fiatarone, M. A., Blumberg, J. B., and Cannon, J. G. (1991) Vitamin E protection of exercise-induced oxidative damage in young and elderly subjects. FASEB J. 5: A919.

Meydani, S. N., Cathcart, E. S., Hopkins, R. E., Meydani, M., Hayes, K. C., and Blumberg, J. B. (1986) Antioxidants in experimental amyloidosis of young and old mice, in: Fourth International Symposium of Amyloidosis. Glenner, G. G., Asserman, E. P., Benditt, E., Calkins, E., Cohen, As., and Zucker-Franklin, D., eds. Plenum Press, New York, pp. 683–692.

Odeleye, O. E., Watson, R. R. (1991) Health implication of the n-3 fatty acids. Am. J. Clin. Nutr. 53: 177–178.

Pincemail, J., Deby, C., Camus, G., Pirnay, F., Bouchez, R., Massaux, L., and Goutier, R. (1988) Tocopherol mobilization during intensive exercise. Eur. J. Appl. Physiol. 57: 189–191.

Pubelle, P., Chantreuil, J., Bensadous, J., Blotman, F., Simon, L., and Crastes de Paule, A. (1982) Plasma lipoperoxides and aging. Biomed. 36: 164–166.

Quintanilha, A. T., Packer, L., Davies, J. M. S., Racanelli, T., Davies, K. J. A. (1982) Membrane effects of vitamin E deficiency: bioenergetics and surface charge density studies of skeletal muscle and liver mitochondria. Ann. NY Acad. Sci. 399: 32–47.

Samarajski, T., Ordy, J. M., and Rudy-Reimer, P. (1968) Lipofuscin pigment accumulation in the nervous system of aging. Anat. Rec. 160: 555–574.

Simon-Schnass, I., and Pabst, H. (1988) Influence of vitamin E on physical performance. Internat. J. Vit. Nutr. Res. 58: 49–54.

Sumida, S., Tanaka, K., Kitao, H., and Nakadomo, F. (1989) Exercise-induced lipid peroxidation and leakage of enzymes before and after vitamin E supplementation. Int. J. Biochem. 21: 835–838.

Wartanowicz, M., Panczenko-Kresowska, B., Ziemlanski, S., Kowalska, M., and Okolska, G. (1984) The effect of alpha-tocopherol and ascorbic acid on the serum lipid peroxide level in elderly people. Ann. Nutr. Metab. 28: 186–191.

Witting, L. A. (1974) Vitamin E-polyunsaturated lipid relationship in diet and tissues. Am. J. Clin. Nutr. 27: 952–959.

Wu, W-H., Meydani, M., Meydani, S. N., Burklund, P. M., Blumberg, J. B., Munro, H. N. (1990) Effect of dietary iron overload on lipid peroxidation, prostaglandin synthesis and lymphocyte proliferation in young and old rats. J. Nutr. 120: 280–289.

Free Radicals and Aging
ed. by I. Emerit & B. Chance
© 1992 Birkhäuser Verlag Basel/Switzerland

Vitamin C and vitamin E – synergistic interactions in vivo?

J. J. Strain[a] and C. W. Mulholland[ab]

[a]*Human Nutrition Research Group, University of Ulster, Coleraine BT52 1SA, Northern Ireland, and* [b]*Ulster Hospital, Dundonald, Belfast BT16 ORH, Northern Ireland*

Summary. The synergistic relationship between ascorbic acid (vitamin C) and α-tocopherol (vitamin E) in the inhibition of lipid peroxidation has been known for some time and is now well established in vitro systems. The possibility that ascorbic acid may also reduce tocopheroxyl radicals in vivo is a subject of some interest and speculation. Although not all experiments have failed to suggest a synergistic antioxidant interaction, recent data indicate that the postulated synergism between these vitamins might be relatively unimportant compared with other metabolic processes.

Introduction

There is now compelling evidence (reviewed by McCay, 1985; Niki, 1991) to indicate synergism between ascorbic acid and vitamin E in various in vitro model systems. The mechanism for the synergism is the reduction of tocopheroxyl radicals by ascorbate to regenerate vitamin E and has been demonstrated by electron spin resonance (Niki et al., 1982) and pulse radiolysis (Packer et al., 1979). These tocopheroxyl radicals are formed when vitamin E scavenges peroxyl radicals in the lipid phase. Thus this mechanism of regeneration of vitamin E rather than the irreversible oxidation of tocopheroxyl radicals by reaction with further peroxyl or tocopheroxyl radicals to form non-radical products links water soluble ascorbate with the interruption of free radical chain propagation in the lipid phase. Although experimental support for the regeneration of vitamin E by ascorbate has been largely confined to studies in chemical solutions or media containing liposomes, there is also evidence for in vitro synergistic interactions between these vitamins in platelets (Vatassery et al., 1989; Chan et al., 1989) erythrocyte membranes (Van den Berg et al., 1990), low density lipoproteins (Sato et al., 1990) and in the prevention of malignant transformation of cells (Borek, 1991).

The relative importance of the synergistic interactions between vitamin E and ascorbate should depend on the competing reactions of the tocopheroxyl radicals and the proximity of the tocopheroxyl radicals to the lipid-water interface. Indeed with respect to the latter, the high efficiency of regeneration of short-chain homologs of vitamin E by ascorbate in liver microsomes may account at least in part for their greater antioxidant potency compared with α-tocopherol (Kagan et al., 1990). Obviously observing any synergism between ascorbate and vitamin E will be more difficult in vivo due to inter alia the complications of alternative metabolic processes which consume or eliminate these vitamins from the body (Frei et al., 1990; Packer and Landvik, 1990; Jialal and Grundy, 1991; Niki, 1991), or reductions of tocopheroxyl radicals by other means such as enzymatic mechanisms or by glutathione (Packer et al., 1989; Bast and Haenen, 1990). It is not unexpected, therefore, to find that the experimental evidence for an in vivo interaction between vitamin E and ascorbate is much less compelling.

Studies in animals and humans

A majority of workers have also observed synergistic interactions between vitamin E and ascorbate in vivo. For example dietary vitamin C has been shown to enhance tissue levels of vitamin E (Hubra et al., 1982; Bendich et al., 1984) and partially alleviate signs of vitamin E deficiency (Chen and Thacker, 1987). In contrast some workers have reported antagonistic interactions between vitamin E and vitamin C. For example dietary vitamin C can also lower plasma levels of vitamin E and enhance erythrocyte haemolysis (Chen and Chang, 1979; Chen, 1981). Others, however, have concluded that the in vitro interactions between these vitamins do not occur in vivo on the basis of data from ascorbic and dehydroascorbic acids status in rats fed diets varying in vitamin E levels (Behrens and Madere, 1989) and from the biokinetics of dietary RRR-α-tocopherol in guinea pigs fed different dietary levels of vitamin C and vitamin E (Burton et al., 1990). The latter workers utilized an elegant experimental protocol to measure vitamin E turnover in a number of tissues of guinea pigs which varied in both vitamin C and vitamin E status. Reasons given by these workers for the failure to detect any in vivo interactions included the possible existence of a free radical reductase system or the lack of oxidative stress in the animal model.

Few, if any, comparable data exist for humans. Arad et al. (1985) found that dietary vitamin C enhanced tissue levels of vitamin E in premature infants. In a placebo-controlled double-blind supplementation trial conducted on 32 healthy young adults to assess the effect of α-tocopherol or ascorbic acid separately or in combination on the total radical-trapping antioxidant potential (TRAP; Wayner et al., 1987) of plasma, we found no significant differences between the four treatment groups for TRAP values after 28 days (Mulholland and Strain, 1991). This ex vivo investigation, although unable to demonstrate a synergistic interaction between ascorbic acid and α-tocopherol, showed that dietary antioxidant supplementation could increase the free radical trapping capacity of plasma and further that the increase produced was much larger than the molar increase in plasma antioxidant concentration. The latter finding suggested some interaction among the various antioxidants.

Conclusion

The synergistic interaction between vitamin C and vitamin E which readily occurs in vitro has yet to be demonstrated conclusively in vivo. Future research should ideally aim to investigate the effects of multiple combinations of antioxidants at varying doses on peroxyl radical fluxes in healthy animals and in pathologies which have been shown to have an oxidative stress component.

Arad, J. D., Dgani, Y., and Eyal, F. G. (1985) Vitamin E and vitamin C plasma levels in premature infants following supplementation of vitamin C. Internat. J. Vit. Nutr. Res. 55: 395–397.

Bast, A., and Haenen, G. R. M. M. (1990) Regulation of lipid peroxidation by glutathione and lipoic acid: involvement of liver microsomal vitamin E free radical reductase, in: Antioxidants in Therapy and Preventive Medicine, I. Emerit et al., eds. Plenum Press, New York, pp. 111–116.

Behrens, W. A., and Madere, R. (1989) Ascorbic and dehydroascorbic acids status in rats fed diets varying in vitamin E levels. Internat. J. Vit. Nutr. Res. 59: 360–364.

Bendich, A., D'Apolito, P., Gabriel, E., and Machlin, L. J. (1984) Interaction of dietary vitamin C and vitamin E on guinea-pig immune responses to mitogens. J. Nutr. 114: 1588–1593.

Borek, C. (1991) Free-radical process in multistage carcinogenesis. Free Rad. Res. Comms. 12–13: 745–750.

Burton, G. W., Wronska, U., Stone, L., Foster, D. O., and Ingold, K. U. (1990) Biokinetics of dietary RRR-α-tocopherol in the male guinea pig at three dietary levels of vitamin C and two levels of vitamin E. Evidence that vitamin C does not "spare" vitamin E in vivo. Lipids 25: 199–210.

Chan, A. C., Tran, K., Raynor, T., Ganz, P. R., and Chow, C. K. (1991) Regeneration of vitamin E in human platelets. J. Biol. Chem. 266: 17290–17295.

Chen, L. H., and Chang, H. M. (1979) Effects of high level of vitamin C on tissue antioxidant status of guinea pigs. Internat. J. Vit. Nutr. Res. 49: 87–91.

Chen, L. H., and Thacker, R. R. (1987) Effect of ascorbic acid and vitamin E on biochemical changes associated with vitamin E deficiency in rats. Internat. J. Vit. Nutr. Res. 57: 385–390.

Chen, L. H. (1981) An increase in vitamin E requirement induced by high supplementation of vitamin C in rats. Am. J. Clin. Nutr. 34: 1036–1041.

Frei, B., Stocker, R., England, L., and Ames, B. M. (1990) Ascorbate: the most effective antioxidant in human blood plasma, in: Antioxidants in Therapy and Preventive Medicine. I. Emerit et al., eds. Plenum Press, New York, pp. 155–163.

Hruba, F., Novakova, V., and Ginter, E. (1982) The effect of chronic marginal vitamin C deficiency on the alpha-tocopherol content of the organs and plasma of guinea-pigs. Experientia 38: 1454–1455.

Jialal, I., and Grundy, S. M. (1991) Preservation of the endogenous antioxidants in low density lipoprotein by ascorbate but not probucol during oxidative modification. J. Clin. Invest. 87: 597–601.

Kagan, V. E., Serbinova, E. A., and Packer, L. (1990) Recycling and antioxidant activity of tocopherol homologs of differing hydrocarbon chain lengths in liver microsomes. Arch. Biochem. Biophys. 282: 221–225.

McCay, P. B. (1985) Vitamin E: Interactions with free radicals and ascorbate. Ann. Rev. Nutr. 5: 323–340.

Mulholland, C. W., and Strain, J. J. (1991) Effect of antioxidant vitamin supplementation on the peroxyl radical trapping ability of plasma. Proc. Nutr. Soc. 50: 133A.

Niki, E., Tsuchiya, J., Tanimura, R., and Kamiya, Y. (1982) Regeneration of vitamin E from α-chromonoxy radical by glutathione and vitamin C. Chem. Lett., 789–792.

Niki, E. (1991) Vitamin C as an antioxidant. World Rev. Nutr. Diet. 64: 1–30.

Packer, J. E., Slater, T. F., and Willson, R. L. (1979) Direct observation of a free radical interaction between vitamin E and vitamin C. Nature, Lond. 278: 737–738.

Packer, L., and Landvik, S. (1990) Vitamin E in biological systems, in: Antioxidants in Therapy and Preventive Medicine. I. Emerit et al., eds. Plenum Press, New York, pp. 93–103.

Packer, L., Maguire, J. J., Mehlorn, R. J., Serbinova, L., and Kagan, V. E. (1989) Mitochondria and microsomal enzymes have a free radical reductase activity that prevents chromanoxyl radical accumulation. Biochem. Biophys. Res. Comms., 159: 229–235.

Sato, K., Niki, E., and Shimasaki, H. (1990) Free radical-mediated chain oxidation of low density lipoprotein and its synergistic inhibition by vitamin E and vitamin C. Arch. Biochem. Biophys. 279: 402–405.

Van den Berg, J. J. M., Kuypers, F. A., Roelofsen, B., and Op den Kamp, J. A. F. (1990) The cooperative action of vitamins E and C in the protection against peroxidation of parinaric acid in human erythrocyte membranes. Chem. Phys. Lipids 53: 309–320.

Vatassery, G. T., Smith, W. E., and Quach, H. T. (1989) Ascorbic acid, glutathione and synthetic antioxidants prevent the oxidation of vitamin E in platelets. Lipids 24: 1043–1047.

Wayner, D. D. M., Burton, G. W., Ingold, K. U., Barclay, L. R. C., and Locke, S. J. (1987) The relative contributions of vitamin E, urate, ascorbate and proteins to the total peroxyl radical trapping antioxidant activity of human blood plasma. Biochim. Biophys. Acta 92: 408–419.

Free Radicals and Aging
ed. by I. Emerit & B. Chance
© 1992 Birkhäuser Verlag Basel/Switzerland

The threshold of age in exercise and antioxidants action

Abraham Z. Reznick[a], Eric H. Witt[b], Michael Silbermann[a] and Lester Packer[b]

[a]*Department of Morphological Sciences, The Bruce Rappaport Faculty of Medicine and the Rappaport Family Institute for Research in the Medical Sciences, Technion-Israel Institute of Technology, Haifa, Israel, and* [b]*Department of Molecular and Cell Biology, University of California, Berkeley, California 94720, USA*

Summary. Physical activity and exercise are important factors in determining the quality of life in old animals and humans. With age there is a slow but significant reduction in muscle mass and ability to perform certain physical activities. This may be due to changes with the age of muscle composition and protein turnover, as well as decrease of trophic influence in neural control of muscles of old individuals. Exercise in general was shown to improve muscle performance even in old age. However a concept of threshold of age in exercise was advanced forward in the 1970s. Accordingly, the idea was that for a given exercise of a particular duration and intensity there is a certain age beyond which this exercise may not have a positive influence, but can become detrimental to the exercising animal or human.

Recent studies on the effect of antioxidants such as Vitamins C and E and selenium have shown that these agents could decrease the free radical associated muscle damage caused by extensive exercise. Thus, administration of these antioxidants especially vitamins C and E may reduce the oxidative damage due to exercise, and may alter the threshold of age by delaying it to an older age.

Introduction

Aging is a complex process which has both intrinsic and extrinsic causes that affect the basic processes of life. Environmental factors such as nutrition, exercise, and way of life can influence basic aging processes directly or indirectly by affecting other intrinsic random and/or programmed processes that in turn may influence the aging phenomenon (Reznick et al., 1989).

In essence, these environmental factors can be various environmental stresses and insults. For example, physical exercise can be considered such an environmental stress which involves recruitment of physiological, biochemical and energetic resources of our body. Indeed, in many studies physical exercise has been shown to improve the physical ability of trained individuals. However, the effect of exercise on aging systems remained controversial and thus required further discussion.

Exercise and aging

With the advance of age, the body composition is changing slowly and steadily. Studies on various parameters of body composition by many researchers showed that the greatest percentage of change in body composition is in the skeletal muscle mass which is reduced by 30–40% at the age of 75 (Evans, 1980; Fryer, 1960). Specifically, the decrease in muscle mass is probably due to a decrease of mainly type II (white-un-aerobic) fibers which are reduced by 40% at the age of 70 in humans, compared to 25-year-old people (Larsson et al., 1978). This reduction of muscle mass is usually accompanied by reduced functional capacity and reduced strength of muscles. However, exercising old people for 3 months for strength training caused a relative increase in the type II muscle fiber area (Aniansson and Gustaffsson, 1981). Similar results were obtained by Silbermann et al., (1983), when old mice were exposed to endurance training for 10 weeks. The cause or causes of this age-related decline of skeletal muscle mass was attributed mainly to deficiency of neural input for the muscle with age, which resulted in a decrease of tropic influence of neurons at the neuromuscular junction (Larsson, 1982). Other causes that were discussed in the literature were decline in muscle regenerating capacity and reduced oxygen diffusion to muscle mitochondria with age (Evans, 1980).

It has been shown that protein metabolism and turnover is slowed down in aging animals (Lavie et al., 1982). Similarly, Golden and Waterlow showed that there is a significant reduction in whole body protein turnover with age (Golden and Waterlow, 1977). The decrease of protein turnover and especially muscle protein turnover in aging people (Young et al., 1981) could reduce the ability of aging muscles to respond to the stress of exercise. Thus, Evans (1980) studied the influence of physical exercise on protein turnover in young and middle aged people (52 ± 1.4 years) and showed that despite reduced muscle mass, physical exercise could increase protein turnover in these two age groups. Unfortunately, this study was not conducted on older people.

The concept of threshold of age in exercise

In a paper published in 1972, Edington et al. have formulated for the first time the concept of a threshold of age in exercise (Edington et al., 1972). In this and subsequent studies (McCafferty and Edington, 1974) these researchers have put forward a concept that says that for a given exercise of a particular duration and intensity, there is a certain age beyond which this exercise may not have a positive influence but can become detrimental to animals and similarly to humans. Similar observations were formulated by Goodrick (1980). Indeed, young rats trained from a young age had a longer survival rate than their control counterparts (Edington et al., 1972; McCafferty and Edington, 1974; Goodwrick, 1980), but after a certain age this effect of exercise was diminished. For example rats that started to exercise at the age of 20 months had a shorter survival rate than their control counterparts.

Studies conducted by Reznick and Steinhagen-Thiessen and their coworkers in the 1980s on mice trained at various ages, confirmed the existence of such a threshold. In several organs of old animals a short-term training of 5–7 weeks caused a marked reduction of key enzyme activities in old animals. Thus, measuring the activities of superoxide-dismutase, catalase, creatine phosphokinase and aldolase showed a reduction of 10-50% in old mice that trained for short periods of 5–10 weeks (Steinhagen-Thiessen et al., 1981; Reznick et al., 1982; Reznick et al., 1983; Steinhagen-Thiessen and Reznick, 1987; Steinhagen-Thiessen et al., 1983; Silbermann et al., 1987). This was not observed in young animals trained for the same periods. On the other hand, initiation of a training program at young (5 months) or middle age (15 months) did not result in reduction of enzyme activities due to exercise. On the contrary some of the old animals trained throughout life had levels of activity comparable to their young counterparts. Comparative studies on the morphology of muscles of bones of these animals corroborated the biochemical studies (Silbermann et al., 1983; 1987).

Exercise and antioxidants action

In a recent review (Witt et al., 1991), the effect of antioxidants on exercise inducing oxidative damage has been discussed. A great deal of evidence has been accumulated that shows that exercise can cause free radical formation in muscles and livers of trained animals (Witt et al., 1991; Davies et al., 1982). Indeed, reports on increased lipid peroxidation in exercise have appeared consistently in the literature. A more recent work demonstrated that exercising a rat to exhaustion increased the rate of formation in muscles of carbon centered and hydrogen free radicals (Simon-Schass and Pabst, 1988).

In general, antioxidant nutrients include vitamins E and C, the mineral selenium and coenzyme Q10 which occur naturally in our body. Naturally, one would expect that supplementation or deficiency of these nutrients may affect the exercise-induced oxidative damage in muscles. Indeed, selenium deficiency has been found to cause an elevation in pentane production in the breath of rats, a common and accepted indication for increase of lipid peroxidation (Dillard et al., 1978). However, when vitamin E in adequate amounts was supplemented to

426

these rats, it would abolish the increased pentane production in sele-
nium-deficient rats. In another study, vitamin E deficient rats exhibited
a sixfold increase of pentane production than animals that were supplemented with vitamin E (Dillard et al., 1978), and provision of 10,000 IU
of vit. E per kg diet reduced very impressively the carbon centered and
hydrogen radicals in muscles of rats that were exercised to exhaustion
(Hiramatsu et al., 1991).

In most studies supplementation of vitamins E and C supported the
notion that antioxidants may have a protective effect against exercised-
induced free radical damage. Studies in humans taking 600 mg of
vitamin E for two weeks showed a decrease of lipid peroxidation during
exercise (Dillard et al., 1978a). Another study on mountain climbers
showed that those individuals who were supplemented with vitamin E
did not exhibit the increase of breath pentane seen in unsupplemented
climbers (Simon-Schass and Pabst, 1988).

Recently, the authors of this report examined the effect of exercise to
exhaustion on rat muscle protein oxidation as measured by the carbonyl
assay. Control animals supplemented with vitamin E had about 30–
40% lower levels of protein carbonyl than the animals exercised to
exhaustion (unpublished data). Very interestingly, exercise to exhaustion in non supplemented animals caused a 17% increase in protein
carbonyl levels compared to sedentary controls. Supplementation of
vitamen E to rats exercised to exhaustion reduced the increase of
carbonyl levels in protein to only 4–5%, compared to 17% in the
unsupplemented animals (see above).

Conclusions

The wealth of data on the effect of vitamin E and other antioxidants
on reducing the exercise-induced oxidative damage, is increasing. Data
on the effect of vitamin E provision on physical performance of old
people is not clear. Most studies to date were performed on young
animals and humans. It is high time that such studies should also be
conducted on aging populations. If indeed antioxidant manipulation
has a positive effect on reducing the oxidative damage due to exercise,
one would predict that antioxidants can neutralize or delay the detrimental effects seen in exercising aging individuals. If this is the case,
proper supplementation of antioxidants could possibly alter the
threshold of age by raising it to a greater longevity of life.

Aniansson, A., and Gustaffsson, E. (1981) Clin. Physiol. 1: 87–98.
Davies, K. JK. A., Quintanila, A. T., Brooks, G. A., and Packer, L. (1982) Biochem. Biophys.
    Res. Commun. 107: 1198–1205.
Dillard, C. J., Litov, R. E., and Tappel, A. C. (1978) Lipids 13: 396–402.

Dillard, C. J., Litov, R. E., Savin, W. M., Dumelin, E. E., and Tappel, A. L. (1978a) J. Appl. Physiol. 45: 927–932.

Edington, D. W., Consmas, A. C., and McCafferty, W. B. (1972) Exercise and longevity: Evidence for a threshold of age. J. Gerontol. 27: 341–346.

Evans, W. J. (1980) Exercise and Muscle Metabolism in the Elderly, in: Nutrition and Aging. Munro, H. and Hutchinson, A., eds. Academic Press, New York, pp. 179–191.

Fryer, J. H. (1960) Fedn. Proc. Am. Soc. Exp. Biol. 19: 327–335.

Golden, M. H. N., and Waterlow, J. C. (1977) Clin. Sci. Mol. Med. 53: 229–288.

Goodrick, C. L. (1980) Gerontology 26: 22–27.

Hiramatsu, M., Velasio, R. D., and Packer, L. (1991) Biochem. Biophys. Res. Comm. 179: 859–864.

Larsson, L., Sjodin, B., and Karfsson, J. (1978) Acta Physiol. Scand. 103: 31–39.

Larsson L. (1982) in: The Aging Motor System. Prozzoto, F. J., and Malete, G. J., eds. Praeger Inc., New York, pp. 60–67.

Lavie, L., Reznick, A. Z., and Gershon, D. (1982) Biochem. J. 202: 47–51.

McCafferty, W. B., and Edington, D. W. (1974) Gerontologia 20: 44–48.

Reznick, A. Z., Steinhagen-Thiessen, E., and Gershon, D. (1982) Biochem. Med. 28: 347.

Reznick, A. Z., Steinhagen-Thiessen, E., Gellerssen, B., and Gershon, D. (1983) Mech. Ageing Dev. 23: 253.

Reznick, A. Z., Steinhagen-Thiessen, E., Gershon, D., and Silbermann, M. (1989) in: Molecular Biology of Stress. Breznitz, S. and Zinder, O., eds. Alan R. Liss Inc., New York, pp. 223–240.

Silbermann, M., Finkelbrand, S., Weiss, A., Gershon, D., and Reznick, A. (1983) Muscle and Nerve 6: 136–142.

Silbermann, M., Weiss, A., Reznick, A. Z., Eilam, Y., Szydel, N., and Gershon, D. (1987) Comp. Gerontol. 1: 45.

Simon-Schass, I., and Pabst, H. (1988) Int. J. Vitam. Nutr. Res. 58: 49–54.

Steinhagen-Thiessen, L., Reznick A., and Hilz, H. (1980) Mech. Ageing Dev. 12: 231.

Steinhagen-Thiessen, E., Reznick A., and Hilz, H. (1981) Mech. Ageing Dev. 16: 363.

Steinhagen-Thiessen, E., Reznick A. Z. (1987) Gerontology 33: 14–19.

Witt, E. H., Reznick, A. Z., Viguie, C. A., Starke-Reed, P., and Packer, L. (1992) J. Nutr. 122: 766–773.

Young, V. R., Gersovitz, M., and Munro, H. N. (1981) in: Nutritional Approaches to Aging Research. Moment, G. B. ed. CRC Press, Boca Raton, Florida, pp. 48–75.

Free Radicals and Aging
ed. by I. Emerit & B. Chance

Antioxidant therapy in the aging process

Guilherme Paulo Deucher

Clinica Guilherme Paulo Deucher, Rua. Borges Lagoa 1231, 2° andar, CEP 04038, São Paulo, Brazil

Summary. A total of 1,265 patients with age-related diseases such as diabetes, arthritis, vascular disease and hypertension as well as 1,100 persons in diminished health without apparent disease, were treated with the metal chelator EDTA and antioxidants such as vitamin C, E, beta-carotene, selenium, zinc and chromium. Good results were observed in the majority of patients. This is encouraging for the initiation of controlled clinical trials.

> *"Every man desires to live long, but no man would be old."*
>
> Jonathan Swift

Introduction

The role of oxygen-derived free radicals in the intrinsic aging process (Harman, 1986; Cutler 1991) as well as in age-related degenerative (Bulkley, 1983; Vladimirov, 1986; Pryor, 1987) and autoimmune diseases (Emerit, 1986) seems to be well established. This implies the use of free radical scavengers and antioxidants for prevention and therapy of age-related diseases. Hopefully also human life span may be prolongated or at least the periods of illness and suffering during the last years of human life may be shortened. Generation of the highly cytotoxic hydroxyl radical via metal-catalized Fenton reactions has been claimed to be a mechanism in oxyradical-related tissue damage (Halliwell et al., 1985), and the removal of "free" iron has been proposed as an appropriate treatment (Povoa, 1985; Halliwell, 1987; Deucher, 1987; 1988).

For these reasons, we have been using antioxidant therapy (AOT) in combination with chelation therapy (CT) for the past seven years in a large number of aged patients. The results obtained in several disease processes and also in aged persons without apparent disease are communicated in the following issue.

Treatment

Metal chelation (Fe2+ Pb2+ Hg2+, others) was obtained by administering a dose of 1.5 g of EDTA in each of 20 treatments. EDTA was chosen as the chelating agent, since it is widely used and well tolerated (Levine, 1979). It forms stable, soluble and non-toxic compounds with most metals (Petersdorf, 1983).

Metal chelation was combined with daily oral supplements of vitamin C (3 g), vitamin E (400 mg), beta-carotene (15 mg) selenium (100 μg), chromium (200 μg) and zinc (35 mg). The mean duration of chelation plus antioxidant therapy (CT + AOT) was 60 days. The clinical situation was reevaluated two to six months after the beginning of treatment.

Effects of chelation therapy alone

The efficacy of this treatment is documented by the increased excretion of heavy metals in the urine. As one can see in Table 1, iron as well as lead showed a significantly increased excretion after one intravenous dose of 1.5 g EDTA given as a perfusion for 2 hours. The 150 patients of this study group had not been professionally exposed to these metals.

Diabetes . . . . 4 patients
Arthritis . . . . 8 patients
Peripheral vascular disease . . . . 30 patients
Coronary heart disease . . . . 10 patients
Hypertension . . . . 9 patients
No apparent disease . . . . 89 patients
$\overline{150}$

In 28 of these 150 patients, the spontaneous chemiluminescence of whole blood could be studied before and 1 month after 12 consecutive perfusions of 1.5 g EDTA. There was a marked decrease in chemiluminescence, as may be seen in Table 2.

Table 1. Atomic absorption spectroscopy study. Excretion of Pb and Fe in 24-hour urine before and after 1.5 g EDTA in 150 patients

	Before $n = 150$	After $n = 150$	Increase	p^*
μg Pb	54 ± 3[*]	243 ± 9	+350%	<0.001
μg Fe	537 ± 27	1878 ± 89	+250%	<0.001

*Student's paired t test; **mean ± standard error; female $n = 59$; male $n = 91$; mean age = 53 years.

Table 2. Effect of 12 consecutive perfusions of 1.5 g EDTA on whole blood spontaneous chemiluminescence in 28 patients (cpm × 1000)

Before CT $n = 28$	After CT $n = 28$	Decrease	p^*
248 ± 24**	117 ± 14	−53%	<0.001

*Student's paired t test; **Mean ± standard error; Reference value: up to 40,000 cpm.

Increased heavy metal excretion and decreased chemiluminescence were correlated in these patients with clinical improvement.

Effects of combined CT and AO therapy

A total of 1,265 patients with age-related diseases such as diabetes, arthritis, peripheral or coronary vascular disease and/or in consultation for high blood pressure, were treated according to the treatment schedule given above.

	Aging process – Diabetes	
$n = 84$	$(m = 44)$	Mean age = 61 years

Insulin dependent 45.2%
One or more oral antidiabetic drugs. 54.8%
Clinical results: 90% improved. Decreases in necessity for these drugs ranged between 50% and 100%.

	Aging process – Arthritis	
$n = 170$	$(m = 58)$	Mean age = 56 years

Pain in one or more articulations 100%
Decrease in mobility. 100%
Inflammation or swelling 60%
Deformity. 45%
90% improvement in pain, mobility, inflammation and swelling. No improvement in deformity.

Aging process – Peripheral vascular disease

Fontaine grade[a]	n	Mean age (years)	Clinical improvement
Claudication (50 to 300 m)	520 $(m = 325)$	65	78% Increase in deambulation (500 to 5000 m)
Rest pain	48 $(m = 32)$	68	70% Pain decrease
Ischemic ulcers	55 $(m = 40)$	68	65% Ulcer cicatrization

[a]Fontaine et al. 1958; Guilmot et al. 1989.

	Aging process – Coronary heart disease	
$n = 215$	$(m = 162)$	Mean age = 63 years

Clinical results: 70% improvement. Disappearance of angina pectoris.

Aging process – Hypertension

$n = 173$	$(m = 124)$		Mean age = 65 years

Effect of CT + AOT on blood pressure in hypertensive patients

	Blood pressure (mm Hg)		
	Before*	After**	p
Systolic (mean)	165.0 ± 1.6	149.0 ± 1.5	<0.01
Diastolic (mean)	94.0 ± 1.0	86.0 ± 0.8	<0.01

*Using one or more antihypertensive drugs; **Without using antihypertensive drugs.

In addition to these patients, a total of 1,100 aged persons received CT + AOT for various complaints, without apparent disease. They showed improvement in their general health condition and were able to engage in the same activities they enjoyed several years before, for instance a game of tennis.

Aging process – No apparent disease		
$n = 1,100$	$(m = 445)$	Mean age = 46 years
General fatigue		100%
Pain		40%
Depression		40%
Reduced memory		29%
Irritability		24%
Headache		18%
Reduced sexual capacity		12%
Other symptoms		30%

Some good results were observed using the same therapy in cerebral vascular accidents, Parkinson's disease and certain autoimmune disorders. The number of these patients was limited, and therefore the results are not included here.

On the other hand, our preliminary results in patients with Alzheimer's disease were not encouraging, but imply further study.

Some examples of patients in which common therapies failed, even vascular surgery, had their ulcers healed with uncommon speed after AOT + CT. This is a small sample showing that this therapy could be important at least in the treatment of pain and/or saving of limbs. (Figs. 1 to 10).

Conclusions and prospectives

Good results were obtained with chelation therapy and antioxidant supplementation in patients with age-dependent diseases or diminished health conditions due to advanced age. This therapeutic approach appears to be appropriate for retardation of the aging process. The clinical reevaluations cannot be described in detail in this short review. However modification of blood pressure values and blood glucose levels in patients with hypertension and diabetes respectively should be convincing evidence for the efficacy of the treatments. The fact that these patients could reduce or even completely abandon their intake of anti-hypertensive or antidiabetic drugs is of great importance. Also the decreased need for analgesic and antiinflammatory drugs in patients with arthritis is an important observation, since most drugs have side effects which become more serious the longer the treatment. Chelation therapy and antioxidant supplementation should therefore find their

432

M.G.P. – m – 45 years

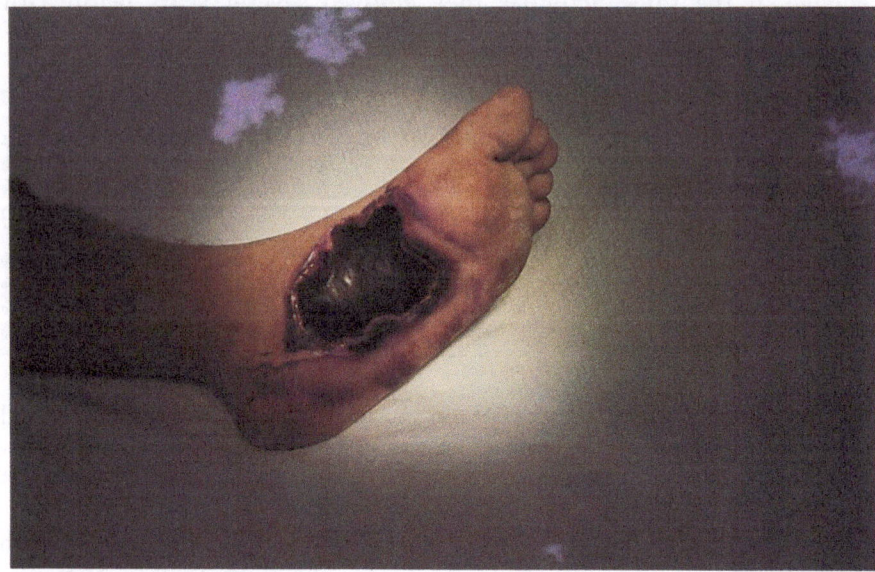

Figure 1.

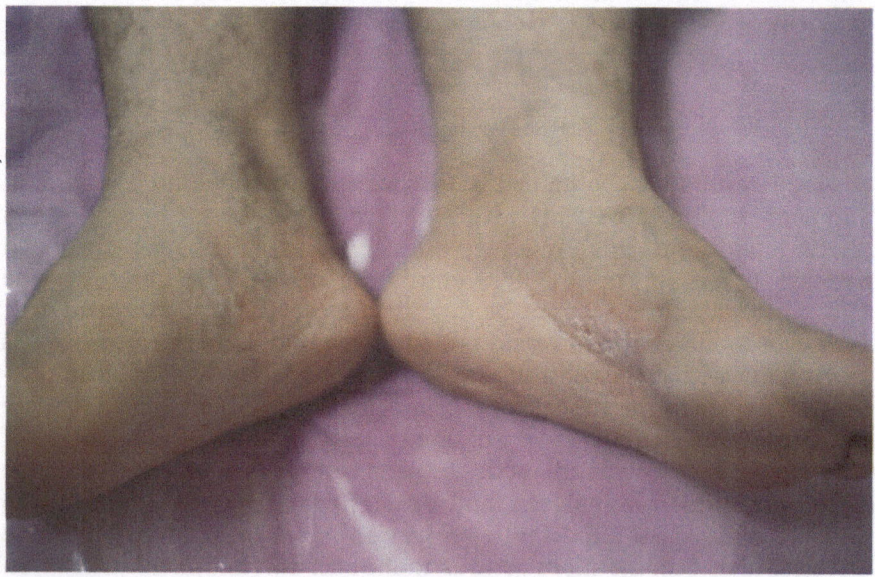

Figure 2.

Scleroderma and ischemic gangrene left leg

I.M. – f – 32 years

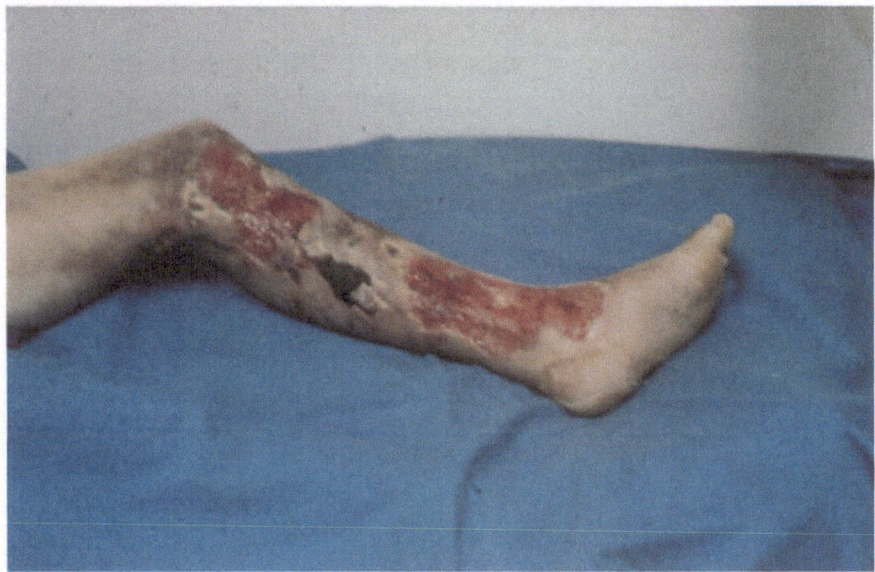

Figure 3.

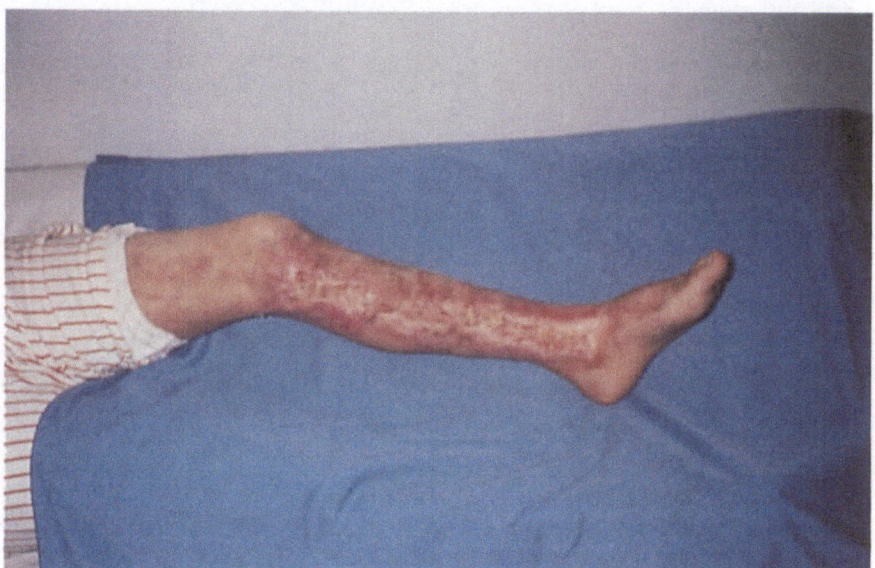

Figure 4.

Extremely painful Martorell's ulcer (hypertensive patients) Patient No. 1

A.Y. – f – 67 years

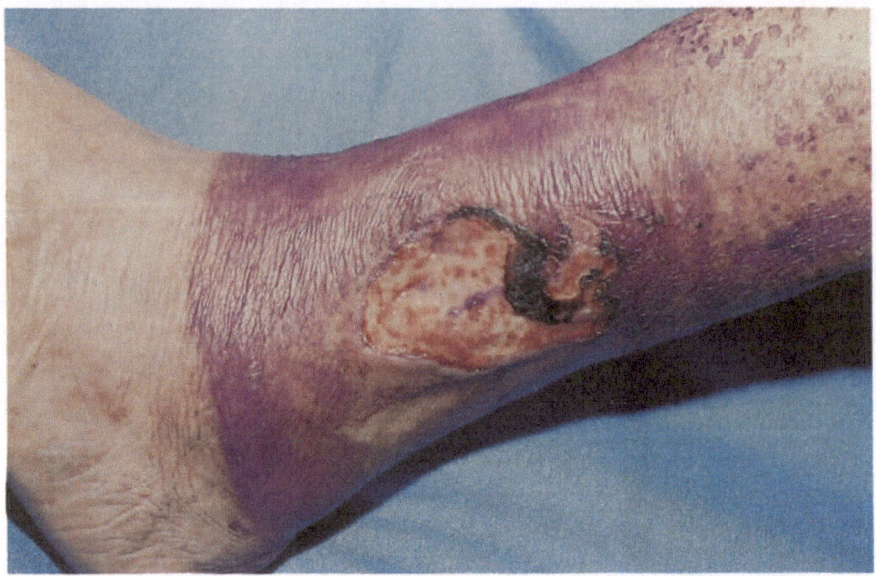

Figure 5.

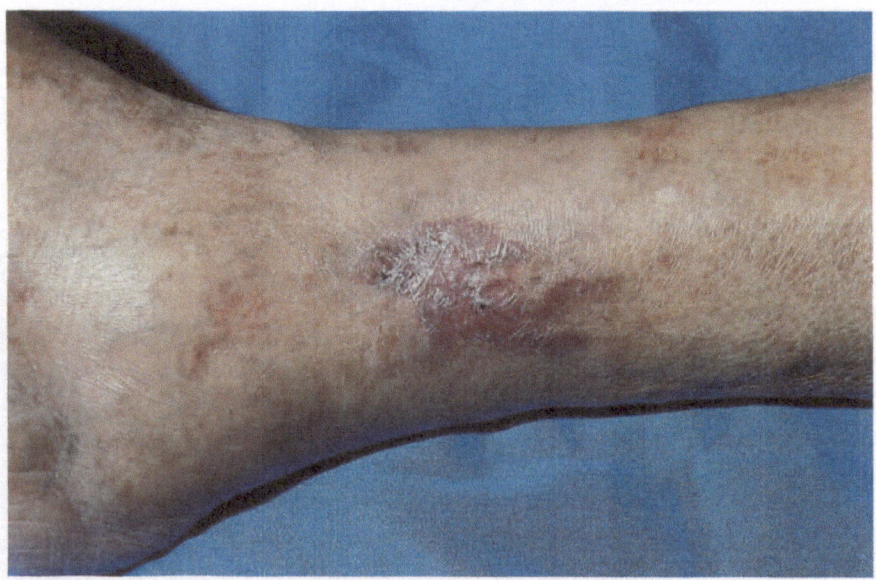

Figure 6.

Patient No. 2

S.T. – m – 59 years

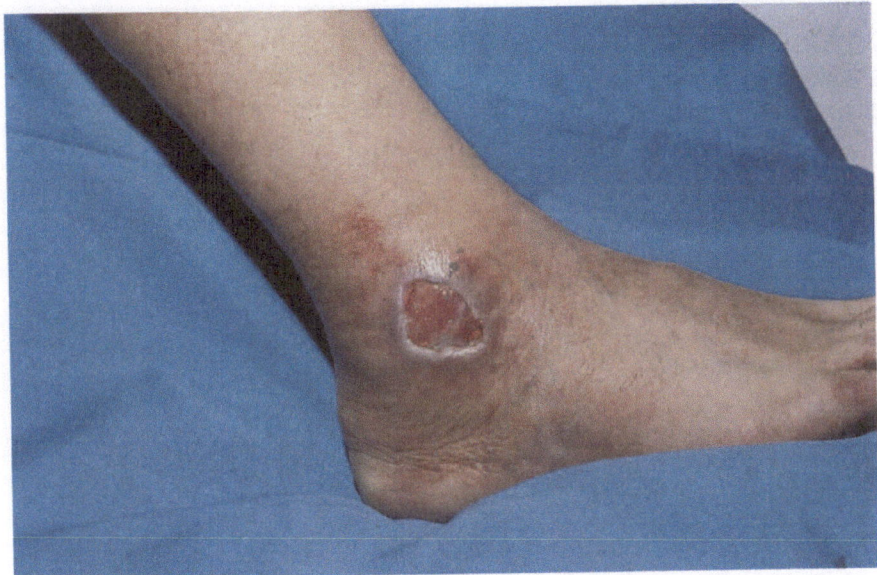

Figure 7.

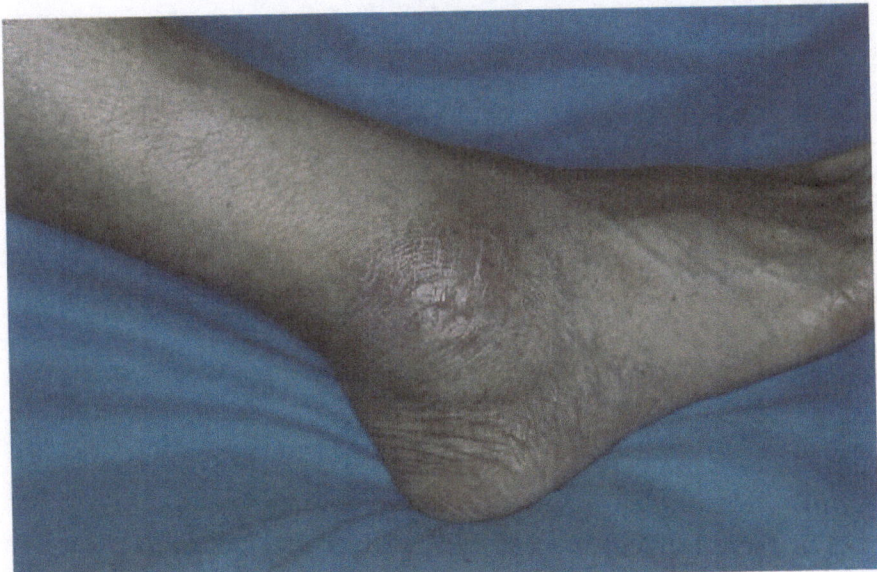

Figure 8.

436

Patient No. 3

J.B.C. – m – 57 years

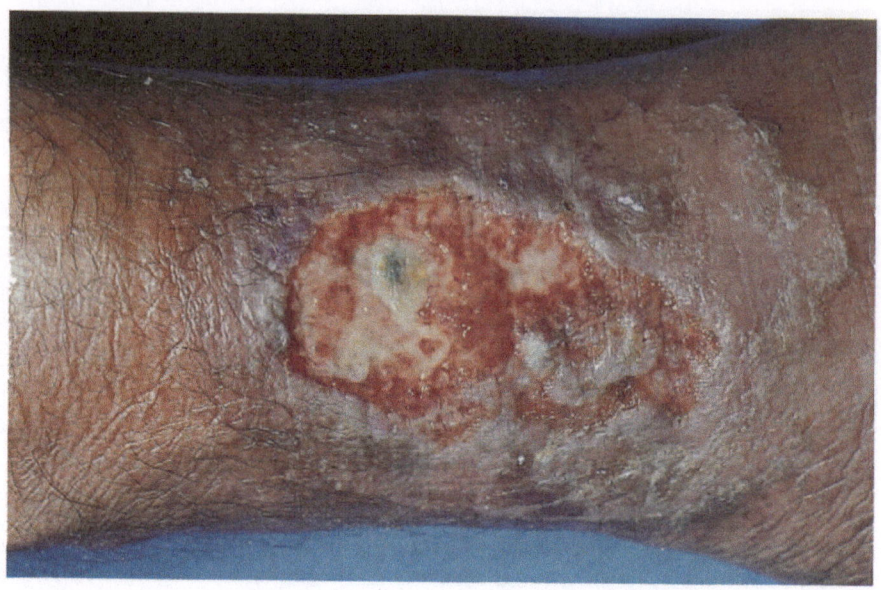

Figure 9.

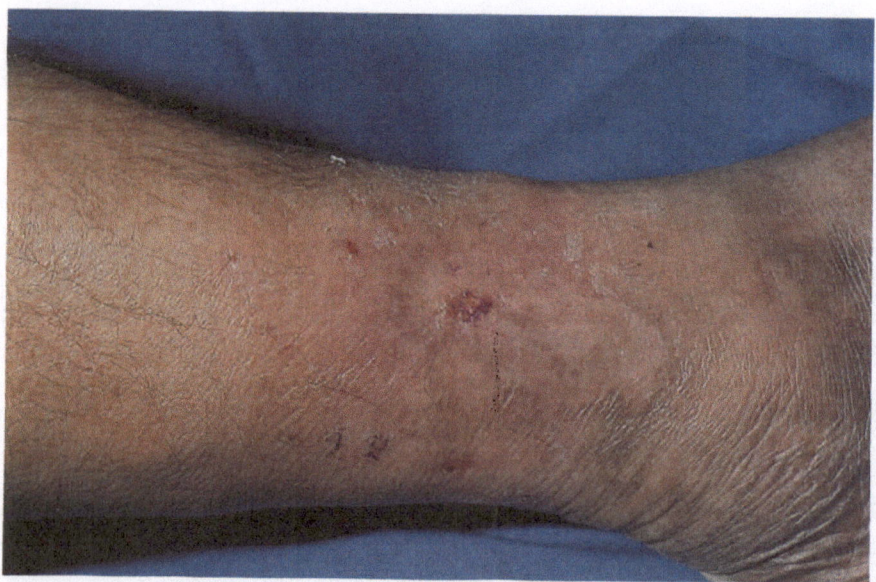

Figure 10.

place in long-term strategies for the treatment of degenerative disease and the aging process.

The results presented here should be considered as preliminary and have to be confirmed by controlled clinical trials. They are clearly encouraging, since they are based on a large number of observations.

Appropriate nutrition and life style, ingestion of sufficient amounts of oral antioxidants and proper excretion of accumulated toxic heavy metals are necessary for good health, retardation of aging and prevention of age-related diseases. Decreasing the periods of illness and suffering, increasing well-being and health – these are the major objectives of scientists, physicians and of all mankind.

Acknowledgments. The author is indebted to Drs. Daud Abuchalla and Helion Povoa Filho for the extremely helpful discussions and for their knowledge, to Dr. José de Felippe Junior for the statistical analysis and to Ms. Tânia Maria Guilherme Sisdelli for preparing the manuscript.

Bulkley, G. B. (1983) The role of oxygen free radicals in human disease processes. Surgery 94: 407–411.

Cutler, R. G. (1991) Antioxidants and aging. Am. J. Clin. Nutr. 53: 373S–9S.

Deucher, G. P. (1987) Chelation des métaux lourds, radicaux libres et maladies humaines. Angiologie 63(4): 1776–7.

Deucher, G. P. (1988) Quelação de metais pesados, radicais livres e doenças humanas. F. Med. (BR) 96(4): 193–194.

Emerit, I. (1986) Oxygen-derived free radicals and DNA damage in autoimmune diseases, in: Free Radicals, Aging and Degenerative Diseases. John E. Johnson Jr. et al., eds. Alan R. Liss, Inc., New York, pp. 307–324.

Fontaine, R. et al. (1958) Le traitement des oblitérations artérielles, in: Actualités Médico – Chirugicales. Masson Éditeurs, Paris, p. 44.

Guilmot, J. L. et al. (1989) Semiologie clinique, in: Artériopathies Athéromateuses des Membres Inferieures. Rouffy. J. et al., eds. Masson, Paris, pp. 175–187.

Halliwell, B. et al. (1985) The importance of free radicals and catalytic metal ions in human diseases. Molec. Aspects Med. 8: 89–193.

Halliwell, B. (1987) Oxygen radicals and metal ions: Potential antioxidant intervention strategies, in: Oxygen Radicals and Human Disease. Ann. Intern. Med. 107: 526–545.

Harman, D. (1986) Free radical theory of aging: Role of free radical in the origination and evolution of life, aging and disease processes, in: Free Radicals, Aging and Degenerative Diseases. John E. Johnson Jr. et al., eds. Alan R. Liss, Inc., New York, pp. 3–49.

Levine, W. G. (1979) The Chelation of Heavy Metals. International Encyclopedia of Pharmacology and Therapeutics. Pergamon Press, Great Britain, p. 112.

Petersdorf, R. G. et al. (1983) Harrison's Principles of Internal Medicine. McGraw-Hill Book Company, New York, p. 1274.

Povoa, F. H. (1985) Radicais livres e quelação. F. Med. (BR) 91: 81–86.

Pryor, W. A. (1987) The free radical theory of aging revisited: A critique and a suggested disease-specific theory, in: Modern Biological Theories of Aging. Warner, H. R. et al., eds. Raven Press, Great Britain, pp. 89–111.

Vladimirov, Y. A. (1986) Free radical lipid peroxidation in biomembranes: Mechanism, regulation and biological consequences, in: Free Radicals, Aging and Degenerative Diseases. John E. Johnson Jr. et al., eds. Alan R. Liss, Inc., New York, pp. 141–195.

Acknowledgements

Many of the speakers of the meeting in Paris accepted to contribute to this book by a review resuming the knowledge in their particular research field. We would like to thank all these colleagues for their cooperation.

We also gratefully acknowledge the support to this conference provided by: The American Aging Association; Bergaderm, France; Chauvin Blache, France; Fondation Ipsen, France; Henkel, Germany; Hoffmann-La Roche, Switzerland; Institut de Recherches Internationales Servier, France; L'Oréal, France; Lutsia, France; Nestlé, Switzerland; Rhone-Poulenc, France; Smith-Kline-Beecham, France.

The Editors

Subject Index

BIRKHÄUSER

LIFE SCIENCES

Experientia Supplementum

Cell Motility Factors

Edited by
I. D. Goldberg
Long Island Jewish Medical Center, NY, USA

1991. 225 pages. Hardcover. ISBN 3-7643-2569-0 (EXS 59)

Contents:

Molecular Analysis of Amoeboid Chemotaxis: Parallel Observations in Amoeboid Phagocytes and Metastatic Tumor Cells – Adhesion Systems in Embryonic Epithelial-to-Mesenchyme Transformations and in Cancer Invasion and Metastasis – Neutrophil Chemotactic Factors – Purification and Characterization of Scatter Factor – Purification, Characterization and Mechanism of Action of Scatter Factor from Human Placenta – Scatter Factor Stimulates Migration of Vascular Endothelium and Capillary-Like Tube Formation – The Cellular Response to Factors which Induce Motility in Mammalian Cells – The Role of E-Cadherin and Scatter Factor in Tumor Invasion and Cell Motility – Heterogeneity Amongst Fibroblasts in the Production of Migration Stimulating Factor (MSF): Implications for Cancer Pathogenesis – Cell Motility, A Principal Requirement for Metastasis – Tumor Cell Autocrine Motility Factor Receptor – Interleukin-6 Enhances Motility of Breast Carcinoma Cells – Interleukin-6 Stimulates Motility of Vascular Endothelium – Computer Automation in Measurement and Analysis of Cell Motility *In Vitro.*

Contributors:

S.A. Aznavoorian, M.E. Beckner, J. Behrens, M.M. Bhargava, W. Birchmeier, B. Boyer, W. Carley, A. Coffer, J. Condeelis, M.A. Donovan, P.G. Dowrick, I. Ellis, U.H. Frixen, E. Gherardi, I.D. Goldberg, D. Grant, A.M. Grey, L. Harvath, R. Hofmann, A. Howell, S. Jaken, J.G. Jones, A. Joseph, H. Kleinman, Y. Li, L.A. Liotta, D. Liu, P.M. Luckett, I.R. Nabi, B. Palcic, M. Pendergast, M. Picardo, A. Raz, E.M. Rosen, G. Rushton, M. Sachs, E. Schiffmann, J.H. Schipper, A.M. Schor, S.L. Schor, J. Segall, P.B. Sehgal, E. Setter, S. Silletti, I. Spadinger, M.L. Stracke, I. Tamm, J.P. Thiery, G. Thurston, A.M. Valles, R.M. Warn, H. Watanabe, K.M. Weidner.

Please order from your bookseller or directly from:
Birkhäuser Verlag AG
P.O. Box 133
CH-4010 Basel / Switzerland

Orders from the USA or Canada should be sent to:

Birkhäuser Boston Inc.
c/o Springer Verlag New York Inc.
44 Hartz Way
Secaucus, NJ 07096-2491 / USA

Birkhäuser

Birkhäuser Verlag AG
Basel · Boston · Berlin

BIRKHÄUSER
LIFE SCIENCES

Agents and Actions Supplements

Drugs in
Inflammation

Edited by

Parnham, M. J.,
Nattermann Research Centre, Cologne, Germany,
Bray, M. A.,
Ciba-Geigy AG, Basel, Switzerland,
Van den Berg, W. B.,
Univ. Hospital, Nijmegen, The Netherlands)

1991. 264 pages. Hardcover. ISBN 3-7643-2504-6 (AAS 32)

The search for more efficacious and better tolerated anti-inflammatory drugs has received a new impetus in recent years with advances in our understanding of new mediators (e.g. cytokines and tachykines) and previously poorly understood processes such as synovial cartilage breakdown. This book presents invited review lectures and research communications given at the inaugural symposium of the International Association of Inflammation Societies in June, 1990. The topics include the latest developments in basic research and approaches to anti-inflammatory drug therapy, con–centrating on minimization of side effects, analgesics, treatment of synovial damage and modulation of cytokine release.

Please order through your bookseller or directly from:
**Birkhäuser Verlag AG, P.O. Box 133
CH-4010 Basel / Switzerland
Fax ++41 / 61 721 79 50**

Orders from the USA or Canada should be sent to:
**Birkhäuser Boston Inc.
c/o Springer Verlag New York Inc. 44 Hartz Way
Secaucus, NJ 07096-2491 / USA
Call Toll-Free 1-800-777-4643**

Birkhäuser

Birkhäuser Verlag AG
Basel · Boston · Berlin